Studies in Systems, Decision and Control

Volume 501

The series "Studies in Systems, Decision and Control" (SSDC) covers both new developments and advances, as well as the state of the art, in the various areas of broadly perceived systems, decision making and control–quickly, up to date and with a high quality. The intent is to cover the theory, applications, and perspectives on the state of the art and future developments relevant to systems, decision making, control, complex processes and related areas, as embedded in the fields of engineering, computer science, physics, economics, social and life sciences, as well as the paradigms and methodologies behind them. The series contains monographs, textbooks, lecture notes and edited volumes in systems, decision making and control spanning the areas of Cyber-Physical Systems, Autonomous Systems, Sensor Networks, Control Systems, Energy Systems, Automotive Systems, Biological Systems, Vehicular Networking and Connected Vehicles, Aerospace Systems, Automation, Manufacturing, Smart Grids, Nonlinear Systems, Power Systems, Robotics, Social Systems, Economic Systems and other. Of particular value to both the contributors and the readership are the short publication timeframe and the worldwide distribution and exposure which enable both a wide and rapid dissemination of research output.

Indexed by SCOPUS, DBLP, WTI Frankfurt eG, zbMATH, SCImago.

All books published in the series are submitted for consideration in Web of Science.

Kofi Kissi Dompere

Theories of Knowing and Knowledge Production

An Organization of Spaces of Functors and Categorial Converters

Springer

Kofi Kissi Dompere
Department of Economics
Howard University
Washington, DC, USA

ISSN 2198-4182 ISSN 2198-4190 (electronic)
Studies in Systems, Decision and Control
ISBN 978-3-031-42862-3 ISBN 978-3-031-42863-0 (eBook)
https://doi.org/10.1007/978-3-031-42863-0

This Springer imprint is published by the registered company Springer Nature Switzerland AG
The registered company address is: Gewerbestrasse 11, 6330 Cham, Switzerland

DEDICATION

This monograph is dedicated to all professors, scholars, researchers and supporters working to understand information-knowledge system as an economic production-consumption process under the space of general input-output dualities, where neuro-decision-choice actions take place over the space of real cost-benefit dualities relative to the general space of processes and technologies as the space of functors and categorial converters and its relationship to information-knowledge economy, and the space of identification-transformation dualities and how these processes and technologies as neuro-decision-choice development are affected by human conditions of curiosity, creativity, courage, hope, despair, failure and success over the general space of incentive-disincentive dualities conformally mapped onto the space of cost-benefit dualities relative to the space of ignorance-knowledge dualities over the space of existence-nonexistence dualities.

To Ms. Caroline Gillespie, Dr Paige of New Castle, Pa., the Lanes of Kittanning of Pa, Mr. Dick Gregory of Washington, D.C, and all traditional thinkers of the adinkra symbolism and its conceptual and philosophical system, the people who worked to make my staying and studies in USA tolerable over the space of human existence. Finally, to a great friend Dr Anthony Akoto Osei and his contributions to Ghana and Dillard university in New Orleans, USA.

ACKNOWLEDGEMENTS

Organization-disorganization dualities relate to composition-decomposition analytics in the space of neuro-decision-choice actions. The composition-decomposition analytics relate to input-output processes which establish the space of production-consumption dualities in the space of rational-irrational dualities. The nature of the space of rational-irrational dualities depends on collectively acceptable rules over the space of thinking-reasoning dualities. Irrespective of how it is conceived and interpreted, it always affects all areas of human thought and the derived laws of thought. The laws of thought are the products of the space of thinking-reasoning dualities which is also shaped by the derived laws of thought to guide all neuro-decision-choice behaviors toward actions in managing the social set-ups in time and over time in the space of production-consumption dualities within the multiplicity of areas of knowing over the space of problem-solution dualities requiring curiosity, creativity, innovation, hope, despair, success, failure chance-taking and resource commitments. It is within this dualistic-polar framework that a general theory of production-consumption duality must be conceived as an interdisciplinary and multidisciplinary enterprise over the space of principle of opposites and the principle of non-satiation. In this respect, I express gratitude to thinkers with the vision to interdisciplinary and multidisciplinary research, teaching and learning with view of enhancing the human intellectual heritage through global collaboration and the development of dynamic framework of research universities to reduce national and international cognitive capacity limitations. I thank all those involve in the development of fuzzy paradigm of thought to help our understanding of our agreements on collective thinking and reasoning as well as the understanding of the collective efforts in relations to human understanding of uncertainty, identity and change and their relational connectivity to the space of life-death dualities where such understanding will help to set peace against war, nonviolence against violence in our socio-natural space within the space of relativity.

Special thanks always go to friends, critics and those who have taken interest and time to read my monographs, whose positive and negative encouragements respectively have made my works in research, teaching, and learning enjoyable. My thanks also go to all my graduate students in my courses in economic theory, mathematical economics and cost-benefit analysis who raised some interesting questions regarding the applications of the fuzzy techniques and logic of thinking and reasoning in the space of relativity with inexact information structure and the use of cost-benefit methods and techniques in understanding neuro-decision-choice actions over the spaces of individual-collective dualities and democratic-nondemocratic dualities. I accept all responsibility for all controversial ideas and errors.

PREAMBLE

(QI)*The optimist is right. The pessimist is right. The one differs from the other as the light from the dark. Yet both are right. Each is right from, his own particular point of view, and this point of view is the determining factor in the life of each. It determines as to whether it is a life of power or of impotence, of peace or of pain, of success or failure.*

The optimist has the power of seeing things in their entirety and in their right relations. The pessimist looks from a limited and a one-sided point of view. The one has his understanding illumined by wisdom, the understanding of the other is darkened by ignorance. Each is building his world from within, and the result of the building is determined by the point of view of each. The optimist, by his superior wisdom and insight, is making his own heaven, and in the degree that he makes his own heaven is he helping to make one for all the world beside. The pessimist, by virtue of his limitations, is making his own hell, and in the degree that he makes his own hell is helping to make one for all mankind.

You and I have the predominating characteristics of an optimist or predominating characteristics of a pessimist. We then are making, hour by hour our, own heaven or our own hell; and in the degree that we are making the one or the other for ourselves are we helping make it for all the world beside.

The word haven means harmony. The word hell is from the old English "hell", meaning to build a wall around, to separate; to be "helled" was to be shut off from. Now if there is such a thing as harmony there must be that something one can be in right relations with; for to be in right relations with anything is to be in harmony with it. Again, if there is such a thing as being "helled", shut off, separated from, there must be that something from which one if held shut off or separated. [R7.32, p.13-14]

(QII) *Freedom is control over ourselves and over external nature which is founded on knowledge of necessity. Such knowledge also requires that we know what chance factors enter into the process with which we are concerned, and the probabilities characterizing their operation, so that we can (a) control the operation of chance and (b) take precautions to meet its operation in so far as we do not control it, as a result of which the whole process is directed to desired end* [R17.11b. p193].

(NQIII) *Judgments of probability express our expectation of the occurrences of accidents. According to some theories, probability is purely subjective, in the sense that a judgment of probability is an expression of nothing but our own subjective uncertainty or lack of knowledge. But on the contrary, the idea of probability reflects an objective reality - or rather, one aspect of objective reality- namely the operation of accidental causes in a whole sequence of events or in an aggregate of instances. This is just as much an objective reality as the operation of a single cause on a single occasion, which is not a subject of probability*. [R17.11 p.193].

(QIV) *The distinction between uncertainty and risk is very often referred to in economic theory. But no formal representation of expectations has been suggested so far, in which risk and uncertainty could be identified each of them separately. Pure risk arises in those situations in which we cannot formulate any further questions in the evidence which, if answered, would make us change the probability weights which we have already attached to different outcomes. Pure uncertainty arises when none of the questions which we can formulate have been answered. Different hypotheses are often suggested to explain behaviour in either of these two situations. In fact, however, the elements of uncertainty and of risk combine always one with the other, and there is no way in which they could be separated* [R24.22, p.58].

(QV) *The interpretation of probability as a measure of the degree of confirmation has also been criticised on the ground that cases in which it is possible to identify outcomes equally confirmed by the evidence are rather rare. In general, probability in the classical sense cannot be ascertained by means of the operation in terms of which it is defined. The probability of death of an Englishman of a particular age cannot, for instance, be derived from the available evidence by means of the Principle of Insufficient Reason. The fact that there may be rules of logic which we cannot perceive is not of much help either. If we do not know how long people live and how often they die at each particular age, we simply do not know what the probability of their death actually is* [R24.22, p.59].

(QVI)*The probable is something which lies midway between truth and error (HOMASINS, 1688); an assertion, of which the contrary is not completely self-contradictory or impossible, is called probable (Reimarus). Kant says: 'That which, if it were held as truth, would be more than half certain, is called probable.' Perhaps, after these examples, someone may wish to know what modern philosophy has contributed to this subject. I quote literally from Robert Eissler's 'dictionary of philosophic Concepts (1910). Probability, in the subjective sese, is a degree of certainty which is based on strong or even overwhelming reason for making an assertion. . . . In the objective sense, the probable is that which is supported by a number of objective arguments. . .. There are differing degrees of probability and these depend upon the kind and number of reasons or facts upon which the assertion or conclusion of probability is based.*

To consider now a familiar, modern source, Webster's 'New International Dictionary gives the following definition of probability' Quality of state of being probable; reasonable ground for presuming, likelihood; more narrowly, a conclusion that is not proof but follows logically from such evidence as available; as, reports devoid of all probability, to establish probability of guilt. [R24.35, pp. 1-2].

(QVII) Let us go back to the beginning. We face today the same old questions that philosophy has always faced:

"What are Humans, that they can understand the World?

And what is the World, that Humans can understand it?"

Or alternatively:

"How is it that human thought and practice are adapted

To the tasks of dealing with the world? And what it is

The nature or measure of that adaptation?"

[R3.31, p25.]

(QVIII) *If we are to heal the wounds created by the Cartesian split, and integrate humanity with nature, it will follow as a result that human actions, too, are performed 'within' the world of natural processes; and the older philosophical barriers separating rationality from causality will have to be dismantled along with all the others. A world of nature into which humanity has been reintegrated will no longer be an impersonal, mechanistic world. Rather, it will be a world within which the human reason itself is a causally efficacious agency, within which – as the ancient recognized but the philosophers of the seventeen-century denied- we have the elbowroom that we require to exercise the autonomy that is chief mark of our humanity* [R3.11, p.35].

(QIX) *In psychological research, we are seeking the invariants of human behavior; but in seeking them, we cannot ignore the complexity of the relation between the underlying system and that behavior. In any cognitive activity, a strategy or program intervenes between the physiology of the brain and the actual human performance-- just as it does between computer hardware and computer behavior. This is one reason why computer programing languages provide such appropriate formalisms for stating psychological theories. Our experiments and our theories must pay explicit attention to this intervening layer of strategy or programs.* [R8.27b, p.89.]

(QX) *This lesson from industrial psychology has not yet been applied very frequently to experimentation in cognitive psychology. In running experiments on problem-solving tasks, for example, it is not all common to train subjects in a particular specific method for doing the task, even in experiments that aim at measuring speed of task performance. Instead, subjects are given rather general instructions about the task requirements, and perhaps some practice examples, and are then allowed to work out their own methods of approach. Under these circumstances, at least four distinct sets of factors – the characteristics of the problem environment, the physiological characteristics of the subjects (e.g., shot-term memory limits), then previous learning and experience, and the particular strategy adopted—are all confounded in the data. The situation is made worse when data are averaged over many subjects, who may be using a variety of different methods to perform the task* [R8.27b, p.90].

(QXI) *Physical systems are designed and built to perform certain defined functions. Submarines, aircraft, and spacecraft must navigate in their respective environments to accomplish their objectives, whereas an electric power system network must meet the load demands. In order to* *determine whether a system is performing properly, and ultimately to control the system performance, the engineer must know what the system is "doing' at any instant of time. In other words, the engineer must know the state of his system. In navigation, the state consists of position and velocity of the craft in question; in an ac electric power system, the state may be taken as voltages and phase angles at network notes. Physical systems are often subject to random disturbances, so that the system state may itself to be random. In order to determine the state of his system, the engineer builds a measuring device (e.g., sextant, radar, voltmeter) and takes "measurements" or "observations" on his system. These measurements are generally contaminated with noise caused by the electronic and mechanical components of the measuring device.* [R24.16b, p.1].

(QXII a)*A definition is usually interpreted as a logical device THAT INTERESTS ONE; (a) they distinguish, find, or construct the object that interests one; (b) to make the meaning of a term already current in science more precise, and also to form the meaning of a newly introduced term. Since the knowledge of the distinctive features of objects and of the meanings of the terms in question entails the formation of concepts of the objects defined, logicians often speak of definitions of concepts in question rather than of definitions of objects and meanings of terms.* [R3.33, p.9]

[QXII b] *A nominal definition is a definition which serves (a) to formulate in an explicit form the meaning of a term already current in the language of science or a natural language (including symbols of the t artificial languages of science) .... (b) to fix the meaning of a term newly introduced into a natural language or the language of science. In this case a new meaning for a new sign expression is introduced. (c) to record the fact that term of the definiendum (Dfd) and of* *the definiens (Dfn) denote identical object. (d) to introduce a new term as a mere abbreviation of another, usually more complicated, expression. Here we ignore temporarily the meaning of the Dfd and Dfn terms and consider them simply as sign forms....*

A real definition is one that specifies and unambiguously distinguishes the object in question from other objects of a given domain [R3.33, pp,13-14].

REFLECTIONS ON THE QUOTATIONS

These quotations as the preamble to this monograph need some explanations as to their relevance to the development of the organizational analytics of functors and categorial converters and their relational continuum and unity in the space of ignorance-knowledge dualities, The quotations relate to the principle of opposites, concepts and phenomena of certainty, uncertainty, probability, problem-solving processes in knowing and information-knowledge processes and neuro-decision-choice psycho-economic states. The quotation (QI) relating on states of optimism and pessimism reflects principle of opposites of the space of optimism-pessimism duality with relational continuum and unity, where the optimistic characteristic disposition has a corresponding pessimistic characteristic disposition specified in degrees of an actions such that as the characteristic disposition of pessimism falls the characteristic disposition of optimism rises. In this situation, both positions are true and false under conditions of neuro-decision-choice determination in the space of relativity in relational continuum and unity, where contradictions are accepted with epistemic conditionality and not in the space of absoluteness in relational separation and disunity, where contradictions are not accepted.

Quotations (III, IV, V, VI) present some analytical difficulties in the concepts and phenomena of probability and expectations in various definitional and contextual settings. The essence and the values of these quotation are to show epistemic defectiveness of the concepts and phenomena of probability and expectations. The concept. The phenomenon of probability is cognitive formation in the space of neuro-decision-choice formation in the space of information-knowledge variety production as input-output processes in the general space of human actions over the space of fife-death dualities. The phenomena and concept of probability do not arise in a epistemological vacuum. They have a source within the epistemological space, where varieties must be identified and known under cognitive capacity limitations in terms of vagueness, inexactness, and volume incompleteness in their characteristic dispositions and not in the ontological space, where varieties are in the space of *what there is* (actual) and *what would be* (potential) with exactness or precision and volume-completeness in their characteristic dispositions. In this respect, while it is meaningful for one to speak of the phenomenon and concept of epistemological probability, it is not meaningful to speak of phenomenon and concept of ontological probability. The phenomena and concepts of possibility and probability are analytically related and reside in the space of imaginations which is derived from the space of potentials, The rise of imagination is due to variety *uncertainty* which is derived from an *epistemic distance* between the variety *ontic characteristic disposition* as an identity with information-knowledge equality and hence certainty from the ontological space and the *variety epistemic disposition* as a derived identity from the epistemological space with information-knowledge inequality and hence uncertainty. The collection of the ontic variety certainties and epistemic variety uncertainty constitutes the space of certainty-uncertainty dualities which then gives rise to the space of imaginations containing the possibility and probability spaces which house necessity and freedom respectively in the general space of cognitive actions where freedom may be related *free will* under social philosophical consciencism over the general space of human behavior.

The studies of possibility and probability have two sides of philosophy with its paradigm of thought and logical laws, and mathematics with corresponding paradigm of thought and logical laws. Philosophical studies deal with questions regarding the phenomena, concepts, sources and interpretations of probability possibility, their interdependencies and their relationships with statistical sciences, neuro-decision-choice processes, and games over the spaces of subjective-objective and qualitative-quantitative dualities in the space of relativity. Mathematical studies deal with probability measures, forms of probability functions and their uses in all areas of science and nonscience. What the mathematical probability measures are not explicitly revealed. The general difficulty in all human endeavors is how to explain and understand human cognitive capacity limitations, the source conditions of imaginations and their relationships to curiosity, creativity, hope, despair over the space of problem-solution dualities, giving the space of mind-body structure, and its link to the space of success-failure dualities, where possibility space containing necessity and probability space containing freedom emerge. The development of mathematical-philosophical system of understanding of possibility-possibility dualities must explicitly relate the space of imagination. The space of imagination must then be connected to the space of certainty-uncertainty dualities and how this space of certainty-uncertainty dualities arises in the space of all human endeavors. Uncertainty is present in all experiential information in the space of acquaintance relative to the space of knowing, information-knowledge accumulation, and its applications.

The philosophy and mathematics of theories of information and knowledge must discuss the similarities and differences between them, where these differences and similarities must apply over all categories contained in the epistemic space with relational and information fields. While uncertainty is found in all areas of scientific and non-scientific areas of knowing, it is useful to understand that uncertainty is not possibility or probability but relate to them in outcomes of knowing as intellectual investment flows and information-knowledge accumulation as the production of intellectual stocks in the space of variety acquaintance-description dualities. Information and data are not the same, they must be distinguished and explicated. Statistical and data analyses are to provide us an epistemic conditionality of zones of acceptance in the spaces of true-false dualities and correct-incorrect dualities in relation to the spaces of problem-solution and question-answer and ignorance-knowledge dualities since cognitive capacity limitations place neuro-decision-choice actions in the space of error-correction modes within the socially infinite self-correction information-knowledge system, where new variety characteristic dispositions are encountered in the acquaintance space.

Errors and mistakes are not ignorance. Errors and mistakes are the results of incorrect applications of either a macro-paradigm of thought with the corresponding macro-algorithm or micro-paradigms of thought with corresponding micro-paradigms or both. In this respect, one can speak of possibility space of error-mistake processes over the space of correct-incorrect dualities over the space of neuro-decision-choice actions relative to the space of sureness-unsureness dualities, where principle of doubt leads creativity and innovation in thinking-reasoning processes.

Uncertainty is not a quantitative estimation of error in data as derivative of information over the space of acquaintance. Data is recording numbers of occurrences of counting occurrences of varieties of interest in the space of acquaintance, where incorrect counting is error, and incorrect recordings are mistakes. Deficiencies in observations of characteristics in the variety characteristic dispositions in the acquaintance space are due to cognitive capacity limitations of multiplicity of forms that affect epistemic variety dispositions compared to ontic variety dispositions. The errors in counting and mistakes in recording generate first order incorrectness that is not due to cognitive limitations. The errors and mistakes in applications of paradigms of thought and algorithms over the space of quality-quantity dualities while they may lead to wrong outcomes are not uncertainties even though they may enhance the degrees of uncertainties, where uncertainties are conditions in the pace of cognitive penumbra but not in the space of error-non-error and mistake-non-mistake dualities. Their contributions to outcomes in the space of neuro-decision-choice actions. In all these conditions, a theory of uncertainty must be developed and separated from the theory of imagination, the theory of possibility and the theory of probability. The relationships among these three theories and theory of imagination in the space of knowing and information-knowledge production must be examined by philosophical-mathematical representations and analyses over the space of qualitative-quantitative dualities.

The theory of possibility must be related to the theory of variety identification that houses initial conditions that give rise to the development of the theory of necessity. The theory of probability must be related to the theory of transformation that houses freedom and sufficient conditions contained in the social philosophical consciencism which contains all aspects of culture of knowing in the space of thinking-reasoning dualities, conditions of national personality, national exceptionalism, containing provisional and incomplete knowledge relative to epistemological time, where all information-knowledge acceptances are always under epistemic, conditionality in the space of claims. The epistemic conditionality is governed by cognitive capacity limitation, elements of errors, mistakes, and fuzzy-stochastic structures.

Quotations (VII, VIII, IX) relate to some difficulties, demands, and needs in Knowing and information-knowledge processes to human existence and how the knowing and knowledge production define the conditions to maintain the process and continuity of life. The knowing and information-knowledge processes relate to human understanding of nature and the management of cognition which relates the infinite spaces of neuro-decision-choice processes and relationality over the space of relations in the spaces of thinking-action and individual-collective dualities to overcome elements in the general space of problem-solution dualities that presents elements in the spaces of joy-sorrow, happiness-sadness, wellness-sickness, cure-decease dualities and such conditions in the space of life-death dualities within the general space principle of opposites where thinking-reasoning actions are to seek balances in variety dualities and opposites

Quotations (X, XI) is reflections on the continuity and interdependent relationships among the space of knowing and information-knowledge production, the space of problem-solution dualities, the space of neuro-decision-choice actions, the space of success-failure dualities and the space of input-output dualities. Knowing and information-knowledge development are seen in terms of elements in space of problem-solution dualities, where in general, every variety problem has a corresponding variety solution in degrees of correctness and over the infinite space of in the space of past-future dualities, and where every solution generates new variety problems, and the easiness to which new solutions may be obtained will depend on what is the nature of the preceding variety solution and how it is obtained.

Quotation (XII) points to the logical nature, structure and conditions of definition and the theory of definition in the development of knowing, variety information-knowledge elements over the space of nominal-real linguistic dualities in order to establish distinction, similarities, differences and commonness over the space of source-destination dualities in all human-human, human-machine and machine-machine communications. All varieties of states processes are defined by their characteristic dispositions, where group formation, category formation and set formation in the epistemic field and relational field are established by characteristic dispositions that the constructs of dualities and polarities, synonyms, and antonyms under the general principles of opposites. In this way, the concept and phenomenon of information find expressions and meanings in the space of characteristics and communication of information find expressions and meanings in the space of signals.

Finally, the quotations in this preamble provide us with some foundations in understanding the source, concepts and phenomena of uncertainty and its relation to information and knowledge in the space of human actions as seen in the spaces of question-answer dualities, problem-solution dualities and ignorance-knowledge dualities, necessity-freedom dualities. The way these essential concepts and phenomena are introduced over the space of ignorance-knowledge dualities affects the general understanding over the space of identification-transformation dualities under cognitive-capacity limitations and the space of curiosity, creativity, hope, despair, and similar elements in the space of quality-quantity dualities. The understanding of the general behavior of the space of identification-transformation dualities offers important pathways to the developments of theories of socio-natural dynamics over the space of quality-quantity dualities, such as theories of socioeconomic development, social evolution and environmental-quality maintenance under decision-choice rationality.

PREFACE

ABSRACT

The *preface* introduces the motivation for writing the monograph, what it is about and how the conceptual framework of the monograph is conceived and connected to other previous monographs by the same author on information, knowledge, data, evidence, and decision-choice actions. The preface introduces the basic reflection on the development of the monograph follow by discussions on functors and categorial converters and their relationship to processes and technologies. This is followed by discussions on the space of goals, objective, and vision over the space of individual-community dualities, the spaces of certainty-uncertainty dualities, thinking-reasoning dualities in relation to contradictions, conflicts under the general principle of opposites in the process of knowing and generation of anomalies. The preface is ended with the structure and organization of the monograph. The preface provides an organization of paremiology of the essential spaces and sub-spaces contained in the development of the epistemic discussions for understanding of the paths of discussions.

I: BASIC REFLECTIONS ON THE STRUCTURE OF THE MONOGRAPH

This monograph on the theory of organizational analytics is about the organization of instruments of transformation made up of functors and categorial converters in conformal mapping in the spaces of input-output dualities and cost-benefit dualities. It is an analytical continuity of my examination and development of the theories of information and knowledge as economic production and not as communication. At the level of statics, it is about variety identification of property and relation dispositions in the space of quality-quantity dualities to define the initial conditions. At the level of dynamics, it is about variety transformation of property and relation dispositions in the space of quantitative and qualitative motions. The importance of this approach is to provide a framework for analysis of the concepts and phenomena of thinking, reasoning, algorithms, creativity, curiosity, hope failure and paradigms of thought in knowing as production of intellectual investment flows and information-knowledge accumulation as production of intellectual capital stocks. These provide us the boundary conditions of strengths and limitations of organization over the space of problem-solution dualities and ignorance-knowledge dualities. The theory of knowing as intellectual investment flows and the theory of information-knowledge as the theory of intellectual capital stocks is the most general theory of production-consumption dualities. It is also the theory of input-output dualities which contains the theory of supply-demand dualities relative to supply chain and demand chain over the space of fundamental-applied dualities. All these are dualistic-polar structures with one important element connecting them in relational continuum and unity, and this important element is the space of production-consumption dualities containing the spaces of parent-offspring and processor-successor dualities .

The theory of intellectual investment flow contains the theory of investment in commodity production in the space of opened-closed economies. The theory of intellectual capital stocks contains the theory of capital stock-flow dynamics in commodity production in closed-opened economies [R13.10] [R13.11]. Here, we can conceive of the theory of the choice of the path of intellectual production to maximize the present value of social production-consumption path with given initial intellectual capital stock. The general specification will depend on the input-output parameters where the input-output parameters will vary over societies relative to social preferences and space of goals, objectives, and visions. The present value of the social production-consumption will always be related to the general space of goals, objectives and visions constrained by social preference order and social philosophical consciencism. The inputs of the functor or categorial converter will always be labor and intellectual capital stocks. This approach of a theory of optimal intellectual investment and an optimal information-capital accumulation will be taken in another monograph, where the introduction of functor and categorial converter provide a mean to introduce curiosity and creativity in knowing to abstract an optimal epistemic model with efficient management and control.

The development of the general theory of intellectual capital composed of the theory of intellectual investment and the theory of intellectual capital stock requires the organizational analytics of the basic organization and the corresponding superstructure within the basic-superstructure duality. The organizational analytics will then allow the development of epistemic anomalies that arise in the process of linking the epistemological space to the ontological space and reconciling the differences between variety ontic dispositions and variety epistemic dispositions as seen within the space of ontic-epistemic dualities. The ontic-epistemic differences must lead to critical analyses of the concepts and phenomena of onticity and epistemicity and their relationships to the concepts and phenomena of cognitive capacity limitation, certainty, uncertainty, imagination, and reality in the enterprise of knowing in the epistemological space. It must then be shown as to how imagination generates possibility and probability, and how reality connects to the actual. From the conditions of imagination, it must be shown as to how necessities and initial conditions are related to possibility and freedom and sufficient conditions are related to probabilities and how the space of goals, objectives and visions is embedded in the space of imaginations and how the space of imagination is connected to the space of potential inherent in the ontological space which contains the space of actuals, the varieties of which are defined in universal diversity and unity. The processes and the paths through which the ontological diversities are established by the corresponding ontological characteristic dispositions and the epistemological diversities are established by the correcting epistemological characteristic dispositions which are abstracted by cognitive agents with cognitive capacity limitations from ontological signal dispositions must be shown, where the processes and the paths are defined in space of categorial converters or functors.

The epistemic connectivity in the space of imagination-reality dualities requires the discussions of effects of curiosity, creativity, hope, despair, success, failure and how realities are abstracted from imaginations and the imaginations from potentials through the space of functors and categorial converters. The theory of intellectual investment flows and the theory of intellectual capital stocks under the principle of opposites composed of dualities and polarities must discuss the dualistic-polar structures of varieties in relation to conflicts over the space of internal-external dualities to bring about internal and external motions and transformations. The transformations as journeys among duals of dualities and the poles of polarities require the discussions on stopping rule in the dualistic-polar crossing in the sense that when does one cross zone of ignorance to the zone of knowledge. The epistemically dualistic-polar crossing over the space of problem-solution dualities may be formulated as fuzzy decision-choice problem, solved by fuzzy optimization techniques, and analyzed by fuzzy logic reasoning.

II: PROCESSES, TECHNOLOGIES, FUNCTORS AND CATEGORIAL CONVERTERS

All cognitive journeys with the space of identification-transformation dualities require motions. The motions are defined in terms of processes and technologies which act on varieties for identification and transformation over the space of ignorance-knowledge dualities. The motions are defined over the space of qualitative-quantitative dualities in terms of properties, relations, and both. Each variety motion is composed of a process which defines the thinking as the algorithm for the cognitive journey and a technology as the use of the algorithm to accomplish the cognitive journey from one zone of a dual to the zone of corresponding dual in the same duality. The activities over the space of identification-transformation dualities are the works of neuro-decision-choice actions though processes and technologies, the finish line is always a change in either relation or property. The space of thinking-reason dualities must then be related to the space of categorial converters or the functor space, where the elements in these spaces are composed of specific processes and technologies and related to variety behaviors in the space of variety relation-property dualities.

The functor space and the space of categorial converters come to us as natural space and social space, where there are natural functors and categorial converters, and social functors and categorial converters. The space of natural categorial converters and the functor space constitute the primary categories of the space of processes and technologies as identities in the ontological space containing the space of actual-potential dualities. The space of social categorial converter and the functor space constitute derived categories of space of processes and technologies as derived identities in the epistemological space containing the space of imagination-reality dualities. The natural functors and categorial converters connect activities within the space of actual-potential dualities transforming variety actuals to variety potentials and variety potentials to variety actual in natural existence in ontological space within the ontological time of the space of ontological beginning-end dualities. The social functors and categorial converters connect activities within the space of imagination-reality dualities transforming variety realities to variety imagination and variety imaginations to variety realities in social existence in epistemological space within the epistemological time of the space of epistemological past-future dualities where the derived processes and technologies are the creations of cognitive actions through the space of thinking-reasoning dualities that connects the space of potentials to the space of imaginations and then to the space of reality transforming variety imaginations to variety actuals in comparison to the variety of what there is the actual. The institutions of knowing must be studies as organizations of institutions of production of knowledge through thinking to generate knowledge about processes and technologies and reasoning through the utilization of the knowledge for variety identifications and transformations over the space of theory-applied dualities that must be connected to input-output and construction-destruction processes, where the static processes are about variety identification and dynamic processes ere about variety transformation. The decision-choice activities generate the spaces of functors and categorial converters to mimic the natural processes and technologies for social transformations, where all knowing are about variety identifications and all neuro-decision-choice actions are about variety transformations.

III: THE SPACES OF GOALS, OBJECTIVES, VISIONS, AND INDIVIDUAL COMMUNITY DUALITIES

In general analytics, human actions through decision-choice actions are defined in the space of problem-solution dualities that is contained in the space of imagination-reality dualities directed to solving problems. The problems are generated in the space of realities in comparison with *what there is* and *what would be*. The solutions are generated from the space of imaginations. We are, here, confronted with the questions as to the concept and phenomena of problem and solution in identity and distinction. The problem-solution identity and distinction find meaning, and definitions in cost-benefit relativity which must be examined in the space that contains goals, objectives and visions, which is defined in the space of imaginations, where the cost-benefit conditions of variety imaginations are compared with cost-benefit conditions of variety realities, where the benefit conditions of goals, objectives and visions are assessed to be greater than variety realities as seen in the space of individual-community dualities that houses conflicts in preferences. The journeys over the space of imagination-reality must be connected to the functor space and then the space of realities [R17.17] [R17.18].

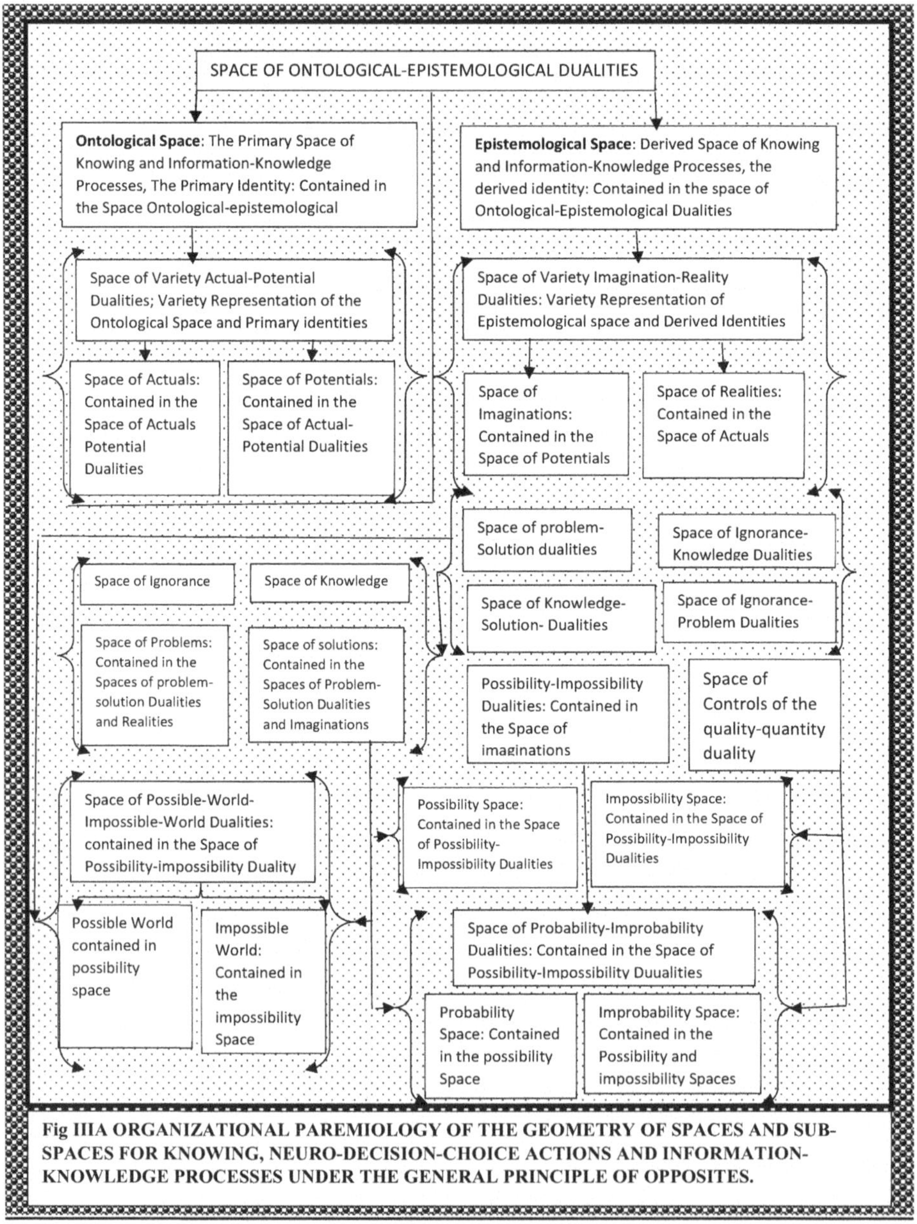

Fig IIIA ORGANIZATIONAL PAREMIOLOGY OF THE GEOMETRY OF SPACES AND SUB-SPACES FOR KNOWING, NEURO-DECISION-CHOICE ACTIONS AND INFORMATION-KNOWLEDGE PROCESSES UNDER THE GENERAL PRINCIPLE OF OPPOSITES.

IV: THE SPACE OF CERTAINTY-UNCERTAINTY DUALITIES AND NEURO-DECISION-CHOICE ACTIONS

The analysis of the conditions of goals, objective and vision leads us to raise the questions regarding the spaces that contains them and the conditions that constrain their realization over the space of success-failure dualities in relation to the varieties in the functor space and the space of categorial converters under the ruling social philosophical consciencism. The constraints in the success-failure are abstracted from the space of certainty-uncertainty dualities. In this respect, there is a need for a theory of uncertainty as contained in the theory of certainty-uncertainty duality that will show the concept, phenomenon, meaning, and source of uncertainty and its relation to imagination, possibility, and probability with clear definition of the phenomena of these concepts. The failure to clear analyses of the concept, phenomena of uncertainty, potential, imagination, possibility, probability, their sources, and their relational sequential interdependence connectedness has brought some analytical confusions in the theory of neuro-decision-choice processes. Part of these confusions are discussed in [R8.16b] It is a substantial epistemic confusion when discussions are undertaken of different kinds of probability without possibility and their relationships to the simple concepts of possible and probable and what gives rise to these linguistic phenomena. The possible is cognitively internal and non-resource-based and probable is cognitively external and resource-based over the space of neuro-decision-choice actions. The problem with the development of the theory of probability is found on the path of its development and how the relationship of possibility and probability is not discussed.

The concept of likelihood used in defining probability and possibility is a conceptual problem in the sense that, how is likelihood connected to uncertainty and as an input into the neuro-decision-choice processes over the input-output dualities. Is probability a cognitive outcome or is cognition a probability outcome? In other words, is probability generated by cognitive capacity limitations of neuro-decision-choice processes or are cognitive capacity limitations in neuro-decision-choice processes generated by probability? Another way of looking at the analytical confusion is to ask the question as to the relationships among uncertainty, imagination, possibility, probability, and cognition, and in what sequential order do they appear in human cognitive actions in the space of parent-offspring processes. The monograph provides some directions to the solutions to these important questions by developing an introduction to the theory uncertainty different from the theories of possibility and probability. The conception of different forms of probability and uncertainty makes no sense when we do not know what is are the concepts of uncertainty, probability, and their sources of within human experiential information. The introduction of concepts of subjective probability, objective probability, ontic probability, epistemic probability, revealed probability defined in terms of probability is not analytically useful without clear definition of probability and its source. Similarly, the introduction of different types of uncertainty expands and intensifies the system of difficulties of clarity and understanding of human actions under cognitive capacity limitations in relation to varieties in the space of necessity-freedom dualities, anticipation-expectation dualities. We must always keep in mind the sequential logical development of phenomenon, concept, definition, and measurement, where phenomenon gives rise to concept and then to definition and measurement over the space of quality-quantity dualities in the space of ignorance-knowledge dualities.

V: THE SPACE OF THINKING-REASONING DUALITIES

The journeys in the spaces of epistemological dualities and polarities are undertaken with varieties in thinking-reasoning dualities. The elements in the space of thinking-reasoning dualities must be related to the varieties in the space of ignorance-knowledge dualities and their space of functors and categorial converters. The structure and conditions of the thinking-reasoning dualities must be studied and analyzed in terms of paradigms of thought and how the paradigms of thought affect the cognitive journeys and the transversality conditions in the space of ignorance-knowledge dualities. The elements of thinking-reasoning dualities must be related to curiosity, creativity, over the spaces of hope-despair dualities and success-failure dualities. The paradigms of thought must be structured to account for mega-algorithm that guides a set of meta-algorithms, their developments and use over the space of thinking-reasoning dualities relative the elements in the space of problem-solution dualities. The behaviors of elements in the space of thinking-reasoning dualities derive its foundation from the social philosophical consciencism as well as influence the dynamics of social philosophical consciencism and intergenerational connectivity. The behaviors of the elements in the space of thinking-reasoning dualities set the paths and directions of human progress and telescopic future by creating negation and negation of negation over the space of destruction-construction dualities contained in the space of imagination-reality dualities.

VI: CONTRADICTIONS, CONFLICTS AND THE PRINCIPLE OF OPPOSITES

The whole of elemental behaviors over the space of ontological-epistemological duality is centered on varieties which are defined by the principle of opposites composed of dualities and polarities. Every duality is composed of opposing duals for contradictions and conflicts. Every polarity is composed of opposing poles for contradiction and conflict. The varieties reside in dualistic-polar games. Every pole of polarity has a residing duality. The identity and the behavior of every pole of polarity is established and determined by the nature and the identity of the residing duality. Based on the residing dualities the polarities are transformed. Contradictions are discounted in classical paradigm of thought which accepts exact game space. Contradictions are accepted by fuzzy paradigm of thought which accepts fuzzy game space. Analytically, every game construct is reducible to a dualistic-polar stricture. The general game space presents three scenarios of dualities and polarities. They are: 1) *positive-sum game* (win-win) which is generated by *creatively destructive conflicts* in the space of contradictions 2) *negative-sum game* (lose-lose) which is generated by *uncreatively destructive conflicts* in the space of contradictions and 3) *zero-sum game* (win-loss) which is generated by *neutral* (creative-uncreative) *conflicts* in the space of contradictions. The phenomena and the concept of game and game space are not only important in understanding in knowing and information-knowledge development, but they encompass everything in similarity, commonness, differences, and distinction in variety dynamics that must be made explicit without which the understanding socio-natural dynamics will be impossible in our socio-natural environment. It is useful to note that the fuzzy paradigm of thought provides a way to deal with lexical elasticities in degrees of belonging and substitutions within synonyms and antonyms in communication under the principle of opposites.

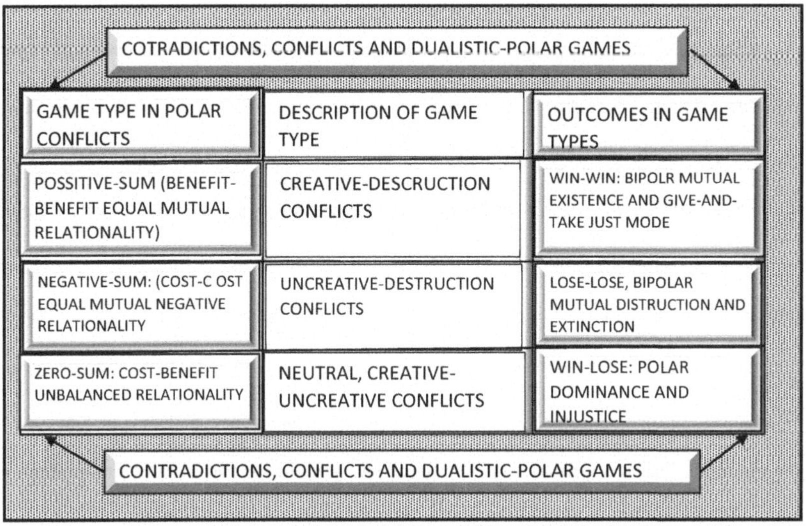

Figure VI.1: Table of Game Space, Names, Structures and Outcomes

Over the space of theoretical and analytics dualistic-polar development, all the relevant phenomena and concepts must be defined in terms of their characteristic depositions over the relevant dualistic-polar game spaces for transformation over the space of internal-external dualities. The relevant spaces must be defined as collections of dualistic-polar varieties for identifications and transformations. All the ontological varieties must be established through variety characteristic dispositions that belong to the ontic dual. All-knowing varieties must be established through variety acquaintance with the ontological signal dispositions from the ontological variety characteristic depositions under cognitive capacity limitations to form the epistemic dual. The ontic dual and the epistemic dual unite to form ontic-epistemic duality, the differences of which give rise to epistemic distance, where the variety epistemic distance gives rise to variety uncertainty and the collection of these variety uncertainties constitutes the space of uncertainties. The space of uncertainties gives rise to space imaginations which must be abstracted from the space of potential contained in the ontological space The conceptually analytical ideas contained in this approach are : 1) there is an explicit definition of information to establish variety identities and categories, 2) there is an explicit definition of uncertainty in relation to information, 3)there is a defined and stated source of variety uncertainty as cognitive dependent from ontic-epistemic disparities, 4) the is a statement of definition and source of variety imagination and space of imaginations, 5) there is an explicit discussion on how variety uncertainty gives rise to variety imagination, 6) there are explicit discussions on how variety imaginations are derived from variety potentials contained in the ontological space, 7) the variety imagination is shown to give rise to variety possibility and variety probability, 8) the variety possibility and variety probability are shown to be generated by neuro-decision-choice actions to deal with cognitive capacity limitations and uncertainties over the space of problem-solution dualities, 9) the space of problem-solution dualities through the space of certainty-uncertainty dualities is discussed to give rise to the space of ignorance-knowledge dualities from the space of acquaintances, 10) the space of ignorance-knowledge dualities gives rise to the space of true-false dualities , 11) the space of true-false duality gives rice to the space of acceptance-rejection dualities, 12) and the space of acceptance-rejection dualities gives rice to the space of realities as knowledge contained in the space of actuals.

All the dualistic-polar games are actionized by elements as processes and technologies in the space of categorial converters and functor space. In this respect, possibilistic reasoning and probabilistic reasoning must be seen as reverse processes in examining the length of the epistemic distance. It is in this context that statements and outcomes have fuzzy-stochastic conditionality in acceptance of knowledge that may enter as input into the space of neuro-decision-choice actions and obtain outputs as information. In this respect the classical definition of knowledge as "*justified true belief*' must be altered into *justified true characteristic disposition (JTCD)*, where cognitively conformal mapping from the space of ontological signal will help to make possible the equality between ontic variety characteristic dispositions and epistemic variety characteristic dispositions. The ontic variety characteristic dispositions are the identities of *what there is,* and the epistemic characteristic dispositions are the claimed identities of *what there is.* The theory of knowledge is the theory of production of the claims of *what there is* and the verifications of the claims relative to *what there is* from the space of knowledge by acquaintance (exponential information) to the space of knowledge by description (abstracted information by paradigm of thought).

VII: THE MONOGRAPH

The dualistic-polar journeys over the space of variety opposites are carried on by socio-natural processes and technologies. These processes and technologies are defined in the space of categorial converters and functor space relative to the space information-decision-choice-interactive processes over the space of identification-transformation dualities, In the epistemological space, errors and mistakes occur in the space of true-false dualities and anomalies arise in the space of basic-applied dualities in journeys over the space of ignorance-knowledge dualities. The monograph is about the development of theory of organizational analytics in functor space and the space of categorial converters. The introduction of functors and categorial converter is to provide a general framework for the understanding of socio-natural production system as residing in the space of input-output dualities, where the space of inputs is conformally mapped onto the space of outputs which is then conformally mapped onto the space of inputs. The instruments of mapping are functors and categorial converters which are divided into natural functors and categorial converters, and social functors and categorial converters for actions over the space of identification-transformation dualities.

The creations of natural functors and categorial converters are natural processes induced from the organization of activities defined in the space of ontological time of the space of beginning-end duality, where the natural decision agent is to be determined. The natural functors and categorial converters are organized to operate with information-knowledge equality in the space of certainty and recklessness. in the information-knowledge equality is in the sense that *what there is,* is *what there is,* and *what there would be* is *what there would be* within the ontological time of the space of beginning-end duality. The creations of social functors and categorial converters are social processes induced from the activities defined in the space of epistemological time of the space of past-future duality. The decision-choice actions by neuro-decision-choice agents are then based on the past-present conditions contained in the space of epistemological time of the space of past-future dualities. The social functors and categorial converters operate with information-knowledge inequality with uncertainties and risks. In the sense that what is claimed may not be *what there is*, and what is claimed to happen may not be *what would be* within the epistemological time of the space of past-future dualities. The conditions of the creations of social functors and categorial converters and the demands for their uses to close the ontic-epistemic gap relative to elements in the space of certainty-uncertainty dualities call for the phenomena of the space of imaginations derived from the space of potentials. The organizational analytics lead to the understanding that within the space of imaginations are developed possibility, probability, curiosity, creativity, hope and other such phenomena which come to influence the performances of the functors and categorial converters over the space of success-failure dualities as the functors and categorial converters are used to effect the outcomes of goals, objectives, and visions which are developed in the space of imaginations to enter the space of realities contained in the space of actuals .

The organizational analytics of social functors and categorial converters relate to all epistemological varieties, where the nature and behavior of every socio-natural variety are governed by the principle of opposites under the general space of dualistic-polar games with continual functor and categorial-converter actions under cognitive capacity limitation producing negation, and negation of negation in the spaces of identification-transformation dualities and certainty-uncertainty dualities connected to the spaces of problem-solution dualities and ignorance-knowledge dualities in relational continua and unity which are generated by the elements in the space of goals, objectives, visions and preference order relative to the space of cost-benefit dualities. The conditions of the dualistic-polar game space are such that at the level of identification, every variety characteristic disposition is a closed set for its identity over the space of relation-property identification; and at the level of transformation, every variety characteristic disposition is an open set for change over the space of relation-property dualities and hence every variety characteristic disposition resides in the space of closed-open dualities for identification and transformation. Over the general space of knowing and information-knowledge production, the theoretical framework must show how cognitive agents relate the elements in the space of dualistic-polar structures to the varieties in the spaces of relation-property dualities, identification-transformation dualities relative to conditions of varieties in the space of goals, objectives and vision within the corresponding space of problem-solution dualities, where the connecting varieties of these spaces in continua and unity are established by varieties in the space of social categorial converters or functor space induced by neuro-decision-choice actions under the ruling philosophical consciencism relative to the elements in the space of property-relation dualities.

It must be noted that the outcome of any functor action or categorial-converter action is information to be processed into knowledge as input into the space of neuro-decision-choice actions. The theory of organizational analytics of the functors and categorial converters presents a framework that allows the analysis and integration of the roles of thinking, reasoning, curiosity, creativity, hope, despair, and similar phenomena into the dynamics of the space neuro-decision-choice actions over the spaces of problem-solution dualities and ignorance-knowledge dualities. Several interesting elements in the space of question-answer dualities and primary-derived dualities are then discussed such as innovations and discoveries in the space of categorial converters and the functor space relative to the spaces of fundamental-applied dualities and ignorance-knowledge dualities contained in the space of imagination-reality dualities.

Some questions for discussions center around: 1) the understanding of how functors operate over the space of problem-solution dualities, 2) the understanding of the relationships among the organizational analytics the space of categorial converters and the functor space and the mathematics of conformal mapping 3) the understanding of natural and social categorial converters and functors in the processes of knowing relative to the space of incentives relative to the space of cost-benefit analysis, 4) the understanding of the relationships among, paradigm of thought, thinking, reasoning and their developments over the space of neuro-decision-choice actions, 5) the understanding of the space of ontic-epistemic dualities relative to the space of certainty-uncertainty dualities and the source of epistemological uncertainties in human actions. , 6) the understanding of the relationship among epistemological uncertainty, potential, imagination, and reality, 7) the understanding of imagination, possibility, probability, and risk in human action, 8) the understanding of the success-failure dualities relative the spaces of problem-solution dualities and ignorance-knowledge dualities in the space of information-neuro-decision-choice systems, 9) the understanding of relational connectivity among the space of categorial converters, functor space, information-knowledge stocks and social philosophical consciencism, as guides to neuro-decision-choice actions of elements in the space of individual-collective dualities within the space of thinking-reasoning dualities and 10) the understanding of the relationships among preference orders, organizational analytics of the elements in the space of categorial converters and functor space and the space of cost-benefit dualities as housing the space of human-decision-choice incentives where cost may be viewed as suffering and benefit may be viewed as joy in an abstract space of pain-pleasure dualities and 11) how together, all these conditions provide an integrated framework for the understanding of social dualistic-polar dynamics in neuro-decision-choice space relative to natural dualistic-polar dynamics in the ontological space.

VIII: ORGANIZATION OF THE MONOGRAPH AND THE CHAPTER SUMMARIES.

The monograph is organized with a preamble, a prologue, a preface, an acknowledgement and eight chapters with epilogue. Each chapter deals with a specific set of analytical difficulties within the development of the theory of organizational analytics of the space of functors and the space of categorial converters in support of the understanding of variety dynamics, social progress and neuro-decision-choice action through the development of the theories of problem-solution dualities and polarities, information-knowledge dualities, ignorance-knowledge dualities and entropy under fuzzy information and paradigm of thought for the general understanding of the principles of opposites, contradiction tension, and conflict over the dualistic-polar game space over the space of identification-transformation dualities.

There is a *preamble to the monograph*. The preamble is composed of series of quotations from important works of different authors in philosophy, science, mathematics, economics, decision-choice theories, and other relevant areas to the information-knowledge questions about the general science and art of knowing under cognitive capacity limitations that generate questions over the spaces of certainty-uncertainty dualities and projected to the space of doubt-surety dualities. The preamble is used to initialize the idea that all information-knowledge systems are generated by knowing actions irrespective of the area of human endeavor through neuro-information-decision-choice processes over the space of problem-solution dualities within the ontological-epistemological duality. Again, the objective of the preamble is to introduce the reader to the conflict zones of knowing and to point out the epistemic zones of irreducible disagreements in knowing as intellectual investment and information-knowledge system as intellectual capital accumulation enterprise.

The *prologue* is analytically structured to follow the contents of the preamble. The prologue introduces a set of essential epistemic propositions in the theoretical system on categories, epistemic field, epistemic-field embedding, relational field and information fields and their relationships to categories of intellectual investment flows and information-knowledge accumulation in intellectual capital stocks. The prologue is used to set the stage of the development of the theory. Based on these essential propositions, the concepts of methodology, techniques, and methods for discussing the propositions are defined, explicated, and analyzed to provide conditions of their differences and commonness in theoretical and applied analytics. The techniques and methods are toolbox collections in any methodology of interest for knowing, thinking, and reasoning over the space of the problem-solution dualities. The prologue is also used to introduce the concepts of exact and inexact information structures and how they relate to the space of absoluteness and the space of relativity to affect the methodology of thinking and reasoning over the space of decidability-undecidability dualities. The prologue points to the path of the theoretical journey and the parameters of containments and deviations regarding the applicable areas over the space of question-answer dualities and polarities to establish diversity and unity principles in knowing and science.

The *preface*: the prologue is followed by a preface which introduces the motivation and conditions for writing the monograph, what it is about and how the conceptual framework of the monograph is connected to other previous monographs by the same author on information-knowledge processes and general conditions of categories of knowing. The general conceptual frame makes explicit the structure of the theory of problem-solution duality and polarity and how it forms the foundations of knowing and information-knowledge accumulation in all areas of cognitive activities over the space of fundamental-applied dualities in knowing. The analytical connectedness of these monographs is found in the consistent use of the principle of opposites, inexact information, fuzzy paradigm of thought with its mathematics, logic and corresponding toolbox of techniques and methods of analysis, where knowing is seen as an intellectual investment flow in a discovery of knowledge and information-knowledge is seen as an intellectual capital accumulation of discoveries, providing us with factories of production and consumption in the general input-output space and decision-choice actions over the space of problem-solution dualities. Together, these monographs present the meaning and essence of my works, research agenda and possible contribution to organization of knowing, research, and teaching, where information-knowledge stock-flow disequilibrium dynamics is seen as an economic production and related to continual neuro-decision-choice dynamics over the space of problem-solution dualities that resides in temporary equilibrium states and permanent disequilibrium dynamics. In a sense, the interrelated works provide foundations for explaining the continual process of transformation of varieties in terms parent-offspring replacement processes at the levels of both fundamental and applied knowledge on the dynamics of emergence of the new and the disappearance of the old, where the end is in the beginning and the beginning is in the end in continuum and connectivity as well as taking a critical look at the logic of economic theory of production in which the production of intellectual investment flows and intellectual capital stocks are based.

Chapter 1 is about organizational analytics of neuro-decision-choice actions over static-dynamic dualities and their relevance and effectiveness over the space of problem-solution dualities under conditions of knowing and information-knowledge processes within the space of identification-transformation dualities. It is about how organization is essential in connecting neuro-decision-choice actions over the space of problem-solution dualities to knowing and information-knowledge production at the levels of the individual and the collective, at the level of domestic and international, corporation and competition, production, and consumption over the space of cost-benefit dualities defined over the economic, political, and legal structures of social set-ups. It relates conditions of efficiencies to neuro-decision-choice outcomes over the space of success-failure dualities as well as connecting the fundamental information-knowledge system to the applied information-knowledge system and all decision-choice activities of human endeavors over the space of progress-retrogress dualities. The importance, here, is to examine the effects of social organization on decision-choice actions over the space of problem-solution dualities [R8.16a] and the epistemic field with relational and information fields [R8.16b] in relation to knowing and the stock-flow disequilibrium dynamics of the information-knowledge system at the levels of variety identifications and transformations and their importance to the spaces of input-output, supply-demand and production-consumption dualities. The concept of primary-derived organizations, institutions and the space of problem-solution dualities are discussed and related to neuro-decision-choice activities in knowing and information-knowledge accumulation and how they are influenced by social philosophical consciencism that has taken hold of the social set-up.

Chapter 2 provides the analysis of the relationships of organizational analytics of information-knowledge systems on basic structure, superstructure, and the space of basic-superstructure dualities in the development in knowing and the production of information-capital accumulation and their connectivity to the spaces of input-output and cost-benefit dualities. Two sets of theories of explanatory and prescriptive structures are introduced and discussed in the knowing and information-knowledge processes which are connected to descriptive and predictive structures through the functor space and connected to the space of imagination-reality dualities. An introduction to a theory of certainty-uncertainty dualities is advanced to provide explanatory conditions and sources of uncertainties which give rise to the space of imagination containing possibility and probability spaces and hence the rise of possibility and probability theories. Discussions are made on basic-superstructure components of organization of knowing and information-knowledge systems, where definitions and comparative analysis of policy and decision-choice action are presented and linked to relationality, and organicity of the space of problem-solution dualities. The roles of individuals and the collectivity in the organization of problem-solution dualities are examined in the space of individual-community dualities in relation to conflicts and conflict resolutions over the space of cost-benefit dualities. The roles of institutions of research, teaching and learning are examined, leading to the discussions on the goals, objectives, and research in the organization of knowing and information-knowledge production, where we discuss learning with and without supervision in relation to thinking and reasoning. The chapter is concluded with the discussions on the development of special institutions with global collaborations, and competitions over the epistemic field, where the space of relevant alternatives may be expanded, and the space of irrelevant alternative may be reduced with global collaboration and creative competition in discovery of elements in the space of categorial converters or the functor space under the guidance of the ruling social philosophical consciencism.

Chapter 3 deals with the theory of epistemic anomalies in the organization of knowing and information-knowledge production, where such anomalies are introduced, defined, and explicated for understanding in theories and applications over all areas of knowing under characteristic-signal analytics within the epistemic field supported by the relational and information fields. These anomalies in knowing and information-knowledge developments are explained to arise within the space of imagination-reality dualities contained in the space of actual-potential dualities as we journey between the space of imaginations and the space of realities through thinking and reasoning connecting the knowledge by acquaintance to knowledge by description with the use of an organic paradigm of thought and micro-paradigms of thought. Some anomalies are paradigm-derived, some are observationally derived, some are intentionally derived, and some are unintentionally derived. The space of imagination-reality dualities, the space of imaginations and the space of realities are then discussed in relation to the conceptual links to phenomena of possibility and probability which give rise to the spaces of possible world, impossible world and possible-world-impossible world dualities and then linked to the anomalies. From the space of imagination, we discuss the concepts, meanings and roles of creativity, curiosity, hope and courage in relation to the space of cost-benefit analysis within the space of identification-transformation dualities. Using fuzzy paradigm of thought, conditions of optimality [R6.3] are derived for hope and despair under dualistic-polar structures, where in general the phenomena of thinking and reasoning are shown as actions to balance forces within opposites and polarities to abstract the best cost-benefit varieties in human endeavor relative to some elements in the space of goals, objectives, and visions. Directly and indirectly, the forces between opposites generate the space of problem-solution dualities under the principles of opposites in all social spaces. The chapter presents the needed foundations for further algebraic analytics to make explicit the rise and nature of various anomalies in relevant spaces within the epistemic field.

Chapter 4 of this monograph discusses the roles of creativity and curiosity in knowing as the production of intellectual investment flows and information-knowledge accumulation as production of intellectual capital stocks in the in the space of problem-solution dualities and polarities within the space of imagination-reality dualities as contained in the space of actual-potential dualities relative to the space of ignorance-knowledge dualities within the space of ontic-epistemic dualities relative to the space of ontological-epistemological duality. The understanding of the roles of creativity and curiosity are related to the spaces of courage-fear dualities, hope-despair dualities, and success-failure dualities. Definitions of these concepts are presented, and their contents explicated with corresponding operator functions. Comparative analytics are undertaken to analyze their interdependencies and their relative roles and degrees of efficiencies in the space of ignorance-knowledge dualities and problem-solution dualities to show the strengths and limits of organization in neuro-decision-choice actions over the space of human behaviors under the principle of opposites. Through comparative organizational analytics in varieties and categories, we show the relational connectivity in the space of ontological-epistemological dualities in knowing, differences, similarities of *ontological information-time processes* and *epistemological information-time processes* in reference to the ontological beginning-end duality of *Bennu-Anoma-Gyenyame* conditions and epistemological past-present duality of Sankofa conditions in the space of ontological and epistemological dualities where the essentiality of symmetry and supersymmetry are discussed under the principle of opposites relative to the space of relativity-absoluteness dualities with illustration by epistemic geometric representation with the corresponding mathematical spaces in terms of axiological structures of varieties in the space of real-nominal dualities relative to the space of cost-benefit dualities relative to the space of cost-benefit dualities.

Chapter 5 is devoted to the analysis of fuzzy paradigm of thought as an organic paradigm for thinking and reasoning and in the development of fuzzy meta-paradigms for the analysis of the conditions in the space of ignorance-knowledge dualities. From the cognitive framework of fuzzy paradigm of thought the technique and method of fuzzy optimization are used to analyze the optimal conditions of varieties in the space of ignorance-knowledge dualities, where variety ignorance and variety knowledge are mutually constraining for abstracting the optimality which is then used to establish and define the transversality conditions in the journey from the zone of ignorance to the zone of knowledge in specifying the variety identity for knowing. The conditions of optimality are used to decompose the zones of knowledge and ignorance for specifying the optimal knowledge and optimality fuzzy decomposition for neuro-decision-choice actions over the space of the problem-solution dualities. Discussions are made with examples on the use of fuzzy Numbers for optimizing ignorance-knowledge relational and the use of the conditions of optimization for decomposition of the space of ignorance-knowledge dualities into the zone of ignorance and the zone of knowledge. The cumulative results of the chapter are the development of a theory of ignorance-knowledge dualities that allows the specification for the conditions of zones of ignorance and knowledge for confidence claim of variety identities with fuzzy conditionality as known where the fuzzy paradigm of thought is coordinated with the principle of opposites with relational continuum and opposites.

In chapter 6, we explore certain concepts of conceptual and epistemic anomalies such as paradox, phantom, possible world, impossible world, imagination, fantasy, curiosity and creativity in relation to reality, actuality, potentiality and other such analytical structures that emerge in knowing and information-knowledge accumulation. These concepts are placed in a system of decision-choice actions in relation to elements in the spaces of decidability-undecidability dualities and input-output dualities. The epistemic concern here, is how these anomalies relate to the space of possibility-probability dualities that houses the spaces of necessity-freedom dualities and risk-benefit dualities as one journeys between variety ignorance-and variety knowledge in ignorance-knowledge dualities in the space of intellectual investments and information-capital accumulations over the space of individual-collective dualities. The discussions on the anomalies are then linked to the discussions of categorial conversion in [R17.15] and to the discussions of social philosophical consciencism in [R17.16], where the contents in philosophical consciencism include the anomalies, conceptual weapons that generate structural conditions in the spaces of justice-injustice, war-peace, and other dualistic-polar structures. The space of imagination-reality dualities is introduced as an analytical space and shown to be a sub-space of the space of actual-potential dualities. The space of imagination is shown as a derived space from the space of the potential and them shown to contain the possibility and probability spaces, while the space of realities is shown to be contained by the space of the actuals. The space of imagination-reality dualities is argued to generate all the intellectual anomalies, the structures of which are discussed and to examine the effective integration of hope, courage, curiosity into creativity and innovations in knowing and information-knowledge processes.

Chapter 7 follows chapter 6. It deals with the theory of epistemic anomalies, where such anomalies are introduced, defined, and explicated for theory and applications in all areas of knowing under characteristic-signal analytics. These anomalies in knowing and information-knowledge developments are explained to arise within the space of imagination-reality dualities as we journey between the space of imaginations and the space of realities through thinking and reasoning connecting the knowledge by acquaintance to knowledge by description with the use of an organic paradigm of thought and micro-paradigms of thought within the organic. Some anomalies are paradigm-derived while some are cognitive-capacity-limitation derived. The space of imagination-reality dualities, the space of imaginations and the space of realities are then discussed in relation to the conceptual links to possibility and probability which then give rise to the spaces of possible world, impossible world and possible-world-impossible world dualities and then linked to intentional and unintentional anomalies and the development of social philosophical consciencism to further define the social framework of neuro-decision-choice actions and interpretations of input-output results over all relevant spaces . From the space of imaginations, we discuss the concepts, meaning and roles of creativity, curiosity, hope, courage in relation to the space of cost-benefit analysis in the space of identification-transformation dualities over all relevant spaces. Using fuzzy paradigm of thought, optimality conditions are derived for hope and despair under dualistic-polar conditions where in general thinking and reasoning are to balance forces within opposites and polarities to abstract the best cost-benefit balances. The chapter presents the needed foundation for further algebraic analytics to make explicit the various anomalies in relevant spaces.

The epilogue of this monograph provides a summary of analytical connectivity of the principle of opposites composed of dualities and polarities, epistemic fields, and fuzzy paradigm of thought in the development of knowing as the production of intellectual investment flows and information-knowledge accumulation as the production of intellectual capital stocks. The principle of opposites places the system in the space of dualistic-polar games to cross the transversality conditions where the rewards of the game are variety identity and polar dominance. The conditions of functors and categorial converters are presented and linked to curiosity, creativity, hope, and despair with suggestions for further research.

PROLOGUE

ABSTRACT:

The prologue of this monograph is used to indicate the entry points into the development of this monograph concerning the theory of organizational analytics in the functor space and the space of categorial convertors as establishing the theoretical and applied framework for examining socio-natural processes and technologies in transforming intra-categorial and inter-categorial variety characteristic dispositions as well as continuity to the theory of info-statics and the theory of info-dynamics in the spaces of problem-solution-dualities and ignorance-knowledge dualities, where every variety solution is an identification-transformation action, every variety problem is an identification-transformation action, every variety knowledge is as identification-transformation action and every decision-choice action is a variety identification-transformation action in the space of ignorance-knowledge dualities. The monograph is about the development of organization of functors and categorial conversions that will include mental qualities of curiosity, creativity in knowing and knowledge as production and consumption over the space of problem-solution dualities. Toward this, the prologue introduces some essential concepts and phenomena such as curiosity, creativity, and others through function actions in discussions of decision-choice activities over the spaces of ignorance-knowledge dualities and problem-solution dualities. In this respect, dualistic-polar game of hope-despair duality is analyzed as an example in the use of fuzzy paradigm of thought under principle of opposites. The prologue extend to the structures of the spaces of functors and categorial converters containing conditions of fundamental dimensions of curiosity, creativity, hope-despair dualities anomalies and the structure of thinking-reasoning dualities within the general theory of knowledge as a theory of intellectual actions over the spaces input-output and production-consumption dualities composed of six sub-theories of 1) information theory, 2) theories of languages and Definition, 3) descriptive theory, 4) explanatory theory, 5) predictive theory, 6) prescriptive theory and 7) quality-control theory.

1): ENTRY POINT AND EPISTEMIC CONNECTIVITY

In every theoretical development, irrespective of what category of the space of problem-solution dualities, it is always useful to indicate the entry point of its development. This indication will show us the dualistic-polar journeys that are being undertaken through the space of problem-solution dualities relative to the spaces of primary-derived dualities and variety ignorance-knowledge dualities. The dualistic-polar journeys are called epistemic processes with a general structure and categorial particularities. The general structure relates to the accepted macro-paradigm of thought as principle of cognition defined over the space of thinking-reasoning dualities. The categorial peculiarities relate to category-specific micro-paradigms of thought, where the micro-paradigms are developed by thinking as category specific algorithms to guide reasoning activities within categories to transform variety problems into variety solutions under the conditions of the accepted macro-paradigm of thought which is also developed by thinking. The epistemic process is the understanding of intellectual investment flows and information-capital accumulation in this monograph as a continuation of series of my works in presenting the theory of knowing and information-knowledge system as neuro-decision-choice activities over the spaces of production-consumption and input-output dualities, where knowing is the production of intellectual investment flows to update information-knowledge accumulation as the production of intellectual capital stocks. This epistemic structure presents knowing and information-knowledge accumulation as two interdependent processes over the space of production-consumption in infinitely disequilibrium dynamics as self-correction system just as the theory of physical investment and capital in the enterprise of production and consumption of goods and services. This epistemic structure must be grasped by all involved in the theory of knowledge if we are to understand and relate economic concept of scarcity where scarcity is defined through the elements in the space of categorial converters and the functor space relative to the spaces of identification-transformation and ignorance-knowledge dualities. In other words, scarcity is always defined within economics in terms of processes and technology given the varieties in the ontological space.

The prologue points to some essential concepts and phenomena, and their fuzzy semantics structures relative to the classical semantic structures that will be central to the development of the theory of organizational analytics of functors and categorial converters and their requirements for applications over the space of identification-transformation dualities in the institutional setting and departmental zones of knowing and knowledge development over the space of problem-solution dualities for basic and applied conditions in the theory of intellectual investment-capital processes through the neuro-decision-choice activities, where curiosity, creativity, hope, despair and many such conditions must be integrated to understand the conditions in the spaces of success-failure dualities and risk-riskless dualities. In this respect, reflections on the roles that curiosity plays in knowing and creativity plays in efficiencies in knowing, hope plays in motivating curiosity in choice, and success-failure conditions enhance creativity must be undertaken by all involved in the space of problem-solution dualities. Similarly, reflections must be undertaken on how curiosity, creativity, hope, despair, goals, objectives, and visions operate over the spaces of imaginations, ignorance, and knowledge to generate the space of possibility-probability dualities, where the concept and nature of knowledge economy is introduced for future development through the operations of functors and categorial converters. In the space of imagination as abstracted from the space of potentials, every cognitive agent resides in the space of potential as a contributor to knowledge. To enter the space of realities every resident in the space of potential must exercise his or her curiosity as a knowledge activist in the social setting and not passive and blind user of elements contained in the social philosophical consciencism. The activist must always exercise creativity for progressive change and always confine in setting hope against despair over the space of success-despair dualities for human progress. The intellectual capital-investment processes must integrate the knowledge activist as the center in all activities in the epistemological space containing all spaces and subspaces relative the space of categorial converters and the functor space, the contents of which are the results of the knowledge activist. Furthermore, the mappings induced by functors and categorial converters must maintain the conformal relation with the elements in the space of cost-benefit dualities as the space of incentive-disincentive dualities [R5.13] [R5.14].

2: ON THE STRUCTURES OF THE SPACES OF FUNCTORS AND CATEGORIAL CONVERTERS

The functors and categorial converters serve as cognitive vehicles of neuro-decision-choice agents between imagination and reality to overcome the adverse conditions in the spaces of possibility-impossibility and probability-improbability, fear-fearless, and hope-hopeless dualities over the epistemological space. The space of functors and categorial converters is divided into natural and social sub-spaces. The space of natural functors and categorial converters correspond to the activities in the ontological space. The space of social functors and categorial converters correspond to the activities in the epistemological space. The space of natural functors and categorial converters constitute the identity in the space of general universal processes, while the space of social functors and social categorial converters constitutes the derived identity in all human actions. In other words, the information-knowledge of cognitive agents in variety transformations are derived from the understanding of behavior of elements in the ontological space, where the choice of variety transformations including the varieties of knowing are induced by preference defined over the space of cost-benefit dualities relative to the space of production-consumption and input-output dualities.

The sources of natural functors in the ontological space will not be discussed, however, for their morphologies see the works of various aspects of physical, chemical, and biological science in relation to variety behaviors over the space of identification-transformation dualities. We shall discuss the nature and forms of social functors and categorial converters and the role they play in the construct of the theory of intellectual investment-capital process viewed in the space of production-consumption and input-output dualities. The approach here in additions to my previous works on the theories of information and knowledge in the space of production-consumption, input-output, demand-supply chain dualities is indirectly a critique of classical theories of information, knowledge, and knowledge economy. It is also criticisms of classical theories of decision as well as neoclassical theories of production and consumption. In this approach, the theory of information-knowledge process is viewed in terms of an integrated combination of production, quality control and consumption in the space of input-output dualities. Much of the classical works on the theory and philosophy of knowledge taking the concept and philosophy in terms of justified true belief (JTB) is centered on quality control where discussions are made of conditions of verification, falsification, collaboration, and others in the same category that demands the definition of variety knowledge and a system of tests on what is claimed to be known and the defined identities around the dubious definition of JTB. The classical theories and philosophy on information, unlike those on knowledge, on the other hand has no definition of the concept and phenomenon of information and assumes that the concept and phenomenon of information is known and understood. Much of the theory and philosophy is devoted to information storage, communication, computability, and retrieval in terms of data over the space of source-destination dualities.

The organic approach that is pursued and suggested here must have clear definitions of information, data, knowledge, and fact in distinction and must combine production, quality control and consumption, where the structure of the universe in relation to knowing as a production of intellectual-capital stock-flow process must be viewed as cognitive transformations from the ontological space to the epistemological space, as well as intra-epistemological transformations, where the space of the neuro-decision-choice activities serve as the ultimate connectors to the development of human information-knowledge stock-flow as the intellectual heritage of humanity constrained by all forms of cognitive capacity limitations over the space of fundamental-applied dualities, where the cognitive capacity limitations will vary from different social geography to another social geography [R8,16b]. The distribution of the degrees of information-knowledge stocks and the knowing process over the space of human geography will establish the distribution of sub-spaces in the space of categorial converters and the functor space relative to specific and general categories in the epistemic field with relational and information fields The same organic distribution of the degrees of information-knowledge stocks and knowing processes will also establish the types of social philosophical consciencism that will prevail and the corresponding distribution of national personalities that will induce the space of national types of exceptionalism.

Functors and categorial converters in the space of knowing as production of intellectual investment flows and knowledge accumulation as production of intellectual capital stocks have the following epistemic representations:

2.1) *Conditions of the Foundational dimensions of the curiosity*: The basic dimension of the knowing is cognitive journey to ascertain the characteristic dispositions of varieties that reveal the variety identities from the ontological space which contains the space of actual-potential dualities with contents of varieties to the epistemological space which contains the space of imagination-reality dualities, where the cognitive journeys are defined in the space of ignorance-knowledge dualities and demand consistent curiosity as an initiating functor or categorial converter in the search of relevance (broadly defined) over the space of the unknowns in the space of actual-potential dualities to connect the ontic characteristic dispositions to the epistemic characteristic dispositions. The analytical process must integrate and examine the role of curiosity at all levels of knowing, where curiosity connects the space of potential to the space of imaginations that allows the connectivity within the spaces of necessity-freedom and initial-sufficient dualities.

2.2) *Conditions of Creativity in the Space of Categorial Converters and functor Space:* The discovery of an element in the space of categorial converters and the functor space and the degrees of efficiencies that they operate over the space ignorance-knowledge dualities depend on creativity over the space of individual-collective dualities for the connecting cords between the space imaginations and the space of realities in the space of variety dualistic-polar structures setting knowledge against ignorance. The dualistic-polar connectivity between the space of actual-potential dualities in the ontological space and the space of imagination-reality dualities is established by social factor of creativity in complement to curiosity which then affects the discovery of functors or categorial converters and their innovations in connecting imaginations to potentials and transforming imaginations to realities through the space of construction-destruction processes over the space of variety disequilibrium dynamics.

2.3) *Conditions of the Space of Hope-Despair Dualities*: The varieties in the hope-despair dualities are established by superimposition over the spaces of courage and creativity, where the elements in the hope-despair dualities are intimately connected to the space of incentive-disincentive dualities over the space of dualistic-polar games, creative-destructive functors and categorial converters that bring negation and negation of negation over the space of variety input-output dualities. The actions of functors and categorial converter must pass through the space of success-failure dualities where the outcomes are conformally mapped onto the space cost-benefit dualities under the conditions in the space of hope-despair dualities. It is the conditions of variety hope, defined over success-failure dualities, that provide the required impetus to act, enhance and bring about increasing knowing and intellectual capital-investment stock-flows. It is also the conditions of variety despair, also defined in the space of success-failure dualities, that diminish the required impetus to act and bring about diminishing knowing and intellectual capital-investment stock-flows. The roles of hope-failure dualities as important factors must be integrated over the space of neuro-decision-choice actions over the space of problem-solution dualities

2.4) *Conditions of the Space of actions of functors and Categorial Converters*: In the space of general and specific principle of opposites with relational continua and unity, the variety imagination and variety potentials are destroyed and transformed by functors called destructive functor or categorial converter for negations. The space of potentials is the collection of all dualistic-polar varieties, where the characteristics of the potentials dominate the characteristics of the actuals in dualities and polarities. The space of potential is ontologically composed of *what there would be*, the telescopic future. within the space of present-end dualities which is equipped with ontological creative-destructive functors and categorial converters. The ontological creative-destructive functors and categorial converters are unknown and will be available for known when future varieties are transformed to present as telescopic present and present varieties are transformed into potential over the space of ontological beginning-end processes. Neuro-decision-choice activities are to acquire information-knowledge about the ontological functors and categorial converters in the space of ontological destructive-constructive dualities through courage, creativity and hope to transform them into corresponding epistemological functors and categorial converters in space of epistemological destructive-constructive dualities, where undesirable variety varieties are destroyed by epistemological destructive functors and categorial converters and transformed into the space of potentials. In this respect, the space of ontological functors and categorial converters is the primary identity and the space of epistemological functors and categorial converters is the derived identity through the space of thinking-reasoning dualities that connects the epistemological space to the ontological space by connecting the space of imagination to the space of potentials as well connecting the space of realities to the space of actuals.

3) THE SPACE OF THINKING-REASONING DUALITIES, SOCIAL FUNCTORS AND CATEGORIAL CONVERTORS IN THE EPISTEMOLOGICAL SPACE

The epistemological space is the space in which cognitive agents operate to accesses information and develop knowledge about conditions of varieties and categorial existence in the ontological space regarding what there was and what there is and use the acquired information as inputs into multiplicity of neuro-decision-choice actions through the space of thinking-reasoning dualities connected to the spaces of destruction-construction dualities, input-output dualities and production-consumption dualities under continual variety transformations to satisfy needs and wants in the social space of life-death dualities. The space of thinking-dualities is connected to all spaces and subspaces contained in the epistemological space. The elements in the space of thinking-reasoning dualities create actions of thought and practices in the space of imagination-reality dualities that ensure the idea that thought without action is empty in the space of realities and action without thought brings blind forces in the space of reality in that the people of actions are present-oriented while people of thought are future-oriented, while the principle of opposites connects them in relational continuum and unity. In this respect, any theory of knowledge as a serious cognitive representations of intellectual investment-capital processes must be explicit with the role that elements in the space of thinking-reasoning dualities play in connecting cognitive awareness about ontological varieties and social varieties through acquaintances in the space of acquaintances under a dynamic complex system of cognitive capacity limitations and how they form organic connectors between the epistemological space and the ontological space.

The cognitive awareness in the epistemological space produces information-knowledge conflicts and inequality in such a way that the variety awareness generates epistemic characteristic dispositions of epistemological varieties that may substantially deviate from the ontic characteristic dispositions of the ontological varieties creating variety ontic-epistemic disparities in the identification process over the acquaintance space leading to transformation decisions with information-knowledge differences in the space of cost-benefit dualities. As such, information and knowledge in the epistemological space are not the same but interdependent in their characteristic dispositions over the spaces of thinking-reasoning dualities and ignorance-knowledge dualities.

Similarly, data, fact and evidence have different characteristic dispositions but conceptually dependent in their definitional structures and intimately connected to information-knowledge characteristic dispositions in the domain of intellectual investment-knowledge stock-flow processes in production and quality control, where thinking-reasoning actions are driving force of claims over the space of true-false dualities connected to the spaces of acceptance-rejection dualities and acquaintance-description dualities giving us knowledge by acquaintance which is connected by thinking-reasoning process with paradigm of thought over the space of micro-macro dualities to give us knowledge by description where the paradigm of thought constrains criteria for quality control.

Knowledge by acquaintance is an experiential information produced by reflections as variety epistemic characteristic dispositions in the epistemological space derived from the ontological signal dispositions relative to ontic characteristic dispositions in the ontological space. The ontic characteristic dispositions reside over the space of absoluteness with information-knowledge equality and hence irrefutable in the space of thinking-reasoning dualities. The epistemic characteristic dispositions reside over the space of relativity with information-knowledge inequality and hence claims are refutable in the space of thinking-reasoning dualities in the sense of principle of opposites, where every claim contains true-false duality in relativity over the space of dualistic-polar games. The space of ontological *what there was, what there is* and *what there would be* resides in the space of absoluteness as an identity connected to the space of beginning-end dualities of *ontological time,* where the ontological time is connected to ontological processes and defined in the space of beginning-end duality. The space of epistemological *what there was, what there is* and *what there would be* resides in the space of relativity as derived identity connected to the space of past-future dualities of the *epistemological time*, where the epistemological time is connected to epistemological processes and defined in the space of past-future dualities. It must be noted that there are important differences and commonness between ontological processes and epistemological processes, between ontological time and epistemological time, between the space of ontic characteristics and the space of epistemic characteristics and between the space of ontological *what there was, is and would be* and epistemological *what there was, is and would be* and the space of ontological beginning-end duality and epistemological past-future dualities. These differences point to the spaces of certainty-uncertainty and unknowability-unknowability dualities.

The structure as presented here allows the derivations of the conditions and phenomena of the concepts of the space of certainty-uncertainty dualities, epistemological uncertainty, epistemological certainty, imagination relative to potential, possibility-impossibility dualities, probability-improbability dualities, success-failure dualities, doubt-surety dualities as neuro-decision-information-interactive concepts, where the concept of information has an explicit definition as characteristic dispositions which are the contents of information that fix the identities of varieties, and supported by signal dispositions as identity carriers in the space of ontological-epistemological dualities on the basis of which the family of ordinary languages (FOL) and family of abstract languages (FAL) are developed as representations of real and nominal varieties with action connectivity over the space of source-destination dualities, all by the system of thinking-reasoning processes. In this respect, the concept of information as appears in different category of thinking and communications in theory and practice is not fluid and amorphous in the space of thinking-reasoning dualities. The concept and phenomenon of information are one and indivisible. The concept and phenomenon of characteristic disposition are intricate categories of information in accordance with varieties and categorial varieties over the epistemic field containing families of categories as input-output dualities in the space of thinking-reasoning processes in establishing distinctions and commonness with differences and similarities in the general space of production-consumption dualities.

Available to us are two types of information. There is ontological *information* which is always equal to *ontological knowledge* as represented in the space of ontic characteristic dispositions of varieties in perfect identities. There is also the *epistemological information* also known as *knowledge by acquaintance* or *experiential information* which is derived in the space of acquaintance from variety *ontological signal dispositions*, which are always unequal to the variety ontological knowledge, to obtain *epistemic characteristic dispositions* of varieties in imperfect identities. Experiential information or knowledge by acquaintance is composed of empirical information and axiomatic information in the spaces of exact-inexact, vague-clear, and fundamental-applied dualities. A general theory of knowledge must consist of conditions of intellectual investment-capital stock-flow and source-destination processes through the space of thinking-reasoning processes, where the elements operate over the spaces of production-consumption dualities and input-output dualities with explicit inclusion of the effects of thinking-reasoning processes in the use of information as input to produce knowledge as output which then serves as input for further production over the space of ignorance-knowledge dualities connected to the space of varieties with variety distinctions and commonness over social preferences and under the principle of non-satiation, where thinking-reasoning processes connect the mind-matter visible-invisible and source-destination conditions to the spaces of problem-solution dualities to develop functors and categorial converters over the space of destruction-construction dualities. The functors and categorial characteristic converters must be seen as cognitive conformal mappings from categories into the space of ignorance-knowledge dualities. The approach here points to a critical objection of classical definition of knowledge as a *justified true belief* (JTB) in the definition resides in the space of confusion, since this definition has no defined connection to information which is also resides in state of definitional confusion without any explicit characteristic disposition. If knowledge is justified by true belief, then how is this definition relates to concepts of data, fact, evidence, and information?

The variety epistemic characteristic dispositions are derived from the ontological signal dispositions of the same varieties from the space of actuals as projected by the ontological characteristic dispositions onto the ontic space. The derivation process is a functor-action mapping from the space of ontic characteristic dispositions to the space of epistemic characteristic dispositions, where the variety epistemic characteristic dispositions go through refinements by another functor action within the space of ignorance-knowledge dualities, where the refinements are undertaken with thinking-reasoning processes to obtain knowledge with epistemic conditionality which must be defined and specified to include transversality conditions from the zone of variety ignorance to the zone of variety knowledge .

4) ONTOLOGICAL TIME AND EPISTEMOLOGICAL TIME.

It is useful to understand the space-time relation over the space of identification-transformation dualities. The space-time structure may be divided into ontological space-time structure and epistemological space-time structure. The ontological space-time is defined as the space of beginning-end duality with fixed endpoints, where both the beginning and end will never be known by any cognitive agent. The epistemological space-time is defined as the space of past-future duality with variable endpoints, containing the space of past-present dualities, also with variable endpoints, which contains the space of knowability. The space of past-future dualities is contained in the space of beginning-end duality. The future structure ia always unknown and constructed from imagination derived from the ontological potential. The ontological time correspond to ontological and natural processes; the epistemological time correspond to epistemological and social processes in the spaces of identification-transformation and static-dynamic dualities, where the ignorance-knowledge processes take place over ontological-epistemological space-time dualities.

5) THE SPACE OF THINKING-REASONING DUALITIES AND LOGIAL ANOMALIES IN THE EPISTEMOLOGICAL SPACE

The elements in the space of thinking-reasoning dualities operate over the space of imagination-reality dualities contained in the epistemological space derived from the ontological space as the identity. The journeys from the space of potentials to the space of imaginations and to the space of realities are undertaken with different functors and categorial converters through the possibility space and then the probability space where the possibility space is generated by qualitative uncertainties and the probability space is generated by quantitative uncertainties. These cognitive journeys are plagued with epistemic anomalies such as errors, paradoxes, phantom problems, ill-posed problems, unsolvable problems which are defectiveness that are results of functor and categorial converter actions in intra-space and interspace crossings with paradigm of thought and thinking-reasoning processes. These defectiveness in the space of categorial converters and functor space must be examined, explained, and explicated within the theories of knowing and information-knowledge stock-flow accumulation as intellectual production-consumption processes over the space of fundamental-applied dualities, where information and knowledge are both inputs as well as outputs. In the organic epistemic process, the anomalies in knowing and information-knowledge production, whether theoretical or applied, are epistemic and arise within the space of acquaintance-description dualities and are generated by the elements in the space of certainty-uncertainty dualities and attributable to paradigm of thought in the space of thinking-reasoning dualities over the space of absoluteness-relativity dualities.

The theories of knowing and information-knowledge accumulation must also indicate the intellectual qualitative characters of the functor and characteristic converters over the space of ignorance-knowledge dualities and the conditions of moving from the space of ignorance to the space of knowledge that will indicate the epistemically dualistic-polar transversality conditions. They must indicate the quality control processes over the input-output dualities as they relate to the spaces of problem-solution dualities and question-answer dualities.

6) COMPONENTS OF THE THEORIES OF KNOWING AND KNOWLEDGE

The theory of intellectual investment flows as the theory of knowing and the theory intellectual capital stocks as the theory of information-knowledge accumulation constitute the *general theory of knowledge* over the fundamental-applied duality. The general theory of knowledge is a theory of mental actions over the spaces input-output and production-consumption dualities composed of six sub-theories which are applicable to all areas of knowing and human actions as intellectual factories of information-knowledge enterprise. The sub-theories are: 1) information theory, 2) theory of languages, 3) descriptive theory, 4) explanatory theory, 5) predictive theory, 6) prescriptive theory and 7) quality-control theory. These theories are seen in terms of identification-transformation variety. identities and behaviors over the space of static-dynamic dualities.

6.I) *Information Theory*: The information theory presents definitional characteristics of information and identities of varieties contained in source-destination dualities in the space of ontological-epistemological dualities. The information theory must provide an explicit definition of information in terms of meaning and essence in presenting variety identities and corresponding phenomena to reveal variety similarities, commonness, and differences over the space of identification-transformation dualities to establish initial conditions of knowing and category formation. The variety identities, phenomena, similarities, differences, and commonness must be seen in terms of characteristic dispositions defined in the space of quantity-quality dualities. Information theory must include the sub-theories of characteristics and signals as input-output processes in the space of source-destination duality, such that a general theory of communication within the space of ontological-epistemological dualities contained in the information theory is also viewed in the spaces of production-consumption and input-output dualities. In this way the information theory must explain the nature of acquaintance space, production of *experiential information* or knowledge by acquaintance and must account for the division of experiential information into *empirical information* and *axiomatic information* and their usefulness in the intellectual investment-capital stock-flow processes in the space if ignorance-knowledge dualities. The collection of all information theories constitutes the *space of information theories*. The philosophy of information studies the foundations of the origins of the concept and phenomenon of information and types of the nature of information and the efficiencies of the methods, techniques, and paradigm of thought for the production of the theory over the space of imagination-reality dualities contained in the space of actual-potential dualities within the space of ontological-epistemological duality containing the spaces of ontic-epistemic dualities, identification-transformation dualities and static-dynamic dualities relative to ontological-epistemological space-time structures. The philosophy of information must examine representation and storage in the space of digital-analog duality relative to the space of discrete-continuum dualities. The philosophy must examine how the information theory deals the complexities of characteristic dispositions of categories and categorial relations in the epistemic field to account for differences in multiplicity of varieties, categorial varieties, relations of varieties and categories over the spaces of intra-inter dualities that give rise to concepts of knowledge, data, facts, ideas, and evidence in their characteristic dispositions within the infinite space of error-correction dualities contained in the infinite space of imagination-reality dualities in the space of discrete-continuous dualities. The collection of all philosophies of information theory is the space of philosophy of information.

6.II)*Theory of languages and definitions*: The theory of languages and definitions is a logical nature, structure and condition of definition and the theory of definition in the development of knowing, variety, information-knowledge elements over the space of nominal-real linguistic dualities in order to establish distinctions, similarities, differences and commonness over the spaces of representations and source-destination, coding-decoding dualities in all human-human, human-machine and machine-machine dualities in all communications-in terms of visual, audio, nonvisual and non-audio, print and non-print. The objective of theory of language is the establishment of communicational elements in the space of coding-decoding dualities relative the space of characteristic-signal dualities in relation to source-destination dualities, where variety identities are coded as signal dispositions and must decoded into characteristic dispositions to reveal variety identities. The theory of definition is to encode the characteristic dispositions as signal dispositions of variety identities for decoding in the space of source-destination dualities for any language of communication. The theory of language may then be viewed as philosophy of language linguistics, representations of variety dispositions, methods, and analytical techniques, with definitions representing variety identities over the spaces of real-normal dualities, vagueness-clarity dualities, and exactness-inexactness dualities relative and connected to the space of variety dispositions abstracted from the space of characteristics from ontological-epistemological space-time structures. The categories of variety dispositions present diversities which are linked together in unity of communications and understanding over the spaces of source-destination dualities and static-dynamic dualities by accepted operating rules called grammar. The collection of all *theories of languages (theories of definitions)* constitutes the space of theories of languages (theories of definitions) and the collection of philosophy of theories languages (theories of definitions) and sub-theories constitutes the space of philosophy of theories language (theories of definitions).

6.III) *Descriptive Theory:* The descriptive theory must describe conditions of varieties and categories and fix the identities from the contents of information in constructing the space of real-nominal dualities for use in the space of source-destination dualities in the epistemological space. The descriptive conditions must relate to variety characteristic dispositions for identity recognitions. The descriptive theory fixes the linguistic characters of the meaning and essence of the content and phenomena of varieties in the space of epistemological varieties. It describes the variety identities and behaviors in terms of characteristic dispositions with the contents and signals from the information theory. The descriptive theory must contain sub-theories of ontological coding and epistemological decoding in the space of representation of phenomena and identities to produce experiential information and epistemic characteristic dispositions for epistemic varieties in the space of quality-quantity dualities. The sub-theory of ontological codding and epistemological decoding constitute the general theory of languages which may be divided into a family of ordinary (natural) languages (FOL) and the family of abstract languages (FAL) that are used to store and transmit the contents or characteristic dispositions in the space of epistemic varieties as abstracted from the information theory. The theory must relate to all categories and departments in the epistemic field accounting for differences in multiplicity of varieties, categorial varieties, relations of varieties and categories over the space of intra-inter dualities that gives rise to knowledge, data, facts, ideas, visions, goals, and evidence as conceptual sequence of information derivatives in the infinite space of error-correction processes over the space of ignorance-knowledge dualities. In other words, a descriptive theory must deal with the conditions in the epistemic field with relational and information fields defined over the space of varieties to make explicit the variety identities in the communicational space of source-destination dualities where the explicit identities find meaning in the space of characteristic disposition. The collection of all descriptive theories constitutes the space of descriptive theories with relational continuum and unity. The philosophy of descriptive theory studies the foundations of the origins of the concept and phenomenon of variety representation, meaning and phenomenon of means of representation in relation to forms of languages in the space of visual-nonvisual dualities, audio-nonaudio dualities and other forms over the space of source-destination dualities in characteristic dispositions, paradigm of thought, the efficiencies of methods, techniques, and logic of the theory over the space of constructionism-reductionism dualities of the relative to the space of imagination-reality dualities contained in the space of actual-potential dualities within the space of ontological-epistemological duality, and within the spaces of identification-transformation dualities, static-dynamic dualities, vagueness-clarity dualities discrete-continuous dualities and exactness-inexactness dualities relative to ontological-epistemological space-time structures. Descriptive theory moves from information theory, containing both experientially empirical and axiomatic characteristic dispositions, to produce conditions for fixing epistemic identities in variety, differences, and commonness in characteristic dispositions for category formations over the space of definitional real-nominal duality in the space of languages. The collection of all descriptive theories constitutes the space of descriptive theories and the collection of all philosophies of descriptive theories constitutes the space of philosophy of descriptive theories.

6.IV) *Explanatory Theory:* The explanatory theory presents logical connectivity of variety and categorial internal structures and variety and categorial external structures to explain the intra-variety, intra-category, inter-variety and inter-category activities in the space of behavioral relations over past-present dualities with the contents from the descriptive theories over the space of identification-transformation dualities generating explanations to establish initial conditions and necessities inherent in the characteristic dispositions of varieties and categorial varieties within the space of identification-transformation dualities. The explanatory conditions deal with state of variety characteristic dispositions contained in the space of state-process dualities to establish conditions of intertemporal comparisons relative to the space of functors and categorial converters for actions over the space of input-output dualities contained in the spaces of actual-potential dualities and imagination-reality dualities, and cost-benefit dualities contained in the space of imagination-reality dualities. The philosophy of explanatory theory studies the foundational meaning and action of the concept and phenomenon of variety explanation from the dispositional results of descriptive theories, their relevance in the general theory of knowledge and how the explanatory processes are affected by meanings , phenomena and means of representation in relation to forms of languages, of visual, audio, nonvisual, non-audio, the paradigm of thought, efficiencies of its methods, techniques, and logic for processing the results from information theory and the descriptive theory over the space of imagination-reality dualities contained in the space of actual-potential dualities within the space of ontological-epistemological duality and within the spaces of identification-transformation dualities, static-dynamic dualities, vagueness-clarity dualities and exactness-inexactness dualities relative to ontological-epistemological space-time structures. Explanatory theory deals with the roles of explanatory functor actions and categorial-converter actions as conformal mapping on variety dispositions as well as categorial dispositions in essence and behavior. Explanatory theory moves from information theory, containing experientially empirical characteristic dispositions of epistemic varieties, to descriptive theory, producing results for fixing epistemic identities in essence and behavior over the space of definitional real-nominal duality, to produce the results of variety behavior and essence as to fixing the understanding of the initial conditions and necessity of varieties over the space of identification-transformation dualities. The collection of all *explanatory theories* constitutes the space of explanatory theories, and the collection of philosophy of explanatory theories and sub-theories constitutes the space of philosophy of explanatory theories.

6.V) *Predictive Theory:* The predictive theory presents a logical relation between telescopic present and telescopic future in terms of variety and categorial behavioral consistency and change of variety characteristic disposition based on conditions of explanatory theories, predictive functor actions and characteristic-categorial-converter actions on the space of variety past-present dualities with inputs of experientially *empirical information* as presented from the results of information theory and processed by explanatory theory . In this way, variety identities are mapped onto variety transformation over the space of static-dynamic dualities with functor variety and categorial internal structures and variety and categorial external structures to predict the intra-variety, intra-category, inter-variety and inter-category future dispositions in the space of variety behavioral relations and identities with the results from the explanatory theories over the space of identification-transformation dualities generating results that will establish variety future telescopic conditions and sufficiency inherent in varieties and categorial varieties within the space of identification-transformation dualities. The predictive conditions deal with comparative states of present-future varieties contained in the space of state-process dualities to establish conditions of intertemporal comparisons, relative to the space of predictive functors and categorial converters for actions over the space of input-output dualities and variety present-future telescopic identities based on empirically processed knowledge by descriptions. Prescriptive theory moves from information theory, containing experientially empirical characteristic dispositions, to descriptive theory, producing results for fixing epistemic identities over the space of definitional real-nominal duality, to explanatory theory of variety behavior and essence, to obtain the results of predictions. The philosophy of explanatory theory studies the foundational meaning and actions of the concept and phenomenon of variety explanation from the results of descriptive theories, their relevance in the general theory of knowledge, and how the explanatory processes are affected by meanings phenomena and means of representation in relation to forms of languages, of visual, audio, nonvisual and non-audio, and a paradigm of thought with its efficiencies of methods, techniques, and logic for processing the results of variety characteristic dispositions from information theory and the descriptive theory over the space of imagination-reality dualities contained in the space of actual-potential dualities within the space of ontological-epistemological duality with active reference to the spaces of identification-transformation dualities, static-dynamic dualities, vagueness-clarity dualities and exactness-inexactness dualities relative to ontological-epistemological space-time structures. The collection of all *predictive theories* constitutes the space of predictive theories, and the collection of philosophy of predictive theories and sub-theories constitutes the space of philosophy of predictive theories.

6.VI) *Prescriptive Theory:* The prescriptive theory presents logical relations between telescopic present and telescopic future in terms of undesirable variety characteristic dispositions and categorial behaviors in the space of present realities and change them into desirable variety characteristic dispositions in the future reality based on conditions of descriptive theories relative to axiomatically experiential information, prescriptive functor actions and characteristic-categorial-converter actions on the space of variety present-future dualities with a movement from potential space to the space of imaginations containing possibility space and probability sp[ace. In this way, present undesirable variety identities are conceptually mapped onto desirable variety identities over the space of static-dynamic dualities with functor actions to transform unwanted variety and categorial internal structures and variety and categorial external structures containing necessity and freedom based on preference order over the space of cost-benefit dualities to prescribe transformation rules for the present undesirable intra-variety, intra-category, inter-variety and inter-category dispositions to achieve the desirable variety dispositions in in the space of identification-transformation dualities with explicit neuro-decision-choice actions in the space of behavioral relations with the results from the axiomatic information in information theory and descriptive theory bypassing the conditions of explanatory theory over the space of identification-transformation dualities generating to generate rules of actions to establish desired variety future telescopic conditions in the space of necessity-freedom dualities where necessity is inherent in the possibility space and sufficiency is inherent in *probability space* in varieties and categorial varieties within the space of identification-transformation dualities. The prescriptive conditions deal with comparative states of present-future varieties contained in the space of state-process dualities with subjective criteria and neuro-decision-choice actions to establish conditions of intertemporal comparisons relative to the space of prescriptive functors and categorial converters for actions over the space of input-output dualities and variety present-future telescopic identities from the space of imaginations and space of realities. The philosophy of predictive theory studies the foundational meaning and actions of the concept and phenomenon of variety identities from the results of descriptive theories on the basis axiomatic information, their relevance in the general theory of knowledge, and how the predictive processes are affected by meanings, phenomena and means of axiomatic representations in relation to forms of languages, of visual, audio, nonvisual and non-audio, and a paradigm of thought with its efficiencies of methods, techniques, and logic for processing the results of variety characteristic dispositions from information theory and the descriptive theory over the space of imagination-reality dualities contained in the space of actual-potential dualities within the space of ontological-epistemological duality with active reference to the spaces of identification-transformation dualities, static-dynamic dualities, vagueness-clarity dualities and exactness-inexactness dualities relative to ontological-epistemological space-time structures. The collection of all *prescriptive theories* constitutes the space of prescriptive theories, and the collection of philosophy of prescriptive theories and sub-theories constitutes the space of philosophy of prescriptive theories.

6.VII) *Quality-Control Theory.* Quality-control theory moves through the general space of all theories to execute tests of claims over the spaces of true-false dualities and acceptance-rejection dualities in relations to information theory, descriptive theory, explanatory theory predictive theory and predictive theory as seen from the space of neuro-decision-choice actions from the space of imaginations abstracted from the potential space. The quality-control tests are required since the space of imaginations and its sub-spaces are the result of conditions of the space of certainty-uncertainty dualities resulting from the space of ontic-epistemic dualities, where the epistemic varieties reside in the space of information-knowledge inequality at the acquaintance space and ontic varieties reside in the space of information-knowledge equality such that the ontic varieties constitute identities and the epistemic varieties are derivatives giving rise to epistemic distances that produce uncertainties. These epistemic distances are the foundation of the rise of the space of imaginations containing possibility and probability sub-spaces. The possibility and probability spaces are sequentially derived from the space of acquaintance to give rise to possibility, probability, statistical theories, and general theory of entropy over the space of quality-quantity dualities for the development of the quality-control theory. The quality-control theory is about the tests of efficiencies, broadly defined, on how the neuro-decision-choice actions perform over the space of identification-transformation dualities in terms of identities and behavior and the degrees of closeness of the epistemic claims to ontic identities.

In the space of variety identification, the tests are performed on the correctness of characteristic dispositions for variety identities that enter the space of epistemic characteristic disposition in terms information, behavior, and essences. The tests thus involve claims in information theories, descriptive theories, and explanatory theories. In the space of variety transformations, the tests are performed on correctness of transformations of variety characteristic dispositions in terms of future behavior and essences. The tests thus involve claims in predictive theories and predictive theories in all areas of knowing. The philosophy of quality-control theory studies and examine the foundations of meaning and actions of the concept and phenomenon of variety tests applied to conditions of information-knowledge claims from the results of all theories covering all categories contained in the epistemic field. It must examine forms of claims and their conditionality in the space of possibility-probability dualities and necessity-freedom dualities, paradigms of thought with its efficiencies of methods, techniques, and logic to arrive at the claims in accounting for elements in the spaces of certainty-uncertainty dualities, vagueness-clarity dualities, and exactness-inexactness dualities relative to the space of imagination-reality dualities contained in the space of actual-potential dualities. It is important to note that the quality control theory unifies all areas categories of knowing. The collection of all *quality-control theories* constitutes the space of quality-control theories, and the collection of philosophy of quality-control theories and sub-theories constitutes the space of philosophy of quality-control theories.

7): REFLECTIONS ON THE THEORIES

All these theories are common to all categories of knowing and information-knowledge accumulation, where knowing is the production of intellectual investment flows and information-knowledge accumulation is the production of intellectual information-capital stocks as input-output system relative to the general space of problem-solution dualities for the establishment of infinite intellectual stock-flow disequilibrium system under cognitive limitations which give rise to *uncertainties*, the elements of which are established in the space of characteristic-signal dualities from the space of varieties, where the variety uncertainties give rise to imaginations to reduce uncertainties. The theories are developed in the space of imaginations relative to ignorance-knowledge dualities from the space of actual-potential dualities to deal with uncertainties from the space of ontic-epistemic dualities, where uncertainty gives rise to *imaginative thinking* which then gives rise to *possibilistic thinking* with anticipations in the space of hope-despair dualities relative to the space of potentials as well as gives rise to the *probabilistic thinking* with expectations in the spaces of success-failure and risk-riskless dualities relative to resource space in the ontological space and social philosophical consciencism of epistemological space. The relational structure of the theories is presented as a paremiological geometry in Figure 7.1.

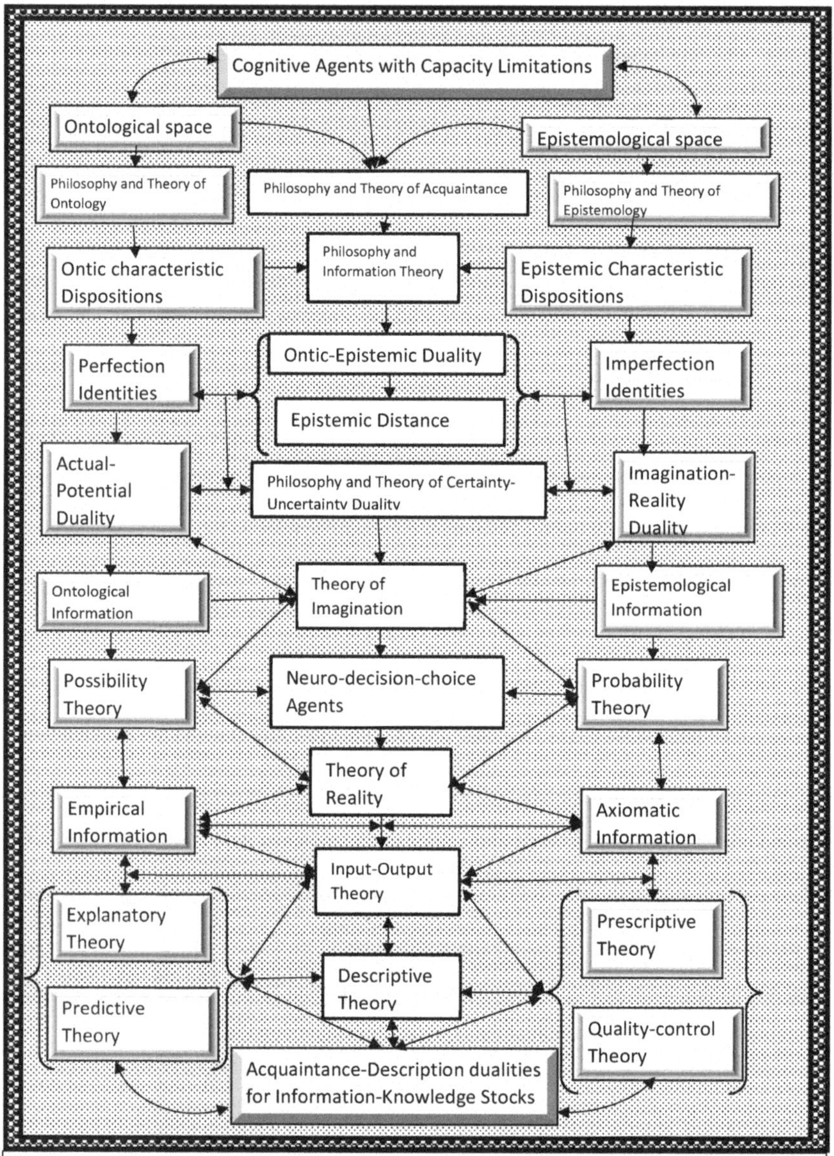

FIGURE 7.1: Paremiological Geometry of a System of Theories in an Acquaintance-Description Dualities for Knowing and Information-Knowledge Production.

The space of human actions must be understood. From the space of acquaintance within the space of ontological-epistemological dualities, we have the ontic-epistemic duality which gives rise to variety epistemic distances generated by cognitive capacity limitations. The variety epistemic distances give rise to the space of certainty-uncertainty dualities within the space of ignorance-knowledge dualities relative to the space of human actions. The variety uncertainties from the space of certainty-uncertainty dualities give rise to *imaginative thinking* within the epistemological space connected to the ontological potential. The imaginative thinking gives rise to conditions of *possibilistic thinking* with anticipation in the space of hope-despair dualities affected by the conditions of past-present dualities, as well as gives rise to conditions of *probabilistic thinking* with expectation in the spaces of success-failure and risk-riskless dualities relative to the conditions of resource space. The resource scarcity and the need to reduce uncertainties leads to the development of paradigm of thought and *rational thinking* in the space of acquaintance-descriptive and true-false dualities relative to the space ignorance-knowledge dualities to improve the knowledge inputs in the space of neuro-decision-choice actions relative to cognitive capacity limitations of all forms including fuzziness, volume-limitation, geographical limitation and many such limitationality and limitativeness. The theory of knowledge composed of knowing as production of intellectual investment and information-knowledge accumulation as intellectual capital stock is composed of all the relevant sub-theories, and then traced to cognitive capacity limitations related to inputs of neuro-decision-choice actions.

The knowing as the production of intellectual investment flows and the information-knowledge accumulation as production of intellectual capital stocks take place over the space of identification-transformation dualities, fundamental-applied dualities to deal with varieties in the space of ignorance-knowledge dualities in all categories in the epistemic field where the methods, techniques of executing the knowing and information-capital accumulation are theories which take forms of specificities that are categorial dependent. Every practice has a theory behind it as a framework of action and every theory has a practice behind it as thought in the framework of thinking and reasoning. Theories projects thoughts which contain ideas, plans, suggested rules of explanation, reasoning, and prescriptions for practice in terms of inputs and directions of actions. Practices are actions, the results of which become outputs in response to theories. In this way, both knowing, and information-knowledge accumulation find expressions over the space of thought-practice dualities, where every theory expands the space of practices, and every practice expands the space of theories in the space of problem-solution dualities and question-answer dualities relative to the spaces of blindness-emptiness dualities and ignorance-knowledge dualities. In this way, every theory is an organization of knowledge by acquaintance or experiential information to present a hypothesis or group of hypotheses to arrive at knowledge by description relative to the length of the epistemic distance in the space of ontic-epistemic dualities over the space of methodological constructionism-reductionism dualities. As discussed, the concept and practice of experimentations in some categories of knowing are to improve observations and exactness of characteristic dispositions in the acquaintance space to reduce variety epistemic distances in the variety ontic-epistemic dualities.

The development of any theory with sub-theories in the space of methodological constructionism within the space of theoretical independence-interdependence dualities takes place in the space of imaginations by the use of instruments abstracted from the space of thinking-reasoning dualities to connect the potential space to the space of realities and then to the space of actuals through forms of test by collaboration, verification, falsification and statistical methods in the space of methodological constructionism. The conditions of the space of imaginations are essential to the understanding of neuro-decision-choice activities in the space of knowing that connects information-knowledge processes of past, present and future (Sankofa-anoma information-time trinity) and sets the architect of human creation, progress and telescopic future in the space of possibility-probability dualities by overcoming forms of cognitive capacity limitations over the space of negative-positive dualities within the space of imagination-reality dualities. The space of imaginations contains the space of possibility-probability dualities, the conception of possible worlds and the possibility space which houses the necessity and anticipations, as well as the probability space, which houses freedom and expectations. The development of the general theory of knowledge must connect thinking-reasoning processes in knowing and information-knowledge production and conditions of the space of necessity-freedom dualities to conditions of the space of possibility-probability dualities. The space of possibility-probability dualities must also be connected to the space of anticipation-expectation dualities in dualistic-polar structures which houses, anomalies in the space of thinking-reasoning dualities relative to all journeys within the spaces of ignorance-knowledge dualities, problem-solution dualities, and success-failure dualities within the space of imagination-reality dualities.

The general conditions of curiosity, courage, hope, creativity, necessity, and freedom as affecting thinking and reasoning in the knowing processes must be examined within the theory of knowing and the theory of information-knowledge systems and how anomalies, ill-posed-problems, unsolvable problems, un-computable problems arise in space of thinking-reasoning dualities. In this respect, one is not only interested in the definition of information, knowledge and their justification, but one is also interested on how the information-knowledge items are produced and justified and linguistically related to produce other related concepts and phenomena. The understanding of the roles played by the neuro-qualitative elements in thinking and reasoning will provide important conditions for the development of social frameworks to enhance the efficiencies and effectiveness of the functors and categorial converters by integrate incentive dynamics within the spaces of cost-benefit dualities and input-output dualities relative to social philosophical consciencism that has taken hold over the space of individual-collective dualities of societies in the selections from space of categories of problem-solution dualities, social resource in the stock-flow processes as intellectual investments and capital stocks.

The rise of anomalies such as paradoxes, phantom problems, ill-posed problems, unsolvable problems, news, faked news, misinformation, and disinformation must be examined, explained, and explicated within the theories of knowing and information-knowledge accumulation as intellectual production-consumption processes with conformal functor and categorial-converter mappings over the space of fundamental-applied dualities, where information and knowledge are both inputs as well as outputs. The anomalies in knowing whether theoretical or applied are the results of neuro-decision-choice actions. They arise within the space of knowledge by acquaintance-description dualities and are attributable to paradigm of thought and cognitive capacity limitations through as system of theories and practices over the space of relativity-absoluteness dualities. The theories of knowing and information-knowledge accumulation must also indicate the intellectual quantity and qualitative motions over the space of ignorance-knowledge dualities and the conditions of moving from the space of ignorance to the space of knowledge to indicate how the functor and categorial converter actions reach the zone of epistemic transversality conditions. They must indicate the tracking and quality control processes over the input-output dualities. The relevant varieties in the space of question-answer dualities are discussed in the prologue of the monograph [R8.16] and will not be repeated here.

Similarly, discussions on paradigms of thought, principle of opposites, the conditions of methodological and analytical toolboxes, defining the nature of the functors and categorial converters must be discussed relative to paradigm shifting over the space of mega-meta dualities over the spaces of problem-solution dualities, question-answer dualities and cognitive capacity limitations that generate the space of certainty-uncertainty dualities, relative to the spaces of cost-benefit dualities, and incentive-disincentive dualities, with their relational effects on knowing and information-knowledge accumulation over the space of production-consumption dualities. Within the space of imagination-reality dualities, the theory of knowledge is seen not simply as discussions on and finding the meaning of knowledge and what ideas and conceptions constitute knowledge and its test of validity in the space of ontic-epistemic duality. The theory of knowledge is about the theory of information-knowledge accumulation as the production of intellectual capital stocks with functor and categorial-converter actions over the space of quality-quantity dualities within the space of general input-output dualities supported by the theory of knowing as a production of intellectual investment flows from the space of imaginations to update the intellectual capital stocks in the space of realities, where the fundamental and applied variety dispositions are connected by dualistic-polar structures with mutual give-and-take dynamic sharing relations for continuum and unity.

The categorial-converter and functor actions as process-technological functions act as gradual processes in mapping variety dispositions in the zones of ignorance into variety dispositions of zones of knowledge and zones of penumbral regions of conflicts in the space of neuro-decision-choice actions over the space of individual-collective dualities in human endeavors in the past-present telescopic processes in the space of knowability-unknowability dualities for present-past and present-future telescopic processes. The successes of the constructs and practices of these theories over the spaces of static-dynamic dualities and identification-transformation dualities will always depend on cognitive capacity limitations and the social philosophical consciencism that holds social personalities and national personalities, national exceptionalism over the space of individual-collective dualities and national-international dualities defined over the general space of neuro-decision-choice processes.

The term cognitive capacity limitation as applied in these discussions is in reference to development and utilization of the contents in the social philosophical consciencism in the development of individual personalities and exceptionalisms and national personalities and exceptionalisms. The cognitive capacity refers to individual and collective upper bounds in information acquisition, storage and processing giving cognitive abilities over the spaces of problem-solution dualities, question-answer dualities and ignorance-knowledge dualities in relation to a family of forms of knowing such as pleasure-pain, joy-sorrow, happiness-sadness, new-old life-death, oneness-dividedness, composition-decomposition, unity-disunity, organization-disorganization dualities and other dualistic-polar existentiality. Cognitive ability is innate to all humans from which cognitive capacity are developed in the space neuro-decision-choice processes in relation to the space of curiosity, creativity, hope, despair and many such characteristics over the space of quality-quantity dualities. In all these, what the system theory of knowing as a continual production of intellectual investment flows and the theory information-knowledge accumulation as production of intellectual stocks needs is to plant seeds of curiosity, creativity knowledge production, pursuit of knowing in the social philosophical consciencism for the youths and the youths will grow with cognitive capacity for exploring the zone of knowledge by avoiding the zone of ignorance. In this respect, the theory of knowledge provides thought for action to avoid the state of emptiness and action is guided by thought to avoid the state of blindness, supporting the general principle of neuro-decision-choice actions as always balancing forces between opposites to achieve benefit over cost in in the space of varieties, where the theories of information and production are viewed in terms the economic theories of production-consumption dualities and input-output dualities in the space of visible-invisible dualities relative to the space of property-relation dualities and polarities.

A PROPOSAL

1. Statement of Aims and Rationale behind the Monograph

a) The *aim* of this monograph is to provide organizational analytics of space of categorial converters, functor space and how these spaces connect to the organization of knowing as the production of intellectual investment flows and information-knowledge accumulation as the production of intellectual capital stocks as a theory of production-consumption duality and input-output dualities over the space of neuro-decision-choice actions constrained by cognitive capacity limitations in dynamic complexities. The goal is extended to the development of a framework for analytical understanding of transformations in the space of dualistic-polar structures where inputs become outputs, output become inputs, problems become solutions and solutions become problems in static-dynamic disequilibrium processes under actions of functors and categorial converter in the space of matter-energy equilibrium, where the natures of functor and categorial actions allow the analyses of dualistic-polar games, the results of which are dualistic-polar dominance in variety identities such that there are explanations of relational connectivity and unity between the ontological space and the epistemological space, where the functors and categorial converters are socio-natural technologies and processes.

The resulting framework is a monograph in support of the development of a theory of knowledge as a combined sub-theory of *intellectual investment flows* and a *sub-theory of intellectual capital stocks*. The supporting goal of the monograph is to place neuro-decision-choice actions as central to the development of the theory of knowledge which reveals social systems as knowledge economy in the epistemological space. The placement of the neuro-decision-choice actions allows the examinations and discussions of the role of creativity, curiosity, courage and other qualitative characteristics over the space of thinking-reasoning duality of human activities in the space of fundamental-applied dualities, where the goal is the development of a general framework to an epistemic approach to the understanding of conditions human existence over the space of socio-natural dualities relative to the space of knowability-unknowability dualities that gives rise to the power of imagination, the audacity of knowing, the courage of action, the sadness of failure, the indestructability of hope, the joy of success, the existence of non-satiation, the living essence of thinking, the force of reasoning, the conception of vision, the pursuit of goals, the search for meaning of life, the rise of problems and the search for solutions.

b) The *objective* of the monograph is the further advancement of my previous monographs on information-knowledge systems in human actions and capital-investment theories in social production-consumption enterprise. The conditions of the problem-solution processes define the meaning of life, the conditions of thinking-reasoning processes define the foundations of problem-solution processes, the conditions of intellectual capital-investment processes define the foundations of the input-output processes, the conditions of input-output processes form the foundations of supply-demand while the conditions of supply-demand processes form the foundations of production-consumption processes which form the foundations of the essence of human existence, its growth and development which constitute the foundations of civilization, irrespective of how its linguistic elasticity in definition. This is the main objective which is supported by sub-objectives. It is also to offer a framework to examine the logical validity of the classical theory of knowledge [R8.17] [R8.37] [R8.47] [R8.54] and the currently available philosophy and the theories of information and to present a direction where the theories of information and knowledge are viewed in terms of general theories of production-consumption dualities and input-output dualities relative to the conditions of theory of problem-solution dualities. An important supporting objective of the monograph is the development of a framework on functors and categorial converters in complementing and providing philosophical and theoretical continuity of my previous works especially from the theory of fuzzy rationality[R4.7], Cost-benefit analysis and theory of fuzzy decisions: Identification and measurement[.], Cost-benefit Analysis and theory of fuzzy decision: Fuzzy value theory[.],,the theory of info-statics[R17.17], the theory of info-dynamics[R17.18], the general theory of entropy[R21.5] the theory of problem-solution dualities and polarities [R8.16a], the theory of epistemic field [R8.16b], where these theories and interdependent in essential ways with give-and-take sharing modes. Together, these theories have essential objective in presenting a general understanding of neuro-decision-choice actions in terms of problem-solution processes where the problem-solution processes is input-output processes.-The other objective is to show how the functors and categorial converters act as powerful analytical tools in understanding socio-natural transformation over the space of internal-external dualities.

The sub-objectives include is the development of a framework of a theory of epistemic transversality conditions, a theory of certainty-uncertainty dualities to show the sources of uncertainties, and their relation to the rise of the space of imaginations that gives rise to the concept of possibility and probability, hence giving rise to the need for possibility and possibility measures which in the space of imaginations, where the analytical framework gives a process to explain the similarity and differences in the concept and phenomena of qualitative and quantitative uncertainties that give rise to the development of the objective and subjective theories of probability measures, where the objective uncertainties may be traced to the space of acquaintance in terms decisions on variety repetitions and quantitative inexactness and the subjective uncertainties may be traced to decisions vagueness and qualitative inexactness. In this framework the objective is to follow the logical sequence of definition of concepts and phenomena and then pursue measurements as an objective to solve the defective in the theory of probability. The theory of probability is not a theory of uncertainty, and an uncertainty theory is not theory of uncertainty. The supporting objective is related to the development of theory of uncertainty that that laid down the foundations for the development of the theory of possibility and the theory of probability which are sub-theories of the theory of imagination which arises due to uncertainties from the space of ontic-epistemic dualities. The extension of the supporting objectives is to bring into focus that the theory of possibility must help to establish initial conditions and necessity while the theory of probability must help to establish freedom and sufficient conditions in the space of neuro-decision-choice actions and how these conditions are epicenter of neuro-decision-choice actions operating with preference in the space of variety in diversity, similarity, and principle of non-satiation. Another supporting objective is to introduce ontological time and epistemological time and discuss the similarities and differences between ontological beginning-end duality and epistemological past-future duality while present is always neutral.

c) The *rationale* for writing this monograph is derived from multiple epistemic limitations of conditions that are established for the relationships among the concepts of possibility, possible world, necessity, probability. Freedom, imagination, creativity, curiosity, hope, despair, reality, and paradigm of thought over the space of universal history knowing over the space of universal beginning-end dualities that contains the epistemological space of past-present dualities with neutrality of present, where claims of truth are always under fuzzy-stochastic conditionality over the space of possibility-probability dualities that also contains the space of necessity-freedom dualities. The major rationale is supported by sub-rationales of the form:

c1) There is a need to establish a system of space of cognitive functors and categorial converter to show transformations over internal-external dualities which includes the space of ignorance-knowledge dualities, where these spaces may be seen in terms of organization of problem-solution processes in such a way that each space is collection of categories to justify variety problem-solution dualities.

c2) There is also a need to reexamine the concepts of uncertainty, risk, and neuro-decision-choice actions in relation to imagination and reality and their connectivity to human actions, neuro-decision-choice activities, their rationality and creativity relative to national personalities, national exceptionalism and the ruling national social philosophical consciencism in the space of neuro-decision-choice actions relative to the spaces of problem-solution dualities and ignorance-knowledge dualities defined over ontological-epistemological time dualities.

c3) There is an important need to provide a framework for integrating the phenomena of creativity in the production of thinking algorithms, curiosity in a search for information-knowledge variety, hope and successes for reasoning in the use of thinking algorithms over the space of problem-solution dualities and failures and despairs over the space of input-output dualities as connected to the dynamics of human actions in maintenance of life.

c4) There is a need to examine and provide a framework of epistemic analytics for the idea that all knowing make progress over space of curiosity-creativity dualities, where hope defines epistemic actions and the pathways that connect potentials to imaginations and from imaginations to realities and realities to actuals over the space of ontological-epistemological information-time structures.

c5) There is also a need to point out the processes that connect knowing to categories of language developments and to the space of categories of problem-solution dualities, ignorance-knowledge dualities and cost-benefit dualities which serves as important and general analytical toolbox in the neuro-decision-choice actions over the space of fundamental-applied dualities that supports social dynamics over the space of neuro-decision-choice actions in infinite disequilibrium processes for the continual advancements of knowledge , discoveries of technologies and innovations of existing processes.

c6)There is also a need to develop the understanding of a unified framework of the knowing and information-knowledge development for in the enterprise of information-knowledge accumulation and its relationship to the general space of neuro-decision-choice actions and how the framework of information-knowledge-neuro-decision-choice system create the space of functors and categorial converter for the development of artificial intelligent system of communication in the space of human-machine dualities.

d) The *monograph*: Given the stated goal, aim, objective and rationale the monograph is set to provide answers to several questions through the development of the *theory of the theory of organizational analytics in the functor space and the space of categorial converters*, where the functors and categorial converters are composed of categories of processes and technologies that are linked to the space of *science, technology, engineering and mathematics* (STEM) defined over the space of fundamental-applied dualities to establish the organizational dynamics of variety input-output dualities, supply-demand dualities, production-consumption dualities and problem-solution dualities that establish information-knowledge economy. It is important to understand the role of STEM in the space of information-knowledge production, cultural dynamics, and socio-natural transformations, where the efficiencies of social transformations and its relations to science, technology, engineering and mathematics depend on the organizational efficiency of the social setup, and social philosophical consciencism visions containing visions and preferences in the space of individual-collective dualities.

The functors and the categorial converters are organized in categories as primary and derived categories with defined characteristic dispositions which are related to institutional and organizational conditions of knowing and information-knowledge production in a sequential parent-offspring dynamics which are linked to the primary category and derived categories of processes and technologies which are inseparably interrelated in multiplicity of regularity of interdependent rhythms in continuum and unity. With respect to the behaviors of development and behavior of elements in the functor space and the space of categorial converters, the monograph fist presents organizational analytics of theories of knowing and knowledge. This is then followed by the development and discusses on the organizational analytics of the basic-superstructure of social system and intellectual investment flows and intellectual capital stocks as input-output system to the space of neuro-decision-choice system.

The rise of confusions, information-knowledge disparities over the space of ignorance-knowledge is presented as the theory of epistemic anomalies containing discussions of phantom problems, ill-posed problems, unsolvable problems, and other anomalies that may arise in the space of ignorance-knowledge dualities. The search for the sources of these disparities in the space of ignorance-knowledge dualities leads to the introduction of the *onticity*, *epistemicity*, ontic-epistemic duality, creativity, curiosity and hope-despair duality and how their organizationally integrated structures affect the discoveries and innovations in the social functor space and the space of categorial conversions over the space of success-failure contained in the space of imagination-reality dualities and the understanding of behavior of natural functor space and the space of categorial converter and the comparative differences and similarities over the ontic-epistemic dualities. The behaviors of the elements in the space of success-failure dualities are seen in terms epistemic journeys between dualities and polarities, such as journeys from ignorance to knowledge, where the transition from ignorance to knowledge is examined in terms of epistemic transversality conditions which leads to the development and analysis of the theory of epistemic transversality conditions. The need to examine general motions over the space of quality-quantity dualities, linked to the dynamics of the space of internal-external duality, leads to the further reflections of important roles played by the tools of principle of opposites, cost-benefit analysis and fuzzy paradigm of thought over the spaces of relational continuum and unity between the spaces of actual-potential dualities and the imagination-reality dualities contained in the space of ontological-epistemological dualities.

The primary and derive categories of functors and categorial converter are linked to variety behavior over the space of identification-transformation dualities to explain the nexus of intellectual environment, social constructs, governance, and neuro-decision-choice actions over the space of individual-community dualities and the corresponding space of problem-solution dualities. The monograph makes it explicit that social progresses come to us through the cognitive activities over the space of imagination contained in the space of potentials, where the space of imaginations is connected to the space of realities by the varieties in the space of functor space and the space of categorial converters to ensure the continually and infinitely dynamics of the space of individual varieties and categories of varieties where analytically the space of actuals contains the spaces of *what the was* and *what the is*, while the space of potentials contains *what there would be.* The knowing processes through creativity and curiosity come to us by the elements in the functor space and the space of categorial converters that expose the variety characteristic dispositions to form the space of realities, where the contents of the space of realities enter the existing social philosophical consciencism which then guide and help creativity, curiosity, hope, despair over the space of success-failure dualities.

e) The *abstract* provides a general summary of the contents of the book and a thematic structure, where the topical developments of the argument are given in the *table of contents.*

2) ABOUT THE AUTHOR

(See the attachment about the author)

3) THE GENERAL REFLECTIONS ABOUT THE MONOGRAPH

The monograph is a follow up of the theory of epistemic field [R8,16b.] and its connection to the theory of problem solution dualities and polarities[R8.16a] other continuous theoretical works on information, knowledge, and decision in the spaces of production-consumption dualities, input-output dualities, supply-demand dualities over the space of social-natural dualities, where the elements of the ontological space follow natural rhythms and the elements of epistemological space follow social rhythms as derived creativity and courage in the thinking-reasoning dualities from the signal dispositions of natural rhythms over the space of acquaintance [R22.1] [R22.5] [R22.11] [R22.12] [R22.18]. This monograph is devoted to the development of the theory that will support the formation and utilization of categorial converters and functors in the space of identification-transformation dualities and ignorance-knowledge dualities relative to the general space of problem-solution dualities. The development is based on the fundamental idea that neuro-decision-choice actions are the foundation of life which is anchored in the space of problem-solution dualities to from which the path of social history is established by success-failure outcomes. The success-failure outcomes are dependent on the efficiencies of creativity and curiosity in the epistemic connectivity between the space of actual-potential dualities and the space of imagination-reality dualities, where elements in the space of success-failure dualities relate to the elements in the space of risk-riskless duality with conformal mapping onto the space of cost-benefit dualities. The space of risk-riskless dualities is shown to connect to the space certainty-uncertainty dualities and the space of resource-resourceless dualities. The conceptual system of the theory is developed under the fuzzy paradigm of thought over the space of relativity. The space of relativity relates to the principle of opposites. The principle of opposites is composed of duality and polarity. The activities in the space of identification-transformation dualities relate conflicts of opposites which are set in dualistic-polar games where the reward of wining is variety identity dominance. The theory of cognitive transversality conditions help to establish how a journey passes from a zone of duality to another zone of duality as well as the contending zones of conflict in wining claims of the dualistic-polar games and thus defining the stopping point of the game identity dominance. The transformation of characteristics defining the identity of the variety is then shown to be variety destruction-replacement process of the variety, where the transversality conditions are related to quality controls provided by the conditions of entropic tests taking account of conditions in the space of possibility-probability dualities as connected to the space of necessity freedom dualities under the conditions of categorial conversion and social philosophical consciencism, where the experiential information or knowledge by acquaintance is vague and inexact. The analytical tools are drawn from the fuzzy paradigm of thought in processing the variety information to establish the transversality condition and to obtain variety identity.

Generally, the monograph is developed to explore a framework for organizational analytics of functor and categorial converter over the space of static-dynamic dualities that allows the examination of anomalies in the space of thinking-reasoning dualities and how cognitive behavioral elements are integrated into the thinking-reasoning dualities to correct the anomalies, improve the intellectual capital stocks, and maintain the continual dynamics of the social philosophical consciencism. The monography points to space of knowability-unknowability dualities, the benefits, costs, the limits of organization and the paradigms of thought. The organizational analytics are such that each category of functor and categorial converters relates to categories of processes and technologies over the space of identification-transformation dualities for continual variety transformations and dynamics of input-output dualities and infinite problem-solution dynamics in the space of epistemological past-future dualities contained in the space of ontological beginning-end dualities. Every category of functor houses categories of processes and technologies for knowing and utilization. Every space of functors or every space of categorial converters houses a transitional station of processing learning, technological knowing and development through need, curiosity, creativity, hard work and perseverance and hope of the cognitive intra-categorial and inter-categorial journeys in relational continuum and unity over the space of ignorance-knowledge dualities. The transitional stations of the epistemic journey allow critical examinations of factors that affect the continuity of the neuro-decision-choice actions in tactics, strategy, and possible restructuring in the space of error-correction dualities.

4. THE SYNOPSES

A) For the general trust of the book see the abstract below.

B) For the main topics see the table of contents.

5) ORGANIZATION OF THE MONOGRAPH

The monograph is organized with a preamble, a prologue, a preface, an acknowledgement, seven chapters and an epilogue.

The *preamble* is composed of series of quotations from important works of different authors in philosophy, science, mathematics, economics, decision-choice theories, information theory and other relevant areas to the information-knowledge, problem-solution, question-answer dualities and ignorance-knowledge about the general knowing, composed of science and art of knowing and information-knowledge development under cognitive capacity limitations that generate risk-riskless dualities over the spaces of certainty-uncertainty dualities and projected to the spaces of doubt-surety dualities, true-false and decidability-undecidability dualities. In this respect, the preamble introduces the conceptual disagreements and agreements on the idea that all information-knowledge systems are generated by knowing actions with courage, creativity, sweat and disappointments irrespective of the area of human endeavor through neuro-information-decision-choice processes. The information-neuro-decision-choice interactive processes are exercised over the categories of the space of problem-solution dualities defined in relative cost-benefit differences within the space of imagination-reality dualities which is contained in the space of actual-potential dualities within the space of the ontological-epistemological duality as cognitive agents move from the space of acquaintances for knowledge to the space of descriptions of knowledge in the epistemic journeys over the space of variety ignorance-knowledge dualities in search of variety knowledge.

The *prologue* of this monograph is used to indicate the entry point into the development of this monograph concerning the theory of organizational analytics in the functor space and the space of categorial converters over the space of identification-transformation dualities. It is also used to introduce the readers to some important epistemic short comings, linguistic difficulties and criticisms of the existing information theory, theory of knowledge, the phenomenon of knowledge economy and their corresponding philosophies and mathematics. The epistemic short comings and the difficulties find expressions in a lack of clear definitional distinctions among the concepts and phenomena of information, fact, data, knowledge, and evidence which are in many cases interchangeably used in theoretical and applied discourses. Knowledge, fact, data, and evidence are sequentially derived categories of information in knowing and decision. Information is the primary category of knowing. Information is not knowledge, and it cannot be defined as skills acquired through education, understanding, comprehension, theoretical and practical understanding of a subject, and many other conceptual and phenomenal irrelevances. Information is the primary category of knowing while knowledge is a derived category of knowing from information using an accepted paradigm of thought and a system of quality controls. The real varieties are not the inputs into neuro-decision-choice actions. It is the characteristic representations, as either information or knowledge, that enter the neuro-decision-choice actions to produce outputs by transformations.

The monograph discusses the roles of the paradigms of thought over the space of thinking-reasoning dualities and the role of thinking and reasoning as instruments to reconcile dualistic-polar conflicts in the duals of dualities and the poles of polarities. The reconciliations of dualistic-polar conflicts are seen as varieties in the space of problem-solution dualities, where problems are seen as challenges in the space of thinking-reasoning dualities in search of solutions over the space of relativity, the outcomes of which propel the directions of human history. The principle of dualistic-polar structures in continua and unity points to a simple idea that our universal system is an organization of dualities and polarities with continual dynamics of input-output processes linked to supply-demand dynamics over the infinite space of identification-transformation dualities and polarities. It is argued that the fuzzy paradigm of thought is the appropriate thinking-reasoning instruments for the conditions of vague, inexact, and incomplete epistemological information, the opposites in continua and unity in the space of relativity, where the space of relativity admits of contradictions and simultaneous existence of opposites in space and time without which continual dynamics over the space of internal-external dualities are impossible. The monograph extends the arguments to the idea that the difficulties in the conceptual definition and phenomena of information and knowledge are carried to the space of certainty-uncertainty dualities in such a way that the concept and phenomenon of uncertainty are assumed to be known and taken as epistemically granted which then give rise to more thinking complications and reasoning difficulties in the paradigms of thought when the concept of uncertainty is placed under probability measure with no direct association with information and the probability is used in stochastic conditionality in qualifying neuro-decision-choice actions over the space of true-false duality and the space of acceptance-rejection dualities linked to the space of imagination-reality dualities. The concepts, phenomena and sources of uncertainty, possibility, probability are not explicitly defined and specified given rise to further confusion between the concepts of subjective and objective probability.

The journeys in the spaces of epistemological dualities and polarities are undertaken with varieties in thinking-reasoning dualities. The elements in the space of thinking-reasoning dualities must be related to the varieties in the space of ignorance-knowledge dualities and their space of functors and categorial converters. The structure and conditions of the thinking-reasoning dualities must be studied and analyzed in terms of paradigms of thought and how the paradigms of thought affect the cognitive journeys and the transversality conditions in the space of ignorance-knowledge dualities. The elements of thinking-reasoning dualities must be related to curiosity, creativity, over the spaces of hope-despair dualities and success-failure dualities. The paradigms of thought must be structured to account for mega-algorithm that guides a set of meta-algorithms, their developments and use over the space of thinking-reasoning dualities relative the elements in the space of problem-solution dualities. The behaviors of elements in the space of thinking-reasoning dualities derive its foundation from the social philosophical consciencism as well as influence the dynamics of social philosophical consciencism and intergenerational connectivity. The behaviors of the elements in the space of thinking-reasoning dualities set the paths and directions of human progress and telescopic future by creating negation and negation of negation over the space of destruction-construction dualities contained in the space of imagination-reality dualities.

The *preface* introduces the motivation for writing the monograph, what it is about and how the conceptual framework of the monograph is conceived and connected to other previous monographs by the same author on information, knowledge, data, evidence, and decision-choice actions. The preface introduces the basic reflection on the development of the monograph follow by discussions on functors and categorial converters and their relationship to processes and technologies. This is followed by discussions on the space of goals, objective, and vision over the space of individual-community dualities, the spaces of certainty-uncertainty dualities, thinking-reasoning dualities in relation to contradictions, conflicts under the general principle of opposites in the process of knowing and generation of anomalies. The preface ends with the structure and organization of the monograph. The preface provides an organization of paremiology of the essential spaces and sub-spaces contained in the development of the epistemic discussions for understanding of the paths of discussions.

CHAPTER ABSTRACTS

Chapter 1 is about organizational analytics of neuro-decision-choice actions over static-dynamic dualities and their relevance and effectiveness over the space of problem-solution dualities under conditions of knowing and information-knowledge processes within the space of identification-transformation dualities. It is about how organization is essential in connecting decision-choice actions over the space of problem-solution dualities to knowing and information-knowledge production at the levels of the individual and the collective, at the level of domestic and international, corporation and competition, production, and consumption over the space of cost-benefit dualities defined over the economic, political, and legal structures of social set-ups. The chapter relates conditions of efficiencies to neuro-decision-choice outcomes over the space of success-failure dualities as well as connecting the fundamental information-knowledge system to the applied information-knowledge system and all decision-choice activities of human endeavors over the space of progress-retrogress dualities. The importance, here, is to examine the effects of social organization on decision-choice actions over the space of problem-solution dualities [R8.16a] and the epistemic field with relational and information fields [R8.16b] in relation to knowing and the stock-flow disequilibrium dynamics of the information-knowledge system at the levels of variety identifications and transformations and their importance to the spaces of input-output, supply-demand, and production-consumption dualities. The concept of primary-derived organizations, institutions and the space of problem-solution dualities are discussed and related to neuro-decision-choice activities in knowing and information-knowledge accumulation and how they are influenced by social philosophical consciencism that has taken hold of the social set-up.

Chapter 2 provides the analysis of the relationships of organizational analytics to information-knowledge systems on basic structure, superstructure, and the space of basic-superstructure dualities in developing the knowing process, the production of information-capital accumulation and their connectivity to the spaces of input-output and cost-benefit dualities. Two sets of theories of explanatory and prescriptive structures are introduced and discussed in the knowing and information-knowledge processes which are connected to descriptive and predictive structures over the space of imagination-reality dualities. An introduction to a theory of certainty-uncertainty dualities is advanced to provide explanatory conditions and sources of uncertainties which give rise to the space of imagination containing possibility and probability spaces and hence the rise of possibility and probability theories, the concept of possible and impossible worlds. The definitions and sources of uncertainty, possibility, probability, corresponding spaces, and their relational interdependencies are provided. Discussions are made on basic-superstructure components of organization of knowing and information-knowledge systems, where definitions and comparative analysis of concepts of policy and decision-choice action are presented, distinguished, and linked to relationality, and organicity of the space of problem-solution dualities.

The roles of individuals and the collectivity in the organization of problem-solution dualities are then examined in the space of individual-community dualities in relation to conflicts and conflict resolutions over the space of cost-benefit dualities. The roles of institutions of research, teaching and learning are examined, leading to the discussions on the goals, objectives, and research in the organization of knowing and information-knowledge production, where we discuss learning with and without supervision in relation to thinking and reasoning. The chapter is concluded with the discussions on the development of special institutions with global collaborations, and competitions over the epistemic field, where the space of relevant alternatives may be expanded, and the space of irrelevant alternative may be reduced with global collaboration and creative competition in discovery of varieties in the space of categorial converters or the functor space under the guidance of the ruling social philosophical consciencism.

Chapter 3 deals with the theory of epistemic anomalies in the organization of knowing and information-knowledge production, where such anomalies are introduced, defined, and explicated for understanding in theories and applications over all areas of knowing under characteristic-signal analytics within the epistemic field supported by the relational and information fields. These anomalies in knowing and information-knowledge developments are explained to arise within the space of imagination-reality dualities contained in the space of actual-potential dualities as we journey between the space of imaginations and the space of realities through thinking and reasoning connecting the knowledge by acquaintance or experiential information to knowledge by description with the use of an organic paradigm of thought and micro-paradigms of thought. Some anomalies are paradigm-derived, some are observationally derived, some are intentionally derived, and some are unintentionally derived. The space of imagination-reality dualities, the space of imaginations and the space of realities are then defined and discussed in relation to the conceptual links to phenomena of possibility and probability which give rise to the spaces of possible world, impossible world and possible-world-impossible world dualities and then linked to the anomalies, where such anomalies are social but not of nature. From the space of imagination, we discuss the concepts, meanings and roles of creativity, curiosity, hope, and courage in relation to the space of cost-benefit analysis within the space of identification-transformation dualities. Using fuzzy paradigm of thought, conditions of optimality are derived for hope and despair under dualistic-polar structures, where in general, the phenomena of thinking and reasoning are shown as actions to balance the forces within opposites and polarities to abstract the best cost-benefit varieties in human endeavors relative to some varieties in the space of goals, objectives, and visions. Directly and indirectly, the forces between opposites generate the space of problem-solution dualities under the principles of opposites in all social spaces for social input-output dynamics and generate conflicts in the space of positive-negative duality for natural input-output dynamics. The chapter presents the needed foundations for further algebraic analytics to make explicit the rise and structure of various anomalies in relevant spaces within the epistemic field providing further guidance for institutional development of knowing.

Chapter 4 of this monograph discusses the roles of creativity and curiosity in knowing as the production of intellectual investment flows and information-knowledge accumulation as production of intellectual capital stocks in the in the space of problem-solution dualities and polarities within the space of imagination-reality dualities as contained in the space of actual-potential dualities relative to the space of ignorance-knowledge dualities within the space of ontic-epistemic dualities relative to the space of ontological-epistemological duality leading to the introduction of the concepts and phenomena of onticity and epistemicity. The onticity presents conditions of ontological variety characteristic dispositions *(what there is)*, while the epistemicity presents the conditions of epistemological variety characteristic dispositions (*claims of what there is*) as interpreted and derived from ontological variety signal dispositions. The understanding of the roles of creativity and curiosity are related to the spaces of courage-fear dualities, hope-despair dualities, and success-failure dualities relative to the conditions of epistemicity. Definitions of these concepts are presented, and their contents explicated with corresponding operator functions. Comparative analytics are undertaken to show their interdependencies and their relative roles and degrees of efficiencies in the spaces of ignorance-knowledge dualities and problem-solution dualities showing the strengths and limits of organization in neuro-decision-choice actions over the space of human behaviors under the principle of opposites and cognitive capacity limitations. Through comparative organizational analytics in varieties and categories, we show the relational connectivity in the space of ontological-epistemological dualities in knowing, differences, similarities, ontological information-time processes and epistemological information-time processes in reference to the ontological beginning-end duality of Bennu-Anoma-Gyenyame conditions and epistemological past-present duality of Sankofa-anoma conditions in the space of ontological-epistemological dualities, where the essentiality of symmetry and supersymmetry are discussed under the principle of opposites relative to the spaces of relativity-absoluteness dualities and dualistic-polar stability-instability dualities with illustration provided by an epistemic geometric representation with the corresponding mathematical spaces in terms axiological (intrinsic value) structures of varieties in the space of real-nominal dualities relative to the space of cost-benefit dualities in the space of neuro-decision-choice (processes).

In chapter 5, the conceptual frame of anomalies in knowing and information-knowledge production is presented and related to some important elements, such as courage and creativity, in the space of quality-quantity dualities over the spaces of ignorance-knowledge dualities and problem-solution dualities and polarities. Chapter 5 is a continuation with algebraic analytics of anomalies in knowing and information-knowledge systems and the roles that curiosity, creativity, courage, and innovation play in the space of imagination-reality dualities contained in the epistemological space and as a sub-space of the space of actual-potential dualities which is contained in the ontological space, all in relation to the conditions of onticity and epistemicity. The natures of the phenomena of thinking and reasoning relative to creativity, courage, curiosity and knowing are presented, defined, and explicated, where the differences and similarities among them are shown in relation to the neuro-decision-choice actions over the spaces of problem-solution dualities and certainty-uncertainty dualities and dynamic behaviors of varieties in the functor space and the space of categorial converters. The concepts and phenomena of thinking and reasoning are related to possible worlds and the space of realities by the way of comparative analytics of relations among necessity, freedom and creativity which are then examined by way of analytical connectivity through definitions, explications, and analyses. From the analytical connectivity among possibility space and probability space in the space of imaginations to the space of realities as a subspace of the actuals, the concepts and phenomena of phantom problem, ill-posed problem, unsolvable problem, news, fake news, alternative facts, and alternative realities are introduced, defined, explicated, and connected to impact analyses of neuro-decision-choice actions under the general social philosophical consciencism. The social philosophical consciencism generates systemic risks over the space of cost-benefit dualities at generational points, where the dynamics of the social philosophical consciencism help to explain the dynamics and differences in generational preferences, variety relevance and style of social management.

Chapter 6 of this monograph presents discussions on the theory of epistemic transversality conditions which relate to dynamic optimality of epistemic journeys between duals of duality and poles of polarity to obtain stopping rules of the epistemic journeys over the space of dualistic-polar conditions. The epistemic transversality conditions are made up of *epistemic identification transversality* conditions and *epistemic transformation transversality* conditions. The epistemic identification transversality conditions define the epistemic stopping rules for cognitive journeys to obtain the variety identities that reveal the inherent necessity, conditions of being and the knowledge about the varieties at static state in the space of ontic characteristic dispositions, where the varieties are made up of states and processes with properties and relations in the space of identification-transformation dualities to produce varieties in the space of epistemic characteristic dispositions. The transformation transversality conditions define the terminal state of the journey between variety ignorance and variety knowledge, variety problem and variety solution, variety question and variety answer and, variety cost and variety benefit, presenting the conditions when variety knowledge is found for the replacement of variety ignorance or a variety solution is found to replace variety problem or variety answer is found to replace variety question over the spaces of dualities and polarities. The epistemic transversality conditions are related to cost-benefit marginal conditions of knowing as the production of intellectual investment flows and information-knowledge accumulation as the production of intellectual capital stock, where thinking-reasoning processes are the vehicles of the journeys within and between dualities and polarities in the space of dualistic-polar structures. The epistemic transversality conditions are then related to telescopic past, present and future in all neuro-decision-choice processes and used to partition the knowing space into zones of knowledge and ignorance or zones of problems and solutions leading to the discussions of their utility in neuro decision-choice space. Conceptual definitions are provided and their relationships to the principle of opposites with relational continua and unity are discussed and linked to the fuzzy paradigm of thought, where ignorance and knowledge are mutually constraining, general duals of dualities are mutually constraining, poles of polarities are mutually constraining, and the dualities and polarities are also mutually constraining in the space of dualistic-polar structures under the general principle of opposites. The chapter is concluded with discussions linked to *asantrofi-anoma* problem, conditions, and rationality to provide computational understandings of epistemic transversality conditions in cognitive dynamics within ignorance-knowledge and problem-solution situations providing human learning foundations for machining learning. The definition and discussions of the asantrofi-anoma principle and rationality are provided.

Chapter 7 of this monograph presents some important discussions and reflections on comparative analytics about the differences and similarities of the ontological principle of opposites in the ontological space and epistemological principle of opposites in epistemological space. It is argued that the principle of opposites is composed of a system of dualities and polarities. There is a system of *key opposites* of dualities and polarities, and a system of *supporting opposites* of dualities and polarities which are defined and specified. The key aggregate opposites which are common to both the ontological and epistemological spaces are positive-negative dualities, input-output dualities, and identification-transformation dualities, where the ontological space contains the space of actual-potential dualities and epistemological space contains the space of imagination-reality dualities. The supporting opposites are specific to epistemological space. They are specified and explained to be generated by the requirements of neuro-decision-choice processes under individual and collective preferences which are admissible in the social philosophical consciencism over the epistemological space containing the space of imagination-reality dualities and constrained by all forms of cognitive capacity limitations. Among the supporting opposites are problem-solution, war-peace, cost-benefit, minimum-maximum, true-false, acceptance-rejection, and other similar dualistic-polar structures which relate to the category of the key aggregate opposites. Comparative analytics are undertaking to show the similarities and differences of the ontological and epistemological spaces, where organizational analytics allow for the understanding of the differences and similarities to be connected to the space of knowing as intellectual investment flows and information-knowledge accumulation as the production of intellectual capital stocks.

The epilogue of this monograph provides a summary of analytical connectivity of the principle of opposites composed of dualities and polarities, epistemic field, and fuzzy paradigm of thought in the development of knowing as the production of intellectual investment flows and information-knowledge accumulation as the production of intellectual capital stocks. The principle of opposites places the system in the space of dualistic-polar games to cross the transversality conditions where the rewards of the game are variety identity and polar dominance. The conditions of functors and categorial converters are presented and linked to curiosity, creativity, hope, and despair with suggestion of further research.

KOFI KISSI DOMPERE
PROFESSOR EMERITUS, DEPT. OF ECONOMICS
HOWARD UNIVERSITY, WASHINGTON

ABOUT THE AUTHOR:

Kofi Kissi Dompere is a professor emeritus of economics at Howard University. He holds a doctoral degree in economics, master's degree in applied mathematics, master's degree in business administration in finance and bachelor's degree in economics and mathematics, all degrees from Temple University in Philadelphia. He has authored several scientific and scholarly works in economics, philosophy, decision theory and fuzzy mathematics. Some of the articles may be found at the following sites https://howard.academia.edu/KOFIKISSIDOMPERE,
He is the winner of 2006 Cheikh Anta Diop award for excellence in scholarship. He has a number of published monographs including: 1)The Theory of Problem-Solution Dualities and Polarities; Information-Decision-Choice Foundations of the Unity of Knowing and the Unity of Science, www.springer.com, (2021) 2) A General Theory of Entropy: Fuzzy Rational Foundations of Information-KnowledgeCertainty,www.springer.com, (2019), 3) The theory of Info-dynamics: Rational Foundations of Information-knowledge dynamics, www.springer.com, (2018), 4)The Theory of Info-statics: Conceptual foundations of Information and Knowledge, www.springer.com, (2018), 5) The theory of Categorial Conversion, www.adonis-abbey.com, (2017), 6) The Theory of Philosophical Consciencism, www.adonis-abbey.com, (2017), 7) Fuzziness, Democracy, Control and Collective Decision-choice Systems: A Theory on Political Economy of Rent-Seeking and Profit-Harvesting, www.springer.com (2014) 8)Social Goal-Objective Formation, Democracy and National Interest: A Theory of Political Economy Under Fuzzy Rationality,www.springer.com (2014), 9) The Theory of the Knowledge Square: The Fuzzy Rational Foundations of the Knowledge-production System, www.springer.com (2013), 10) Fuzziness and Foundations of Exact and Inexact Sciences(2013), www.springer.com, 11) Fuzzy Rationality: A Critique and Methodological Unity of Classical, Bounded and Other Rationalities(2009) www.springer.com 12) Epistemic Foundations of Fuzziness: Unified Theories on Decision-Choice Processes www.springer.com (2013.), 13) Fuzziness and Approximate Reasoning: Epistemics on Uncertainty, Expectation and Risk in Rational Behavior(20013 www.springer.com, 14) African Union: Pan-African Analytical Foundations, www.adonis-abbey.com , 15) Polyrhythmicity: Foundations of African Philosophy www.adonis-abbey.com 2006, 16) Cost-Benefit Analysis and the Theory of Fuzzy Decisions: Identification and Measurement Theory (2004) (www.springer.com), 17) Cost-Benefit Analysis and the Theory of Fuzzy Decisions: Fuzzy Value Theory (2004) (www.pringer.com), 18) The Theory of Aggregate Investment in Closed Economic Systems (www.greenwood.com), 19) The Theory of Aggregate Investment and Output Dynamics in Open Economic Systems (www.greenwood.com), 20) Epistemics of Development economics: Toward a Methodological Critique and Unity (Co-author with Dr. Ejaz) (www.greenwood.com, 1995). His teaching areas include Economic Theory (Micro and Macro), International Economics (Trade theory, Monetary Theory and Theory of Commercial Policy), Cost-Benefit and Project Analysis, Mathematical Economics and Econometrics at all levels. His current research focus is on the theories of knowing and information-knowledge development through the understanding of the phenomena of problem-solution and question-

answer dualities viewed from the theories of economic production-consumption, input-output processes under the conditions of cost-benefit dynamics, fuzzy paradigm of thought and the principle of opposites, mathematics and epistemics on fuzzy phenomena and their applications to neuro-decision-choice actions, mathematics of politico-legal economy and data analytics. He is a resource person for UNITAR (United Nations) and economic consultant in the areas of cost-benefit analysis, national income accounts, development, and macroeconomics. Resource Person, World Bank, Refereed for Studies on African Economic Policy, and Trade Negotiations. He served as Senior Technical Research Consultant for Cost Benefit Analyst - Zero-Based Regulatory Review Project of Federal Highway Administration (USA) Basic Technologies International, Annandale, VA. He also served as a consultant to consolidation of East and Southern African regional Groupings into a common market, organized by International Development Research Center – Canada. Served as a Technical Consultant Government of Botswana, USAID Washington, DC on National Accounts, Macroeconomic Policy, and Industrial Data Systems Design. Economic Consultant, World Bank Social Dimensions of Structural Adjustment – Ghana. Technical Consultant and African Delegate, 24th Session, United Nations Statistical Commission, New York.

He is the founder of The Research Institute of Information Decision and Economic Sciences (RIIDES) in which he serves as a Distinguished Research Professor. His partial research impact on the scientific and scholarly community may be assessed on http://worldcat.org/identities/lccn-n93122486/ and the partial list of monographs may be see on https://www.abebooks.com/book-search/author/kofi-kissi-dompere/ and also on https://www.semanticscholar.org/author/K.-K.-Dompere/2608523 . He is an active participant of discussions on fundamental problems in physics. Under RIIDES, he is working on the frontiers of information-knowledge and decision-choice systems with monographs of 1) The Theory of Organizational Analytics, The Functor Space, The Space of Categorial Converters and the Space of Identification-Transformation Dualities 2) The Unity of Knowing and The Unity of Science: Information-Decision-Choice Foundations in the Space of Production-Consumption Dualities.3) The Theory of Imperial and Market Incumbencies, Competitive fringes and Neocolonial States Under Fuzzy Rationality.

He is the producer and host of a radio program "African Rhythms and Extensions" on WPFW 89.3fm in Washington D.C., USA, www.wpfwfm.org .

KOFI KISSI DOMPERE

PROFESSOR EMERITUS OF ECONOMICS

DEPARTMENT OF ECONOMICS

HOWARD UNIVERSITY

WASHINGTON, D.C., 20059

E-mail: kdompere@howard.edu, cc: domperekofi494@gmail.com , Tel. #: 202-529-4223

Contents

CHAPTER 1 THE THEORY OF KNOWING AND THE THEORY OF KNOWLEDGE: ORGANIZATIONAL ANALYTICS 1

CHAPTER 2 THE INFORMATION-KNOWLEDGE SYSTEMS AND THE ORGANIZATIONAL ANALYTICS OF THE BASIC-SUPERSTRUCTURE DUALITY 35

CHAPTER 3 THE THEORY OF EPISTEMIC ANOMALIES IN THE PRINCIPLES OF DIVERSITY AND UNITY OF KNOWING AND SCIENCE: Organizational Characteristic-SignalAnalytics 111

CHAPTER 4 ONTICITY, EPISTEMICITY, CREATIVITY, CURIOSITY AND HOPE-DISPAIR DUALITY 177

CHAPTER 5 THE THEORY OF EPISTEMIC ANOMALIES WITHIN THE PRINCIPLES OF DIVERSITY AND UNITY IN KNOWING AND SCIENCE: Organizational AlgebraicAnalytics 247

CHAPTER 6 THE THEORY OF EPISTEMIC TRANSVERSALITY CONDITIONS OVER THE SPACE OF IGNORANCE-KNOWLEDGE DUALITIES AND POLARITIES 315

CHAPTER 7 FURTHER REFLECTIONS ON THE PRINCIPLE OF OPPOSITES, THE VARIETY SPACE AND KNOWING IN THE SPACE OF ONTOLOGICAL-EPISTEMOLOGICALDUALITY 377

EPILOGUE 411

APPENDIX: GLOSSARY OF NON-ENGLISH TERMS 421

BLURB 431

INTERDISCIPLINARY AND MULTIDISCIPLINARY REFERENCES 433

Index 521

CHAPTER 1

THE THEORY OF KNOWING AND THE THEORY OF KNOWLEDGE: ORGANIZATIONAL ANALYTICS

ABSTRACT

Chapter 1 is about organizational analytics of neuro-decision-choice actions over static-dynamic dualities and their relevance and effectiveness over the space of problem-solution dualities under conditions of knowing and information-knowledge processes within the space of identification-transformation dualities. It is about how organization is essential in connecting decision-choice actions over the space of problem-solution dualities to knowing and information-knowledge production at the levels of the individual and the collective, at the level of domestic and international, corporation and competition, production, and consumption over the space of cost-benefit dualities defined over the economic, political, and legal structures of social set-ups. It relates conditions of efficiencies to neuro-decision-choice outcomes over the space of success-failure dualities as well as connecting the fundamental information-knowledge system to the applied information-knowledge system and all decision-choice activities of human endeavors over the space of progress-retrogress dualities. The importance, here, is to examine the effects of social organization on decision-choice actions over the space of problem-solution dualities [R18.16a] and the epistemic field with relational and information fields under comparative analytics of fuzzy and classical paradigms of thought [R18.16b] in relation to knowing and the stock-flow disequilibrium dynamics of the information-knowledge system at the levels of variety identifications and transformations and their importance to the spaces of input-output, supply-

K. K. Dompere, *Theories of Knowing and Knowledge Production*, Studies in Systems, Decision and Control 501, https://doi.org/10.1007/978-3-031-42863-0_1

demand and production-consumption dualities. The concept of primary-derived organizations, institutions and the space of problem-solution dualities are discussed and related to neuro-decision-choice activities in knowing and information-knowledge accumulation and how they are influenced by social philosophical consciencism that has taken hold of the social set-up.

1.1 INTRODUCTION

The focus of this monograph is to examine the effects of social organization on individual-collective efficiencies of decision-choice outcomes on the space of problem-solution dualities and their relationship to the outcomes of knowing and information-knowledge production under social investments that will allow analytical integrations of human curiosity, creativity, success-failure dualities and hope-despair dualities into the process of knowing, and information-knowledge production irrespective of the area of knowing. The analytical integration is relative to the spaces of problem-solution dualities, cost-benefit dualities, and ontic-epistemic dualities from which an introduction of the theory of certainty-uncertainty duality is advanced using functors and categorial converters under the fuzzy paradigm of thought, category theory, principle of opposites and the theory of epistemic fields [R8.16a] [R8.16b]. The understanding of socio-natural activities in the space of input-output processes with diverse inputs, diverse outputs and diverse relations requires some form of organization of diversities into a unity or an organization of partitions of the unity into diversities with relational connectivity. In general, the purpose of organization is to relate deferent resources in input-output processes in an orderly framework for the achievements of specific goals and objectives as to how cognitive agents may know through decision-choice activities over the space of problem-solution dualities. Organization is required where there exist diversities or an organized partition into diversities if the unit is too large for clarity. We, thus, speak of composition and decomposition, organization, and disorganization in management structures over the space of problem-solution dualities relative to the space of ignorance-knowledge dualities. The goal of the organizational analytics is related the epistemic works to the question as how societies should be organized and how should the economic, political, and legal institutions be arranged so they can serve the efficiencies of neuro-decision-choice actions over the space of problem-solution dualities and the general information-knowledge input-output dynamics.

The continual process of stock-flow of information-knowledge system in disequilibrium dynamics over the space of input-output dualities requires a continual organization and composition of microunits into oneness, wholeness and unity and as well as dividedness, and decomposition into micro-units for details with reorganization and integration to meet the continual expansion of the space of ignorance-knowledge dualities induced by identifications and transformations on objects in the space of varieties, where there is always a decision-choice action of epistemic transversality conditions of variety identities and variety transformations in the space of organization-disorganization dualities. The epistemic transversality conditions are important in all areas of dualistic-polar dynamics in knowing when one can claim one dual over the other dual or one pole over another pole in identity of knowing in the space of ignorance-knowledge (*onim-sua-ohu*) dualities relative to the knowing of identifications and transformations of variety identities and the enveloping (genetic) paths of varieties where *onim* means ignorance, *ohu* means knowledge and *sua* is a connecting cord of learning action that must meet the cognitive transversality test conditions.

In this respect, the culture of organization should inculcate in the evolving social philosophical consciencism that knowing takes place through neuro-decision-choice actions over the space of problem-solution dualities in slow measures over the space of ignorance-knowledge dualities under success-failure processes, avoiding dogmatism and mental decay which are generated from the house of cognitive imbecilities and intellectual laziness. Similarly, it must be understood that every objective reality contains subjective reality and hence provisional that must be verified in the space of the actuals with multiplicity of varieties under the principle of opposites.

The multiplicity of variety identities and transformations is such that the relationships between the neuro-decision-choice activities and cognitive behaviors must be collective as well as organized to increase the effectiveness of individual and collective knowing outcomes through the individual and collective neuro-decision-choice actions over the space of problem-solution dualities, the outcomes of which must help to explain the path of *social progress, empire dissolutions, war and* peace and many important factors that affect the individual and collective conditions over the space of sadness-happiness dualities and the evolving culture and civilization. The concept of social progress is a troublesome one with many interpretations as the interpretations are induced by subjective preferences in the space of individual-community dualities, where the individual subjective preferences and the community collective preferences, over the space of variety real cost-benefit dualities, may be in conflicts and non-inter-supportive. It is these conflicts of preferences over the space of individual-community dualities that give rise to disinformation, misinformation, deceptive practices and information-knowledge propaganda over the neuro-decision-choice space relative to the space of problem-solution dualities and the need for organization with rules and constraints to help the stability and commonness among individuals and the collective in distributions of the real cost-benefit dualities of individual and collective neuro-decision-choice actions under the general principle of opposites.

The analytical usefulness and power of the principle of opposites must be understood in knowing and information-knowledge accumulation and the concepts of language vagueness, lexical elasticity, and indecision. By the principle of opposite, each opposite is both parent and offspring in the space of micro-macro dualities, where every micro-variety serves as macro-variety under decomposition and every macro-variety serves as micro-variety under composition and organization into unity within the space of nothingness-somethingness dualities. The idea that every micro-variety serves as a macro-variety simply implies that in reductionist process any micro-unity may be subdivided in an infinitely reduction process to the state of nothingness relative to the space of natural technologies over the space of visible-invisible dualities, where visibility infinitely vanishes into nothingness and invisibility takes residence in somethingness in the space of dualistic-polar transformation games by natural processes, technologies and laws relative to *transversality conditions* in all dualistic-polar structures. The analytical usefulness and power of the principle of opposites with relational continuum and unity is complemented and sharpened with the use of fuzzy information and fuzzy paradigm of thought over the space of relativity under decompositions and compositions. The concepts of visibility, invisibility, micro and macro are always in relation to technologies and their evolutions at the space of acquaintance.

1.2 ORGANIZATION, SPACE OF PROBLEM-SOLUTION DUALITIES AND NEURO-DECISION-CHOICE SYSTEMS

The starting point of the theory of problem-solution dualities and polarities in this monograph is the view that the ultimate determining factor of the *meaning of life* is decision-choice actions over the space of problem-solution dualities, and the *essence of life* finds expressions in the manner in which the decision-choice actions are exercised over the space of variety problem-solution dualities to simultaneously produce variety knowing and information-knowledge system and use the results as inputs into further neuro-decision-choice activities to produce means and needs for human existence in dualistic-polar continuum and unity over the space of variety identification-transformation dualities. Here, the idea of knowledge for knowledge's sake finds no legitimacy in our concept of information-knowledge-decision-choice interactive systems.

Information and knowledge, whether intentionally or unintentionally acquired, serve as input into neuro-decision-choice actions to accomplish goal-objective elements toward a vision as abstracted from the space of imaginations contained in the space of potentials as a subspace of actual-potential dualities. Information and knowledge are varieties within the space of social production-consumption dualities which is completely defined and established in the space of variety identification-transformation dualities, The identification space provides us with identity information regarding *what there was* and *what there is* in the space of knowability containing states and processes which are connected to the space of cost-benefit dualities for preference evaluations for defining the necessity and anticipation. The transformation space provides us with conditions of functors and categorial convertors regarding *what varieties* must be destroyed and *what varieties* must be constructed from the space of imaginations within the space of input-output dualities with dualistic-polar interchanges under the preference assessment scheme over the space of cost-benefit dualities. The information-knowledge accumulation at any time point comes to us as fundamental and applied information-knowledge stock-flow systems which are available as input service flows for present and future generational and intergenerational neuro-decision-choice actions depending on the timepoint needs in relation to human activities and generational preferences as projected from the ruling social philosophical consciencism and organized over the spaces of input-output and cost-benefit dualities, where all human actions are defined and established over the space of identification-transformation dualities regarding variety property and variety relations over the space of static-dynamic dualities, where the power of curiosity and creativity find dynamic expressions in the space of problem-solution dualities and polarities.

The set of inter-generational results of meaning and essence of life constitutes the enveloping of human progress and the history of the evolving civilization. It is the intimate continual interactions and connectedness of neuro-decision-choice actions and the problem-solution dualities, as the defining factors of the meaning and essence of life, that provide the factors for establishing the *principles of unity and diversity* of knowing and science organized under information-knowledge and neuro-decision-choice systems. In fact, the evolving information-knowledge system is the evolving problem-solution system that defines the character of different generations and different nations with different organizational structures as induced by the contents of social philosophical consciencism. The contents of the social philosophical consciencism is also governed by the general principles of opposites with dualities and polarities over the space of negative-positive dualities that contains the space of main-fringe dualities of ideas, where the fringe may enter general thinking as the main stream and main-stream ideas may enter into the fringe where every negative disposition and every positive disposition find residence in the space of cost-benefit dualities in relational continuum and unity in the space of oneness-dividedness dualities where such oneness and dividedness must be organized into useful structures to obtain the understanding relational continuum and unity in the general space of give-and-take sharing modes for mutual existence and destruction.

The adherence to diversity in categories of knowing is a strategic dividedness and the adherence to unity of categorial diversities is a strategic oneness with a multiplicity of connectedness in all neuro-decision-choice actions over the space of problem-solution dualities while the use of diversities in the space of unity-disunity dualities is an efficient approach to information-knowledge stock-flow processes under the presence of cognitive capacity limitations of neuro-decision-choice agents over the space of individual-collective dualities, where efficient organization of individuals into useful collectivity may reduce the constraints imposed by the individual cognitive capacity limitations, where the organization of individual cognitive efforts into collective cognitive effort is an important strategy in the knowing process as a production of intellectual investment flows. The organization strategy is a creation of conditions of neuro-decision-choice efficiencies within the spaces of individual-collective dualities in terms of cooperation over the space of give-and-and sharing modes as well as encouraging individual competitive spirits through the systems of incentive-disincentive and reward-punishment dualities over the general space of cost-benefit dualities. To say that the meaning and essence of life are defined in the space of problem-solution dualities is also to say that the meaning and essence of life is defined in the space of identification-transformation duality under epistemological time structure.

The organization takes the form of creating categorial diversities through organized dividedness with relational continua and categorial and oneness with relational unity in give-and-take sharing modes. The organized dividedness into categorial diversities is the emphasis on individuality within the wholeness. The organized oneness into categorial unity is the emphasis on the rational interdependence over the space of mutual construction-destruction dualities over the individual-collective relational existence as well as on give-and-take sharing processes over the space of categorial success-failure dynamics of neuro-decision-choice activities over the epistemic field and the space of problem-solution dualities. The efficiencies of categorial diversities and unity over the space of variety problem-solution dualities require inter-categorial and intra-categorial arrangements into an organization of institutions of neuro decision-choice actions over the space of categorial varieties in relational inter-dependencies and relational give-and-take sharing processes for dualistic-polar universal systems, where every neuro-decision-choice action, as a policy or non-policy, is powered through or guided by an institutional channel to a problem-solution duality, and where every result, whether success or failure, is a part of the intellectual investments information-knowledge flows to update the intellectual capital accumulations of information-knowledge stock at each disequilibrium state.

In this respect, to understand human history, one must understand its relational structure with neuro-decision-choice actions over the space of problem-solution dualities, the results of which must be related to an enveloping of neuro-decision-choice outcomes in the conflict zones of socio-natural processes, where the effects of natural outcomes dominates human cognitive capacity under multiplicities of natural cognitive capacity limitations to deal with the varieties in the space of the problem-solution dualities, the morphological processes of which establish the foundationally relational dynamics of neuro-decision-choice actions, social progress and increasing knowing of socio-natural varieties where the space of natural cognitive capacity limitations may be reduced or expanded by social organization over the spaces of input-output and production-consumption dualities. In this framework, all cognitive activities are defined and examined over the spaces of production-consumption and input-output dualities as translated over the space of cost-benefit dualities with conflicting preferences over the individual-collective dualities, and where every neuro-decision-choice action is either a change in relation, property, or both over the space of organized-disorganized dualities.

Within the space of oneness-dividedness dualities, a question arises as to whether the human collective cognitive capacity can be expanded by reducing the cognitive capacity limitations through forms of individual and collective organizations. Every variety in nature is under the indestructible laws of organization of intra-variety diversities, and every variety seeks its destruction when its internal conditions disobey the indestructible laws of organization, while it maintains its growth and development when its internal conditions and intra-varieties obey the laws of organization to strengthen its integrity to take advantage of its internal relational unity of varieties, its positive externalities and fight against its negative externalities over the space of cost-benefit dualities. Under the principle of the laws of organization, a question, therefore, arises as to whether there are limits to the utilizing power of the laws of organization as they relate to socio-natural varieties in identities and transformations where needs and wants find residence and organized in the space of production-consumption dualities. If there are limits, how are the limits defined and established, and what factors help to establish the limits for each variety and place constraints on the space of neuro-decision-choice activities within the space of success-failure dualities? The epistemic interest, here, is not on the development of the theory of organization even though this may be implicit in the understanding of the relationships among decision-choice actions, problem-solution dualities, and information-knowledge systems over the space of general production-consumption and input-output dualities in terms of allocative efficiencies.

The understanding of the concept of limits and non-limits of the laws of organization and the factors that determine them are extremely important also to the understanding of variety transformation from the new to the old, from old to the new, and from life to death and death to life, from the seed to plant and from the plant to the seed and from problems to solutions and solutions to problems under the principle of opposites, where there are evolutionary games between variety duals of dualities and the variety poles of polarities for mutual destructions and mutual existence. In other words, how are the laws of organization set, by who, and manifest themselves within the principle of opposites in relation to the internal dualistic-polar conditions of the socio-natural variety structures, where nature resolves the variety conflicts directly over the spaces of actual-potential dualities and polarities through natural technological processes, and society resolves variety conflicts indirectly with cognitive capacity limitations over the space of imagination-reality dualities and polarities contained in the space of actual-potential dualities and polarities by mimicking nature's technological processes through knowing and information-knowledge accumulation and applications organized within the space of success-failure dualities relative to the space of certainty-uncertainty dualities?

Any variety is destroyed when its intra-variety organization of contents and parts are dismantled to compromise the integrity of the variety and transform it to a new variety. In this respect, the limits of organization vary over the morphologies of variety's internal arrangements, integrity of its constituent parts, intra-variety relationality and the strength of competitive effects of conditions of its externalities. The important point that must be considered in understanding the limits and non-limits of the laws of organization involves in the understanding of the key intra-variety aggregate parts that hold the other minor parts into the relational unity of the organism. The key intra-variety parts are essential in maintaining the organizational efficiencies of the variety's internal conditions. The relational conditions of the intra-variety parts are related to intra-vertical and intra-horizontal sharing of risks of the organizational dis-integrity. What is organization and its characteristic disposition? An answer to this question requires an analytical definition and explication of organization to set the boundaries of its use in the space of source-destination dualities and polarities involving knowing, neuro-decision-choice actions and information-knowledge production, where organization may involve decomposition recompositing and composition over the space of problem-solution dualities.

DEFINITION 1.2.1: Organization

Organization is an arrangement of intra-variety diversities to create a complex network of intra-variety relationships of internal parts to maximize the positive give-and-take sharing conditions to ensure the integrity of the whole for the general survival of its diverse parts and contents as well as minimize the negative give-and-take sharing modes of antagonistic relations of mutual destruction and to strengthen its immune system against internal and external advisories in order to maintain the health of the variety. It is also decomposition of wholeness into manageable relational parts over the space of problem-solution dualities for the efficiencies of the neuro-decision-choice actions over the space of fundamental-applied dualities.

NOTE 1.2.1

Some notes of analytical clarity of the definition will be helpful since this definition may seem to be different from those that are commonly available. The role of definitions in the source-destination communications is very important for massage interpretation, understanding and as inputs into decision-choice actions [R3.26] [R3.82] [R10.11]. This definition is abstracted from an understanding of the external and internal conditions of varieties. The internal conditions refer to internal parts and their give-and-take sharing relations, refer to as the *internal relationality* (an example is the human body), The decomposition, re-composition and composition refer to the relational understanding of the variety existence in the space of internal-external dualities over the space of problem-solution dualities and polarity. The set of the external conditions refers to the environment, with negative and positive give-and-take sharing relations, that surrounds the variety and is referred to as *external relationality*. A strong internal relationality strengthens the varieties' internal-external transformation games. The internal give-and-take sharing flow conditions must be fair, equitable, just, and dutiful with respect to intra-variety communications of information-knowledge flows. Any organization must be seen in the space of individual-community duality to take advantage of the benefits of unity, while maintaining the individualities in diversities in such a way as to fight the destructive nature of extreme individual preferences against the collective preferences within the unity. The translation of the net organizational advantages to the conditions of the principles of unity of knowing and science implies developing a socially powerful relationality of give-and-take sharing modes of categorial problem-solution dualities as defined in different disciplines. The success of neuro-decision-choice actions over one category or discipline may point to advantageous directions for neuro-decision-choice activities in other categories of problem-solution duality under the laws of unity, diversity, and relationality in a well-structured organization to help create a useful relational field to complement the epistemic field, where the limits of organization depend on the nature of the relational field as revealed through the information field.

1.2.1 Limits and Efficiencies of Organization in the Space of Problem-Solution Dualities and Neuro-Decision-Choice Systems

The intra-varieties are made up of different parts with different roles and with different give-and-take sharing conditions with complex communication networks for the growth and development as well as the maintenance of the organizational entity over the space of problem-solution dualities. The different parts may be analytically group into key parts as *intra-macro varieties* and the supporting parts as *intra-micro varieties* at equal levels of essentiality if the variety is to function *efficiently*. The intra-macro varieties are the essential *organizational pillars* while the intra-micro varieties are the essential *connecting supports* for the integrity of the variety. The conditions of the relationality of the intra-macro varieties and the intra-micro varieties set the *organizational limit inferior* and the *organizational limit superior* for conditions of growth, development, and survivability of the variety.

The organizational limit inferior and limit superior are determined by the conditions of relational field for the internal stability of the unit. The stronger the intra-relationality the greater is the non-vulnerability of both the organizational limits, while the weaker the intra-relationality the greater is the vulnerability of both the limits. The limits move closer and closer as the intra-varieties become tighter and tighter. The organizational limit inferior and limit superior are relevant in understanding the diversity and unity of knowing and science, where there are no categorial isolations but categorial relations of cooperation of intra-diverse forms in the sense that disciplines do not exist in individual isolation but in individual cooperation over the space of problem-solution dualities to take advantage of all possible real benefits of relational unity through give-and-take sharing processes reduce the real cost of disunity and increase the real benefit of unity of diversities under relational continua.

The intra-variety cost-benefit structure may be seen from the point of view that the conditions of individual isolationism and disunity imply no information-knowledge give-and-take sharing relations and non-cooperative decision-choice actions of the diversities over the space of problem-solution dualities. This is what is meant by saying that the category is internal decision-choice closed. The conditions of individual cooperation and unity imply collaborative efforts in give-and-take sharing relations and mutual support in decision-choice actions of the diversities over the problem-solution dualities. This is what diversities in unity means and the category is said to be internal decision-choice open. Let us examine the concept of organization relative to knowing, information-knowledge system and decision-choice actions over the space of problem-solution dualities and polarities and relate it to the possible and probable limits of individual and collective decisions and choices. The analytical conclusions are that limits to decision and choice have no limits if individual and collective decisions are related to variety problem-solution duality in never-ending processes of changing needs and wants under the principle of non-satiation where information-knowledge accumulation has no upper bound. The organization limit inferior and limit superior examined over the space of static-dynamic dualities, at any social set-up, relate to the lower and upper efficiencies under the principle of non-satiation, where the lower efficiency relates to conditions of the zone of greater cost over benefit and the upper efficiency relates to the conditions of the zone of greater benefit over cost under the contents of the prevailing social philosophical consciencism.

1.3 THE ORGANIZATION OF KNOWING, SCIENCE, AND PROBLEM-SOLUTION DUALITIES

The dynamics of neuro-decision-choice activities are such that every result of neuro-decision-choice activity is an information-knowledge item about a variety in a property, relation, or both at the static state. Every human activity is an information-knowledge item about a change in property, a change in relation or both at the dynamic state of neuro-decision-choice action. Every decision-choice action is also about either an identification of a variety property, an identification of a variety relation or both which are related to elemental identities in the space of variety problem-solution dualities within the spaces of organization-disorganization dualities contained in the space of visible-invisible dualities. The organization for dealing with neuro-decision-choice activities of varieties of the space of problem-solution dualities in knowing and science is complex at organic level and simpler at categorial level where organized actions take place from complexity to simplicity by decomposition into a system of diversities with or without relational continuum or from simplicity to complexity by composition of diversities welded together by relational give-and-take sharing modes.

At the organic level of knowing, there is the social organization which is the base. Its contents are made up of a collection of all intra-categorial and inter-categorial institutions to facilitate all individual, and collective or social decision-choice activities over the space of problem-solution dualities. At the individual level of knowing, there is the organizational decomposition of the universal oneness into micro units as the basis of social organization of knowing. The contents of any organization of knowing as well as any organization are made up of a collection of units of diversities with or without relational connectivity. In examining organization, it is always useful to keep in focus the unbreakable relations among neuro-decision-choice actions, problem-solution dualities, give-and-take sharing modes, knowing, information-knowledge accumulation and their impacts on change in relation, change in property, where both property and relation are defined over the space of qualitative-quantitative dualities to produce qualitative and quantitative motions of variety identities and transformations.

Here, it is helpful to note an important complication of the structure of the space of problem-solution dualities, where every result of individual and collective neuro-decision-choice action generates information-knowledge items in the degrees of efficiency units of a solution or a problem in the social organization. Every neuro-decision-choice result of a problem-solution duality is an information-knowledge flow to increase the information-knowledge stocks, where the result is in dualistic-polar units. The social organization is simply a network of formal-informal dualities of multi-relational connectivity of cognitive agents behaving over the game space of success-failure dualities contained in the open space of the problem-solution dualities and polarities toward common elements in the goal-objective space through decision-choice actions to set the solution dual against the problem dual in order to reverse the poles of the problem-solution duality by setting the solution pole against the problem pole in the space of relativity where dualities and polarities are always in unbreakable mutual relations over the identification-transformation processes contained in the space of imagination-reality dualities as abstracted from the space of actual-potential dualities.

Given the matter-energy category of existence, the understanding of natural organizations, decision-choice actions of cognitive agents constitute the understanding of the system of production modules over the space of problem-solution dualities as assessed from the cognitive field by neuro-decision-choice agents. The social organizational structure is designed by cognitive agents in line with the understanding of natural organizations to maximize benefit applications of the laws of diversity and unity as they relate to themselves, categorial problem-solution dualities, the diverse variety events, and outcomes in the epistemic field with the corresponding relational and information fields.

In taking advantage of the laws of unity and diversity of knowing and science, there are two socio-natural forces of differential unity that must be brought together. One is diversity and unity of categorial decision-choice skills of cognitive agents over the infinite generations. The other is the diversity and unity of categorial problem-solution dualities over the epistemic field in infinite stock-flow disequilibrium dynamics. The process of knowing through the production of intellectual investment flows and information-knowledge accumulations as production of intellectual capital stocks through the neuro-decision-choice actions over the epistemic field is first to partition the epistemic field into categories of problem-solution dualities and partition the neuro-decision-choice skills into dynamic categories and then coordinate categorial skills with categorial problem-solution dualities for research, knowing, leaning and teaching to abstract information-knowledge from variety identification and transformation problem-solution dualities for knowing investments and information-capital stock system in continual decision-choice actions over the space of problem-solution dualities which constitutes the spaces of production-consumption and input-output dualities.

At the categorial level, there are multitudes of sub-organizations containing different institutional arrangements for use as channel processes of source-destination continual information flows for continual neuro-decision-choice activities over the action space and the space of the problem-solution dualities by creating specific tactics and strategies, pulling together human and non-human resources, and facilitating works on different categorial problem-solution dualities toward specific sets of collective and individual goals. At the level of society as a social organization, the collection of the categorial sub-organizations constitutes the *superstructure* of the social organization. Each categorial sub-organization contains a set of institutional arrangements of unity through which an individual and collective neuro-decision-choice actions are directed to specific set of problem-solution dualities within the category. At the levels of social setups, there are two aggregate types of knowing and problem-solution dualities that sustain the dynamics of the variety problem-solution dualities and the dynamics of the information-knowledge system with its investment and stocks in the space of equilibrium-disequilibrium dualities. They are referred to as *fundamental problem-solution dualities* with a corresponding fundamental information-knowledge system, and the *applied problem-solution dualities* with the corresponding applied information-knowledge system. The two organs of knowledge and social information-knowledge system are intimately connected with multiplicity of give-and-take-sharing relationality, where the fundamental outputs supply the needed inputs for the applied problem-solution processes and the applied outputs supply inputs to the fundamental problem-solution processes for their mutual stock-flow disequilibrium dynamics in self-correction systems defined over the ignorance-knowledge dualities. Under the principle of opposites, the dualistic-polar give-and-take sharing conditions establish the understanding of a shared mutuality of dualistic-polar commonness over the spaces of existence-destruction and present-future dualities relative to the space of cost-benefit dualities for a mutual telescopic future history, where the give-and-take sharing conditions in practice is divided into a *give-and-take sharing principle* and a *give-and-take-sharing rationality* in the space of justice-injustice dualities. In human behavior, the dualistic-polar give-and-take sharing conditions translate into the practice of *each-for-all-and-all-for-each*, *ubuntu*, or *dibi-name-nibi* and shared future as opposed to *pesemakominya* (each for himself). The union of the fundamental structure of knowing and information-knowledge production, and the applied structure of knowing and information-knowledge production constitutes the totality of socio-economic production in unity, where the intersection is non-empty to define the multiplicity of relational continua at all generational timepoints in domestic and international information-knowledge heritage under infinite disequilibrium dynamics with continual production over the space of ignorance-knowledge dualities and error-correction system.

1.3.1 Individual and Collective Existence over the Space of Finite-Infinite Dualities, the Space of Problem-Solution Dualities and the Knowing Process

At the level existence, it is useful to understand the phenomenon of individual life and the phenomenon of generation life over the space of finite-infinite dualities in knowing and information-knowledge production as well as over the space of production-consumption dualities over the epistemological space and input-output dualities over the ontological space. The journey of cognitive agents over the epistemological space is indirect with multiplicities of constraints on social individual, collective, generations and intergeneration. The conditions of this journey relate to individual and generation existence over the space of finite-infinite duality in existence through production and reproduction of real life. The cognitive constraints in knowing and information-knowledge production have differential effects on individual efficiencies and generation efficiencies as we examine then in the spectrum of human existence. The individual life resides in finite space while the generation life resides in infinite space within the epistemological space contained in the space of ontological-epistemological duality that contains infinite variety transformations.

The finite space of life relates to a finite space of individual knowing and information knowledge production over the general space of problem-solution dualities relative to the space of cost-benefit dualities for the individuals while the infinite space of life relates to knowing and information-knowledge production over the general space of problem-solution dualities relative to the general space of cost-benefit dualities for infinite generations. The problem-solution dualities are cost-benefit defined under a system of preferences over organization over the space of imagination-reality dualities contained in the space of actual-potential dualities which is also contained in the space of ontological-epistemological dualities. The individual problem-solution dualities as organized in the space imagination-reality dualities to contain individual goals, objective and vision in the space of imaginations are terminated at the level of knowing while the accomplishments amplify the intellectual capital stocks for intra-generational and inter-generational utilization through a system of information-knowledge storage and transfers, the efficiencies of which depend on the available organizational nature of technologies of storage and transferability. Here, it is useful to note the relative organizational efficiencies of oral traditions, non-verbal traditions, different forms of storage such as writing in books and digital recording and the need of smart materials that increase the stock-flow information-knowledge processes over the generational and intergeneration information-knowledge production. The advantages of individual and collective intellectual production must be organized for knowing and storage within each generation with increasing development of intergenerational transferability. The understanding of the behavioral conditions in the space of individual-collective dualities is extremely important in understanding all actions for decomposition-composition dynamics, unity-disunity dynamics, organization-disorganization dynamics and the derived questions involving the structures of economic, political and legal at the level of social setup in relation to multiplicity of elements in the space of property-relation dualities that generate the spaces of justice-injustice dualities, necessity-freedom dualities, peace-war dualities and many others over the space of ignorance-knowledge dualities, where the behaviors of varieties in these spaces of dualities are decision-choice determined over the spaces of individual and collective thinking-reasoning dualities.

1.3.2 Primary-Derived Organizations, Institutions, and the Space of Problem-Solution Dualities

In terms of knowing, science, non-science and knowledge in every social setup, there is an organization with its institutional arrangements for the decision-choice activities over the space of the fundamental problem-solution dualities for the production and storage of the fundamental information-knowledge system. Similarly, there is a sub-organization of the superstructure with its institutional arrangements for the decision-choice actions over the space of the applied problem-solution dualities to produce applied information-knowledge system. The sub-organization with its institutional arrangements that houses and transmits decision-choice activities over the space of the fundamental problem-solution dualities is the collection of institutions of learning, teaching, research, and other forms of education that are either privately or publicly undertaken to provide the *fundamental information-knowledge system* to guide social practices. These institutions will be called the *primary category of institutions* of the social set-up at the level of the superstructure and in a more explicit form as the *primary category of superstructure institutions*. The organization with its institutional arrangements that houses and transmits decision-choice activities over the *applied problem-solution dualities* is the collection of institutions of industry, production of goods and service in varieties, that are either privately or publicly undertaken to satisfy the individual and social consumptions over the space of preference. These institutions will be called *derived category of institutions* of the social set-up at the level of superstructure, and in a more explicit form as the *derived category of superstructure institutions*. The basic structure generates the sub-space of basic problem-solution dualities with a corresponding evolving set of basic institutions while the superstructure generates the sub-space of superstructure problem-solution dualities with the evolving set of superstructure institutions, while the union of the sub-space of basic problem-solutions dualities and corresponding set of evolving institutions and superstructure of problem-solution dualities and the evolving set of institutions constitutes the space of evolving space of problem-solution dualities and the corresponding evolving set of institutions.

The approach, here, in using the concept of problem-solution duality and decision-choice action in establishing the *principles of diversity and unity* of knowing and science leads to a situation where the social setup is first partitioned into the *basic organization* with its institutions as the *primary*, and the *superstructure organization* with its institutions as the *derived.* Given the basic organization, the superstructure organization is then partitioned into organization of institutions for fundamental information-knowledge production called *primary category* of super structure institutions of the organic organization, and organization of institutions for applied information-knowledge production called the *derived category* of superstructure institutions at the level of the organic social organization, where the fundamental and applied information-knowledge systems are seen in terms of production-consumption and supply-demand over the space of input-output dualities, in such a way that inputs become outputs and outputs also become inputs in the space of variety destruction-construction dualities under a systems of technological processes and quality controls over the space of input-output dualities, where a theory is about the development of thinking algorithms and application is about the use of the thinking algorithms in the spaces of identification-transformation, problem-solution and production-consumption dualities contained in the space of imaginations.

The organizations and supporting institutions of fundamental and applied production systems are mutually dependent and creating, constituting the elements in the space of production-consumption dualities as well as sub-space through decomposition of the general space of input-output dualities of epistemological space as well as the space of destruction-construction dualities. The organizations and their institutions are the products of decision-choice actions over the space of problem-solution dualities in relation to elements in the goal-objective space under some individual and collective preference scheme in selection of elements from the space of problem-solution dualities. Interestingly, the building of organization and supporting institutions also belongs to the problem-solution dualities as well as subject to decision-choice actions and all possible cognitive capacity limitations. It is within this context that the principle of opposites with duality and polarity finds definitional understanding. The recently emerging fields of neuroeconomics, neuroscience neurochemistry, neurobiology, epigenetics, engineering-economic system and many others which belong to neuro-decision-choice sciences are attempts to understand cognitive actions on decision-choice activities over the problem-solution dualities, dynamics of the input-output of information-knowledge system for variety identifications and transformations in the space of sequencing the contents of variety characteristic dispositions and the organizational principles of diversity and unity in relation to the space of cost-benefit dualities for human progress in the space of socio-natural destruction-construction dynamics under the matter-energy equilibrium. The structural connectedness of different organizations and institutions are presented as an epistemic geometry in Figure 1.3.2.1.

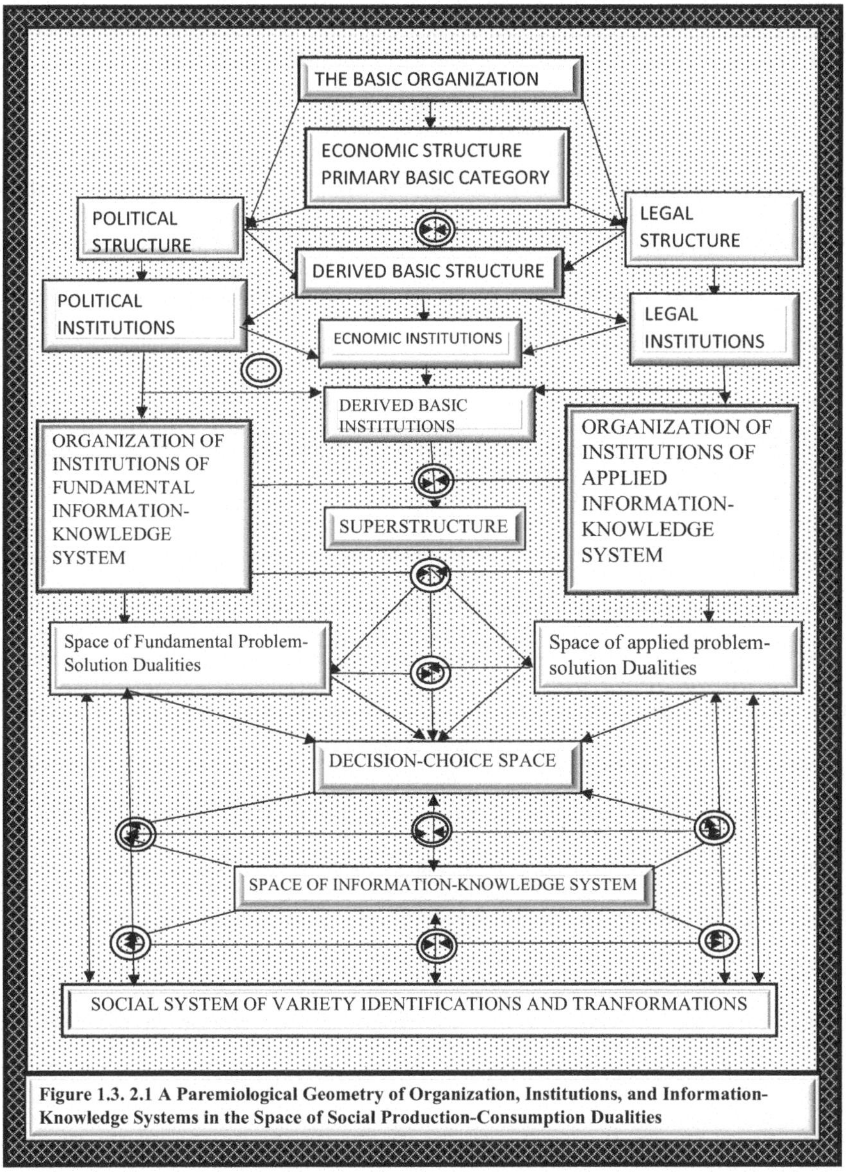

Figure 1.3. 2.1 A Paremiological Geometry of Organization, Institutions, and Information-Knowledge Systems in the Space of Social Production-Consumption Dualities

1.3.3 Organizations, Institutions and the Social Philosophical Consciencism

The understanding of the diversity and unity of knowing and science requires an understanding of the organicity of the social structure and its guidance of the neuro-decision-choice actions with varying degrees of cognitive capacity limitation within the space of individual-community dualities as organized from the space of imaginations contained in the space of imagination-reality dualities. Organization is continually dynamics which is about structuring and restructuring of variety properties and relational give-and-take-sharing conditions among individual diversities to maximize the net benefit of collective relationality over the space of negative-positive dualities toward accomplishments of some elements in the space of goals and objectives as envisioned and organized. The conditions of relational field are cost-benefit distributions of collective and individual social actions over the space of production-consumption dualities. Every organism involves production-consumption distribution processes for continual existence over the space of parent-offspring dualities. The space of production-consumption dualities has a dynamic structure that is continually being transformed by individual and collective neuro-decision-choice processes under conflicts and cooperations. It is here that the idea that the people are the backbone of national greatness acquires meaning and essence. The people are the knowing, where information and knowledge are of the people who are the producers and consumers of the information-knowledge in the general space of input-output dualities, where the utility of information is determined by cognitive agents in relations to decision-choice actions over the elements in the space of problem-solution dualities under the prevailing social philosophical consciencism.

Social philosophical consciencism contains the evolution of the contents of the collective mind, where the varying aspects of the set of contents of the social philosophical consciencism constitute the varying contents of the individual mind that is unknown by other decision-choice agents. The individual mind, in this respect, is an empty luggage of the brain at birth and is gradually filled with the contents from the evolving social philosophical consciencism in an encounter with the social set-up and its members, where languages and information-knowledge items are developed as proven and unproven ideas overt the space of good-bad dualities. The individual contents are abstracted using the *theory of information-knowledge filtering relative to individual goals and objectives* where such goals and objectives may conflict with the collective goals and objectives. The theory of information-knowledge filtering is not about elements in the space of true-false dualities. The filtering from the social philosophical consciencism is about information-knowledge input for accomplishing goals and objectives.

The dynamic structure of the social organism is made up of *basic structure* and *superstructure* toward the exercise of individual and collective neuro-decision-choice actions over the space of variety problem-solution dualities for continual transformations over the space of variety production-consumption dualities changing properties and relations. The *basic structure* of the social organization is composed of the *economic structure* with the category of economic institutions, *political structure* with category of political institutions and *legal structure* with the category of legal institutions. In terms of the basic structure, the economic structure and its institutions constitute the primary category of the basic organization, while the political and legal structures and their institutions constitute the derived categories of the basic organization. The basic structure of the social organization houses the primary category of institutional arrangements for neuro-decision-choice actions over the primary category of the basic problem-solution dualities. Let us always keep in mind the analytical importance of primary-derive dualities and parent-offspring dualities to understand the conditions of telescopic past and telescopic forward in relation to the present in the analysis of conditions of the space of variety unity-disunity dualities with internal diversities and epigenic institutional processes. There are some characteristics of the basic structure and the superstructure that are useful to examine and know for the understanding of the organization of knowing as production of intellectual investment flows and information-knowledge accumulation as production of intellectual capital stocks over the space of static-dynamic dualities.

The superstructure of the social organization houses the derived category of institutional arrangements for neuro-decision-choice actions over the derived category of super-structural problem-solution dualities. At the level of the basic structure, the efficiencies of the neuro-decision-choice actions at all states and processes demands valid and relevant information-knowledge inputs that must be acquired by cognitive agents through decision-choice actions over the superstructure problem. In this respect, the basic structure and the superstructure are in relational continuum and unity through a system of give-and-take sharing activities under stock-flow processes of information-knowledge disequilibrium dynamics in other word they constitute a space of basic-superstructure dualities. The results of applied decision-choice become inputs into theoretical decision-choice process for corrections and further claims over the spaces of true-false and acceptance-unacceptance dualities, where theories create foundations of practices, and the results of practices create conditions of theory refinements. In this respect, where the fundamental and applied information-knowledge structures are interdependent and relationally united in the general knowing and information-knowledge production under disequilibrium dynamics with the general power of individual and collective decision-choice actions as abstracted from the guidance of social philosophical consciencism.

Given the primary category of the superstructure institutional arrangements, there are two sets of sub-institutional arrangements that may be identified for total information-knowledge production at each point of time in support and maintenance of the existence of any social setup whether primitive or non-primitive, where the information-knowledge production may be seen as an input-output process of the basic organization containing sub-organizations with corresponding sets of supporting institutions in relational unity defined in a dynamics of give-and-take-sharing structures for relational continua. Each of these sets of institutions is infinite in possibilities with lower bounds and no upper bounds, just like the space of problem-solution dualities is infinite with no upper bound in the sense that every problem must be seen in terms of *what it is not* (solution) as well as *what it is* (problem) and similarly, every solution must be seen as *what it is not* (problem) and *what it is* (solution) within the institutional arrangements for the channels of decision-choice actions over the decision-choice organizing space of problem-solution dualities.

The analysis of general decision-choice activities over the space of the variety problem-solution dualities, also as the space of variety (*what-was-not*)-(*what-was*) dualities or variety (*what-there-is*)-(*what-there-is-not*) dualities or the variety of (*what-would-be-not*)-(*what-would-be*) dualities to produce information-knowledge system, must acknowledge the existence of category of *individual problem-solution dualities* and the category of *social problem-solution dualities* with corresponding decision-choice actions and their inter-categorial and intra-categorial interactions under the distribution of differential preferences of the spaces of variety cost-benefit dualities and individual-collective dualities as facilitated by the corresponding set of institutions under the general principle of non-satiation over the space of varieties. The preference space of individuals, collective and society is molded and shaped by the contents of the social philosophical consciencism at generational timepoints.

The social organizations and corresponding institutions are shaped by the social philosophical consciencism operating over the space of imaginations transforming variety imaginations to variety realities with the elements defined over the space of negative-positive dualities, where every negative variety and every positive variety are defined in the space of the general cost-benefit dualities, where the distribution of cost-benefit dualities in the space of individual-collective dualities induces continual conflicts. Here, social organization that defines institutions is to reconciled conflicts in preferences in the spaces of individual-collective dualities, inter-individual structures, and intra-individual structures. as well as reconciliations dualistic-polar conflicts in opposites. Each set of institutional arrangements for the individual and collective decision-choice actions in the spaces of fundamental and applied problem-solution dualities is infinite and clouded in the spaces of certainty-uncertainty, vagueness-preciseness exactness-inexactness and doubt-surety dualities due to cognitive capacity limitations at the level of formations of *knowledge by acquaintance* and at the level of formations of *knowledge by description* through a paradigm of thought over the space of micro-macro dualities.

The neuro-decision-choice activities over the space of problem-solution dualities and polarities to produce information-knowledge system are also defined over the spaces of (*what-there-was*)-(*what there was not*) (telescopic past varieties) dualities, (*what-there-is*)-(*what-there-is-not*) (telescopic present varieties) dualities and (*what-there-would-be*)-(*what-there-would-be-not*) (telescopic future varieties) dualities under continually institutional and generational changes over the space of dynamic variety quality-quantity dualities. These dualities induce different categories of knowing in science and non-science as well as requiring different cognitive skill sets to reveal their identities and transformations through the spaces of acquaintance and paradigm in static and dynamic states. Each category of knowing has fundamental problem-solution dualities and corresponding fundamental information-knowledge system over the space of static-dynamic dualities. Additionally, it has an applied problem-solution dualities and corresponding applied information-knowledge system in static and dynamic states. The organizational structure for the fundamental sector is different from that of the applied sector but in relational continuum and unity. Let us examine the similarities and differences. Additionally, there is no information-knowledge system if there are no problem-solution dualities, and that the conditions of the space of variety awareness-unawareness and action-inaction dualities belong to the space of problem solution dualities. Similarly, the space of progress-nonprogress dualities vanishes when there is no space of problem-solution duality that provides the foundations of neuro-decision-choice actions of cognitive agents in a complex relationality over the relational field under the ruling social philosophical consciencism.

1.3.4 The Organization of Knowing, Science and Fundamental Problem-Solution Dualities.

Organization of information-knowledge development is about the development of institutions and coordinate them with relevant categories of problem-solution dualities through which neuro-decision-choice actions are implemented with supporting information. The goals and objectives of the neuro-decision-choice actions over the space of the fundamental problem-solution dualities to generate the fundamental information-knowledge system are to provide inputs into the neuro-decision-choice activities over the space of applied problem-solution dualities to produce applied information-knowledge system as inputs into individual and collective neuro-decision-choice activities over the spaces of variety production-consumption dualities and supply-demand dualities within the input-output process. The most important thing to keep in mind, in the whole of information-knowledge system over the space of static-dynamic dualities, is the understanding of the role of neuro-decision-choice activities in all areas of individual and collective existence and the dualistic-polar role of information-knowledge systems that act simultaneously as inputs and outputs of individual and collective neuro-decision-choice actions to guide the knowing activities over the space of problem-solution dualities and shape the enveloping of individual and social transformations over the space of production-consumption dualities relative to the space of the input-output dualities.

Let us keep in mind that our awareness of varieties and the conditions of socio-natural environment is impossible without information through which knowledge is derived. Similarly, our ability to distinguish among varieties, use the varieties and transform varieties depend on knowing as a flow, and information-knowledge accumulation as a stock which are produced in the superstructure of the social basic organization for its self-maintenance. Information-knowledge productions are input-output processes that involve continual accumulation of variety conditions for identities, and variety transformations are input-output processes that involve destruction-construction processes of varieties and continual expansion of the available information-knowledge system, as problems are identified and solved to generate new problems requiring new solutions.

The organic goal of the social basic organization is to provide relevant resources for institutional arrangements through which the neuro-decision-choice activities of the social superstructure may be organized to assist the functioning of the basic organization. The social *basic organization*, as indicated, is composed of *economic structure* as sub-organization that contains the economic institutions for individual and social economic decision-choice actions and policies in support and maintenance of production of life, *the political structure* that contains the political institutions for political decision-choice activities in reconciling conflicts and the distribution of decision-making power over the space of individual-collective dualities and policies and *legal structure* that contains the legal institutions for legal decision-choice activities and policies in regulating preference behavioral actions over the space of individual-collective dualities. The economic, political, and legal neuro-decision-choice actions generate information-knowledge flows to increase the accumulation of the information-knowledge stocks. Let us keep in mind that the information-knowledge flows through knowing are *intellectual investments* while the information-knowledge stocks are *intellectual capital accumulations* from decision-choice activities of knowing and further information-knowledge production-consumption dynamics.

The fundamental information-knowledge conditions on the economic, political, and legal structures, at static and dynamic behaviors, are under the knowing process of social science which is also under decision-choice activities over the space of *social problem-solution dualities*. Similarly, the conditions of fundamental information-knowledge system on the superstructure with a corresponding space of variety problem-solution dualities are under social science. Every social system finds its existence, development, and progress over the space of basic-superstructure dualities. The decision-choice activities in the space of basic-superstructure dualities are in relational continuum and unity, directed to deal with the elements in the fundamental-applied problem-solution dualities to produce the *fundamental information-knowledge system* and *applied information-knowledge system* for continual social transformation toward states of potentially changing individual, social and generational preferences under the principle of variety non-satiation over the conflict zones of spaces of individual-community dualities and generational parent-offspring dualities. The organization of knowing and applied problem-solution duality will be treated separately even though they are always in relational continuum and unity. The development of social organization must adhere to all the elements of relational continuum and unity under the general laws of opposites that define the existential identities and transformations of varieties and categorial varieties in the epistemic field. Let us keep in mind that static stability and equilibrium are not permanent characteristics of the life of an organization. Every living organization resides on a temporary equilibrium path and permanently on a disequilibrium path.

The variety temporary equilibrium states are in continual internal restructuring under dualistic conflicts generating the conditions within the space of construction-destruction dualities. The variety permanent disequilibrium states define polar conflicts for continual transformations for change of internal property and external relations under the guidance of social philosophical consciencism. Every social organization is defined over the space of production-consumption dualities where the goal of the organization is reconciling conflicts in the space of production-consumption dualities through the neuro-decision-choice management of the elements in the spaces of input-output and destruction-construction dualities through the space of supply-demand dualities, where the neuro-decision-choice activities may be organized under either the conditions of market, non-market institutional arrangements or both in response to conflicting preferences over the space of individual-collective dualities within intergenerational dynamics with changing intragenerational preference structures and shifting of intergenerational preference structures over the general space of cost-benefit dualities and the principle of non-satiation over the space of actual-potential dualities from which the space of imagination-reality dualities is abstracted by the neuro-decision-choice actions, where the space of imagination houses the spaces of possibility, necessity, anticipation, probability, freedom, and expectation in the general space of neuro-decision-choice actions over the space of thought-action dualities relative to the space of success-failure processes.

CHAPTER 2

THE INFORMATION-KNOWLEDGE SYSTEMS AND THE ORGANIZATIONAL ANALYTICS OF THE BASIC-SUPERSTRUCTURE DUALITY

ABSTRACT

Chapter 2 provides the analysis of the relationships of organizational analytics of information-knowledge systems on basic structure, superstructure, and the space of superstructure dualities in the development of knowing and the production of information-knowledge accumulation and their connectivity to the spaces of input-output and cost-benefit dualities. Two sets of theories of explanatory and prescriptive structures are introduced and discussed in the knowing and information-knowledge processes which are connected to descriptive and predictive structures over the space of imagination-reality dualities. An introduction to a theory of certainty-uncertainty dualities is advanced to provide explanatory conditions and sources of uncertainties which give rise to the space of imagination containing possibility and probability spaces and hence the rise of possibility and probability theories. Discussions are made on basic-superstructure components of organization of knowing and information-knowledge systems where definitions and comparative analysis of policy and decision-choice action are presented and linked to relationality, and organicity of the space of problem-solution dualities. The roles of individuals and the collectivity in the organization of problem-solution dualities are the examined in the space of individual-community dualities in relation to conflicts and conflict resolutions over the space of cost-benefit dualities. The roles of institutions of research, teaching and learning are examined, leading to the discussions on the goals, objectives, and research in the organization of knowing

K. K. Dompere, *Theories of Knowing and Knowledge Production*, Studies in Systems, Decision and Control 501, https://doi.org/10.1007/978-3-031-42863-0_2

and information-knowledge production, where we discuss learning with and without supervision in relation to thinking and reasoning. The chapter is concluded with the discussions on the development of special institutions with global collaborations, and competitions over the epistemic field, where the space of relevant alternatives may be expanded, and the space of irrelevant alternative may be reduced with global collaboration and creative competition in discovery of elements in the space of categorial conversions or the functor space under the guidance of the ruling social philosophical consciencism.

2.1: INTRODUCTORY REFLECTIONS

The analytical approach taken in the development of a system of sub-theories within the general comprehensive theory of knowing and information-knowledge process is that all human behaviors are defined in the space of production-consumption dualities contained in destruction-construction dualities abstracted from the space of input-output dualities as reflected in the space of cost-benefit dualities, where the nature of social cost-benefit distribution gives rise to the spaces of justice-injustice, conflict-harmony, war-peace, violence-nonviolence dualities and many such dualities in the epistemological space in which humans operate. The cost-benefit distribution shifts with shifting intra-generational and inter-generational preferences just as shifting nature of generational exceptionalism in line with the nature of the ruling social philosophical consciencism. In this framework, there are two sets of theories that are of interest over the space of organizational basic-superstructure dualities. They are the *set of explanatory theories* and the *set of prescriptive theories* which are logically connected to information and definition theories with concept formations over the space of real-nominal dualities, descriptive theories, quality-control theories, and predictive theories with reality formation. Additionally, these theories are established over the space of imagination-reality dualities in relation to how functors and categorial convertors may be and are created to move variety imaginations to variety realities and how are these theories interrelated. The set of descriptive and predictive theories are considered as sub-theories in the set of explanatory theories. The set of explanatory theories and sub-theories finds residence in the space of realities contained in the space the actuals while the set of prescriptive theories find residence in the space of imaginations contained in the space of potentials. The space of the set of theories of explanatory-prescriptive dualities is contained in the space of actual-potential dualities in such a way that all the theories are defined in the space of variety identification-transformation dualities relative to the space of ignorance-knowledge dualities. These two sets of theories involve knowing as a production of intellectual investment flows and the information-knowledge accumulation as a production of intellectual capital stocks in the activities of the basic and superstructure organizations of the social setup. The explanatory and prescriptive theories reside as dualistic-polar relational structures and yet have independent existence over the space of ignorance-knowledge dualities.

Descriptive conditions are generated by neuro-decision-choice processes as outputs to provide information outcomes on variety identities in the space identification-transformation dualities that become inputs into explanatory processes of the variety identities and behavior, the outcomes of which provide inputs into predictive conditions regarding a movement from the space of realities contained in the space of actuals to the space of potentials from which the space of variety imaginations is abstracted and organized by neuro-decision-choice actions to produce varieties over the space of variety old-new dualities. The explanatory theories are based on the outputs of descriptive theories while predictive theories are based on the outputs of explanatory theories in the neuro-decision-choice journey within the space of ignorance-knowledge dualities, from the space of realities through the space of actuals to the space of potential through the space of imaginations and to the space of realities as they are connected to the space of ontic-epistemic dualities under cognitive capacity limitations and transversality conditions over the space of ignorance-knowledge dualities. The analytical pathways and the spaces of production and reproduction of knowing and information-knowledge accumulations of actual-potential dualities and imagination-reality dualities provide instructions and conditions for the understanding of the rise of spaces of certainty-uncertainty dualities, doubt-surety dualities and the of conditions of possibility-impossibility dualities, probability-improbability dualities, possible-impossible-world dualities, probable-improbable-world dualities and success-failure dualities over the space of neuro-decision-choice actions relative to the space of production-reproduction dualities as expressed over the space of identification-transformation dualities.

2.1.1 Reflections on Explanatory Theories

The explanatory theories and sub-theories are directly supported by knowledge by acquaintance in the general space of acquaintance broadly defined. The knowledge by acquaintance of varieties is the outputs of neuro-decision-choice actions as experiential information about types of varieties over the space of identification-transformation dualities. The outputs of which become inputs into information-processing modules in the explanatory theory development, where the tasks of the explanatory theories are to explain and understand variety behaviors in the space of identities regarding their existence and property in the space of relations creating initial conditions for transformation of the variety properties and relations from either the space of realities contained in the space of actuals to the space of potentials or the space of imaginations contained in the space of potentials to the space of realities contained in the space of actuals, where the space of actual-potential dualities constitutes the ontological space and the space of imagination-reality dualities constitutes the epistemological space. The prescriptive theories are directly supported by axiomatic information which may be connected to the experiential information from the space of acquaintance but not necessarily derived from the conditions of explanatory theories. The axiomatic information from the space of acquaintance becomes inputs into the transformation decisions to create rules of actions to change variety properties and variety relations from either imagination to actual or from actual to potentials under cost-benefit conditions. The space of imagination-reality dualities constitutes the epistemological space and linked to the space of actual-potential duality of the ontological space by the space of acquaintance in the space of source-destination dualities for communication between the elements within the space of ontological-epistemological dualities. The communication process within the space of source-destination dualities is variety information defined in terms of characteristic-signal disposition where the characteristic disposition is the contents, and the signal disposition is the message in codes constituting the ontological information that is received at the acquaintance space as the destination for decoding. The decoded signal disposition reveals the contents of the messages as epistemological information in terms experiential information or knowledge by acquaintance which is then associated with variety identity.

The epistemological information is seen as either experiential information or axiomatic information. The axiomatics conditions just like the explanatory conditions define the initial and necessary conditions as neuro-decision-choice information inputs for the understanding of variety transformations in the space of variety property-relation dualities, where the predictive conditions are about ontic behaviors and prescriptive conditions are about epistemic behavior. Both the experiential information and axiomatic information have similarities and differences in terms of their sources, productions, identities, acceptances and uses and the roles they play in the organization of knowing and information-knowledge dynamics over the space of basic-applied dualities within the space of ignorance-knowledge dualities.

In the space of knowing-knowledge dualities, the explanatory theories are derived by organization of neuro-decision-choice actions from the space of the actuals to provide conditions on variety identities in the space of realities contained in the space of variety property-relation dualities in continuum and unity, where the explanatory theories provide refined knowledge as knowledge by description by cognitive agents and about the environment of their existence in terms of *what there was* and *what there is* contained in the past-present telescopic structures, and how the past informs the present and the present informs the past connectivity. In this framework, the set of explanatory theories is knowledge by description which belongs to the category of organized explanatory sciences relative to the elements contained in the space of ontic-epistemic dualities, the objective of which is to increase the knowledge contents of variety identities and relations in the social philosophical consciencism to guide social practices over the space of thought-action dualities. The explanatory theories are about establishing the necessary and initial conditions for predicting categorial conversions over the space of actual-potential dualities, where the conditions of explanation and prediction become organized inputs for neuro-decision-choice activities over the space of problem-solution dualities, where the organizations contain relevant institutions through which cognitive actions are exercised over the elements in the space of problem-solution dualities as they relate to the elements in the space of ignorance-knowledge dualities, where the explanatory conditions become explanatory-based prescriptive conditions. In the case of a system of explanatory theories the epistemic process moves from description concerning concepts and language to explanation concerning behavior to prediction concerning variety occurrence and then to prescription concerning property-relational changes. Over the epistemological space, the experiential information establishes *a priori* knowledge basis as input into neuro-decision-choice actions while knowledge by description establishes *a posterior* information basis as input into neuro-decision-choice actions. In the whole process of cognitive actions, we have explanatory knowledge base on the conditions of experiential information regarding variety identity anchored in the space of identification-transformation dualities. The explanatory theory provides us with the knowledge of variety identity and behavior within the space of identification-transformation dualities.

2.1.2 A Reflection on Prescriptive Theories

Now let us have a reflection on prescriptive theories as an organized system of knowing as a production of intellectual investment flows and information-knowledge accumulation as a production of intellectual capital stocks. Over the space of ignorance-knowledge dualities, prescriptive theories are derived by neuro-decision-choice actions from the space of potential through the space of imaginations to provide rules for actualizing the potential as imagined to be actualized as elements in the space of realities, where the conditions of actualization pass though the spaces of possibility-impossibility and probability-improbability dualities within the space of imagination as generated by epistemic distances from the space of ontic-epistemic dualities. As has been discussed and explained, the ontic characteristic dispositions present complete variety characteristic disposition and perfect variety identities as to *what there was, what there is* and *what there would be*. The epistemic characteristic dispositions present incomplete variety characteristic dispositions and imperfect variety identities as to *what there was, what there is* and *what there would be*. The differences between the ontic and epistemic of variety characteristic dispositions constitute the epistemic distances that generate uncertainties to affect the coordination and effectiveness of neuro-decision-choice actions. As such, the coordination in the space of problem-solution dualities becomes defined over the space of success-failure dualities as well as over the space of individual-community conflicting preferences relative to the conditions over the space of cost-benefit dualities, the elements of which function as incentives which are the foundations of variety ranking and actions over the space of imagination-reality dualities containing all activities that are and may be defined and established in the epistemological space.

The set of the prescriptive theories may be divided into two in terms of information support. A sub-set of the prescriptive theories derives its information support as a posterior knowledge from the conditions generated and organized around the results of the explanatory theories and hence may be referred to as *explanatory-information-support prescriptive theories* to provide rules for variety transformations. Another sub-set of the prescriptive theories derives its information support as a prior knowledge from axiomatic information which is directly abstracted and organized from the space of acquaintance as self-evident truth and hence may be referred to as *axiomatic-information-support prescriptive theory* to provide rules for variety transformations. The explanatory-information-support prescriptive theories provide rules to actualize varieties described, explained, and accepted as belonging to the space of actuals and may be repeated such as reengineering in the different areas of which they do not yet belong in the space of realities, where the solution is knowledge of a variety, and the problem is how to actualize the varieties. The explanatory-information-support prescriptive theories are intra-categorial conversions with known functor in the sense that the organized changes in properties and relations take place in the same space of realities contained in the space of actuals, where the varieties within the space of realities are duplicated with known functors for variety characteristic dispositions. Under the explanatory-information-supported prescriptive theories, varieties are transformed within the space of realities.

The axiomatic-information-support prescriptive theories provide rules to actualize varieties conceived in the space of imaginations, contained in the space of potentials, and accepted to meet the test conditions of possibility and probability as they are transformed to varieties in the space of realities contained in the space of actuals. The axiomatic-information-support prescriptive theories are inter-categorial conversions where varieties are conceptualized in the space of imaginations as variety potential, and where new functors are created to transform varieties from space of imaginations to the space of realities in the sense that changes in properties and relations take place between different spaces in the process of knowing and knowledge production, where new varieties are conceptualized in thew space of imaginations as a subspace of the space of potentials and conceptualized over the space success-failure dualities contained in the space of realities. Examples of prescriptive theories include theories of planning, theories of architecture, theories of engineering of all forms. The test of validity of prescriptive theory is in its success in practice through the space of certainty-uncertainty dualities in bringing the variety imagination into the space of variety realities and survivability in the space of actualities, where the prescriptive conditions become explanation conditions and for further practices. In the case of a system of prescriptive theories the epistemic process moves from axiomatization to description to prescription to explanation and then to prediction. The prescriptive theory presents us with the knowledge about tactics, strategies, and steps of transformations of variety identities and behaviors The organizational structure of the relational connectivity of these theories over the spaces of ontological-epistemological dualities, ontic-epistemic dualities, actual-potential dualities, imagination-reality dualities, and ignorance-knowledge dualities is shown as an epistemic geometry in Figure 2.1.1

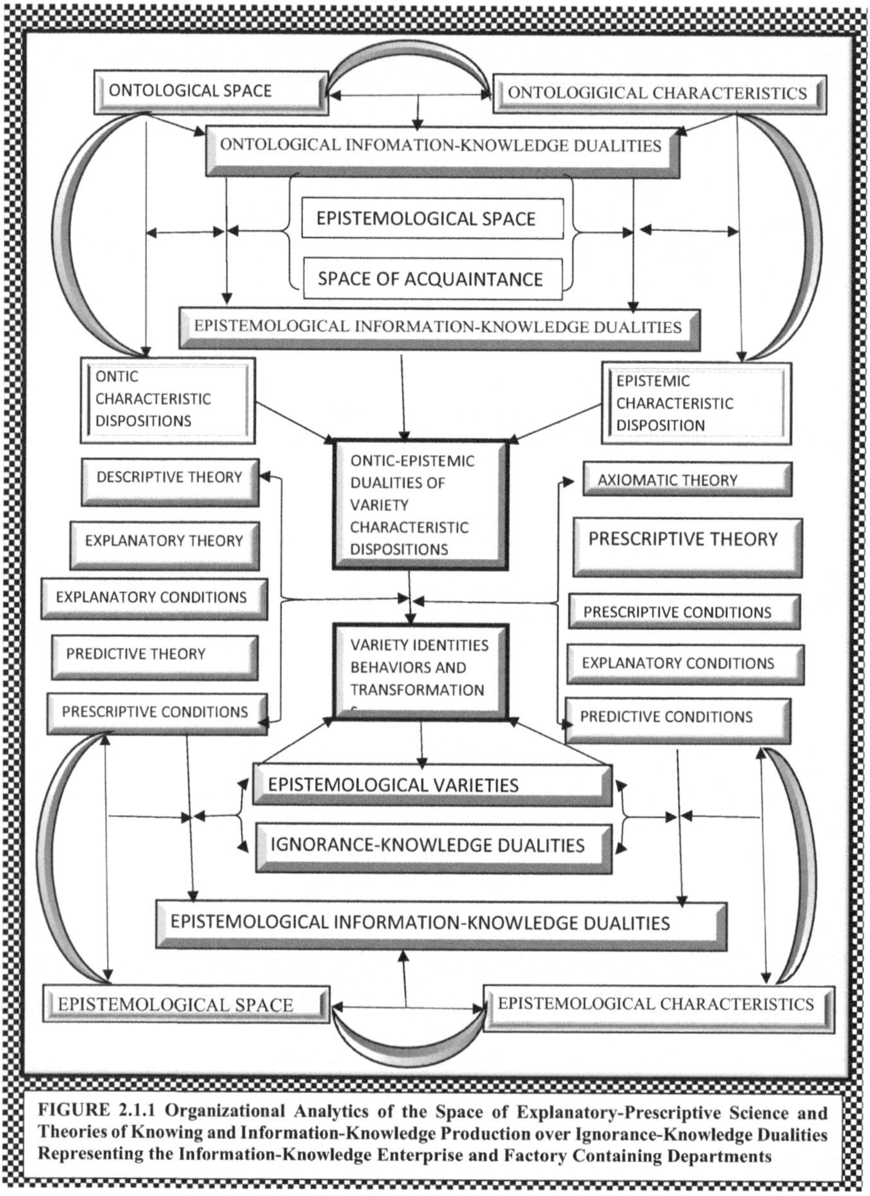

FIGURE 2.1.1 Organizational Analytics of the Space of Explanatory-Prescriptive Science and Theories of Knowing and Information-Knowledge Production over Ignorance-Knowledge Dualities Representing the Information-Knowledge Enterprise and Factory Containing Departments

2.1.3 Introductory Notes on Certainty, Uncertainty, and the Corresponding Spaces

It is useful to note that every variety problem has a set of variety solutions from which an optimal solution may be abstracted in a constrained space over the goal-objective space with conflicting elements in preference schemes within the space of individual-collective dualities containing the space of success-failure dualities. The space of success-failure dualities draws its residence from the space of decidability-undecidability dualities which draws its existence from the space of doubt-surety dualities that generates expectations and hesitations over the space of inaction-action dualities. The question is: Where does the space of doubt-surety dualities derives its existence within the general space of neuro-decision-choice actions over the space of ignorance-knowledge dualities? What are the meanings of surety, doubt, failure, success, ignorance, knowledge, decidability, undecidability, incentive, disincentive, and related ones in relation to neuro-decision-choice actions over the space of general human behavioral activities and how do they help in establishing the paths of individual and collective progresses over the space of social success-failure dualities relative to the space of general problem-solution dualities? What is the epistemic nature of neuro-decision-choice actions and are the corresponding phenomena and the concepts defined over the space of relativity or over the space of absoluteness?

In other words, is human knowledge over the space of fundamental-applied dualities within the basic-superstructure organization established in the space of absoluteness or in the space of relativity and what is meaning and essence of this knowledge? Similarly, are the variety identities defined and established in the space of absoluteness or in the space of relativity and how are the identities determined? What differences do they make in terms of the space in which they are defined and established? What are the causes of doubt and surety and how are doubt-surety dualities formed and related to decidability-undecidability dualities within the space of neuro-decision-choice actions, the results of which are defined by the conditions of success-failure dualities in human actions due to the presence of cognitive capacity limitations? Is the space of doubt-surety dualities connected to the space of uncertainty-certainty dualities and what are the sources of uncertainty-certainty dualities? What different kinds of problem-solution dualities may be confronted by neuro-decision-choice agents, in what spaces, what are the information-knowledge requirements, how are they obtained, and do they relate to the space of certainty-uncertainty dualities and in what space is the space of certainty-uncertainty dualities contained and how is it formed? These questions are very important and relate to the whole conditions of human life and existence within the ontological space from which the epistemological space is formed. They also provide a framework to examine the conditions of meaning and essence of information, knowledge in relation to neuro-decision-choice actions as well as providing a framework to examine and criticize classical representations and theories of information-knowledge-decision-choice-interactive processes and social progress through socioeconomic transformations.

The answers to most of these questions require the development of theories of certainty, uncertainty, and space of certainty-uncertainty dualities. The theories of certainty and uncertainty must be separate and different from and yet connected to the theories of imagination, possibility, probability, possible and probable worlds, and corresponding spaces to show how the space of imaginations as a subspace of the space of potentials is central to the varieties in the space of human success-failure dualities and progress-retrogress dualities. Here, the meanings of certainty, uncertainty in the space of ontological-epistemological duality must be related to cognitive capacity limitations over the space of acquaintance-description dualities contained in the space of ontic-epistemic dualities and related to the space of cost-benefit dualities which contains the space of incentive-disincentive dualities.

The conditions on these statements over the space of question-answer dualities and problem-solution dualities are extremely important in our understanding of the input-output processes as demand-supply chain in the whole production and reproduction of real life under the *anoma-kokone-kone* (individual-community) problem, *funtum-mereku-denkyem-mereku* (unity-disunity) conditions, *asantrofi-anoma* (cost-benefit) rationality and the *Sankofa-anoma* (past-present-future) information-time structure in the space of human actions, where success-failure dynamics are the norms of existence in interdependent continua over the epistemological space. The conditions of the *anoma-kokone-kone* problem provides the dualistic-polar conflicts of social relations of individual and community in generation and over generation; the *asantrofi-anoma* conditions and rationality provide value assessments and ranking of varieties in the space of neuro-decision-choice actions on the basis of cost-benefit or input-output dualities; the *Sankofa-anoma* conditions provide us with past-present-future information-time connectivity in the space of forecasting-discounting dualities as input-output structures over the neuro-decision-choice space while the *funtum-mereku-denkyem-mereku* conditions present relational continuum and unity in the space of varieties under the general principle of opposites over the spaces of identification-transformation dualities ; all of which find epistemic reflections in the space of ignorance-knowledge dualities, where overcoming of one variety ignorance generates a new variety ignorance, variety solution generates new variety problems, and variety success generates variety failures placing neuro-decision-choice agents in continual states of ignorance, problems and failure connected to the spaces of joy-sorrow and happiness-sadness dualities[R27.8] [R27.10].

The Sankofa-anoma conditions are epistemologically troubling over the spaces of basic-superstructure, fundamental-applied and identification-transformation dualities when related to the understanding of the existence of the universe in the telescopic past-present-future structure, where the initial conditions and necessity of the beginning dispositions are evasive of cognition and are difficult to abstract and the future is always unknown and freedom and sufficiency are speculatively locked in the space of imaginations contained in the space of the potentials with a conflicting system of freedoms and preferences that are very difficult to organize within the space of individual-collective dualities intra-generationally and intergenerationally toward the universal end dispositions. The inability of cognitive agents to abstract the information-knowledge conditions of the beginning of the universe obscures the destinations of its connectivity to the beginning and the abstraction of the optimal path on the universal journey, except under conditions of continual try-and-error dynamics and success-failure processes over the space of ignorance-knowledge dualities as generated by multiplicities of cognitive capacity limitations over the space of problem-solution dualities.

The need to understand the information-knowledge processes over the space of general-particular dualities in terms of meanings and productions arises from the general input-output character of information-knowledge processes in neuro-decision-choice actions over the space of theory-practice dualities, where practice without thought is blind and thought without practice is empty in organizations over the space of problem-solution dualities over the space life-death dualities. Alternatively stated, conception in imagination without actualization is not a reality and a reality without imagination is not actualizable over the epistemological space of which all the activities of neuro-decision-choice agents take place. In the development of the theories of decision, choice and action, the inputs as either information or knowledge are taken as given, where the task is on the algorithmic representations and solutions of goals and objectives as mutually constraining in some form of dualities within polarities. Types of the decision-choice theories are defined over the overlapping spaces of uncertainty-certainty dualities and exact-inexact dualities that basically represent the space of doubt-surety dualities over the space of penumbral regions connected to all areas of human actions [R6.3].

The algorithmic representations of neuro-decision-choice problem-solution dualities are mapped onto the space of computable-incomputable dualities to abstract solutions where some problem representations in mathematical space may belong to the space of incomputable structures due to cognitive capacity limitations of neuro-decision-choice agents preventing certainty as an absolute and projecting uncertainty in the space of knowing and information knowledge production. The important question is what the meanings and phenomenon of certainty and uncertainty are and how do they arise and enter the space of human actions which contains relevant sub-spaces? The meanings of the concepts and phenomena of certainty and uncertainty and corresponding spaces are assumed as obvious, known, linguistically primitive and understandable in all areas of cognitive actions and behavior in the classical paradigm of thought, where all actions and behaviors are neuro-decision-choice dependent. In this respect, the answer to the question of the meanings of the concepts of certainty and uncertainty, and the system of their implications in the space of neuro-decision-choice activities are important that must lead us to the develop and present a theory of certainty-uncertainty dualities from which theories of uncertainty and certainty may be derived and related to risks in cost-benefit space. In this respect, how are the concepts of data, evidence, belief related to the elements in the space of certainty-uncertainty dualities. The connectivity of data, evidence, and belief must be seen as sequential information concepts and phenomena, the understanding of which is found in the theory of certainty-uncertainty dualities that must be constructed.

2.2 THE THEORY OF CERTAINTY-UNCERTAINTY DUALITIES: SEMANTIC INTRODUCTION AND REFLECTIONS

We shall have a short semantic reflection on the conditions of the theory of certainty-uncertainty dualities. The theory of certainty-uncertainty dualities must help us to define uncertainty and its distinction from certainty, their meanings, and the role they play in the space of cognitive actions and human behaviors. It must also help us to understand the sources of uncertainty from the universal input-output process over the spaces of socio-natural production-consumption dualities and destruction-consumption dualities through the distinction of events in social space contained in epistemological space and natural space contained in ontological space. The input-output conditions of neuro-decision choice system are information and knowledge over the space of identification-transformation dualities contained in the epistemological space. There are two spaces of information and knowledge that must always be considered in understanding the spaces of problem-solution dualities, question-answer dualities, and success-failure dualities over the space of cause-effect dualities. They are the *ontological space* with the space of *ontic variety characteristic dispositions* and *epistemological space* with the *space of epistemic variety characteristic dispositions*. The ontological space is the *primary identity,* and the variety ontic characteristic dispositions are the perfect variety identifications of *what there was* and *what there is* which constitute the *space of variety actuals* and *what there would be* which constitutes the *space of variety potential*. The important defining element of the ontological space is the variety information-knowledge equality, where the variety characteristic dispositions are the same as the corresponding variety signal dispositions over the space of natural identification-transformation dualities. The conditions of variety information-knowledge equality and the equality of variety characteristic disposition with the corresponding variety signal disposition are such that there is always *certainty* of natural behaviors over the space of natural identification-transformation dualities.

The epistemological space is the *derived identity* by cognitive agents from the ontological space through the multiplicities of acquaintances with cognitive capacity limitations and where the variety epistemic dispositions are abstractions from the ontological signal dispositions with imperfect variety identifications of *what there was,* and *what there is* which constitute the space of *variety realities* as derived actuals for past-present telescopic epistemic structures to constitute the epistemic variety characteristic dispositions and *what there would be* which constitutes the space of *variety imaginations* for epistemic telescopic future as abstracted from the space of potential. The important defining element of the epistemological space is the *variety information-knowledge inequality*, where the abstracted variety characteristic dispositions from the ontological signal dispositions through the space of acquaintance constituting the space of epistemic dispositions may differ from the variety ontic characteristic dispositions over the space of cognitive identification-transformation dualities. In this respect, there are two spaces relevant to the understanding of the rise of certainty and uncertainty defined over the space of information-knowledge dualities. They are the *space of ontic characteristic dispositions* with information-knowledge equality and perfect variety identities and the *space of epistemic characteristic dispositions* with information-knowledge inequality and imperfect variety identities. The conditions of variety information-knowledge inequality and the inequality of variety characteristic dispositions with the corresponding imperfect variety dispositions are such that there is always an *epistemic distance* generated in *ontic-epistemic variety characteristic dispositions* with ontic perfection and epistemic imperfection that affect the input-output conditions of social behaviors over the space of social identification-transformation dualities. These differential conditions affect socio-natural input-output outcomes over the general spaces of input-output dualities and production-consumption dualities.

There are other important information-knowledge interpretive differences over the ontological space with input-output process without cost-benefit implications over the space of absoluteness with excluded middle and disunity and epistemological spaces with input-output process with cost-benefit implication over the space of relativity with relational continuum and unity. The knowing as production of intellectual investment flows and information-knowledge accumulation as the production of intellectual capital stocks are organized as neuro-decision-choice activities within the space of basic-superstructure dualities relative to the space of fundamental-applied dualities in support of evolving generational existence. The existence of the space of certainty-uncertainty dualities is derived from the existence of the space of ontic-epistemic dualities also as the space of perfect-imperfect dualities that generates the space of variety epistemic distances the elements of which are the results from the multiplicity of cognitive capacity limitations of neuro-decision-choice agents. In this respect, how are the concepts of data, evidence, belief, and knowledge related to the elements in the space of certainty-uncertainty dualities? In other words, what are the theories of belief, evidence, data and knowledge, doubt, surety, successes, and failure connected to each other and that of certainty-uncertainty dualities over cognitive action space? It must be clear that the concepts of imagination and the space of imaginations are not only derived from the space of certainty-uncertainty dualities they give rise to the concepts of possibility and probability which give rise to the concepts of success and failure and speculative philosophy over the space of cognitive actions as the vehicle that provide us with continual Sankofa-anoma connectivity as information-time continuum and unity. The organizational structure for understanding the path and epistemic continuity intra-generationally and inter-generationally of the development of the theory of certainty, uncertainty and the space of certainty-uncertainty dualities is presented in Figure 2.2.1. We shall turn our attention to organizational analytics of the basic and superstructure of institutions.

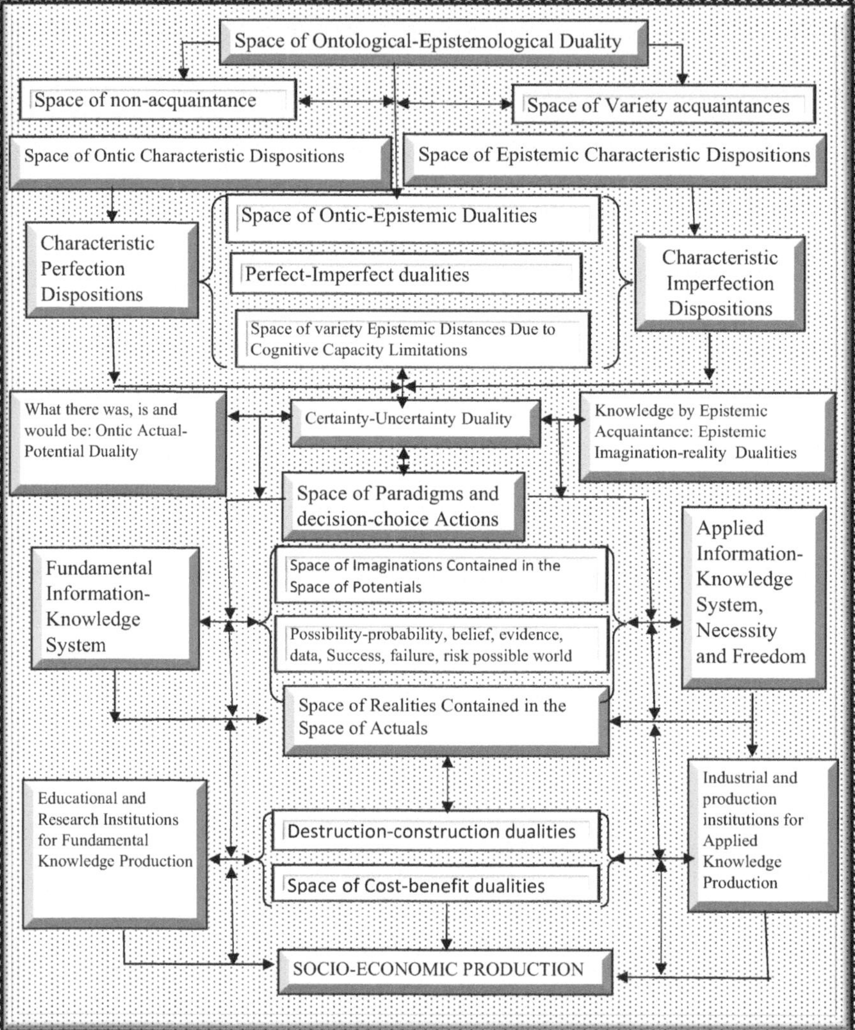

FIG 2.2.1 Organizational Analytics of a Paremiological Geometry of Theories of Certainty, Uncertainty and Certainty-Uncertainty Dualities and their Connectivity over the Space of Cognitive Actions, Information-Knowledge Production in the Space of Human Actions and Quality Control

2.3 THE SUPPERSTRUCTURE AND THE KNOWING PROCESS IN THE SPACE OF THE PROBLEM-SOLUTION DUALITIES AND POLARITIES

To understand the epistemic conditions of organizational basic-superstructure duality of institutional arrangements of neuro-decision-choice actions over the space of problem-solution dualities , it is critically important to understand the neuro-decision-choice actions for the production of the fundamental information-knowledge system and applied information-knowledge system relative to the space of individual-community dualities under cognitive capacity limitations, with a further recognition that the same neuro-decision-choice processes for the production of the fundamental and applied information-knowledge systems are responsible in the productions of variety wants and needs in terms of commodities and services within the space of individual-community dualities. The basic organization provides the framework for both the basic research and applied research to create the interrelated complex know-hows for dealing with *variety identification problem-solution dualities* and *variety transformation problem-solution dualities.*

The neuro-decision-choice actions over the space of identification problem-solution dualities are to reveal new varieties of *what there is* or previous varieties of *what there was* by providing variety information-knowledge dispositions which were previously unknown (this is the identification and explanation). The neuro-decision-choice actions over the space of variety transformation problem-solution dualities are to provide variety information-knowledge dispositions of *what would be,* new varieties through complex systems of neuro-decision-choice activities under individual and social cognitive capacity limitations, where incompleteness and vagueness, as imperfections generate the system of variety certainty-uncertainty dualities with relational continua and unity in knowing. These imperfections the spaced of human actions begin with knowledge by acquaintance with vagueness and volume incompleteness and are amplified by coding representations and the use of the paradigm of thought to create conditions of space of variety doubt-surety dualities. The space of doubt-surety duality leads to the space of decidability-undecidability dualities which requires the development of a criteria, such as general entropy for conditions of true-false acceptances with conditionality defined over the space of hope-despair dualities. The true-false conditionality of production and acceptance is defined in the space of possibility-probability dualities, in relationship to the social space of necessity-freedom dualities within the space of variety identification-transformation conditions involving either *predictability* or *prescriptivity* of variety behaviors.

2.4 THE BASIC ORGANIZATION AND INFORMATION-KNOWLEDGE SYSTEMS

The basic social organization with the primary category of institutions is such that the *economic structure* with its institutions provides the foundation for management and control of the existence of social varieties and their interactions with natural varieties. This is the economic sovereignty distribution of problem-solution dualities over the space of individual-community dualities. The political structure with its institutional configuration provides the foundation for the distribution of individual and collective neuro-decision-choice power for institutional creation, rulemaking, management, and development in the social set-up including individual and collective neuro-decision-choice actions. This is the political sovereignty distribution of problem-solution dualities over the space of individual-community dualities. The legal structure with its institutional foundation is used to endow the social system, with the distributive power of controllability of behavior and legitimacy of the exercise of decision-choice power as well as the creation and enforcement of behavioral rules and regulation of socio-material relations of inter-individual, individual-community, and intra-community structures. This is the legal sovereignty distribution of problem-solution dualities over the space of individual-community dualities. At this point, a distinction must be made between the concept of *decision-choice action* and the concept of *policy* in the social space and the space of knowing, given the institutions of decision-power distribution and conditions of rulemaking and a system of enforcement. The differences and similarities of decision-choice action and policy will be presented through definitions as linguistic varieties relative to social practices.

DEFINITION 2.4.1: Policy

In the space of organization of basic-superstructure dualities, with its set of institutions, a *policy* is an automatic control neuro-decision-choice action in a particular state and time directed to deal with a particular set of problem-solution dualities, irrespective of the variety information-knowledge conditions of that state and time, and irrespective of its dualistic-polar conditions of real costs and real benefits of variety transformations. It is a standardized decision-choice action over the space of problem-solution dualities relative to an occurrence of a defined problem-solution duality or a category of problem-solution dualities for a fixed period. It is said to be individual if the policy is defined over the space of individual problem-solution dualities. It is said to be social if the policy is defined over the space of social problem-solution dualities. The solution is fixes over different versions of a same problem under changing cost-benefit conditions.

DEFINITION 2.4.2: Decision-Choice Action

In the space of organization of basic-superstructure duality with its set of institutions, a *decision-choice action* is a cognitive activity over the space of problem-solution dualities which is always in relation to changing conditions of variety cost-benefit information-knowledge elements of the state and time. It is said to be individual if the action is over the space of individual problem-solution dualities and effect is individual specific. It is said to be social if the action is over the space of social problem-solution dualities where the outcomes have effects on social collectivity.

2.4.1: Comparative Analytics of Policy and Decision-Choice Action Under Information-Knowledge Process Over Problem-solution Dualities

The differences and similarities between the concepts of decision-choice action and policy must be clearly understood. The two are more often confused and interchangeably used over the space of variety problem-solution dualities. Decision-choice action, whether individual or social, is expressed over alternative varieties with differential distribution of cost-benefit dualistic-polar conditions over the space of variety negative-positive dualities. Social or individual policy is in relation to an automatic control decision-choice action over the space of social or individual varieties and fixed for the category of social or individual problem-solution dualities for a time irrespective of the dualistic-polar cost-benefit conditions. The individual policy is an automatic control decision-choice action over the space of private individual varieties and fixed for a category of the same individual problem-solution duality for a given period irrespective of the dualistic-polar conditions of varying cost-benefit structure of a variety or a set of varieties. All policies are neuro-decision-choice actions but not all neuro-decision-choice actions are policies. The policy space is always a sub-space of neuro-decision-choice space. Generally, social policies are created in the primary categories of institutional configurations and practiced in both primary and derived institutional configurations irrespective of the changing cost-benefit configuration. Policy is ideological based with accepted initial solution and benefit irrespective of the changing problem structure and corresponding costs. Decision-choice action which is not policy is not ideologically based and follow mini-max principle over the space of cost-benefit dualities. The effectiveness of scientific practices of the structure of scientific theories must always be devoid of ideological structures that contain problem-solution analysis in the organization of decision-choice actions. When the constraining conditions of policy are transferred to the spaces techniques and methods of knowing, information-knowledge and problem-solution dualities social progress also become constrained in the space of the ideology in the space of policies to victimize the scientific efficiency of organizational analytics.

Both the concepts and practices of decision-choice actions and policies, whether individual or social, find practices in the superstructure of the basic organization. The goal-objective space of the basic organization always exerts preponderating influence on the nature and structure of the goal-objective space of the organizational superstructure, its institutional arrangements, and behavioral conditions over the space of static-dynamic dualities which gives rise to outcomes in the space of problem-solution dualities to assert the directions of in the space of social and institutional progress-retrogress dualities. A policy is a decision-choice action which does not vary with the changing dualistic-polar cost-benefit conditions. Decision-choice activity is a decision-choice action which varies with the changing dualistic-polar cost-benefit conditions. A policy overrules decision-choice actions until it is dismantled. All policies are instituted to act as automatic controllers of decision-choice systems to the extent to which they are in effect, and hence affect some elements over the space of problem-solution duality, knowing activities, intellectual investment, the dynamics of the intellectual capital accumulation and social input-output structure. For example, social policies of environment, biosocial research, medical research and many such structures may restrict or enhance the space of knowing and information-knowledge accumulation over the space of problem-solution dualities.

Both policy and decision-choice action are the products of the social system and studied by social science irrespective of the areas of their applications under the ruling social philosophical consciencism. It is through the policies and decision-choice actions that social sciences, as a category of knowing, affect the natural sciences as a category of knowing as well as well as reveal the inseparable unity of socio-natural sciences in the space of variety knowing through social organizations with departmental and institutional analytics. The fundamental socio-natural information-knowledge system in support of the applied social information-knowledge system is a product of the social system and organized from the social science. Similarly, the development and the growth of natural intellectual capital is under the efficiency and development of intellectual capital of social sciences. The development and growth of intellectual capital of social sciences are facilitated by the gains of intellectual capital of natural sciences. All these involve neuro-decision-choice actions over the space of problem-solution dualities. In a nutshell, neuro-decision-choice actions and the space of problem-solution dualities constitute the principle of unity while categories of characteristic dispositions constitute the principle of diversity and institutional formations. The principle of diversity imposes the techniques and methods while paradigms of thought provide the thinking algorithms of reasoning over the problem-solution space.

The progress of information-knowledge system involving science and non-science is constrained from social rationality and the understanding of organizational rationality and its practices over the space of individual-community dualities as they relate to social policies, and constraints on the space of general decision-choice activities under a preference scheme for the distribution of sovereignty powers over the space of individual-community dualities. All constraints, rules and forms of social controls are imposed on all neuro-decision-choice actions and policies from the legal sector as established from the political structure. While the economic structure is under the control of the neuro-decision-choice activities from the political and the legal structure. The political and legal structures are at the mercy of efficiency of neuro-decision-choice actions over the economic structure, while the economic sovereignty distribution is at the mercy of political and legal structures. The important understanding in the organizational dynamics is that these constraints, rules, controls, and other such constraints on individual and collective neuro decision-choice actions are relational elements which will represent politico-legal constraints that define the socially acceptable domain of neuro-decision-choice practices over the space of problem-solution dualities, and hence intellectual investment and growth of intellectual capital associated with any social set up.

Social policies and neuro-decision-choice actions affect all members of social organization of knowing over the economic, political, and legal structures. The effects of individual policies and neuro-decision-choice actions are individual-specific except in the case of externalities that may be negative or positive. The effects of political and legal actions on the distribution of neuro-decision-choice sovereignty are such that individual body sovereignty as human right may be striped to create slavery, control women reproductive freedom and many individual freedoms to shape cost-benefit distributions over the spaces of input-output and production-consumption dualities. The political-legal constraints and the denial of human rights in relation to individual body sovereignty is not possible in the state of isolated individual existence. The political-legal constraints and the denial of human rights to individual body sovereignties become effective in societies under organization with different ethics and conceptions of human nature related to the space of input-output dualities relative to the space of cost-benefit dualities as organized over the space of categorial converter or the functor space where the political-legal processes may enhance or retard collective neuro-decision-choice over the space of problem-solution dualities and sovereignties over the space of individual-collective dualities.

In the process of knowing through the exercises of either individual or collective decisions over the space of problem-solution dualities, politico-legal constraints may restrict some selections of problem-solution dualities from the space of the problem-solution dualities to affect the path of information-knowledge disequilibrium dynamics. There are historical accounts of such politico-legal actions restricting the space of research, problem-solution dualities, reading and many such conditions on the principles of religious and ideological beliefs coded in laws and social policies. Some Social policies, however, may enhance and open the gates of efficiencies for the development of the fundamental information-knowledge system by promoting increasing neuro-decision-choice actions as well as expanding the toolbox of research, teaching, and learning over both the spaces of individual and social problem-solution dualities. A policy is a special type of decision-choice action as it has been defined; it is also decision-choice determined under cognitive capacity limitations, in such a way that a policy may come to constrain some effective neuro-decision-choice actions over the space of variety cost-benefit dualities. The evaluations of relative cost-benefit conditions are guided by the contents of the social philosophical consciencism that contains legal rules, proven and unproven knowledge, misinformation, disinformation, propaganda and other deceptive elements that affect preferences over the space of individual-collective dualities relative to the space of problem-solution dualities as organizes by the existing conditions of social philosophical consciencism defined over the space of cost-benefit dualities.

The study of neuro-decision-choice and policy efficiencies over the space of individual-collective dualities and their correct applications within the space of ignorance-knowledge dualities is always the subject matter of social science over the space of socio-natural problem-solution dualities where the problem-solution dualities are organized to achieve efficient outcomes. It is, here, that social science of the so-called inexact sciences come to influence the development of information-knowledge system of natural sciences which are claimed to be exact science, the results of which may shape the study of social sciences over the space of problem-solution dualities. Let us keep in mind always that the input-output relations of neuro-decision-choice actions and the space of problem-solution dualities under some preference scheme over the space of variety cost-benefit dualities to generate information-knowledge system as self-exciting and self-correction system are infinite disequilibrium processes over the epistemic field with relational and information fields in relational diversity, continuum, and unity. Often, people fail to understand the social-science foundations of natural sciences as well as underestimate the social-scientific contributions to the development and practice of natural sciences and the corresponding disequilibrium system of processes and technologies that is created. Similarly, it must be understood that scientific and technological information-knowledge stocks are inputs into decision-choice systems for the managements and control of social organizations. The information-knowledge stocks constitute a system of input-output elements in widening and deepening our socio-scientific knowing and the progress. The disequilibrium system of information-knowledge stocks must also be subjected to organizational analytics to abstract the most efficiency path in support of the system of social dynamics in creation and usage of processes and technologies. It is through this eye of seeing that the social understanding of engaging science, technology, engineering, and mathematics (STEM) find theoretical and application meaning and importance over the general social organizing activities of disequilibrium system of information-knowledge stock-flow dynamics over the space of finite-infinite dualities. The system of organizing activities and its efficiencies are also subject to science, technology, engineering, and mathematics in all forms of knowing. Science (S) is to discover the knowledge of the working mechanism of natural processes and technologies (T), engineering (E) are to activate the social applications of the discovered knowledge of nature and its organizing power, and mathematics (M) is to place the knowledge in the space of exactness-inexactness dualities relative to conditions of thought and action in the space of social variety transformations such that thoughts always have actions, and actions are always backed by thoughts over the space of theory-application dualities and polarities in cognitive existence. We must always keep in mind that a program of STEM is a system of cognitive processes to operate over the natural cost-benefit dualities and polarities to organize preparedness for natural disasters and take advantages of natural forces over the space of natural transformations through learning and knowing and their application over the production-consumption dualities.

2.4.2 Relationality, Organicity, Problem-Solution Dualities and Information-Knowledge Systems

As discussed, social-science-non-social science duality on one hand, and science-non-science duality on the other hand exist in dualistic-polar conditions to generate decision-choice input-output processes of diversity, relational continuum and unity of knowing, and unity of science or diversity, separation, and disunity of knowing and science. The relational dualistic-polar conditions, diversity and unity point to the analytical idea that one cannot talk of unity without acknowledging the existence of diversity and relational connectivity. Every unity is in relation to diversities and every diversity is in relation to unity under the principle of individual-community duality or inductive-deductive duality in either relational continua and different give-and-take individual sharing relations and connectivity or relational separation and no-give-and-take individual sharing relations and non-connectivity within the social organization, where there are multiplicities of individual-to-individual give-and-take-sharing relations and individual-to-social give-and-take-sharing relations or an absence of such over the spaces of acquaintance and problem-solution dualities for learning, teaching and research to produce the total information-knowledge system at every point of time in the inter-dependent dynamics of the basic-superstructure organizations and sub-organizations and corresponding dualities.

The essential point of the problem-solution dualistic-polar approach to the principle of diversity and the principle of unity is simply to make clear the foundations of knowing and that the whole enterprise on information-knowledge system must be such that learning, teaching and research are understood in terms of organization of neuro-decision-choice actions within the space of problem-solution dualities. The dualistic-polar relational continuum of individual and community is the *anoma-kokone-kone problem* of the Adinkra tradition of the African system of thought (conflict zones of individual-community relations), where from the individual we identify and understand the group and from the group we identify and understand the individual in their inter-supportive processes in the sense that the individual is simultaneously a cost and a benefit to the community, and the community is simultaneously a cost and a benefit to the individual as seen from the basic-superstructure organizational analytics in dualistic-polar configurations with relational continuum and unity within the spaces of general problem-solution dualities and ignorance-knowledge dualities [R27.9][R27.10].

The organizational analytics in dualistic-polar configurations with relational continuum and unity within the spaces of general problem-solution dualities and ignorance-knowledge dualities generate organizational limit superior and limit inferior in terms of efficiencies as limits of organization for information-knowledge production where such limits may be associated with cognitive capacity limitations and the contents of the social philosophical consciencism ruling at time point. By translating the conditions of limit superior and limit inferior of organization into the space of problem-solution dualities and into the development of information-knowledge system, one may understand the community successes by understanding the individual successes within the community and the structure of the collective relationality within the organization. Similarly, one may understand the individual successes by understanding the community organizing efforts and the nature of collective relationality within the organization.

The families of categories of individual-community dualities, problem-solution dualities, and the corresponding family of categories of institutional arrangements exist in interdependencies of relational continua and unity to facilitate the efficiencies of interdependent neuro-decision-choice actions for variety diversities, categorial diversities and unity in knowing and science under a real cost-benefit rationality and real cost-benefit distributional complexity that constitute the effects on the space of social outcomes. The needed efficiency of individual-to-individual sharing relations relative to the organization of social problem-solution dualities over the space of real cost-benefit dualities calls for an effective organization of human and non-human resources with continually institutional creations which must be consistent with the dynamics of the space of social problem-solution dualities as well as coordinating and harmonizing conflicts of individual neuro-decision-choice actions over the general space of collective-individual problem-solution dualities.

The superstructure through which individual and collective neuro-decision-choice activities take place to affect the basic-structure and the superstructure itself has two categories of *derived category* of institutional arrangements through which the individual and the collective neuro-decision-choice actions are undertaken. One *derived category* of institutional arrangements is the collection of derived institutions through which the individual and collective neuro-decision-choice actions are directed to the *space of fundamental problem-solution dualities and polarities* to develop the fundamental information-knowledge system as an input into various neuro-decision-choice activities over the space of applied problem-solution dualities. The other *derived category* of institutional arrangements is the collection of institutions through which the individual and collective neuro-decision-choice actions are directed to the *space of applied problem-solution dualities and polarities* to develop an applied information-knowledge system as an input into neuro-decision-choice activities over the space of individual and social production-consumption dualities.

The fundamental information-knowledge system and the applied information-knowledge system have dualistic-polar relation with relational continuum, mutually creating and in unity. Since institutions are the channels through which neuro-decision-choice actions are transmitted, they are comparable to the veins of the human body while the neuro-decision-choice actions act like blood for distributive implementations over the action space acting as input-output systems of production and maintenance of continual existence. The structural relationality of the input-output dynamics of organization and institutional arrangements are given in Figure 1.3.2.1 as paremiological geometry in Chapter 1. Since the paremiological geometry presents an intellectually proverbial system of information-knowledge development in relation to social organization, some cognitive dotting and t-crossing must be done to understand the nature of organization, institutional arrangements, and cognitive activities of humanity over the action space of dualistic-polar games under cognitive capacity limitations.

In all individual and social neuro-decision-choice activities over the space of production-consumption dualities, there are dualistic-polar relations between the basic and the superstructure of the social organization for the development, working mechanism and continual transformation of the social set-up regarding individual and social progress through neuro-decision-choice activities over the space of variety problem-solution dualities, where every progress is traced to outcomes of individual and collective neuro-decision-choice actions or a set of actions on setting variety solutions against a variety problems in an organizational environment that may be enhancing or non-enhancing to the process of goal-objective attainment under a preference scheme over the space of individual-collective dualities. There are also dualistic-polar relations and unity between the fundamental problem-solution dualities with fundamental information-knowledge system and the applied problem-solution dualities with applied information-knowledge dualities. Finally, there are dualistic-polar relations between the fundamental and applied information-knowledge systems with the spaces of variety production-consumption dualities, the variety supply-demand dualities, variety input-output dualities and variety cost-benefit dualities producing production, consumption, supply, demand, input and output chains in relational continua and unity of organizational analytics.

The problem-solution dualities, the information-knowledge systems and neuro-decision-choice systems are relationally interdependent in continua and unity to sustain the static and dynamic conditions of any social set-up, where there is multiplicity of conflicts within the space of individual-community dualities, where the conflicts are generated by differences in individual preferences and resolutions are generated by organizations over the space of cost-benefit dualities. The role of organization, in all respects, is to manage the conflict-resolution dualities in the information-knowledge system as seen from the space of individual-community dualities. Let us keep in mind that every variety is dualistic-polar in existence harboring conflicts between the duals and the poles for transformations from *what it is* to *what it would be* or from *what it is* to *what it is not*. In understanding the arguments of unity and disunity of knowing, science, and the progress of knowing and information-knowledge production. we must keep in mind the conditions of backward telescopic knowing and forward telescopic knowing relative to current conditions of knowing and link them to identification and transformation problem-solution dualities and the continually evolving space of problem-solution dualities, where problems are generated by solutions and solutions generate new problems in infinite dimensions that give meaning and essence to continuity of life and infinite information-knowledge disequilibrium dynamics.

Let us keep in mind that the fundamental and applied information-knowledge system constitutes a global input-output foundation of universal variety identifications and transformations through neuro-decision-choice actions over the conflict zones of the space of individual-community dualities, where every individual simultaneously constitutes a real cost as well as a real benefit to the society, and the society also constitutes a real cost as well as a real benefit to the individual in all aspects of human activities for information-knowledge system's development as an input-output system into activities over the space of problem-solution dualities, where the cognitive activities over the space of problem-solution dualities are nothing but social production with continual dynamic input-output structures. Here, the dynamics of information knowledge production is seen as part of the dynamics of social organization of institutions of production in relational continuum and unity, where techniques and methods vary over different categories of knowing as well as different categories of problem-solution dualities. The general understanding is that the space of knowing whether science or non-science is part of the space of problem-solution dualities and that the space of problem-solution dualities is part of the space of knowing, where every knowing creates conditions of ignorance, and every solution creates conditions of new problems. In other words, knowledge resides in ignorance and vice versa, and similarly, solution resides in a problem and vice versa over the space of static-dynamic dualities

The dynamics of the Information-knowledge system as a system of input-output processes provide explanations to decision-choice-action of classification over the space of outcomes of knowing. It also helps to explain substantial investment in espionage in the global system of fundamental and applied information-knowledge accumulation in the conflict zones of international competition in the space of military-economic dualities for global dominance and control. Let us keep in mind that neuro-decision-choice activities over the space of problem-solution dualities produce three organic outcomes of 1) fundamental information-knowledge system as an input into the applied neuro decision-choice module, 2) applied information-knowledge system as an input into transformation neuro-decision-choice modules 3) transformation-action modules on matter-energy varieties into articles of variety production-consumption dualities for needs and wants under a cognitive preference scheme in the global order of relational continua and unity to maintain the information-knowledge disequilibrium process, just like the natural food-chain is an organic input-output process that maintains matter-energy equilibrium. The understanding of the socio-natural dynamics is that there is the mater-energy equilibrium where matter-energy structure cannot be destroyed or created but can be transformed into varieties. There is the information-knowledge disequilibrium, where information cannot be destroyed but with continual variety transformations and accumulation expanding the space of ignorance-knowledge dualities.

2.5: THE SUPERSTRUCTURE ORGANIZATION AND THE INFORMATION-KNOWLEDGE SYSTEMS

We have reflected on the basic social organization, its institutions, knowing and information-knowledge production relative to the space of problem-solution dualities. Let us keep in mind that knowing, whether in a category of science, category of non-science, category of social science or category of non-social science is about organization of institutions through which neuro-decision-choice actions are carried on over the *space of identification problem-solution dualities* to uncover variety characteristic dispositions that reveal past, and present variety identities with variety future imaginations. It is also about neuro-decision-choice actions on the *space of transformation problem-solution dualities* to uncover processes that will bring about changes in variety characteristic dispositions that reveal variety present-future identity differences where the neuro-decision-choice actions are about the present variety conditions and reflections on past variety conditions. Here, every information-knowledge structure is about either variety identity, variety transformations or both and the corresponding socio-natural technological processes defined in terms of functor action or categorial-convertor action. The space of variety identification-transformation problem-solution dualities is infinite that endows the space of knowing with infinite possibilities from the space of imaginations with foundations generated from the conditions of the space of ontic-epistemic dualities containing epistemic distance from which the spaces of certainty-uncertainty dualities and success-failure dualities emerge from the conditions of the space of neuro-decision-choice actions within the superstructure organization. In this infinite space, the required neuro-decision-choice actions for organic knowing are far beyond the strength and cognitive capacity of an individual neuro-decision-choice effort, even if the lifetime of the individual existence is infinite. This constraint of every individual neuro-decision-choice effort over the space of problem-solution dualities establishes an *individual cognitive capacity limitation* in the space of knowing, where the constraints of cognitive capacity limitations may be relaxed through the institutions contained in the superstructure organization.

Similarly, the required neuro-decision-choice actions for organic knowing are far beyond the strength and collective cognitive capacity of the total social neuro-decision-choice of generational efforts over the generational lifetime even if the generational lifetime is extended to infinity. The conditions of individual, collective, intergenerational, and geographical knowing through the constrained neuro-decision-choice actions over the general space of problem-solution dualities and polarities establish *social cognitive capacity limitations* and *generational cognitive capacity limitations* in the space of knowing, productive capacity and amplified by geographical limitations of every society. These cognitive capacity limitations expand the space of problem-solution dualities and polarities requiring increasing superstructure organization and institutional arrangements and rearrangements for neuro-decision-choice actions in their reductions. The path to the reduction of these capacity limitations is through judicious utilizations of organizational wisdom and rationality that must restructure the general conditions of individual-community relationality, intergenerational relationality and international relationality in terms of preference distributions over the space of variety problem-solution dualities under the principles of *each-for-all-and-all-for-each,* in the structure of *ubuntu* and *dibi-mame-nibi*, where humanity and the principle of collectivity are the center of human neuro-decision-choice activities over the action space for shared telescopic future, but not under the principle of each-for-himself (*pesemakuminya*), where the *principle of selfishness* is the center of neuro-decision-choice activities in human existence, personality formations and variety transformations within the institutions contained in cost-benefit distribution in the superstructure organization with no vision of dualistic-polar give-and-sharing mode for common telescopic future.

When we speak of the intellectual heritage of humanity, we are pointing to the output of collective human efforts, the distribution of outcomes was not and is not constrained by ideology, conditions over the spaces of war-peace dualities, national wickedness against other nations, useless competition, imperial destructions and human-right abuses, conditions of profit-loss dualities, hierarchical structure of national prestige, imperial dominance, colonial and neocolonial wickedness, sanction-policy destructions and enslavement atrocities. When we speak of national intellectual capital stocks, we speak of collective and individual efforts, the accumulation of which is the national collective intellectual capital stocks as the foundation of control and management of the social set-up and its progress, where social management and control are neuro-decision-choice dependent with information-knowledge capital service inputs. When we speak of the relationship between thinking and reasoning over the space of the problem-solution duality, we speak of thinking as an organized algorithms and reasoning as organized processes of utilizing the thinking algorithms on a particular problem-solution duality such that the same thinking can or cannot be used to solve problems that may emerge from the same thinking. The problem-solution duality emerging from the replacement of the previous problem-solution duality may or may not be replaced using the same thinking. Anew thinking may be required for reasoning to relate the actual variety to the potential variety. The rise of the new thinking may be constrained by cognitive capacity limitations in the space of individual-collective dualities and organizations that house institutional pathways of cognitive actions in dualistic-polar structures of continuum and unity in the space of imagination-reality dualities.

The individual cognitive capacity limitations in all aspects of human endeavors may be reduced through inter-individual collaboration, teaching, learning with supervision, collective give-and-take sharing modes and collective actions under some form of organization with rules of cost-benefit distribution of outcomes from the neuro-decision-choice actions in the space of variety input-output dualities. The generational cognitive capacity limitation, giving the reduction of individual cognitive capacity limitations, may also be reduced through inter-generational transmission of generational information-knowledge systems with techniques and methods of information-knowledge development and storage. The degrees of reduction in individual cognitive capacity limitations depend on organizations within the social setup at each point of time. This is the inter-individual organization for an increased knowing as intellectual investment flows and increased knowledge as intellectual capital stocks. The reduction in the inter-generational cognitive capacity limitations depends on the organization of inter-generational transmissions of information-knowledge systems for foundations of inter-generational knowing as information-knowledge flows and intergenerational information-knowledge as stocks, connecting static and dynamic states of knowing. This is the inter-generational organization for foundations in the increased inter-generational knowing as intellectual investment flows and as foundations for an increased inter-generational information-knowledge as intellectual capital accumulation and stocks. Finally, there is the national cognitive capacity limitations that are created by geographically special limitations that can be reduced by international intellectual cooperation through a complex system of organizations of different institutional forms at the levels of research, teaching, learning with international neuro-decision-choice actions over the general problem-solution decision-choice space, where humanity is at the center of existence to reconcile conflicts in national preferences and domestic institutions to increase international efforts over the space of general problem-solution dualities, where all organizations and institutions must be created over the space of dualities and polarities.

2.5.1 The Superstructure Organization and the Individual and Social Problem-Solution Dualities and Polarities

At this point, it is useful to examine the role and importance of the superstructure of the social organization over the individual-community dualities in the development of the fundamental socio-natural information-knowledge system through the individual and the collective neuro-decision-choice actions in support of the applied information-knowledge system for variety creations and transformations to meet social needs and wants under the socio-psychological principle of non-satiation over the space of production-consumption dualities. The nominal and real concepts of individual-community dualities find expression in all aspects of social existence. The foundations of the conditions of the space of individual-community dualities are the individual labors with defined health and information-knowledge dispositions to establish the qualitative and quantitative dispositions respectively. An individual living in an isolation needs only intra-individual organization of self in all activities in the maintenance of existence, where cost-benefit relations in neuro-decision-choice actions are internal to the individual as expressed over the space of individual production-consumption dualities with multiplicities of limitations of space and time that tighten the constraints on individual progress in all spaces of behavior. The individual cognitive capacity limitations constrain the effectiveness of individual neuro-decision-choice actions over the space of individual-collective problem-solution dualities for generating extensive knowledge by acquaintance or experiential information as well as the derivation of knowledge by description from the experiential information. The conditions of intra-individual organization are limiting in possibilities that nature offers, close the boundaries of decision-choice actions to an individual-self at the level of socio-natural acquaintance as well as become insufficient in a non-isolated or group existence, where the space of variety possibilities is spun by conditions of the space of certainty-uncertainty dualities. The boundaries of individual cognitive capacity limitations of decision-choice activities over all problem-solution dualities in intra-individual organization become relaxed by individual cooperations as well as expand the space of individual cost-benefit dualities for neuro-decision-choice actions over the space of problem-solution dualities in relational continuum and unity.

The individual cooperation, under organization with rules and regulations, while relaxing some aspects of cognitive capacity limitations on intra-individual neuro-decision-choice actions over the space of general problem-solution dualities, creates dependency on give-and-take sharing process through inter-individual cooperation in neuro-decision-choice actions. The individual give-and-take processes create just and unjust distribution of various aspects human behavior over the space of cost-benefit dualities relative to the space of problem-solution dualities through the rise of possible conditions of free lunch, broadly defined. This just-unjust distribution over the space of cost-benefit dualities creates conflicts over the spaces of inter-individual and individual-community dualities within a social collectivity as well as creates intra-national and international conflicts. The rise of the space of just-unjust dualities within the space of individual-community dualities, where a nation may be considered as individual and the world may be considered as a global community, relative to the space of cost-benefit dualities, creates social and international and intra-national tensions, animosities and instabilities from the practice of each for himself or herself under the principle of non-satiation requiring an organization of collective decision-choice actions to find the degree of justice acceptable by the social collectivity with the understanding that such a degree of *distributive justice* may vary at each point of time and over different generations with a corresponding degree of distributive injustice with practices or non-practices of reconciliations over the space of justice-injustice dualities under some forms of organization, international cooperation and the practice of restorative justice through diplomacy and international give-and-take sharing modes.

The organization for coordinating the neuro-decision-choice activities of individuals over the inter-individual and individual-community dualities and with institutions through which individual and collective actions function is the superstructure organization of the neuro-decision-choice activities to promote *each-for-all-and-all-for-each* over the spaces of individual and social problem-solution dualities and polarities in relational continuum and unity to generate the fundamental and applied information-knowledge systems through inter-individual and individual-collective actions. The organizational objective is to pull together different individual cognitive skills as contained in human labor, rearrange, and weld them together into a collective whole for inter-communications to increase the epistemic efficiencies of the collective neuro-decision-choice actions over the space of the problem-solution dualities and polarities within the epistemic field with relational and information fields [R18.16], [R18.18]. For this organizational objective to be accomplished, it must be supported by an objective of distributive justice to create an acceptable degree of relational stability over the space of cost-benefit dualities. The effectiveness of this strategic approach is to reduce the space of individual cognitive capacity limitations and widen the effectiveness of the space of collective neuro-decision-choice actions over the space of the problem-solution dualities, leading to an increasing stock-flow processes of the information-knowledge disequilibrium system to provide the elements in the space of question-answer dualities. The degree of success of this approach depends on the degree to which the center holds the individual parts together so they may not fall apart just like the organization of living bodies with relational continuum and unity. In other words, the degrees of individual-collective success depend on the organizational effectiveness within the relational field as projected by the information field. The organizational effectiveness is found in the degrees of individual sovereignty participation in the space of individual-collective dualities relative to the collective sovereignty in the social set-up.

2.5.2 The Importance of Organization over the Space of Inter-individual and Individual-Community Dualities

The fundamental information-knowledge system, the applied information-knowledge system, the variety identification, and the variety transformations constitute the complete social production of results of private and public decision-choice activities over the problem-solution dualities for the understanding of social variety identities and transformations over the economic, political, and legal structures. The needs and justifications for inter-individual organizations or intra-generational organizations, inter-generational organizations, intra-national organizations, and international organizations find some epistemic supports from economic theory of specialization as applied to all aspects of individual and social preference ordering of production and consumption of socio-economic varieties. The justification is anchored in the space of cost-benefits dualities relative to neuro-decision-choice actions in the space of individual-collective dualities towards an individual or collective goal or set of goals as well as in the space of intergenerational dualities in preservation and continual growth of the heritage of universal stock-flow of information-knowledge system.

Generally, a need for an organization with rules and regulations arises when there are diversities projecting individualities with differential preferences, abilities, neuro-decision-choice limitations, *limitationality* and *limitativeness* in forms of give-and-take processes of individual and collective existence, and where relational unity of the diversities will raise the efficiency in all forms of operation, expand the collective space of real benefits and reduce individual real costs in the space of individual and collective cost-benefit dualities. The concepts of limitation, limitationality and limitativeness are seen in terms of information-knowledge structure over the points of neuro-decision-choice actions, given the information-knowledge processing capacity over the space of quantitative-qualitative dualities. The need for such organizations finds expressions over the epistemic field of categorial varieties with relational and informational fields. It also finds expression over human population with its own relational and information fields. The central role of organization is to pull together conditions of the relational field and organize them into relational continua and unity to improve individual and collective efficiencies over the space of problem-solution dualities through specialization and coordination over the space of individual-collective dualities. The neuro-decision-choice limitations, *limitationality* and *limitativeness* are simply the *cognitive capacity limitations* that place upper bounds in the space of individual and collective information collection, storage processing and interpretation of results.

An inter-individual organization produces collective actions that extends the domain and effectiveness of individual neuro-decision-choice actions over the space of personal and social variety problem-solution dualities. It is also the means by which individuals may learn, teach and contribute to collective neuro-decision-choice actions for grater successes in expanding the information-knowledge system and the collective can contribute to individual neuro-decision-choice actions over the spaces of individual and social problem-solution dualities with increasing degree of individual and social acceptance in the spaces of success-failure and true-false dualities in relation to input-output processes of the dynamics of information-knowledge systems over the space of certainty-uncertainty dualities. In this way, organization may be seen also as a strategy to reduce the riskiness in the space of certainty-uncertainty duality in knowing, information-knowledge production and their applications to variety transformations to fulfil social needs and wants by pulling labor resources together in a collaborative domain of give-and-sharing modes.

The space of the general problem-solution dualities is the space of question-answer dualities and by logical extension is the space of information-knowledge system which is also by logical extension is the space of production-consumption dualities. The space of production-consumption dualities is by logical extension, the space of input-output dualities which is also by nominal and real extensions the space of destruction-construction dualities, and by further nominal and real extension, is the space of cost-benefit dualities in all aspects of socio-natural variety processes and transformations which constitute socio-natural varieties and categorial varieties in the epistemic field with relational and information fields at the epistemological space. Thus, we have conceptual inter-relational spaces for decision-choice actions of cognitive agents acting individually and collectively in support and defense of general existence. The paremiologically relational transformations of the space organization are shown in Figure 2.5.2,1.

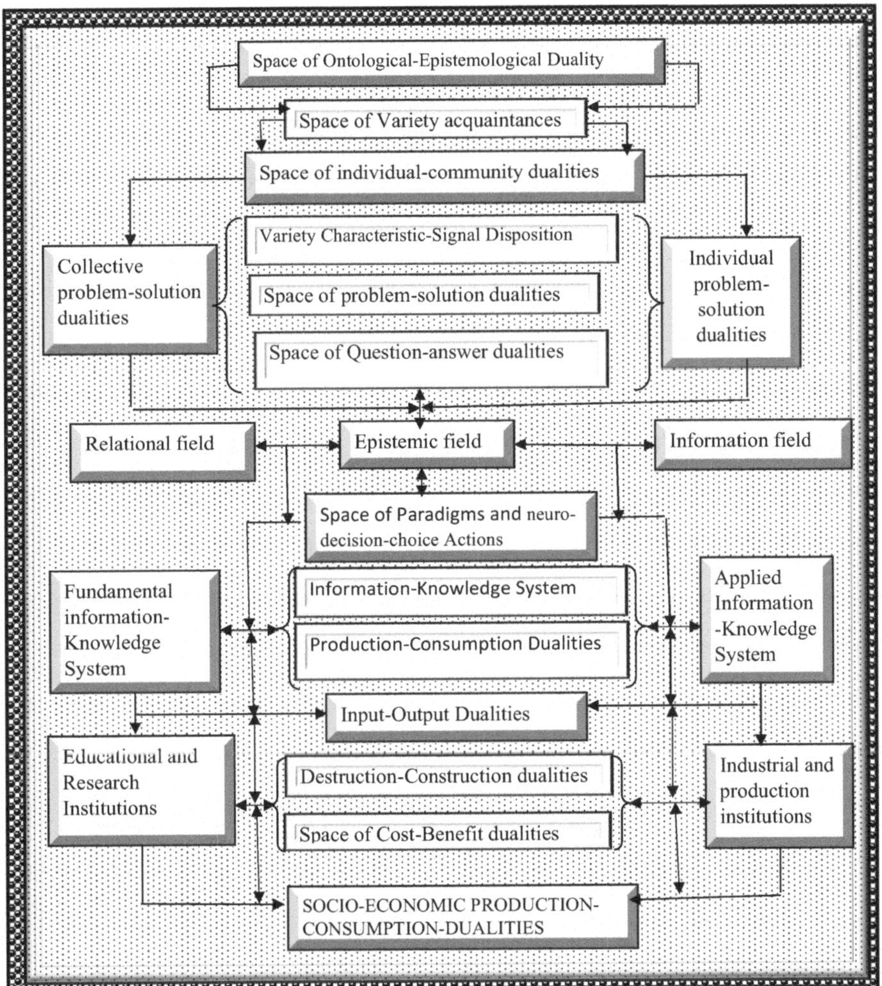

Figure 2.5.2.1 Organizational Paremiology of Unity of Knowing and Science for the Production of the General Information-Knowledge System over the Space of Production-Consumption Dualities as a Design of Intellectual Factories

In general, therefore, the activities of cognitive agents take different morphologies projecting different spaces for individual and collective decision-choice actions over the same epistemological space. These differences in structure and form of neuro-decision-choice actions are projected by differences of individual and collective preferences as induced by variety real cost-benefit dispositions. The relationships between varieties and preferences are very complex in that cognitive preferences will not exist without varieties which may exist without preferences. Preferences do not create the existence of varieties, they, however, facilitate production of varieties in the space of input-output dualities, while varieties tend to create preferences given cognitive neural networks. The distribution of the epistemic morphologies depends on the nature of the socio-natural environment as defined over the space of individual-community dualities and the corresponding needs and wants that are established by individual and collective preferences in competitive and cooperative struggles, where there are continual conflicts between individual and collectives in the space of individual-community dualities relative to the varieties in the space of cost-benefit dualities, where inputs are both costs and benefits and outputs are both benefits and costs.

The different morphologies of individual and collective actions may be classified as either primary category of morphologies or derived category of morphologies over the same epistemological space establishing parent-offspring successor process of existence, where every category serves as a primary to a reality or a possibly derived morphology. In this respect, the space of problem-solution dualities constitutes the primary conceptual space from which all other spaces are derivatives by nominal and real extensions over the space of ignorance-knowledge dualities, given the ontological-epistemological duality, variety actual-potential dualities, the space of acquaintance, the ontic-epistemic dualities, and imagination-reality dualities. A category is said to be a primary if its existence is necessary for another category to emerge, while a category is said to be a derived one if it emerges from a primary category and its characteristic disposition is necessary and sufficient to identify the primary category. The primary category is said to be a parent and the derived category is said to be an offspring in the space of input-output dualities and categorial dynamics. Organizational analytics must relate the technique of input-output process, the parent-offspring methods, primary-category dualistic-polar actions to understand knowing and science in the conceptual power of the space of unity-disunity dualities with internal diversities as essential structures of socio-natural environment in relation to decision-choice actions under conflicting preferences over the space of real cost-benefit dualities defined over the space of problem-solution duality relative to the space of question-answer dualities.

In all the spaces and processes, there are general purposes of organization and sub-organizations. The general purposes of organization and suborganizations are driven and maintained by conditions within the space of real cost-benefit dualities given the distribution of individual and collective preferences. The conditions within the space of cost benefit dualities take deferent forms at different levels of individual, collective, generational, national, and international decision-choice actions over the space of problem-solution dualities. Every neuro-decision-choice action over the space of problem-solution dualities is a variety transformation of relation or property or both as well as a flow of information-knowledge production to be added to the information-knowledge stock. Every natural action is also a flow of information and expansion of possible area of knowing through acquaintance. Every natural action over the space of destruction-construction dualities expands the areas of the paces of problem-solution dualities, knowing and information-knowledge capital accumulation, where organizational collective rationality provides the guidelines for uniting information-knowledge diversities into information-knowledge unity over the space of ontological-epistemological dualities.

At the level of acquaintance, the role of organization is to provide an optimal fuzzy decomposition of the space of the problem-solution dualities into socio-natural categories of problem-solution dualities as seen from the relational and information fields. The fuzzy decomposition assures a unity of the space of problem-solution dualities in a relational continuum for knowing and then provides dividedness by a criterion of relational similarities and differences as established by neuro-decision-choice actions into categories of common problem-solution dualities. At the level problem-solution dualities into knowing and the development of information-knowledge stocks, the purpose of organization is to take advantage of the idea of relational interdependencies of categorial give-and-take-sharing structures of internal-external dualities of the dynamics of the space of variety problem-solution dualities, different cognitive skills and epistemic abilities in specificities and generalities. Here, there is intra-categorial organization of internal varieties, and inter-categorial organization of categories and external varieties.

The natures of these categorial organizations are defined by the internal and external negative-positive dualistic-polar dispositions, while the internal dualistic-polar conditions reside in the space of internal real cost-benefit dualities to define the natures of individual varieties as seen by cognitive agents, and the external dualistic-polar conditions reside in the space of external real cost-benefit dualities to define variety differences and establish the variety comparisons and rankings based on external real cost-benefit rationality over neuro-decision-choice systems. Every organization is about institutional arrangements and rearrangements to reconcile conflicts of the duals in dualities and the poles in polarities under dualistic-polar conditions, where the conflict resolutions are desired neuro-decision-choice results within the space of goals, objectives, and visions. Let us keep in mind that the dualistic-polar space is composed of category of natural dualistic-polar structures which constitutes the primary category with primary principle of opposites and the category of social dualistic-polar structures which constitutes the derived category with derived principle of opposites, where the social processes mimic the natural preprocesses through the applications of the results of knowing and information-knowledge stocks over the space of fundamental-applied dualities as defined over the space of imagination-reality dualities contained in the space of actual-potential dualities in dualistic-polar relational continuum and unity. It must be understood that the category of *natural dualistic-polar* structures does not contain the cost-benefit dualities, which is not part of natural conditions of existence. The category of social dualistic-polar structures contains all the natural dualistic-polar structures in addition to cost-benefit dualities which are neuro-decision-choice evaluations of intra-variety usefulness and comparative analytics of inter-variety usefulness over the spacc of problem-solution dualities supported by a set of comparative social sub-dualities and sub-polarities which are not contained in the set of natural dualities and polarities. The conditions of cost-benefit dualities form an important distinction between the natural and social categories of dualistic-polar structures.

Particularly, within the categorial dividedness, an organization is always a means of abstracting from diversities of useful institutions into relational unity for expanding all spaces of cost-benefit dualities for the advancement of increasing efficiencies of neuro-decision-choice techniques and methods over the individual and the social collectivity under the principle of *each-for-all and all-for-each* toward a common element in the goal-objective space. At the space of individual-community dualities, organization expands the space of individual opportunities and strengths by providing a multi-rationality and multi-relationality in cooperation and increasing the individual and collective neuro-decision-choice powers over the space of problem-solution dualities, where the results from the space of problem-solution dualities increase the stock-flow disequilibrium conditions of variety information-knowledge systems. At the level of complexity, the organization involves decomposition in manageable entities in the space of problem-solution dualities. It also involves composition under similarity principles in organization and utilization of some epistemic toolboxes.

The conditions of social diversities and the conditions of social organization find expressions over the space of cost-benefit dualities and the true-false acceptance power under a system of individual and collective neuro-preference schemes. The conditions of natural diversities and natural organizations find expressions over the space of negative-positive dualities in the transformation power in functors and categorial convertor over natural schemes of conflicting dualistic-polar structures. The cost-benefit dualistic-polar conditions are applicable only in the space of conflicting preferences over the space of individual-collective dualities in social set-ups with organized rules and regulations for creating harmony in the social space of production-consumption dualities and input-output dualities. There are simultaneous existence of real costs and real benefits to every diversity in the social space, where every variety input disposition, variety output disposition, variety negative disposition, and variety positive disposition reside in the space of cost-benefit dualities, the distribution of which generate conflicts, animosities, hatred, war, violence and many other such varieties requiring organization of institutions over the space of justice-injustice dualities as may be revealed by the available information-knowledge conditions. These are the internal real cost-benefit dualistic-polar dispositions that provide epistemic justifications of real cost-benefit assessments of varieties in the neuro-decision-choice processes. There are also simultaneous real costs and real benefits to organization, decomposition, and unity. Organization, while extending opportunity set for individual neuro-decision-choice actions over the space of problem-solution dualities, may also constrain individual neuro-decision-choice creativities through rules and regulations such as legal codes for neuro-decision-choice behavior by restricting the socially allowable area of the space of problem-solution dualities within the space of individual-community dualities embedded in the space of imagination-reality dualities with relational continuum and social unity as organized from the space of actual-potential dualities containing the space of ontic-epistemic dualities .

The set of cost-benefit conditions in the organization of the space of individual-collective dualities is the *anoma-kokone-kone* problem, where the individuals are simultaneously real costs and real benefits to the community which is also real cost and real benefit to the individuals in the space of individual-community dualities of the social organism, where the individuals operate with constrained sovereignty imposed by the politico-legal restrictions. Such organizational restrictions are not applicable to neuro-decision-choice actions of individuals in isolation operating with unconstrained neuro-decision-choice sovereignty even though the individual operates with substantial cognitive capacity limitations and restricted area over the space of acquaintance but with unlimited possibility space from the infinite potential space without legal codes of artificial constraints imposed by collectivity and organizationally relational rationality. It is here that we may speak of socially unconstrained individual neuro-decision-choice rationality in isolation and socially constrained individual neuro-decision-choice rationality in a social set-up The *anoma-kokone-kone* problem arises in all social set-ups and is connected to the space of necessity-freedom dualities over the space neuro-decision-choice actions as established over the space of problem-solution dualities, where the space of necessity-freedom dualities is connected to the spaces of possibility-impossibility dualities and probability-improbability dualities. Individuals in isolations are deprived of real benefits of community organization but are also deprived from dealing with the real costs of community organization, where both necessity and freedom find meanings over the organizational space of real cost-benefit dualities contained in the space of possibility-probability dualities which resides in the space of variety imaginations contained in the space of potentials.

2.5.3 The Spaces of Individual-Community Dualities, Conflict-Resolution Dualities, Tension-Change Dualities, Cost- Benefit Dualities and Organizations

It has been pointed out that all conceivable conceptual spaces of knowing are transformational derivatives from the space of problem-solution dualities as the primary category of space of analytical and epistemic standing of the fundamental and applied information-knowledge systems, given the space ontological-epistemological duality with relational continuum and socio-natural unity. Each space is an organized fuzzy decomposition into categories of dividedness and diversities and then reorganized into relational continuum and unity to create an epistemic field with relational and information fields, where each category is a set of problem-solution dualities with common characteristic dispositions. The intra-categorial problem-solution dualities are internally organized for neuro-decision-choice actions on variety identifications and transformations. Each category is an area of knowledge search, research, teaching and learning as well as has internal institutional arrangements that allow neuro decision-choice actions of cognitive agents to be coordinated with specific category of problem-solution dualities for continual developments of information-knowledge systems at disequilibrium states in infinite process. It is also analytically useful to note that the space of the problem-solution dualities, just like the space of information-knowledge systems, is always on disequilibrium dynamics in support of the stock-flow disequilibrium dynamics of the information-knowledge systems. The only thing taken to be in equilibrium state is the universal matter-energy disposition from which varieties are continually under parent-offspring disequilibrium dynamics in the sense that the mater-energy disposition is active under dualistic-polar-game conditions for either variety maintenance or change relative to relations, properties, or both.

Similarly, the inter-categorial problem-solution dualities are externally organized for neuro-decision-choice actions on the understanding of categorial interconnectedness, relational continuum, destruction-construction process and global unity in relation to different variety identifications and transformations to establish the general global unity of knowing and information-knowledge systems, where the neuro-decision-choice actions are based on variety inter-categorially external real cost-benefit dualistic-polar conditions. The general objective of organization of the superstructure, therefore, is to integrate and manage inter-categories in a relational unity of inter-categorial institutions relative to the results and effects of areas of research, teaching and learning as well as coordinating the inter-categorial internal institutional arrangements in a manner that will allow neuro-decision-choice actions of cognitive agents to be distributed and coordinated across categories of problem-solution dualities for the continual development of information-knowledge systems at disequilibrium states for intergenerational transmissions of information-knowledge systems. Let us also keep in mind that social change is about reconciling conflicts in variety preferences in real cost-benefit conditions over the space of individual, community, national, generational, and international neuro-decision-choice actions.

The nature and structure of the epistemic field with relational and information fields lead to the establishment of categories of department of research, teaching and learning for a continual epistemic probe to develop the fundamental information-knowledge system for each category and the unity of all categories of the fundamental information-knowledge systems for the possible space of applications to either create new social varieties or innovate existing social varieties or to correct errors of neuro-decision-choice actions over the space of problem-solution dualities. At the levels of research, teaching and learning, these categories are called disciplines and departments, and housed under institutions of schools, faculties, colleges, and other names at higher level of education to produce specialized skills with the view of exercising effective neuro-decision-choice actions over the space of the fundamental problem-solution dualities. The categories of applied information-knowledge system are embedded into the epistemic field such that the fundamental information-knowledge system acquires a position of primary state of knowing while the applied information-knowledge system acquires a position of derived category of knowing in give-and-take sharing processes over the space of variety applied-fundamental dualities.

The epistemic field with relational and information fields is thus composed of sub-fields and is socially organized to contain the fundamental epistemic field as the primary category of knowing for generating the fundamental information-knowledge system, and the applied epistemic field with relational and information fields as an organically derived category of knowing for generating the applied information-knowledge system as an neuro-decision-choice input for social variety transformation in the space of goods and services, while the results of applied information-knowledge system become inputs for refining the claims of the fundamental information-knowledge system in the space of acquaintance-description dualities. The results of all neuro-decision-choice actions over the space of problem-solution dualities are continually incorporated into the social philosophical consciencism that becomes the framework of general social space of all neuro-decision-choice actions in the space of individual-collective dualities. It is useful to note that the social philosophical consciencism contains intergenerational methods of intergenerational knowing and information-knowledge stocks, where such stocks include misinformation disinformation, proven knowledge, unproven knowledge, deceptions, propaganda and many such elements that affect perceptions over the space of cost-benefit dualities and conceptions of their distributions over the space of individual-collective dualities.

The objective of organization, here, is to establish and manage the institutional conditions and control of activities in the space of fundamental-applied dualities for individual and collective neuro-decision-choice actions in continual variety transformations of old forms into new forms and new forms into old forms in the space of production-consumption dualities with continual variety destruction and construction. The socially organizational complexity in knowing is such that there is a system of organizations and organizational embedding corresponding to a system of neuro-decision-choice activities of cognitive agents over the space of problem-solution dualities as revealed from the space of ontological-epistemological duality relative to the spaces of preference ordering, goal-objective varieties, and the corresponding real cost-benefit dualities.

The organization of problem-solution dualities into diverse categories, institutions of knowing and information-knowledge production is also an organization of socio-natural varieties into categories of cost-benefit dualities over the space of decision-choice actions, where every problem-solution duality has a corresponding real cost-benefit duality which gives rise to problem-solution duality and the corresponding question-answer duality relative to a goal-objective element(s). The efficient organization of neuro-decision-choice actions over categories of problem-solution dualities and the corresponding categories of cost-benefit dualities demands an organization of societal population into categories of specialties with specialized skills to be coordinated with the categories of problem-solution dualities and the corresponding cost-benefit dualities. This is what is referred to as economics of specialization in space of socioeconomic production-consumption dualities over the space of imagination-reality dualities.

The existence of categories of problem-solution dualities with the corresponding categories of real cost-benefit dualities in the socioeconomic space over the space of identification-transformations dualities and the need for specialized skills demands a fuzzy partition of cognitive agents with different skills and the development of different specialized institutions of research, teaching and learning towards the development of the space of variety fundamental-applied dualities of education in relation to the space of general social production-consumption dualities in a continual disequilibrium process. The utility of the system of organizations and sub-organizations and corresponding institutions and departments finds expressions in the give-and-take sharing relations of categories in that, seemingly problem-solution dualities that are difficult in some categories may become simpler when the problem-solution dualities are seen in terms of inter-categorial transformations to facilitate the neuro-decision-choice processes of setting the variety solutions against the variety problems. In other words, a paradox in one category of knowing may be resolved in another category of knowing. An ill-posed problems in one category of knowing may seem like well-posed problems in another category. Similarly, a problem-solution duality that is difficult to resolve in one category may be resolved in inter-category of knowing. Thinking and reasoning in one category may help in other categories of problem-solution dualities in the input-output processes, where thinking is seen as a set of transformation technologies and reasoning is its set of processes of application in a meta-methodology. It must also be kept in mind that the permanency of our universe is maintained by the elements in the relational field defining a system of give-and-take-sharing conditions which are or may be known through the information field.

2.6 THE INSTITUTIONS OF RESEARCH AND THE EPISTEMIC FIELDS

The epistemic field with its relational and information fields as has been developed in [R18.16] is a family of collection of categories of problem-solution dualities, where neuro-decision-choice activities of cognitive agents are established to relate the conditions of categorial real cost-benefit dualities to conditions of variety identifications, transformation and knowing through neuro-decision-choice actions over the space of the problem-solution dualities mapped onto the space of cost-benefit dualities, where problems and solutions are defined by their relative cost-benefit dispositions in relation to varieties in the space of goals, objectives, and visions over the space of individual-collective dualities. The decision-choice activities are also partitioned into category of thinking and reasoning sub-systems for the development of an institutional framework for knowing, research, teaching and learning for information-knowledge development in terms of specializations and transmissions of general and specific skills over the space of problem-solution dualities. The institutions of research, teaching, and learning, given the available distributions of current and potential specializations, are organized into educational processes for further skill developments and transmissions of know hows of neuro decision choice actions to coordinate different skills of neuro-decision-choice actions within the categories of the problem-solution dualities.

In this respect, what are the goals and objectives of education in terms of research, teaching and learning in support of production and reproduction of real life of individuals and the social collectivity? What is being taught, what is being learnt and what is being researched on from the space of socio-natural varieties? What is the nature of curriculum and curriculum development that will be effective in these continual neuro-decision-choice actions over the categories of the epistemic field with its relational field, information field and the space of static-dynamic dualities?

Should education be viewed in terms of apprenticeship or should it be viewed as the development of thinking and reasoning over the space of problem-solution dualities where principle of doubt of existing information-knowledge items are encouraged. The answers to these questions relate to social goals and objectives of education, research, development of information-knowledge system and neuro-decision-choice activities as they relate to the meaning and essence of life and individual and collective vision of society. Since knowing traverse through a system of spaces of dualistic-polar structures in the spaces of production-consumption and destruction-construction dualities and polarities, we must relate all these to the general space of neuro-decision-choice actions over the space of problem-solution dualities under the creative processes of parent-offspring dynamics.

2.6.1 The Goals and Objectives of Education, Learning and Research in the Social Organization

The goals and objectives of education must be viewed in the space of individual-community duality, where the goals and objectives of the individuals and the society may defer, and the individual goals and objectives are shaped by individual needs through a system of positive and negative incentives as established within the social set-up and related to the space of cost-benefit dualities. In this framework, the individual educational objectives are the acquisition of skills driven by individual self-improvements and benefits relative to costs, where education for education's sake or social progress has no individual neuro-decision-choice relevance and utility. At the level of individual, education may be viewed in terms of an apprenticeship enterprise for skill acquisitions over the space of problem-solution dualities to improve production of varieties and expand the space of the entitlements from the collective production outcomes for individual consumption, where the costs of education are an individual responsibility and education may not be collectively organized. At the general level of society, the objective of education may be viewed in terms of officializing and institutionalizing the apprenticeship processes to transmit skills and information-knowledge conditions to aid individual and collective neuro-decision-choice actions over the space of variety problem-solution dualities over generations. The goal of general education is to teach established and proven skills and accepted information-knowledge items or system of items for individual and collective decision-choice actions over the space of problem-solution dualities in order to teach the methods and techniques to setting variety solutions against variety problems and to develop decision-choice skills to manage contingencies of problem-solution dualities as well as expand the space of collective vision and social progress intra-generationally and inter-generationally. The objective of learning in each category of knowing is to develop know-hows (information-knowledge conditions) of how variety problem-solution dualities are identified, previous solutions were set against the previous problems and how these solutions generated new problems and expanded the social information-knowledge stocks of the intellectual heritage of humanity, where the human intellectual heritage is continually changing information-capital stocks for future generations as decision-choice inputs.

The general goal of social education, therefore, is to create and organize institutions for the development of individual and collective skills for efficient decision-choice operations in terms of general controls and management over the dynamics of the space of variety problem-solution dualities, where the individual zones of cognitive capacity limitations are reduced by the organization of collective give-and-take sharing processes over the space of research, teaching and learning to improve the individual and collective cognitive variety problem-solution replacement skills through thinking and reasoning from the existing information-knowledge stocks and the development of individual and social cognitive capacity of knowing in anticipation of telescopic future. The objective of research within the education is to search and discover new variety identification and transformation problem-solution dualities, and how to set the new solutions against unsolved existing problems and the rise of new problems by understanding variety behaviors and conditions of change over the space of internal-external dualities. The goal of research is to discover new varieties with their characteristic dispositions which constitute the information-knowledge conditions for their identities and transformations, thus increasing the stock-flow dynamics of social information-knowledge accumulation in support of individual and collective existence through neuro-decision-choice actions that may take place through the organization of formal and informal institutional formations and set-ups. The time-point goals and objectives of education are enhanced by intergenerational transmission of information-knowledge stocks to widen the of future generational benefits.

Here, the objective of organization of formal institutions of education is to create elemental arrangements of facilities to bring administrators, learners, teachers, researchers and supporting staffs together in different intra-categorial and inter-categorial areas of knowing under relational continua and unity for continual creation and transmission of information-knowledge stock-flow in the space of investment-capital dualities. The goal is to create a space of institutional continuity of individual-community dualities, generational give-and-take sharing processes, and increasing social capacity for a system of continual development of neuro-decision-choice skills for the activities over the space of problem-solution dualities in the space of human existence. The complementing goal is the development of an institutional guidance of continual creation of input-output processes for sock-flow information-knowledge systems in continual disequilibrium processes over the epistemic field with relational and information fields as inputs into the neuro-decision-choice activities of cognitive agents over the space of ontological-epistemological dualities where the space of social production-consumption dualities find anchorage. It is within this analytical framework that the theory of knowledge, philosophy of education, theory of information, discussions of knowledge economy and many other related theories must find epistemic residence as part of economic theories of production-consumption and input-output dualities.

In the bare essentials, therefore, organizations are the means of achieving collective progresses through institutional creations, controls and managements of different elements in the space of individual-collective dualities under conflicting preferences by collectively setting the real social benefits against the real social costs in variety identifications and transformations, as well as assisting the individuals to efficiently set the individual real benefits against the individual real costs and managing the space of creative destruction in terms destruction to construction and construction to destruction from the space of general matter-energy varieties under dualistic-polar tensions and conflicts. The setting of the real social and individual benefits against the real social and individual costs is also the setting of social and individual solutions against social and individual problems as well as setting knowledge against ignorance with input-output interchanging conditions over the space of identification-transformation dualities. These phenomena are realized in all categorial diversities and unity of the space of problem-solution dualities as the unifying forces of intra-categorial and inter-categorial knowing in creating the disequilibrium dynamics of the stock-flow structure of the universal information-knowledge system.

In brief, the study of production-consumption dualities of information-knowledge system is also the study of general neuro-decision-choice system. It is this relational connectivity of neuro-decision-choice actions and outcomes from the information-knowledge systems that allows the claim that the decision-choice outcomes over the space of certainty-uncertainty duality can be used to study subjective probability measures through utilities, where all variety quantitative and qualitative dispositions are neuro-decision-choice determined under cognitive capacity limitations. Here, objective probability relates to data as information derives from repeated occurrences of incomplete characteristic disposition of a variety in the space of epistemic space contained in the space of ontic-epistemic dualities and hence conceptualized as probable, supported by epistemically empirical conditions, from the possibility space contained in the space of imaginations as derived from the space of potentials. Similarly, the subjective probability unlike the objective probability, relates to neuro-decision-choice abstraction from the conditions of vagueness and volume incompleteness of variety disposition and hence conceptualized as probable, supported by epistemically axiomatic conditions without repetition of observation in the acquaintance space and derived from the possibility space contained in the space of imaginations as derived from the space of potentials [R21.5] [R24.5][R24.22][R24.28]. It is the empirical rate of variety repetition condition over the space of acquaintance for the abstraction of epistemic distance capturing the space of certainty-uncertainty dualities that gives objective probability the conceptual framework of *relative frequency* where technically this repetitive variety cannot be extended to varieties in the space of non-repetitive conditions. It is the neuro-decision-choice conditions with no repetition relation between imagination space and potential space that give subjective probability the conceptual framework axiomatic.

2.6.2 Learning and Research with and without Organized Supervision over the Epistemic Fields

The progress of individual and social knowing and the building of information-knowledge system from neuro-decision-choice actions over the space of the problem-solution dualities are such that the success of education at all levels of knowing with learning and research with supervision, in intra-categories and inter-categories of knowing, must be such that the teacher (supervisor) must initially communicate to the learner the nature of the problem-solution duality and its relationship to the learning assignment and the goal which must be connected to the development of neuro-decision-choice actions in information-knowledge productions and problem-solution processes. Similarly, the development of a research university must be related to the construction of institutions that are related to categorial problem-solution dualities with flexibilities on interdisciplinary relations where the learners are oriented to interdisciplinary methods of research, teaching, language and learning in the space of categorial problem-solution dualities.

The dynamics of relational interdisciplinary approach are motivated by the conditions that within the principle of opposites with relational continuum and unity, a solution from the solution space to either an identification problem or a transformation problem will undoubtedly generate a new variety problem which will require a potential variety solution in terms of variety problem-solution duality. Let us keep in mind that under the principle of opposites, the problem (solution) space is composed of varieties, where the problem (solution) dispositions dominate the solution (problem) dispositions within the space of imagination-reality dualities. The system is constructed under the conditions of relativity and not absoluteness under the *asantrofi-anoma principle* of variety, where every variety is simultaneously composed of inseparable problem (real cost) and solution (real benefit) dispositions in a proportional relativity as viewed over the space of imagination-reality dualities.

A neuro-decision-choice action on a problem-solution duality is simply a transformation of one variety problem-solution duality, where the problem disposition dominates the solution disposition to another variety problem-solution duality, where the potential solution disposition dominates the potential problem disposition relative to an element in the goal-objective space contained in the space of imagination-reality dualities. The new problem-solution duality may be intra-categorial or inter-categorial in nature, requiring an interdisciplinary approach to resolve the new problem-solution conflict which may or may not have an immediate transformation decision-choice action to resolve the dualistic-polar conflict. Examples are found in the space of internalities and externalities of all neuro decision-choice actions that demand rules and regulations from politico-legal actions. The solution-to-problem-and-problem-to-solution dynamics, not only involve all socio-natural varieties as seen over the space of imagination-reality dualities but are easily manifested and observed in the space of socially unstable varieties with continually changing characteristic-signal dispositions, making the nature of social varieties more difficult to track their characteristic dispositions. Let us keep in mind that a gold coin today will be a gold coin tomorrow with the same characteristic disposition. We. however, cannot be sure that a good person today will be a good person tomorrow with the same characteristic disposition. Problems and solutions find definitions in the space of imagination-reality dualities containing the space of cost-benefit dualities and preference schemes expressed over the space of goals, objectives and visions seeking better cost-benefit positions. Problems and solutions do not find definitions and existence in the ontological space containing the space of actual-potential dualities except when it is connected to epistemological space containing the space of imagination-reality dualities, where the space of preferences and cost-benefit conditions are defined and expressed by neuro-decision-choice agents.

The natural variety identities are, however, slower to change in qualitative contents with stable variety characteristic dispositions in revelation at the space of natural varieties. The differential rates of qualitative transformations of social varieties and natural varieties affect the experiential information or the knowledge by acquaintance such that uncertainties in terms of vagueness and volume-limitations as seen from the space of ontic-epistemic dualities are more pronounced in social knowledge by acquaintance than natural knowledge by acquaintance. These differences are very important in understanding the differences in the needed methods and techniques for the knowing process over natural and social varieties. It is here that the studies of social sciences acquire conditions of greater inexactness, and the studies of natural sciences acquire conditions of greater exactness, and the differences between social and natural sciences find expressions over the relative stabilities in their dualistic-polar structures. The differences in the dualistic-polar structures of categories of knowing impose induced differences in the methods and techniques of neuro-decision-choice actions over the epistemic field with relational continuum and unity. It is the existence of these dualistic-polar differences as expressed over categories in the epistemic field with relational and information fields that expresses the distributions of degrees of efficiencies in the use of elements from family of abstract languages (FAL) over the categories such as natural sciences and social sciences. It is here that a lot of paradoxes emerges in the use of classical mathematics and logic in representations and computations of problem-solution dualities in the space of problem-solution dualities.

There are many cases in different areas of research, learning and teaching in social sciences where attempts are made to directly introduce the methods and techniques of natural sciences such as the use of classical mathematics without the critical understanding of the categorial differences over the space of variety quality-quantity dualities. Similarly, the differences between exact and inexact information structures in relation to natural and social phenomena leading to the use of the same methodology or the classical paradigm of thought are not understood. Additionally, the relationships between the classical paradigm of thought with the excluded middle and spaces of absoluteness and relativity in thinking and reasoning are not understood leading to the implicit assumption of exact information inputs in analytical works over non-physical sciences. In this respect, substantially analytical and intellectual compromises are undertaken in non-physical sciences to fit the subject of study into an abstract ignorance of methods and techniques with a set of assumptions that change the characteristic dispositions of the varieties under study, where the learner with supervision carries the same mode of thinking and reasoning in analytics without questioning their foundations and epistemic legitimacy. The inexact information of the category of social sciences presents some analytical difficulties and challenges in using the classical paradigm with its logic and mathematics, where subjectivity and judicious judgments are required in every stage of understanding without excluded middles, where the principles of dualistic-polar shades and fading processes are the foundations of analytical conclusions and acceptance of elements in the pace of truth-falsity dualities with relational continuum and unity without excluded middles.

In the case of learning and research without supervision, the individuals do not have a guidance from a teacher or supervisor but may have the benefit of collective inputs from his or her acquaintances with some aspects of accumulated global intellectual heritage of humanity or transmission of accumulated global information-knowledge system from the previous generation. Generally, the individual or the collective, under learning without a supervision, must learn by try-

and-error process through *learning-by-doing*. In fact, stages of some variety stock-flow of information-knowledge system of the global intellectual heritage in disequilibrium dynamics have been acquired by the methods and techniques of try-and-error processes through learning-by-doing over the spaces of destruction-construction and problem-solution dualities, where the outcome of each variety, at each stage, adds to the learning and the know-how process over the space of success-failure dualities. In this learning-by-doing process, one analytically operates in the space of intense success-failure dualities, where every failure disposition and every success disposition are flows of information-knowledge elements that add to the stock of individual and collective knowing and information-knowledge production to maintain the disequilibrium dynamics of information-knowledge stocks in all social spaces. The expansion of the space of the social information-knowledge systems is obtained by the combined methods and techniques of learning with and without supervision in the transmission of the known and discovery of the unknowns over generations and within generations and where the growth of generational knowledge depends on the efficiencies of methods and techniques of intergenerational information-knowledge transfers conditional on finality of individual and collective life [R.13.8][R13.9] [R13.10] [R13.11]. One may even project the idea that the whole processes of knowing, and information-knowledge production is a learning-by-doing enterprise over the epistemological space, where all goals, objective and visions are contained in the space of imaginations derived from the space of potentials. It is in this respect that theory of learning by doing acquires importance in the space of ignorance-knowledge dualities [R13.8] [R13.10] [R13.11].

2.6.3 The Epistemic Field and Special Institutions for Socio-Natural Information-Knowledge Dynamics

The structure of the epistemic field with relational and information fields provides a guidance in contemporary planned development of information-knowledge systems as a vehicle to effect socio-economic development by unifying the categories of socio-natural information-knowledge systems to create possibilities of neuro-decision-choice transformations over the space of problem-solution dualities moving from the space of variety imaginations to the space of social realities, where all thoughts without practices are social emptiness and all social practices without thoughts is blindness in telescopic social future. In its essentials, therefore, knowledge is produced to guide practice, while information-knowledge stocks are produced as intellectual stocks as inputs to produce thoughts. The planned information-knowledge development and its utilization to effect socio-economic transformation are information-knowledge-based and neuro-decision-choice approaches to the understanding of socio-economic transformations of the society composed of economic, political and economic structures, where the development of information-knowledge system is through information-knowledge and neuro-decision-choice processes from societies considering the deferential cultural dynamics as captured in the continual dynamics of social philosophical consciencism with intergenerational connectivity. From the structure of the epistemic field with relational and information fields, the information-knowledge development must be a system's approach, taking advantages of relational intellectual give-and-take-sharing process of categorial information-knowledge sub-systems and the general unity of knowing and science.

The processes of increasing knowing and the expansion of the space of capital accumulation and inter-generational transmission of stocks will require the development of research universities with inter-dependent curricula, the development of centers of research institutions integrated with universities, the establishment of centers of basic research and applied research, the development of innovation and applicational centers and the development of institutions of industrial utilization of new ideas as well as the back-lock of accumulated ideas, where individual will be equipped with multiple skills with due respect to each category of knowing as one lives in isolated individualism but facilitated by conditions of social collectivity in the space of neuro-decision-choice actions , where objectives of information-knowledge systems are directed to promote the preservation but not to destroy the general and common welfare of humanity by promoting peace over war, nonviolence over violence, love over hate and many of such dualistic-polar structures. The incentive structure to accomplish the development of information-knowledge system in support of common welfare of society and the maintenance and preservation of humanity must be abstracted from the space of individual-collective dualities relative to the space of cost-benefit dualities.

All the institutions of learning and research, teaching and think tanks are geared toward providing information-knowledge inputs into effective neuro-decision-choice actions over the space of problem-solution dualities embedded in the space of cost-benefit dualities which then provides the meaning and essence of individual and collective lives of the society. The social complexity of variety knowing over the space of variety identities and transformations is also the social complexity of neuro-decision-choice actions over the space of problem-solution dualities. The efficiency of the general knowing process requires organizationally fuzzy partitions of the space of the problem-solution dualities into categorial diversities and knowing to create categorial information-knowledge systems on specificities of problem-solution dualities, and reorganization of the fuzzy categorial dividedness into relational connectivity and unity of knowing to create the global information-knowledge system. The fuzzy partition provides sequential problem-solution dynamics that make it easier to organize neuro-decision-choice actions to set variety solutions against variety problems in the same categories, keep track of, manage the emerging problem-solution dualities and develop individual multiple skills in relation to the changing nature of social goal-objective elements, cost-benefit conditions and social philosophical consciencism.

It is here that the economic theory of specialization provides useful answers to categorial dividedness in the epistemic field. The power of organization allows either the integration of diverse categorial information-knowledge systems into a unified totality or the partitioning of the space of problem-solution dualities into new epistemic categories of decision-choice input conditions for critical examination of conditions of decidability, computability, explainability, prescriptability and predictability over the spaces of variety identification and transformation problem-solution dualities, where the conditions of spaces of decidability-undecidability, computability-uncomputability, explainability-unexplainability , predictability-unpredictability and prescriptability-unprescriptibility dualities and polarities may be determined in terms of the conditions of *identifiability* and *transformability* of varieties under destruction-replacement processes, epistemic transversality conditions and the principle of opposites.

2.7 GLOBAL COLLABORATIONS, COMPETITIONS AND ORGANIZATIONS IN KNOWING: A NOTE

This chapter may be ended with a small note on global conditions of collaborations, organizations, and competitions in knowing, information-knowledge development through neuro-decision-choice actions over the space of problem-solution dualities. Just as individuals in a nation have different experiential information so also each national collectivity has different national experiential information that goes to national knowing and development of information-knowledge system. Each generation has its own experiential information, corresponding knowledge and social philosophical consciencism that define its identity, personality, and behavior. Every national collective, generation and individual experiential information structures are under geographical, geomorphological, and historical limitations in addition to a sytem of cognitive capacity limitations that are enhanced by international political arrangements and restrictions. The space of individual national knowing, the development of individual national information-knowledge system and the efficiencies over the space of national problem-solution dualities may be expanded by international collaborations through the development of organization of international institutional arrangements to increase the stock-flow conditions of national information-knowledge system of each participating nation as well as speed up the global intellectual investment flow conditions and stock conditions of information-knowledge capital accumulation.

The international collaboration over the spaces of problem-solution dualities and ignorance-knowledge dualities is not different from international trade and variety give-and-take sharing relations over the space of international production-consumption dualities with organization of exchanges. In the framework of international collaborations, each national entity may benefit from the global organization of collaborative knowing under the principle of give-and-take-sharing processes over the space of just-unjust dualities. The intellectual collaborative process has added benefits in possible tension reduction, peace creation, war reduction, increase in creative competition in the space of dualistic-polar games and increased commitments to maintenance and progress of humanity with shared costs at the space of cost-benefit dualities, where the international collaborations increase the number of benefit varieties over the cost varieties. International collaborations do not mean lack of competitions over the space of problem-solution dualities, where national space of problem-solution dualities may be different and the organization of institutional arrangements over the space of success-failure dualities may enlighten the global system of the spaces of problem-solution dualities with comparative organizational analytics under mutual give-and-take learning culture. The intellectual competition will be exercised in the space of positive-sum games with win-win structures avoiding negative-sum and zero-sum games.

The paths of non-collaborations with envy competitiveness in the space of negative-sum and zero-sum games lead to distributions of international jealousies, animosities and outright dislikes, conflicts, wars, invasions, injustice, counter invasions, imperialism, colonialism, neocolonialism, settler colonialism, international oppression, racism, human atrocities, and lack of international stability, where substantial intellectual investment is devoted to production of knowledge for destruction to destruction but not in the creative destructions in progressive services of humanity. The path to the creative multiplicities of international connectivity may enhance the value and purpose of education at all fronts of knowing and human civilization, where every national social philosophical consciencism is imbued with the meaning and essence of humanity and environmental respect in the space of production-consumption dualities. This path cannot completely take the directions of market and profit-seeking systems of preferences as incentives in the neuro-decision-choice actions. In this framework of international cooperation, we will need a theory of intellectual global corporation and management of cost-benefit distribution just like the theory of international trade and payments. Discussions of the structures of international intellectual collaborations on the space of positive-sum games over the space of theory-practice dualities must seriously be undertaking for all who are interested in conditions of humanity and its progress and natural environment and its survivability in support of humanity where information-knowledge system is an input-output process in socio-natural management and controls with cognitive capacity limitations. International intellectual collaborations in knowledge production are to reduce the national cognitive limitations in knowing, the outcomes of which serve as inputs in terms human actions, just like international trade is production-consumption collaborations to reduce national resource limitations in supply-demand chains, input-output chains to generate the value chain in the space of domestic-international dualities relative to the space of production-consumption dualities over the space of visible-invisible dualities.

CHAPTER 3

THE THEORY OF EPISTEMIC ANOMALIES IN THE PRINCIPLES OF DIVERSITY AND UNITY OF KNOWING AND SCIENCE: Organizational Characteristic-Signal Analytics

ABSTRACT

Chapter 3 deals with the theory of epistemic anomalies in the organization of knowing and information-knowledge production, where such anomalies are introduced, defined, and explicated for understanding in theories and applications over all areas of knowing under characteristic-signal analytics within the epistemic field supported by the relational and information fields. These anomalies in knowing and information-knowledge developments are explained to arise within the space of imagination-reality dualities contained in the space of actual-potential dualities as we journey between the space of imaginations and the space of realities through thinking and reasoning connecting the knowledge by acquaintance to knowledge by description with the use of an organic paradigm of thought and micro-paradigms of thought. Some anomalies are paradigm-derived, some are observationally derived, some are intentionally derived, and some are unintentionally derived. The space of imagination-reality dualities, the space of imaginations and the space of realities are then discussed in relation to the conceptual links to phenomena of possibility and probability which give rise to the spaces of possible world, impossible world and possible-world-impossible world dualities and then linked to the anomalies. From the space of imaginations, we discuss the concepts, meanings and roles of creativity, curiosity, hope and courage in relation to the space of cost-benefit analysis within the space of identification-

K. K. Dompere, *Theories of Knowing and Knowledge Production*, Studies in Systems, Decision and Control 501, https://doi.org/10.1007/978-3-031-42863-0_3

transformation dualities. Using fuzzy paradigm of thought, conditions of optimality are derived for hope and despair under dualistic-polar structures, where in general the phenomena of thinking and reasoning are shown as actions to balance forces within opposites and polarities to abstract the best cost-benefit varieties in human endeavor relative to some elements in the space of goals, objectives, and visions. Directly and indirectly, the forces between opposites generate the space of problem-solution dualities under the principles of opposites in all social space. The chapter presents the needed foundations for further algebraic analytics to make explicit the rise and nature of various anomalies in relevant spaces within the epistemic field.

3.1 INTRODUCTION

In the previous chapters, discussions of organizational analytics were made. The concept of epistemic anomaly was introduced as part in the knowing as production of intellectual investment flows and information-knowledge accumulation as production of intellectual capital stocks as one travels on the path from variety ignorance to variety knowledge over the space of ignorance-knowledge dualities. The cost-benefit conditions of organization of dualistic-polar structures were analyzed and their distributive conditions over the space of individual-community dualities and domestic-international dualities in relation to spaces of justice-injustice, war-peace, imperialism-colonialism, and other such dualities were discussed. Many types of the conceptual anomaly were also introduced. They included misinformation, disinformation, propaganda, paradox, ill-posed problems, unsolved problem in addition to possible world, impossible world, possible-impossible-world dualities. These concepts and phenomena were related to neuro-decision-choice actions over the spaces of problem-solution dualities, question-answer dualities, certainty-uncertainty dualities, and cost-benefit dualities, where some of these anomalies may be either intentional, unintentional or both arising from cognitive capacity limitations. The system of epistemic anomalies is thus composed of intentional anomalies in relation to neuro-decision-choice intentionality, and unintentional anomalies in relation to neuro-decision-choice unintentionality in the enterprise of knowing and information-knowledge development relative to elements in the space of cost-benefit dualities as reflected by cognitive capacities.

The intentional anomalies are non-paradigmatic creations. They arise from deceptions in the game space of human interactions to include misinformation, disinformation, misrepresentation of facts, fake news, alternative facts and realities, propaganda and all kinds of distortions in the communication processes within the space of source-destination dualities through the space of sending-receiving dualities with the objective of influencing the paths of neuro-decision-choice actions of the receiving ends to obtain an advantage in the social relational games within the space of effort-reward dualities with conformal mapping into the space of cost-benefit dualities. The unintentional anomalies are paradigmatic creations to include paradoxes, unsolvable problems, ill-posed problems, problem misrepresentations, phantom problems in the processes of knowing and information-knowledge development to create intellectual capital accumulations, the services of which become inputs into the neuro-decision-choice space where such unintentional anomalies generate risks in space of applications. The unintended anomalies arise from either the structure of the mega-paradigm or from one of the meta-paradigms in the system of meta-paradigms.

All the meta-paradigms are thinking algorithms developed under the guidance of the rules of thought in the mega-paradigm. Both the mega-paradigm and the system of meta-paradigms are neuro-decision-choice related through thinking and may be used in reasoning by any neuro-decision-choice agent as instruments for actions over the space of problem-solution dualities. The laws of thought of the mega-paradigm guide the development of meta-algorithms while the meta-algorithms guide reasoning in the space of problem-solution dualities within the mega-paradigm of thought. The correctness of the development and use of meta-paradigms are verified by the laws of thought of the mega-paradigms, while the success-failure conditions of the meta-paradigms provide conditions for improvements and refinements of the mega-paradigm in the process of knowing. As presented, any meta-paradigm shifting is within the mega-paradigmatic structure and obeys its laws of thought without changing the laws thought of the mega-paradigm. Alternatively, any mega-paradigm shifting alters the laws of thought guiding the development of the corresponding set of meta-paradigms and the methods and techniques of reasoning in the space of problem-solution dualities and the logic of neuro-decision-choice actions within the mega-paradigm. In this respect, the discussions in the literature on paradigm shifting such as one of the Kuhn's paradigm and paradigm shifting in scientific discovery are nothing but meta-paradigms and meta-paradigm shifting within the classical paradigm of thought, with excluded middle and non-acceptance of contradiction in reasoning, which is here considered as discussions on the behavior of meta-paradigms and meta-algorithmic shifts in knowing and science within the classical mega-paradigm [R3.14], [R3.37], [R3.53], [R3.54], [R3.59], [R3.60], [R3.79], [R5.16].

The set of the fuzzy laws of thought, which gives us the fuzzy paradigm containing relational continuum, unity, subjectivity, and acceptance of contradictions, is a mega-paradigm and mega-paradigm shifting away from the classical mega-paradigm. It contains fuzzy meta-paradigms that include the classical mega-paradigm and its meta-paradigms. The fuzzy meta-paradigms give us rules for fuzzy reasoning within the fuzzy mega-paradigm with fuzzy conditionality to admit of contradictions as true values within the thought system of shifting thinking, reasoning, and conditions of neuro-decision-choice actions in relation to paradoxes that may arise in the classical mega-paradigm. All the mega and meta paradigms are about organization of neuro-decision-choice behaviors of cognitive agents in connecting the space of imagination to the space of realities under cognitive capacity limitations which generate qualitative and quantitative uncertainties in all human actions over the space of problem-solution dualities relative to the conditions from the space of ontic-epistemic dualities. The meta-paradigm shifting within a mega-paradigm is an epistemic search process through the space of ignorance-knowledge dualities, problem-solution dualities, and question-answer dualities to find an optimal meta-paradigm for various categories of problem-solution dualities within the epistemic field.

The problem-solution processes are parent-offspring processes, where solutions to previous problems give rise to new problems within categories and between categories. The parent-offspring (predecessor-successor) processes are such that the categorial meta-paradigm may or may not be transferable within categories and similarly, the categorial meta-paradigms may or may not be transferable between categories triggering a search and shifting of meta-paradigm either within categories or between categories under the same mega-paradigm of thought. The shifting of meta-paradigms is also the shifting of meta-algorithms in relation to categorial parent-offspring dynamics over the space of problem-solution dualities within the same mega-paradigm of thought. The mega-paradigms and meta-paradigms are to assist individual and collective cognitive journeys within the space of ignorance-knowledge dualities contained in the space of imagination-reality dualities, where the space of imaginations houses the space of possibility-impossibility dualities, and the space of possibility houses the spaces of probability-improbability dualities and necessity-freedom dualities and anticipation-expectation dualities by organization.

We shall now turn our attention to possibility space and variety designs over the space of real-nominal dualities within the epistemic field.

3.2 ORGANIZATIONAL ANALYTICS ON IMAGINATION, REALITY, POSSIBILITY, POSSIBLE WORLD, CREATIVITY AND ANOMALIES IN KNOWING AND KNOWLEDGE-PRODUCTION SYSTEM.

In order to offer some analytical clarities on the definitions, explications and explanations of the concepts of possible and impossible worlds that have been offered in this monograph, supporting ones in [R8.16a][R8.16b] and the related concepts that may follow, let us examine the foundational concepts of *reality* and *imagination*, *imagination-reality duality* with *anticipation* and *expectation*, *anticipation-expectation duality*, on one hand, *necessity, freedom* and *necessity-freedom duality* on another hand and *risk, reward* and *risk-reward duality* on the other hand relative to the concepts of possibility, possibility space, the spaces of possible-world-impossible-world dualities, problem-solution dualities, possibility-probability dualities and other dualistic-polar structures in terms of knowing and information-knowledge conditions over the spaces of imagination-reality dualities contained in the spaces of actual-potential dualities, identification-transformation dualities, and decision-choice activities over the action space.

All the definitions must be seen in term of their relations to knowing and characteristic dispositions to establish variety identities, similarities and differences which are essential in the development of all languages and communications over the space of source-destination dualities as well as identifications of real varieties. It is here that the system of definitions is divided into real and nominal definitions, where the role of definitions is to establish variety identity, distinction, and commonness for analysis and communications. Every definition is established in the space of characteristics as variety information. All knowing processes and information-knowledge developments are connected to direct and indirect organizations of cognitive actions over the space of individual-collective dualities. In the monograph [R8.16b] we defined the concepts of possibility, impossibility, and the corresponding spaces as well as the space of possibility-impossibility dualities. Additionally, we defined the concepts of probability, improbability, and the corresponding spaces in addition to the space of probability-improbability dualities with relational similarities and differences among them. The possibility space is equipped with functor space or space of categorial convertors that give rise to necessity, initial conditions of change and anticipation. The probability space is equipped with functor space or space of categorial convertors that give rise to freedom, expectations, and sufficiency of variety transformations.

Let us keep in mind that all elements and varieties that may be associated with knowing processes, information-knowledge productions, anomalies, decision-choice mistakes and corrections, and other such structures that take place in cognition and the space of production-consumption dualities with continual interactions that generate epistemic cost-benefit outcomes and propel all dimensions of social system from the space of relativity to the space of absoluteness take place over the dualistic-polar conflicts within the space of actual-potential dualities through the proper sub-space within the space of imagination-reality dualities. In the theory of knowing as intellectual investment flows and the theory of information-knowledge accumulations as intellectual capital stocks, there are many spaces and subspaces in conic structures with relational continua and unity within a system of organizational analytics. These spaces and subspaces must be clearly defined and explicated for distinction and commonness and yet meet the specific and global dualistic-polar conditions under the organicity of the principle of opposites in relational continuum and unity. We shall begin the definitions and explications with the concepts of imagination, reality and the space of imagination-reality dualities and the corresponding spaces in verbal and categorial forms with the understanding that the whole of neuro-decision-choice actions are established over the space of imagination-reality dualities, where variety imaginations are transformed into variety realities and variety realities are transformed into variety potentials through production-consumption processes over the spaces of input-output and individual-community dualities.

DEFINITION 3.2.1A: Imagination and the Space of Imaginations (Verbal)

An *imagination* is a variety characteristic disposition formed in terms of an abstract thinking power to construct cognitive images of varieties from the space of actual-potential dualities that exist in the cognitive space of the imagining agent with *anticipation* and *necessity* (possibility) and unknown to exist in the space of actuals in terms of perceived reality and not encountered in the space of acquaintance, and where the imagined varieties reside only in the mind of the imagining agents while the variety signal dispositions may be shared over the space of source-destination dualities as having the *expectation and freedom* (probability) of being brought to the space of current future realities which is a sub-space of the space of actuals within the space of actual-potential dualities. The collection of all variety imaginations constitutes the *space of imaginations* which is contained in the potential space and houses the spaces of possibility-impossibility, probability-improbability, probable-world-improbable-world, and possible-world-impossible-world dualities from which the spaces of possible world, impossible world, probable world, and improbable world may be abstracted with dualistic-polar restrictions and fuzzy conditionality.

DEFINITION 3.2.1B: Imagination and the Space of Imaginations (Categorial Characteristic)

Let the space of imagination be $(\mathfrak{G})$ with a generic element $(\mathfrak{g} \in \mathfrak{G})$, characteristic disposition $(\mathbb{X}_{\mathfrak{G}})$ and a corresponding variety $(\nu_{\mathfrak{g}} \in \mathbb{V}_{\mathfrak{G}} \subset \mathbb{V})$ and the space of realities $(\mathfrak{H})$ with a generic element $(\mathfrak{h} \in \mathfrak{H})$, characteristic disposition $(\mathbb{X}_{\mathfrak{H}})$ and a corresponding variety $(\nu_{\mathfrak{h}} \in \mathbb{V}_{\mathfrak{H}} \subset \mathbb{V})$. The variety $(\nu_{\mathfrak{g}} \in \mathbb{V})$ is said to be variety imagination if $(\nu_{\mathfrak{g}} \in \mathbb{V}_{\mathfrak{G}} \subset \mathbb{V} \ni \mathfrak{g} \in \mathfrak{G} \subset \mathfrak{U} \subset (\mathfrak{U} \cup \mathfrak{A}) = (\mathfrak{U} \otimes \mathfrak{A}))$ with corresponding characteristic disposition structure of the form $\left(z_{\nu_{\mathfrak{g}}} \in \mathbb{Z}_{\mathbb{V}_{\mathfrak{G}}} \subset \mathbb{Z}_{\mathbb{V}} \ni \mathbb{Z}_{\mathfrak{g}} \in \mathbb{Z}_{\mathfrak{G}} \subset \mathbb{Z}_{\mathsf{U}} \subset \mathbb{Z}_{(\mathfrak{U} \cup \mathfrak{A})} = \mathbb{Z}_{(\mathfrak{U} \otimes \mathfrak{A})}\right)$. The varieties in the spaces of realities and imaginations are epistemologically determined with neuro-decision-choice actions over the space of thinking-reasoning dualities while the varieties in the space of actuals $(\mathfrak{A})$ and the space of potentials $(\mathfrak{U})$ are ontologically determined by natural processes such that $(\mathfrak{H} \subset \mathfrak{A} \subset (\mathfrak{U} \cup \mathfrak{A}) = \Omega), \mathfrak{H} = (\mathfrak{H} \cap \mathfrak{A})$ and $(\mathfrak{G} \subset \mathfrak{U} \subset (\mathfrak{U} \cup \mathfrak{A}) = \Omega)$, and $\mathfrak{G} = (\mathfrak{G} \cap \mathfrak{U})$ where $(\omega \in \Omega)$ is the universal object space and $(\nu \in \mathbb{V})$ is the space of all varieties such that $(\mathbb{Q} = (\mathfrak{H} \otimes \mathfrak{G}) = (\mathfrak{H} \cup \mathfrak{G}))$ is the space of imagination-reality dualities with $(\mathfrak{H} \otimes \mathfrak{G})$ presenting the conditions of the principle of opposites, $(\mathfrak{H} \cup \mathfrak{G})$ representing relational continuum $(\mathfrak{H} \cap \mathfrak{G}) \neq \varnothing$ representing a relational continuum with give-and-sharing modes, where the characteristics of every variety reality have some characteristics of variety imagination with information condition of $\mathbb{Z}_{\mathbb{Q}} = (\mathbb{Z}_{\mathfrak{H}} \cup \mathbb{Z}_{\mathfrak{G}}) = (\mathbb{Z}_{\mathfrak{H}} \otimes \mathbb{Z}_{\mathfrak{G}})$ such that $(\mathbb{Z}_{\mathfrak{H}} \cap \mathbb{Z}_{\mathfrak{G}}) \neq \varnothing$ for the fuzzy conditionality under the principle of opposites with give-and-**take**_sharing modes and $\left(\mathbb{Z}_{\mathfrak{H}} \cap \mathbb{Z}_{\mathfrak{G}}\right) \neq \varnothing$ is the condition of relational unity.

DEFINITION 3.2.2A: Reality and the Space of Realities (verbal)

A reality is a variety, the characteristic disposition of which has been abstracted by cognition as variety epistemic characteristic disposition from the ontological signal dispositions at the space of acquaintance to identify its identity in difference and similarity which exist in the space of the actual-potential duality of the general ontological space with some social agreements regarding approvals for intellectual acceptance as information-knowledge of varieties in the space of actual-potential dualities containing the space of imagination-reality dualities, where every variety reality is a social construct from either of space of actuals by identification or space of imaginations by transformation or both through knowing processes. It is said to be variety identity if its disposition resides in the space of identification, while it is said to be variety transformation if its disposition resides in the space of transformation contained in the space of identification-transformation dualities, The collection of all the variety realities constitutes the space of realities. As an element in the space of the actuals, a variety reality is external information-knowledge supported and justified by cognitive agents from the space of acquaintance-description dualities that have successfully passed through the spaces of possibility-probability dualities, necessity-freedom dualities, doubt-surety dualities, certainty-uncertainty dualities contained in the space of imaginations. Similarly as an element in the space of the potential, a variety reality is internal information-knowledge supported and justified by cognitive agents also from the space of acquaintance-description dualities that have successfully passed through the spaces of possibility-probability dualities, necessity-freedom dualities, doubt-surety dualities, certainty-uncertainty dualities contained in the space of imaginations and other such dualistic-polar elements in the space of true-false dualities, where the real variety space is always in a disequilibrium process following the information-knowledge disequilibrium dynamics. Realities are information-knowledge justified in the space of actual-potential dualities, where the internal reality characteristic dispositions (duals) dominate the internal imagination characteristic dispositions (duals) in the space of imagination-reality dualities while the claim of reality always belongs to the spaces of true-false dualities and acceptance-rejection dualities under cognitive capacity limitations.

DEFINITION 3.2.2 B: Reality and the Space of Realities (Categorial Characteristics)

Let the *space of realities* be $(\mathfrak{H})$ with a generic element $(\mathfrak{h} \in \mathfrak{H})$, variety characteristic disposition $(\mathbb{X}_{\mathfrak{H}})$ and a corresponding variety $(v_{\mathfrak{h}} \in \mathbb{V}_{\mathfrak{H}} \subset \mathbb{V})$. The variety $(v_{\mathfrak{h}} \in \mathbb{V})$ is said to be variety reality if $(v_{\mathfrak{h}} \in \mathbb{V}_{\mathfrak{H}} \subset \mathbb{V} \ni \mathfrak{h} \in \mathfrak{H} \subset \mathfrak{A} \subset (\mathfrak{A} \cup \mathfrak{U}) = (\mathfrak{U} \cup \mathfrak{A}))$ such that $\mathbb{X}_{v_{\mathfrak{h}}} = \mathfrak{I}(\mathbb{S}_{v_{\mathfrak{h}}})$ is an abstraction from an ontological variety signal disposition, where $\mathfrak{I}(\bullet)$ is a *signal decipher function* or *decoding function* which can also serve as functor or categorial convertor. The varieties in the spaces of realities are epistemologically determined where the space of varieties is always a small sub-space of the space of actuals through knowing with epistemic conditionality while the varieties in the space of actuals $(\mathfrak{A})$ and the space of potentials $(\mathfrak{U})$ are ontologically determined in the sense that $(\mathfrak{H} \Subset \mathfrak{A} \subset (\mathfrak{U} \cup \mathfrak{A}) = \Omega)$, where $(\omega \in \Omega)$ is the universal object space and $(v \in \mathbb{V})$ is the universal space of varieties such that $(\mathbb{Q} = (\mathfrak{H} \otimes \mathfrak{G}) = (\mathfrak{H} \cup \mathfrak{G}) \subset (\mathfrak{U} \cup \mathfrak{A}) = \Omega)$ defining relational unity with implied conditions $(\mathbb{Q} = (\mathfrak{H} \otimes \mathfrak{G}) \subset (\mathfrak{U} \otimes \mathfrak{A}) = \Omega)$ is the space of imagination-reality dualities under the principle of opposites $\left((\mathfrak{H}\otimes\mathfrak{G})\subset(\mathfrak{U}\otimes\mathfrak{A})=\Omega\right)$ with $\left((\mathfrak{H}\cap\mathfrak{G})\neq\varnothing \text{ and } (\mathfrak{U}\cap\mathfrak{A})\neq\varnothing\right)$, where every variety reality has a supporting variety imagination and vice versa under dualistic-bipolar conditions over the space of static-dynamic dualities in disequilibrium dynamics and conditions of principle of opposites with give-and-take sharing modes, where the symbol $(\Subset)$ represents the concept of very mall sub-space .

DEFINITION 3.2.3A: The Imagination-reality Duality and the Space of Imagination-Reality Dualities (Verbal)

An *imagination-reality duality* is a dualistic-polar variety, the characteristic disposition of which has been abstracted by cognition as variety epistemic characteristic disposition from the ontological signal dispositions at the space of acquaintance to identify its identity in difference and similarity which exist in the space of the actual-potential dualities as contained in the space of the general ontological-epistemological duality with social agreements regarding approvals of the framework for intellectual activities for the understanding of knowing as intellectual investment flows in terms of information-knowledge varieties, where every variety imagination-reality duality is a social construction from the space of actual-potential dualities through knowing processes, where imagination characteristic disposition is also within the reality characteristic disposition as duals for mutual existence and destruction through give-and-take sharing modes. The collection of all such variety imagination-reality elements constitutes the *space of imagination-reality dualities.*

DEFINITION 3.2.3 B: Spaces of Imagination-Reality Dualities (Categorial Characteristic)

The space of imagination $(\mathfrak{G})$ and information support $\mathbb{Z}_{\mathfrak{G}} = (\mathbb{X}_{\mathfrak{G}} \otimes \mathbb{S}_{\mathfrak{G}})$ is a small collection of varieties, constrained by cognitive capacity limitations, from the infinite space of ontological potential $(\mathfrak{U})$ with generic element $(\mathfrak{g} \in \mathfrak{G} \Subset \mathfrak{U})$ as well as containing the space of possibility-impossibility dualities $(((\mathfrak{P} \cup \mathfrak{M}) \subset \mathfrak{G} \Subset \mathfrak{U}))$, $\mathfrak{P}$ is possibility space and , $\mathfrak{M}$ is impossibility space, where the set of the imagination characteristic dispositions $(\mathbb{X}_{\mathfrak{G}} \subset \mathbb{X}_{\mathfrak{U}} \subset \mathbb{X}_{(\mathfrak{U} \cup \mathfrak{A})})$ is contained in the space of potentials $(\mathfrak{U})$ as the imagination dual, the characteristic dispositions of which the complement of the set of reality characteristic dispositions $(\mathbb{X}_{\mathfrak{H}} \subset \mathbb{X}_{\mathfrak{A}} \subset \mathbb{X}_{(\mathfrak{U} \cup \mathfrak{A})})$ is contained in the space of the actuals $(\mathfrak{A})$ as the opposite dual, and where every space of probability-improbability dualities $(\mathfrak{B} \cup \mathfrak{H}) \subset (\mathfrak{P} \cup \mathfrak{M}) \subset \mathfrak{G}$ with $(\mathfrak{B} \cap \mathfrak{H}) \neq \varnothing$ resides in the space of variety imaginations as possibility-impossibility duality with dualistic-bipolar conditions in the space of imagination-reality dualities, where $\mathfrak{B}$ is the probability space and $\mathfrak{H}$ is the improbability space as they have been structured in the previous chapter. Similarly, every real variety in the space of actual-potential realities resides as an imagination-reality duality with dualistic-bipolar conditions in the space of imagination-reality dualities represented as $\mathbb{Q} = (\mathfrak{H} \otimes \mathfrak{G}) = (\mathfrak{H} \cup \mathfrak{G})$ with fuzzy conditionality as $(\mathfrak{H} \cap \mathfrak{G}) \neq \varnothing$ and information support of the form $\mathbb{Z}_{\mathbb{Q}} = \mathbb{Z}_{(\mathfrak{H} \otimes \mathfrak{G})} = \mathbb{X}_{(\mathfrak{H} \cup \mathfrak{G})} \otimes \mathbb{S}_{(\mathfrak{H} \cup \mathfrak{G})}$. An imagination (reality) variety is one with a dominating dual characteristic disposition over the dual characteristic disposition of reality (imagination) in duality as dualistic-bipolar conditions in a simultaneity of existence of opposites with continuum, and give-and-take sharing mode in unity within the same variety.

NOTE 3.2.1: Reality, Imagination, Imagination-reality Duality, and Corresponding Spaces

As an element in the space of the potential-actual dualities, a variety imagination-reality duality is information-knowledge supported, unsupported, justified and unjustified by cognitive agents from the space of acquaintance-description dualities that has successfully passed through the spaces of possibility-probability dualities, necessity-freedom dualities, doubt-surety dualities, certainty-uncertainty dualities, anticipation-expectation dualities and other such dualistic-polar elements in the space of true-false dualities, where the space of imagination-reality dualities is always in a disequilibrium process following the information-knowledge disequilibrium dynamics. The imagination-reality duality has the imagination characteristic decomposition component of the dualities as unjustified over the space of potentials while the reality characteristic disposition component is justified over the space of actuals, where there are internal dualistic-polar conflicts between reality characteristic dispositions (duals) and the internal imagination characteristic dispositions (opposite dual) within the space of imagination-reality dualities.

NOTE 3.2.2: Imagination, Reality, Imagination-Reality Duality, and the Corresponding Spaces: Extensions

The twin concepts of imagination and reality are extremely important in theory of knowing as an intellectual-investment process and information-knowledge production as an intellectual capital-accumulation process over the space of fundamental-applied information-knowledge dualities as well as neuro-decision-choice actions over the space of production-consumption dualities, where variety imaginations feed into the production of variety realities and variety realities nourish the production of variety imaginations in the space of imagination-reality dualities under the power of epistemic creativity within the space of thinking-reasoning dualities that forms a bridge between the space of imaginations and the space of realities, between the space of ignorance and the space of knowledge and between the space of problems and the space of solutions within the space of individual-collective dualities. All human actions are over the space of production-consumption dualities such that all human conceptions with visions, goals, objectives, designs, music, language development, propaganda misinformation, war, peace, disinformation and deceptions, are abstract images formed within the space of imaginations and transmitted onto the space of realities over the space of risk-benefit dualities under fuzzy conformal mappings from the space of problem-solution dualities, where creativity, curiosity, fear, hope, and other such cognitive characteristic dispositions find expressions and meaning over the epistemological space which constitutes the epistemological universe as the knowable universe contained in the ontological universe which is the totality of all existence.

The universe is conceptualized as the ontological universe and epistemological universe. The ontological universe is the space of totality of all existence. The epistemological universe is the space of knowability of the totality of the space of all that exist. The space of unknowability is contained in the space of ontological universe which is the union of the space of knowable universe and the space of unknowable universe. In the knowing process, the space of knowable and the space of unknowable are implicitly dualistic-polar structures that are revealed by try-and-error processes through the methods of learning-by-doing. The ontological universe is in epistemically in equilibrium state and not subject to principle of expansion in a state knowability in the sense that the space of unknowable universe may have internal expansion without the awareness of cognitive agents. The space of the epistemological universe is continually expanding as the epistemological information-knowledge system expands in the space of knowability relative to the ontological information system. The meaning of any claim of an expansion of the universe is simply confined within the epistemological space containing the space of knowability relative to the space of unknowable universe under the general principle of opposites. This general principle of opposites is claimed, here, to operate in all galaxies, all planets and planetary systems for the continual internal existence, dynamic stabilities, static equilibria and planetary ecological balances in the universal unity and continuum.

To examine the relational complexities among imagination, reality, creativity, problem-solution dualities, question-answer dualities, information-knowledge production and social change, let us recall that $\mathbb{Q} = (\mathfrak{H} \otimes \mathfrak{G}) = (\mathfrak{H} \cup \mathfrak{G})$ defines the space of imagination-reality dualities under dualistic-bipolar conditions with a generic element $(\varphi \in \mathbb{Q})$ that allows for the definitions of the spaces of imaginations and realities to follow dualistic-bipolar conditions $(\mathfrak{G} \otimes \mathfrak{H})$ with relational continuum $(\mathfrak{G} \cap \mathfrak{H}) \neq \varnothing$ and unity $(\mathfrak{G} \cup \mathfrak{H})$ with their corresponding characteristic dispositions that reside in give-and-take sharing modes. Let us also keep in mind that all human behavior is through neuro-decision-choice actions in relation to changes in variety relation and changes of variety properties over the space of identification-transformation dualities where the driving force of neuro-decision-choice activities find incentives and disincentives over the space of cost-benefit dualities to allow the space of imaginations to be connected to the space of realities contained in the spaces of actual-potential dualities and ontic-epistemic dualities, where the space of ontic-epistemic dualities provides the conditions for the rise of the space of uncertainties , imagination, possibilities and probabilities. In this framework, possibilistic uncertainties are related to cognitive limitations of vagueness given rise to subjective probability dealt within neuro-decision-choice actions, while probabilistic uncertainties are related to cognitive limitations of volume incompleteness giving rise to objective probability dealt with relative frequency of classical tradition. The ontic variety characteristic disposition is always a closed set with maximal elements. The epistemic variety characteristic disposition is an open set with less than maximal elements [R8.16b], [R13.8] [R13.9], [R17.13], [R13.15], [R17.18].

The space of variety Imaginations is also the space to form goals, objectives, solutions, designs, engineering, planning and visions in the space of individual-collective dualities, where there are *individual goals, objectives, and visions* and *collective goals, objectives and visions* that may reside in the space of conflict-cooperation dualities in relation to the universal goal-objective space for life and survival generating problems over the space of cost-benefit dualities. Problems are not natural; they are social and in relation to the finding the path to achieve an element in the space of goals, objective, and vision under cognitive capacity limitation relative to the space of ignorance-knowledge dualities. The space of design, engineering, planning and architecture, art, and theory is a space of instruments of actions (imagination toolbox space) contained in the space of imaginations to realize elements in the space of goal, objective, and vision (imagination goal-objective space). The goal-objective and toolbox spaces are interdependent and inter-supportive. The struggles to acquire the correct toolboxes relative to a goal-objective element are called challenges while the drive to attain a goal-objective element is called a problem. At the space of goals, objectives and, visions in the knowing, there is the *individual imagination* in relation to individual *vision, goals* and *objectives* that generate the space of individual problems, relative to the conditions of space of individual realities. Similarly, there is the social or collective imaginations in relation to social or *collective visions*, *goals* and *objectives* that generate the space of collective problems relative to the space of collective realities as constrained by philosophical consciencism that has taken hold of the members of the social set-up at various generational time points, where individual and collective creativities and curiosities play important roles in knowing as intellectual investment flows and information-knowledge accumulation as intellectual capital stocks within the space of fundamental-applied dualities over the dynamic epistemic field with the dynamic conditions of the information and relational fields that hold the structures of qualitative and quantitative motions relative to variety properties and relations over the space of individual-collective dualities.

In the epistemic processes of knowing, information-knowledge production, input-output processes and problem-solution dualities and polarities over the space of fundamental-applied dualities, curiosity and creativity play extremely important roles connecting imaginations to realities through the spaces of possibility-probability and necessity-freedom dualities to reality, actual to potential, potential to imagination and to possibility, and from possibility to probability over the epistemic field in the epistemological space, where, possibility is defined in the *cognitive state of being* and probability is define *resource state of being*. The question is: what is creativity and what role does it play over the epistemological space in knowing, information-knowledge production, problem-solution dynamics, and variety transformations? Similarly, what are possible world, impossible world and possible-world-impossible-world dualities and how do they relate to the conditions in the ignorance-knowledge dualities, problem-solution dualities, identification-transformation dualities, thinking-reasoning dualities, and acceptance-rejection dualities?

Again, do probable world, improbable world and the space of probable-world-improbable world dualities exist and how do they relate to the conditions of in the spaces of ignorance-knowledge, problem-solution and identification-transformation dualities, risk-benefit dualities, and other dualistic-polar conditions in the spaces of relationality-unity dualities and rationality-irrationality dualities? These questions will be answered with definitions, explications and analytical expositions that will establish the conditions of curiosity, over the space of individual-collective dualities, and their relationships to knowing, information-knowledge accumulation in the space of diversity-unity dualities for individual and collective progresses within a nation-building and social progress. The knowing is always in reference to the production of intellectual investment flows, while the information-knowledge accumulation is in reference to the production of intellectual capital stocks that go to amplify social philosophical consciencism which provides differential preferences over cost-benefit conditions, generating distributional outcomes over the space of justice-injustice duality to enhance the conditions over the space of conflict-resolution dualities in the cognitive space which has been defined over the space of imagination-reality dualities containing possible world and corresponding duals.

3.3 DEFINITIONS AND EXPLICATIONS OF THE CONCEPTS OF POSSIBLE AND IMPOSSIBLE WORLDS IN THE SPACE OF IMAGINATION-REALITY DUALITIES

We shall now connect the discussions of the epistemic anomalies with relevant phenomena by definitions and explications while keeping in mind that most of the anomalies arise in the space of imaginations $(\mathfrak{g} \in \mathfrak{G})$ with characteristic disposition $(\mathbb{X}_{\mathfrak{G}})$ which houses the possibility and probability spaces and actualized in the space of realities $(\mathfrak{h} \in \mathfrak{H})$, where $\mathfrak{G} \subset (\mathfrak{H} \otimes \mathfrak{G}), \mathfrak{H} \subset (\mathfrak{H} \otimes \mathfrak{G})$ and $(\mathfrak{h}, \mathfrak{g}) \in (\mathfrak{H} \otimes \mathfrak{G})$ as imagination-reality dualities, where the characteristic disposition of every variety imagination dual contains some characteristics of reality dual. Similarly, the characteristic disposition of every variety reality also contains the characteristics of the imagination dual as the internal structure of dualistic-polar existence. Let us also keep in mind that the possibility space is equipped with a functor space $(\mathfrak{f} \in \mathbf{F})$, where $(\mathfrak{f})$ is a generic element, or space of categorial convertors $(\mathfrak{c} \in \mathfrak{C})$ where $(\mathfrak{c})$ a generic element of the space of categorial convertors. . Each functor or categorial convertor is equipped with an anticipation operator from the space of anticipation operators $(\mathfrak{a} \in \mathbf{A})$ such that the possibility space may be represented as a Cartesian tripled structure of the form $(\mathfrak{P} \otimes \mathbf{F} \otimes \mathbf{A})$ with a generic element $(\mathfrak{p}, \mathfrak{f}, \mathfrak{a}) \in (\mathfrak{P} \otimes \mathbf{F} \otimes \mathbf{A})$ or $(\mathfrak{P} \otimes \mathfrak{C} \otimes \mathbf{A})$ with a generic element $(\mathfrak{p}, \mathfrak{c}, \mathfrak{a}) \in (\mathfrak{P} \otimes \mathfrak{C} \otimes \mathbf{A})$ associated with possibilistic reasoning for social transformation with cognitive capacity limitations. The probability space is equipped with a functor space $(\mathfrak{f} \in \mathbf{F})$, where $(\mathfrak{f})$ is a generic element or space of categorial convertors $(\mathfrak{c} \in \mathfrak{C})$ where $(\mathfrak{c})$ is a generic element. It is also equipped with the space of expectation operators $(\mathfrak{e} \in \mathbf{E})$ such that the probability space may be represented as a Cartesian tripled structure of the form $(\mathfrak{B} \otimes \mathbf{F} \otimes \mathbf{E})$ with a generic element $(\mathfrak{b}, \mathfrak{f}, \mathfrak{e}) \in (\mathfrak{B} \otimes \mathbf{F} \otimes \mathbf{E})$ $(\mathfrak{B} \otimes \mathfrak{C} \otimes \mathbf{E})$ with a generic element $(\mathfrak{b}, \mathfrak{c}, \mathfrak{e}) \in (\mathfrak{B} \otimes \mathfrak{C} \otimes \mathbf{E})$ associated with probabilistic decision-choice actions for social transformation with resource capacity constraints. The space of functors and anticipations is associated with necessity and initial conditions, while the space of functors and expectation is associated with freedoms that turn anticipations into realities as have been discussed in a previous monograph [R8.16b]. The concepts of functors, functor spaces and categorial convertor, anticipation operators, expectation operator and the relevant spaces will be defined and explicated in relation to the spaces of possibility-impossibility dualities and probability-improbability dualities which are traceable through the space of epistemic distances to the space of ontic-epistemic dualities which generates the space of certainty-uncertainty dualities.

Let us reflect on the concept and phenomenon of possible world, where under the dualistic-polar conditions over the episteme field supported by relational and information fields, we must relate it to the concepts and phenomena of impossible world and possible-world-impossible-world dualities under the principle of opposites with relational continuum, unity and give-and-take sharing modes. Given the state of matter-energy equilibrium, the spaces of actual-potential duality of universal existence and neuro-decision-choice actions, how will one view the concept of possible world and impossible world within the space of knowing as the production of intellectual investment flows and the concept of information-knowledge accumulation as the production of intellectual capital stock? The concept and phenomenon of possible worlds may be viewed in two ways in terms of ontological process and in terms of epistemological process. Let us examine them separately.

3.3.1 The Concept and Phenomenon of Possible World in Ontological Space

In the respect of the ontological space, the concept and phenomenon of possible world may be seen in relation to the concept and phenomenon of the space of actuals as a sub-space of the space of actual-potential dualities containing telescopic past, present and future, ontological varieties, where an alternative ontological actuals may exist within the space of potentials that could have been actualized through the space of either *natural functors* or *categorial convertors* which contains the spaces of natural processes and technologies for ontological transformational dynamics. The possible world relative to the ontological actual world has natural functors or natural categorial convertors, such that it has no established framework for examining the conditions of variety transformations which will directly move varieties from potential space to replace or add to the existing variety actuals, where such transformations do not pass through the space of imagination-reality dualities containing the spaces of possibility-impossibility dualities, probability-improbability dualities, anticipation-expectation dualities and necessity-freedom dualities which are defined in the cognitive space of neuro-decision-choice actions. These dualistic-polar structures find no meaning and existence in the ontological space except when they are viewed under conditions of epistemological space. The unactualized ontological alternatives relative to the actual resides in the space of potential where the non-actualization may be attributed to unexplainable and cognitively unknown ontological selection processes from the space of potentials containing process and technological potentialities from the beginning disposition. Such unactualized ontological alternatives are infinite, residing in the space of potentials in multiplicity of relational and information modes, where every potential modality has supporting but unknown conditions within the space of actualization-non-actualization dualities. Since the processes in the space of actual-potential dualities deny the existence of the space of possibility-impossibility dualities, it is analytically useful to speak of *potential world* rather *than possible world,* where possibility-impossibility dualities have no conceptual meaning and relevance in ontological dynamics irrespective of whether one is speaking of semantics, non-semantics, pragmatics, or non-pragmatics.

The concept of potential world relates to the concept and phenomenon of *potential universe* relative to the actual universe. The unknown conditions of the structure and form of potential world deny any logical grounding for examining universal and potential alternatives by cognitive agents such that all potential worlds and universes are attributed to the dynamics of the space of beginning-end dualities with logical entries relating to Devine existence where there is always information-knowledge equality with unknown foundations. The information-knowledge equality conditions as inputs in the natural processes over the ontological space are such that *what there was* is *what there was, what there is,* is *what there is* and *what there would be* is *what there would be* from the initial conditions, producing conditions that deny the concepts and phenomena of certainty-uncertainty dualities, benefit-risk dualities, imagination-reality dualities and the space of imaginations containing possibility and probability and corresponding spaces in dualistic-polar structures. The concepts and phenomena of imagination, possibility and probability are generated by variety information-knowledge inequalities and epistemic distances resulting from ontic-epistemic differences in variety characteristic dispositions, which are irrelevant in the ontological space and its variety dynamics. In its bare essential, possible universes are irrelevant alternatives in the cognitive analysis of the space of universal existence and logical analytics.

The epistemic difficulty in conceptualizing any possible universe relative to the space of actuals simply resides in the space of cognitive capacity limitations that denies human conception of the conditions of telescopic past from the space of beginning such that the conception of each destination in relation to end is unknown and illusive in the understanding of the conditions of ontological transformations which deny any meaningful cognitive imagination of possible universe that must be conceptualized to be contained in the space of potentials under Sankofa-anoma trinity in the spaces of time-information and beginning-end dualities. It is the existence of these cognitive capacity limitations, relative to conditions of universal beginning through the telescopic past relative to the present that impose the epistemic requirements that every dynamic system must specify the *entry point* of the theoretical journeys for every element in the space of ignorance-knowledge dualities and the *exist point* within the infinite journey. The entry point shows where one enters the highway of knowing as the production of intellectual investment flows and the information-knowledge accumulation as the production of intellectual capital stocks. The entry point conditions provide some rational communications regarding differences in agreements and disagreements of cognitive agents in the space of ignorance-knowledge dualities. The exist-point conditions provide rational communications regarding what has been taken out, corrected, and added in the understanding of the epistemic journey towards the tranversality line and to test the conditions of crossing from the zone of ignorance to the zones of knowledge. The universal beginning disposition is not available to cognitive agents and hence the end dispositions and the outcomes between the beginning dispositions and the end dispositions cannot be conceptualized before they happen to create *what there was, what there is* and *what there would be*, which are then subject to knowing and information-knowledge accumulation.

3.3.2 The Concept and Phenomenon of Possible World in the Epistemological Space

In respect to the epistemological space, the concept and phenomenon of possible world may be conceived in relation to the concept and phenomenon of the space of realities as a sub-space of the space of imagination-reality dualities containing the telescopic past, present and future of epistemological varieties, where *what there was* may not be *what there was*, *what there is* may not be *what there is* and *what there would* be may not be *what there would be*, and where the epistemological varieties could be realized through the space of *social functors* or *categorial convertors* from the space of imaginations containing the spaces of possibility, impossibility, probability and improbability with dualistic-polar structures. The unsureness conditions within the spaces of telescopic past-present, present-future, and past-future dualities are generated by variety information-knowledge inequalities and epistemic distances produced by differences of ontic-epistemic characteristic dispositions in the epistemological space relative to the ontological space.

The epistemological space is an imperfect derivative from the ontological space which constitutes the identity, just as every ontic characteristic disposition constitutes an identity and every epistemic characteristic disposition constitutes a derivative.

The possible world relative to the epistemological realities has *social functors* or *social categorial convertors*, such that it contains initial conditions, necessity and anticipation for examining the conditions of variety transformations in terms of semantics and pragmatics of being which must move the epistemological varieties from the space of imaginations containing the spaces of possibility-impossibility dualities, probability-improbability dualities, anticipation-expectation dualities and necessity-freedom dualities, which are defined in the cognitive space of neuro-decision-choice actions under a social philosophical consciencism to add to or replace some varieties in the space of realities, where the initial conditions are known and may be compared to other possible conditions in the space of imagination as may be conceptualized from the space of possibilities . The existence of the space of possibility-impossibility dualities provides us with a logical toolbox to speak of *what if* and *what if not* in terms of imaginations containing possibilities, impossibilities, probabilities and improbabilities derived from the space of certainty-uncertainty dualities that may be examined for conditions within the space of necessity-freedom, anticipation-expectation and realities in the space of imagination-reality dualities that houses goals, objectives, aspirations, visions, drives of actions, hope and other such qualitative cognitive characteristics which spin the spaces of problem-solution and answer-question dualities and give rise to modal propositions for evaluating variety realities relative to varieties in the space of actuals contained in the space of actual-potential dualities under the principle of opposites. The *what if* and *what if not* conditions relative to the space of realities is to examine comparative alternative potential variety cost-benefit conditions relative to the existing variety cost-benefit conditions over the space of true-false dualities.

Every possibility space is contained in the space of imagination and equipped with either a functor space or space of categorial converters whether individually or collectively conceived. Similarly, every impossibility space is contained in the space of imagination but is not conceived to be equipped with functor space or space of categorial converters. The possibility, and impossibility in dualistic-polar existence are social concepts and phenomena as derivatives from within the space of imaginations which is also a social formation and a derivative from within the potential space which is a natural concept from within itself and existing as a dualistic-polar continuum connected by cognitive actions to the space of the actuals. The space of actuals is a concept and phenomenon of natural existence with or without the awareness of other ontological varieties.

Both the spaces of actuals and potentials are equipped with different functor spaces or spaces of categorial convertors. Each variety in the universal object set containing the actual and potential characteristic dispositions is equipped with an appropriate functor for its categorial mapping dynamics. The elements of the functor space (space of categorial converters) of actuals directly transform the varieties in the space of actuals under destruction-construction dynamics to the space of potentials. Similarly, the elements of the functor space (space of categorial convertors) of the space of potentials directly transform the varieties in the space of potentials to the space of the actuals without going through the space of imaginations which contains the cognitive conceptions of possibility and probability spaces under dualistic-polar conditions, where the opposites contain the characteristics of each other for continuum and unity. The concepts and phenomena of functor and categorial convertor in this epistemic field require definitions, explications, and clarifications in their uses. The definitions will be given verbally and algebraically that must meet conditions over the space of static-dynamic dualities.

DEFINITION 3.3.2.1A: Functor, Functor Space, Categorial Converter and Space of Categorial Converters in the Organization of Knowing (Verbal)

A *functor* like *categorial convertor* is a process-technological inter-categorial mapping of variety characteristic dispositions transforming a characteristic disposition of a variety in one category and moving it to another category, where the initial category is the parent category(predecessor), and the new category is the offspring category (successor) within the epistemic field. The collection of all the variety functors (categorial convertors) is the *functor space* (space of categorial convertors). The efficiency of the process-technological mapping is called the *functor moment* (categorial converter moment). The process-technological mapping of category to itself is called *endo-functor* or *endo-categorial convertor*. The process-technological mapping that has a process-technological reversibility is called a *dualistic-polar functor* (dualistic categorial convertor) that brings about negation of negation.

DEFINITION 3.3.2.1B: Functor, Functor Space, Categorial Converter and Space of Categorial Converters (Algebraic)

Given the epistemic field $\mathscr{C}$ as a collection of categories $\mathcal{C}_i . \mathcal{C}_i' \in \mathscr{C}\ \forall i \in \mathbb{I}^\infty$ with supporting relational field $(\mathscr{R})$ and information field $(\mathscr{I})$, $(\mathfrak{f})$ is said to be a functor and $(\mathfrak{c})$ is said to be categorial convertor if $\left(\mathfrak{f} : \mathcal{C}_i \to \mathcal{C}_j, \forall i \neq j \in \mathbb{I}^\infty\right)$, and, $\left(\mathfrak{c} : \mathcal{C}_i \to \mathcal{C}_j, \forall i \neq j \in \mathbb{I}^\infty\right)$. The collection of all $(\mathfrak{f}\text{'s})$ is the functor space $(\mathbf{F})$ and the collection of all $(\mathfrak{c}\text{'s})$ is the space of categorial convertors $(\mathfrak{C})$ where $(\psi_{(\mathfrak{f})})$ is the functor moment and $(\psi_{(\mathfrak{c})})$ is categorial-convertor moment. An $(\mathfrak{f} : \mathcal{C}_i \to \mathcal{C}_i, \forall i \in \mathbb{I}^{\infty})$ is called endofunctor $(\mathfrak{c} : \mathcal{C}_i \to \mathcal{C}_i, \forall i \in \mathbb{I}^{\infty})$ is endo-categorial convertor. The $(\mathfrak{f} : \mathcal{C}_i \rightleftarrows \mathcal{C}_j, \forall i, j \in \mathbb{I}^{\infty})$ as negation of negation is called a *dualistic-polar functor* (categorial convertor) and $(\mathfrak{c} : \mathcal{C}_i \rightleftarrows \mathcal{C}_j, \forall i, j \in \mathbb{I}^{\infty})$ as negation of negation is called *dualistic-polar categorial converter.*

NOTE 3.3.2.1: On Functor Spaces and the Spaces of Categorial Convertors

The functor ++spaces of actuals and potentials are the process-technological mappings within the space of actual-potential dualities transforming variety actuals to variety potentials and variety potentials to variety actuals as have been explained in the theories of info-dynamics [R17.18] and categorial conversion[R17.15][R17,16] over the space of technology-process dualities constituting the space of natural transformation actions within the space of natural input-output dualities as conformal mapping relative to the space of parent-offspring dualities in the space of natural input-output dualities. They are the spaces of *natural functors* or *categorial convertors* under transformative forces of natural principle of opposites. The space of potentials with the corresponding functor space constitutes the identity from which the space of imaginations with corresponding functor space is derived. Similarly, the space of actuals with the corresponding functor space constitutes the identity from which the space of realities with corresponding functor space is derived. Similarly, the space of ontic characteristic dispositions with corresponding functor constitutes the identity from which the space of epistemic characteristic dispositions with corresponding functor space is derived. As such, the space of actual-potential dualities with the corresponding functor space of actual-potential dualities is also an identity from which the space of imagination-reality dualities with corresponding functor space is derived. The union of the spaces of actual-potential dualities and imagination-reality dualities constitutes the space of ontological-epistemological dualities $\Omega \cup \mathfrak{E} = \left((\mathfrak{U} \cup \mathfrak{A}) \cup (\mathfrak{H} \cup \mathfrak{G}) \right) = \left((\mathfrak{U} \otimes \mathfrak{A}) \otimes (\mathfrak{H} \otimes \mathfrak{G}) \right)$ that houses the space of ontic-epistemic dualities producing the space of variety epistemic distances which then gives rise to the spaces of certainty-uncertainty dualities, possibility-impossibility dualities and probability dualities under capacity limitations of cognitive agents, where human actions move from imagination to reality.

The possibility space within the space of imaginations may be viewed in two ways within the space of imagination-reality dualities. One way is in relation to varieties in the space of realities relative to telescopic past and telescopic present through alternative varieties that are not in existence but may be semantically examined in cost-benefit space relative to the space of better-worse dualities. The other way is in relation to what cognitive agents can conceptualize and transform in terms of cognitive images of *what is not* to *what there would be*, in telescopic future with cognitive capacity limitations in the space of imagination-reality dualities containing the space of problem-solution dualities relative to conceived goals, objectives and visions as seen over the space of cost-benefit dualities. The possibility and probability spaces have nothing to do with the alternative structures of the space of the actuals (*what there was* and *what there is*). They have something to do with the varieties in the space of realities as derivatives, where the space of actual-potential dualities is an enabler in conception, knowing and information-knowledge production, the outcomes of which are to serve inputs into the space of neuro-decision-choice actions for the control and management of social conditions through activities over the space of problem-solution dualities as defined by the elements in the space of cost-benefit dualities.

The ontological telescopic past and present varieties are identities of matter-energy dualities in equilibrium state while the space of varieties is always in disequilibrium states generating information-knowledge disequilibrium dynamics toward an unknown destination over the space of natural destruction-construction dualities within the ontological space from the space of relativity to the space of absoluteness in terms of varieties in the space of matter-energy dualities in equilibrium states, where all varieties reside in dualistic-polar zone of conflicts for qualitative and quantitative motions relative to properties and relations of categories and categorial varieties as they have been pointed out in [R8.16a] [R8.16b][R17.14]R17.15][R17.16][R17.18].

Given the state of matter-energy equilibrium, the space of actual-potential dualities with the corresponding functor spaces and neuro-decision-choice actions, how will one view the concepts and phenomena of possible world and impossible world in dualistic-polar structures? The concept of possible world may be seen in relation to the concept of the space of actuals as a sub-space of the space of actual-potential dualities containing telescopic past, present and future ontological varieties, where an alternative ontological actual may exist within the space of potentials that could have been actualized through the space of natural functors. It may also be conceived in relation to the concept of the space of realities as a sub-space of the space of imagination-reality dualities containing telescopic past, present and future epistemological varieties, where the epistemic varieties could be realized through the space of social functors from the space of imaginations containing the spaces of possibility, impossibility, probability, and improbability with dualistic-polar structures.

The possible world relative to the ontological actual world has natural functors or natural categorial convertors, such that it has initial conditions for examining the conditions of variety transformations which will directly move from potential space to add to or replace the existing actuals without passing through the space of imagination-reality dualities containing the spaces of possibility-impossibility dualities, probability-improbability dualities. anticipation-expectation dualities and necessity-freedom dualities which are defined in the cognitive space of neuro-decision-choice actions. The possible world relative to the epistemological world of realities has *social functors* or *social categorial convertors*, such that it must have initial conditions, necessity and anticipations for examining the conditions of variety transformations which must move the epistemological varieties from the space of imaginations containing the spaces of possibility-impossibility dualities, probability-improbability dualities, anticipation-expectation dualities and necessity-freedom dualities which are defined in the cognitive space of neuro-decision-choice actions under dynamics of social intergenerational philosophical consciencism to add to or replace some varieties in the space of realities. The transformations from the space of imaginations must always move through the spaces of possibility-impossibility, probability-improbability, and success-failure dualities under *Asantrofi-anoma* conditions of the space of cost-benefit dualities.

DEFINITION 3.3.2.2A: Possible World and Impossible World Relative to Reality (Verbal)

A possible world is a collection of varieties which exist in a cognitive imaginary *state* within the space of imagination with positive *anticipation. necessity* and *initial conditions* and equipped with a *functor* from a *functor space* as a conceptual alternative to an a real world, where the variety characteristic disposition exists in the mind of the imagining agent with a correspondingly signal disposition projected to the public such that its decoded characteristic disposition has multiple interpretational structures such that what can be realized has no *expectation operator* to move it to the space of realities, and hence has a questionable and disputable value within the space of true-false dualities as examined over the space of decidability-undecidability dualities with an increasing indecision of all its implied conditions in the space of doubt-surety dualities in such a way that every possible world has a multiplicity of split-derived identity, where some of the part of the variety characteristic disposition resides completely in the space of pure imagination with characteristic disposition that dominates the part of its reality characteristic disposition in the space of realities within the space of imagination-reality dualities. The impossible world is also a collection of varieties that exist in cognitive imaginary *state* within the space of imagination without *anticipation. necessity* and *initial conditions* and with a weak functor that has a degree of moment that deprives it of its transformative action from the space of functors as well as has no expectation operator from the space of expectation operators relative to an alternative real world. The Cartesian product of the spaces of possible-world and impossible-world constitutes the space of possible-world-impossible-world dualities with a generic element under dualistic-polar restrictions, information structures defined as characteristic-signal disposition and constrained by a fuzzy conditionality with relational continuum and unity under the principle of opposites.

DEFINITION 3.3.2.2B: Possible Worlds and Impossible Worlds (Categorial Characteristic)

A variety is said to be a *possible world*$(\mathfrak{w} \in \mathfrak{W})$relative to the space of world of realities $(\mathfrak{G})$ if its characteristic disposition belongs to the characteristic dispositions of possibility space $(\mathfrak{w} \in \mathfrak{P} \subset \mathfrak{G} \subset (\mathfrak{H} \otimes \mathfrak{G}))$ equipped with a fuzzy functor$(\mathfrak{f} \in \mathbf{F})$ and not in probability space, and hence not equipped with expectation operator. In other words, the space of possible world is a Cartesian product of the form $\mathfrak{W} = \{(\mathfrak{P} \otimes \mathbf{F}) \mid \mathfrak{P} \subset \mathfrak{G}\}$ with generic element $\nu = \mathfrak{w} = (\mathfrak{p}, \mathfrak{f})$ such that

$\not\exists \mathfrak{e} \in \mathbf{E}$ contained in $\mathfrak{W} = (\mathfrak{P} \otimes \mathbf{F}) \subset \mathfrak{G}$ with information support $\mathbb{Z}_{\mathfrak{w}} = (\mathbb{X}_{\mathfrak{w}} \otimes \mathbb{S}_{\mathfrak{w}})$ such that $\mathbb{X}_{v} = \mathbb{X}_{\mathfrak{w}} \subset \mathbb{X}_{\mathfrak{E}} \subset \mathbb{Z}_{\Omega} \ni, \not\exists \mathbf{E} \subset \mathfrak{W}$. The collection of all possible worlds constitutes the possible-world space $(\mathfrak{W})$ with positive possibility and zero probability measures as cognitively assessed. Similarly, a variety is said to be an *impossible world* $(\mathfrak{v} \in \mathfrak{V})$ if its characteristic disposition $\mathbb{X}_{\mathfrak{v}}$ does not belongs to both the possibility space $(\mathfrak{p} \in \mathfrak{P})$ and probability spaces $(\mathfrak{b} \in \mathfrak{B})$, and hence $\mathfrak{V} = (\mathfrak{M} \otimes \mathfrak{H}) \subset \mathfrak{G}$ that is if $(\mathfrak{v} \in \mathfrak{V})$ then $(\mathfrak{v} \notin \mathfrak{Z})$, $(\mathfrak{v} \notin \mathfrak{B})$ and $\mathbb{X}_{\mathfrak{v}} \in \mathbb{X}_{\mathfrak{V}} \cap \mathbb{X}_{\mathfrak{Z}} = \mathbb{X}_{\mathfrak{V}} \subset \mathbb{X}_{\mathfrak{E}} \subset \mathbb{X}_{\mathfrak{U}}$, where $(\mathfrak{M})$ is the impossibility space and $(\mathfrak{H})$ is the improbability space . The collection of all the impossible worlds constitutes the impossible-world space $(\mathfrak{V})$ with characteristic-signal disposition $\mathbb{Z}_{\mathfrak{V}} = (\mathbb{X}_{\mathfrak{V}} \otimes \mathbb{S}_{\mathfrak{V}})$ without positive possibility and positive probability. The space of possible-world-impossible-world dualities $\mathfrak{Z} = (\mathfrak{W} \otimes \mathfrak{V})$ with a generic element $\mathfrak{z} = (\mathfrak{w}, \mathfrak{v}) \in \mathfrak{Z} = (\mathfrak{W} \otimes \mathfrak{V})$ under dualistic-polar restrictions with characteristic disposition $\mathbb{Z}_{(\mathfrak{W} \otimes \mathfrak{V})} = \left(\mathbb{X}_{(\mathfrak{W} \otimes \mathfrak{V})} \otimes \mathbb{S}_{(\mathfrak{W} \otimes \mathfrak{V})}\right)$ may then be conceptually defined where impossible world in comparison with the reality $(\mathfrak{V} \not\propto \mathfrak{H})$ and $(\mathfrak{W} \propto \mathfrak{H})$ where (α) represents relative relation in terms of variations.

To follow the ensuing definitions in relation to possible world semantics and possible world non-semantics, it is useful to present the analytical nature of possibility, possibility space equipped with a functor space, impossibility and impossibility space with non-functor space and probability, and probability space equipped with a space of expectation operators and improbability and improbability space with non-expectation operators in dualistic-polar setting as they appear in the space imaginations, always in relation to the space of problem-solution dualities relative to goals, objectives, visions over the space of cost-benefit dualities under the socio-economic principle of non-satiation and a system of individual and social preferences. The analytically relational structure of possibility-impossibility dualities and probability-improbability dualities may be represented in a tabular form given the space of imagination as in table3.3.2.1.

TABLE 3.3.2.1: The Relational Structures of Spaces of Possibility-impossibility and probability-improbability Dualities within the Space of Imaginations under Dualistic-Polar Conditions

ANTICIPATION → / ↓ EXPECTATION	**Possibility Space** $(\mathfrak{P}\otimes\mathbf{F})$	**Impossibility Space** $(\mathfrak{M})$	**Possibility-impossibility dualities** $\left[(\mathfrak{P}\otimes\mathbf{F})\otimes\mathfrak{M}\right]$
Probability $(\mathfrak{B}\otimes\mathbf{E})$	**Possibility-probability dualities** $(\mathfrak{P}\otimes\mathbf{F})\otimes(\mathfrak{B}\otimes\mathbf{E})$ **COHORT 1**	**Impossibility-probability dualities** $(\mathfrak{B}\otimes\mathbf{E})\otimes\mathfrak{M}$ **COHORT 2**	**Union of the Spaces of possibility-probability dualities and impossibility-probability dualities** $\left[(\mathfrak{P}\otimes\mathbf{F})\otimes(\mathfrak{B}\otimes\mathbf{E})\right]\cup\left[(\mathfrak{B}\otimes\mathbf{E})\otimes\mathfrak{M}\right]$ **COHORT 3**
Improbability $\mathfrak{H}$	**Possibility-improbability dualities** $(\mathfrak{P}\otimes\mathbf{F})\otimes\mathfrak{H}$ **COHORT 4**	**Impossibility-improbability dualities** $(\mathfrak{M}\otimes\mathfrak{H})$ **COHORT 5**	**Union of the spaces of possibility-improbability dualities and impossibility-improbability dualities** $((\mathfrak{P}\otimes\mathbf{F})\otimes\mathfrak{H})\cup(\mathfrak{M}\otimes\mathfrak{H})$ **COHORT 6**
Probability-improbability dualities $(\mathfrak{B}\otimes\mathbf{E}\otimes\mathfrak{H})$	**Possibility space= intersection of the spaces of possibility-probability dualities and possibility-improbability dualities** $\left[(\mathfrak{P}\otimes\mathbf{F})\otimes(\mathfrak{B}\otimes\mathbf{E})\right]\cup\left[(\mathfrak{P}\otimes\mathbf{F})\otimes\mathfrak{H}\right]$ **COHORT 7**	**Impossibility space = intersection of the spaces of impossibility-probability and impossibility-improbability dualities** $\left[(\mathfrak{B}\otimes\mathbf{E})\otimes\mathfrak{M}\right]\cup\left[(\mathfrak{M}\otimes\mathfrak{H})\right]$ **COHORT 8**	**Union of the Spaces of Necessity-Anticipation Dualities and Freedom-Expectation Dualities Relative to the space of Certainty-Uncertainty Dualities in the space of Imagination-Reality Dualities** **COHORT 9**

DEFINITION 3.3.2.2C: Possible-World-Impossible-World Dualities

If a variety $(\nu \in \mathbb{V})$ is a possible world $(\mathfrak{w})$ then given the interactive spaces of possibility-impossibility dualities and probability-improbability dualities, the space of possible worlds $(\mathfrak{W})$, resides in Cohort 3 such that:

1) $(\nu = \mathfrak{w} \in \mathfrak{W}) \Rightarrow (\mathfrak{w} \in \mathfrak{P}), (\mathfrak{w} \notin \mathfrak{B}) \Rightarrow (\mathfrak{z}_\nu = \mathfrak{z}_\mathfrak{w} \in \mathfrak{Z})$ and $\mathbb{X}_\mathfrak{y} = \mathbb{X}_\mathfrak{W} \subset \mathbb{X}_\mathfrak{Z} \subset \mathbb{X}_\mathfrak{E} \subset \mathbb{Z}_\Omega$

2) $\exists \mathbb{Z} \ni \mathbb{Z}_\mathfrak{W} = (\mathbb{X}_\mathfrak{W} \otimes \mathbb{S}_\mathfrak{W}) \subset \mathbb{Z}_\mathfrak{U} = (\mathbb{X}_\mathfrak{U} \otimes \mathbb{S}_\mathfrak{U})$ for possible world information

3) $\mathbb{X}_\mathfrak{w} \in \mathbb{X}_\mathfrak{W} \cap \mathbb{X}_\mathfrak{Z} = \mathbb{X}_\mathfrak{W} \subset \mathbb{X}_\mathfrak{E} \subset \mathbb{X}_\mathfrak{U}$ supported by $(\mathbb{X}_\mathbf{F} \subset \mathbb{Z}_\mathbf{F})$ or $(\mathbb{X}_\mathbf{C} \subset \mathbb{Z}_\mathbf{C})$

In the case of impossible world $(\mathfrak{Y})$, given the space of imagination, the impossible worlds $(\mathfrak{Y})$ reside in COHORTs 2 and 5, where there are no functor space or space of categorial convertors and the space of expectation operators in COHORT 2 with no transformative significance; thus

1) $(\nu = \mathfrak{y} \in \mathfrak{Y}) \Rightarrow (\mathfrak{y} \notin \mathfrak{P}), (\mathfrak{y} \notin \mathfrak{B}) \Rightarrow (\mathfrak{z}_\nu = \mathfrak{z}_\mathfrak{y} \in \mathfrak{Z} \cap \mathfrak{W})$ and $\mathbb{X}_\mathrm{v} \subset \mathbb{X}_\mathrm{Z} \subset \mathbb{X}_\mathfrak{E} \subset \mathbb{Z}_\Omega$

2) $\exists \mathbb{Z} \ni \mathbb{Z}_\mathfrak{Y} = (\mathbb{X}_\mathfrak{Y} \otimes \mathbb{S}_\mathfrak{Y}) \subset \mathbb{Z}_\mathfrak{U} = (\mathbb{X}_\mathfrak{U} \otimes \mathbb{S}_\mathfrak{U})(\mathbb{X}_\mathfrak{U} \otimes \mathbb{S}_\mathfrak{U}) \not\subset \mathbb{Z}_\mathfrak{A} = (\mathbb{X}_\mathfrak{A} \otimes \mathbb{S}_\mathfrak{A})$ for impossible world information

3) $\mathbb{X}_\mathfrak{y} \in (\mathbb{X}_\mathfrak{Y} \cap \mathbb{X}_\mathfrak{Z}) = \mathbb{X}_\mathfrak{Y} \subset \mathbb{X}_\mathfrak{E} \subset \mathbb{X}_\mathfrak{U} \subset \mathbb{X}_\Omega$ unsupported by $(\mathbb{X}_\mathbf{F} \subset \mathbb{Z}_\mathbf{F})$

Where $\mathfrak{U}$ = potential space, $\mathfrak{A}$ = space of atuals, Ω = space of ontology, $\mathfrak{E}$ = epistemological space, $(\mathfrak{H})$ = space of realities $(\mathfrak{G})$ = space of imagination .

In the case of the space of possible-world-impossible world dualities $\mathfrak{Z} = (\mathfrak{W} \otimes \mathfrak{V}) = (\mathfrak{W} \cup \mathfrak{V})$ with relational continuum of the form $(\mathfrak{W} \cap \mathfrak{V}) \neq \varnothing$ given the spaces of possible world $\mathfrak{W}$ and impossible world $\mathfrak{V}$, we have

1) $(\mathfrak{w}, \mathfrak{v}) \in (\mathfrak{W} \otimes \mathfrak{V})$ under dualistic-polar restrictions with characteristic disposition

2) $\mathbb{Z}_{\mathfrak{W}\otimes\mathfrak{V}} = (\mathbb{X}_{\mathfrak{W}\otimes\mathfrak{V}} \otimes \mathbb{S}_{\mathfrak{W}\otimes\mathfrak{V}}) = \mathbb{Z}_{\mathfrak{W}\cup\mathfrak{V}} = (\mathbb{X}_{\mathfrak{W}\cup\mathfrak{V}} \cup \mathbb{S}_{\mathfrak{W}\cup\mathfrak{V}})$;with$(\mathbb{X}_{\mathfrak{W}\otimes\mathfrak{V}} \cap \mathbb{S}_{\mathfrak{W}\otimes\mathfrak{V}}) \neq \varnothing$

3) $(\#\mathbb{X}_{\mathfrak{W}} \ggg \#\mathbb{X}_{\mathfrak{V}}) \Rightarrow$ possible world and $(\#\mathbb{X}_{\mathfrak{V}} \ggg \#\mathbb{X}_{\mathfrak{W}}) \Rightarrow$ impossible world

Where $\ggg \ \Rightarrow$ substentially greater than and hence conditions of dualistic dominance.

4) $(\mathbb{X}_{\mathfrak{W}} \cup \mathbb{X}_{\mathfrak{V}}) \Rightarrow$ unity and $(\mathbb{X}_{\mathfrak{W}} \cap \mathbb{X}_{\mathfrak{V}}) \neq \varnothing \Rightarrow$ dualistic-polar conditions of continuum .

With fuzzy conditionality.

3.3.3: Comparative Analytics on Definitions of Possible World, Impossible world and Possible-World-Impossible-world Dualities

In terms of the definitions, a variety is a possible world if it belongs to the possibility and improbability spaces equipped with a functor space but not with a space of expectation operators. In this respect, a possible world belongs to the space of necessity with initial information conditions for examining the validity of the elements in the possibility space and not in space of freedom. It finds its definitional characteristics in the space anticipation operators without the space of expectation operators. Its relations to the space of risk-reward dualities, true-false dualities are pure semantics on anticipation in terms of speculative events that are drawn from space of imaginations with fuzzy characteristic membership functions defined between zero and one under dualistic-polar conditions, where the constructs of all spaces are undertaken with fuzzy indicator functions as have been discussed in the above chapters of this monograph.

The connectivity of the space of imaginations and the space of realities finds expressions in Cohort 1 where the possibility space is equipped with functor space and the probability space is equipped with the space of expectation operators to establish the space of necessity-freedom dualities and initial-sufficiency dualities. It is only at the space of possibility-probability dualities equipped with functor space and space of expectation operators that the space of realities finds definition and material existence, where imaginations reside in realities and vice versa and neuro-decision-choice actions operate to transform problems into solutions which then generate new problems in the space of parent-offspring processes.

The varieties in the space of possible worlds are both possible and improbable for transformation, where the variety possibilities from the space of imaginations, contained in the space of potentials, through the spaces of uncertainty-certainty and risk-reward dualities also belong to the space of imagination-reality dualities. The space of risk-reward dualities defined in the space of possibility-probability dualities containing the space of anticipation-expectation dualities and necessity-freedom dualities relates to the curiosity and hope which are amplified by creative processes that connect the space of possible-probable worlds to the space of realities contained in the space of actuals to establish the varieties for identities in terms of categorial conversions, similarities, and commonness. On the other hand, the variety impossible world belongs to the space of impossibility-improbability dualities and related spaces, and does not require curiosity and hope with creative actions and is not related to variety transformations into the space of realities and hence into the space of the actuals.

The varieties in the space of impossible worlds are impossible and improbable for transformation from the space of imagination as contained in the space of potentials. In this respect, the important distinguishing factor between possible and impossible worlds is the probability space that provides freedom for decision-choice agents to examine the structure of the varieties of possible world with conditions of necessity for them to be transformed to enter the space of risk-benefit dualities and then to the space of realities. The similarities are expressed by the common conditions of improbability, where both possible world and impossible worlds are deprived off the space of expectation operators and hence freedom of variety transformations to the space of realities. In this epistemic framework for diversity and unity of knowing and science in the universe, the possibility space houses the necessity, anticipation, qualitative risk and the initial conditions for knowing and transformation while the probability space houses the freedom, expectations and sufficient conditions of change and the quantitative risk-benefit structure in the space of cost-benefit dualities within the space of fundamental-applied dualities in the whole space of production-consumption dualities that relates to the spaces of destruction-construction dualities and input-output dualities.

In this epistemic approach to knowing as an intellectual production of investment flows and information-knowledge accumulation as production of intellectual capital-stocks, the pursue of knowledge for knowledge seek makes no sense in the control and management of any social setup since all investments and capital stocks are part of the input-output processes taking advantages of information-knowledge chain in the space of production-construction dualities in the infinite system of destruction-construction dualities and problem-solution dualities within the space of static-dynamic dualities in the infinite universe of completeness as defined by the space of actual-potential duality, where the possible world and impossible worlds are conceptions under cognitive capacity limitations. Every knowing as intellectual investment flow and information-knowledge as capital stock reside in the space of actual-potential dualities in relation to the space of static-dynamic dualities through the space of possibility-probability dualities contained in the space of imaginations and not in the space of realities under dualistic-polar structures and fuzzy conditionality. Every information-knowledge claim whether supported or unsupported resides in the space of realities. It is within this epistemic framework that one must understand that life or social progress is not a stock of problems where the problems can be solved one after the other individually and collectively to the finish and one can have a happiness rally and party in the state of ecstasy.

Analytically, we have the probable world, improbable world, and probable-world-improbable-world dualities with the corresponding spaces under the principle of opposites. Improbability and possibility combine to define the space of possibility-improbability dualities containing necessity without freedom derivable from the social philosophical consciencism which define the possible world within the epistemic field. Similarly, the probability space and possibility space combine to define the space of possibility-probability duality containing necessities and freedoms derivable from the social philosophical consciencism which defines the conditions for the space of realities within the epistemic field. Within the space of imagination-reality dualities, the possible world may be viewed in relativity to the space of telescopic past-present dualities that holds the space of realities containing the claims of *what there was* and *what there is*. It may also be viewed in relation to telescopic present-future dualities that hold the space of imaginations containing *what there would be*.

Viewed relatively to the space of telescopic past-present dualities (the space of realities contained in the spaced of actuals) in terms of possible world, we are only concerned with comparative analytics of *what if*, associated outcomes and cost-benefit conditions in the space of true-false dualities under philosophical semantics without transformations within the space of imagination-reality dualities even if these outcomes of the semantic conditions under modal logic yield greater net benefits assessments. Viewed relative to the space of telescopic present-future dualities (the space of imaginations contained in the space of potentials) in terms of possible world, we are concerned not simply with comparative analytics of *what there is*, relative to *what there would be* and cost-benefit conditions with concerns of *what there was* (telescopic past) as information input into neuro-decision-choice actions to transformation variety imaginations in the spaces of problem-solution and question-answer dualities under the prevailing social philosophical consciencism within the space of imagination-reality dualities, where the emphasis is on neuro-decision-choice activities over the space of problem-solution dualities defined over the space of cost-benefit dualities.

There are three types of possible world that relate to neuro-decision-choice activities over the space of imaginations contained in the space of potentials. There is the *possible world universe* which is the alternative universe that presents us with infinite conceptual alternatives from the space of conceptual possibilities as may be conceived cognitively within the space of imaginations, the discussions of which are undertaken within the space of Divinity, where these conceptual alternative ontological universes are irrelevant alternatives in the epistemological space. There is the *possible world transformation* which encompasses goals, objectives visions, dreams and solutions for variety conversions relating to art, design, and social progress. There is the *possible world semantics* which defines the space of comparative logical games of different structures with cost-benefit conditions relative to the varieties in the space of realities without expectations in terms of *what if* and *what if not*. The type and conditions of possible world under the principle of opposites are presented in Table 3.3.2.2.

TABLE 3.3.2.2: Types of possible Worlds and their Characteristics and Relations to Neuro-decision-choice Actions

RELATIONAL SPACES / TYPE OF POSSIBLE	NAME OF DUALITIES	CATEGORIAL SPACE AND INFORMATION	Neuro-Decision-Choice of Necessity and Freedom
Possible world alternative universe	Space of Actual-Potential Dualities	$\Omega=(\mathfrak{A}\otimes\mathfrak{U}\otimes\mathbf{F}_{\mathfrak{U}})$ WHERE $\mathbf{F}_{\mathfrak{U}}=(\mathcal{P}_{\mathfrak{U}}\otimes\mathfrak{T}_{\mathfrak{U}r})$	Natural Processes, Technology, Necessity and Freedom with non-cognitive actions
Possible world semantics	Space of Possibility-improbability duality	$(\mathfrak{P}\otimes\mathfrak{M}\otimes\mathbf{F}_{\mathfrak{G}})$ Such that $\mathbf{F}_{\mathfrak{G}}=(\mathcal{P}_{\mathfrak{G}}\otimes\mathfrak{T}_{\mathfrak{G}})$	Anticipation, Non-expectation, necessity, and non-freedom for neuro-decision-choice agents
Possible world transformation	Space of Possibility-probability dualities	$(\mathfrak{P}\otimes\mathbf{F}_{\mathfrak{G}}\otimes\mathfrak{B}\otimes\mathbf{E}_{\mathfrak{G}})$ Such that $(\mathbf{F}_{\mathfrak{G}}\subset\mathbf{F}_{\mathfrak{U}})$	ANTICIPATION EXPECTATION, NECESSITY. FREEDOM, INITIAL CONDITIONS AND SUFFICIENT CONDITIONS

3.4 SEMANTIC ANALYTICS ON IMAGINATION, POSSIBILITY, REALITY, POSSIBLE WORLD, ALTERNATIVE REALITY, ALTERNATIVE FACT, CREATIVITY AND ANOMALIES IN KNOWING AND KNOWLEDGE-PRODUCTION SYSTEMS: Extensions.

In order to offer some analytical clarities on the definitions, explications and explanations of the concepts of possible and impossible worlds and their characteristics that have been offered in this monograph and the related concepts that may follow, and how they relate in differences and similarities to their current usages in philosophical, mathematical and linguistic discourse, let us examine the foundational concepts of *reality, imagination*, and *imagination-reality duality* with *anticipation*, *expectation* and *anticipation-expectation duality*, on one hand, *necessity, freedom* and *necessity-freedom duality* on another hand. It is useful to examine the concepts of *risk, reward* and *risk-reward duality* relative to the concepts of possibility, possibility space, the spaces of possible-world-impossible-world dualities, problem-solution dualities, possibility-probability dualities and other dualistic-polar structures in terms of knowing and information-knowledge conditions over the space of imagination-reality dualities contained in the spaces of actual-potential dualities, identification-transformation dualities, and decision-choice activities over the action space, all of which are derived from the space of certainty-uncertainty dualities that is the result from the space of ontic-epistemic dualities relative to the spaces of functors and categorial conversion operators. All the definitions must be seen in term of their relations to knowing and variety characteristic dispositions in the acquaintance space in order to establish variety identities, similarities and differences which are essential in the development of all languages and communications over the space of source-destination dualities, where production and inputs are sources and consumption and outputs are destination in relational give-and-take sharing modes with dualistic-polar negations of negations.

Let us keep in mind that all elements and varieties that may be associated with knowing processes, information-knowledge productions, anomalies, decision-choice mistakes and corrections, and other such structures that take place in cognition and the space of production-consumption dualities with continual interactions that generate epistemic cost-benefit outcomes and propel all dimensions of social system from the space of relativity to the space of absoluteness take place over the dualistic-polar conflicts within the space of actual-potential dualities through the proper sub-space of imagination-reality dualities containing supply-demand dualities locked in the space of input-output dualities. In the theory of knowing as the production of intellectual investment flows and the theory of information-knowledge accumulation as the production of intellectual capital stocks, there are many spaces and subspaces in conic structures with relational continuum and unity. These spaces and subspaces must be clearly defined and explicated for distinction and commonness and yet meet the specific and global dualistic-polar conditions under the organic principle of opposites in relational continuum and unity in the space of relativity containing the space of absoluteness at the extremes, where all production is seen in terms of knowing in the space of input-output dualities , and all knowing is seen in terms of production in the space of input-output dualities such that production is consumption and consumption is production in the space of input-output dualities. We shall begin with imagination, reality and the space of imagination-reality dualities and the corresponding spaces.

In the foundation of the theory of epistemic field with relational and information fields, the space of actual-potential duality is the fundamental conception of the totality of the universe with matter-energy equilibrium state that encompasses the telescopic past (*what there was*), telescopic present (*what there is*), and telescopic future (*what there would be*), where the universe is composed of a system of varieties in dualistic-polar structures in opposites in negative-positive structures and input-output structures generating continual tension within each duality for change and continual conflicts and resolutions for variety transformations of properties and relations producing qualitative and quantitation motions over the space of static-dynamic dualities. This analytical approach is not to see the concept of possible world as conceptual system of the ultimate perfect world as the creator may have constructed or alternative universes that may have been constructed. This approach sees the concept of a perfect universe as a system of varieties, categorial varieties and categorial dynamics moving from the state of imperfections to an unknown state of perfections through continual transformations producing variety disequilibrium dynamics through negation of negation with a build in self-correction and self-restructuring processes under the principle of opposites that produces continual conflicts for continual transformations, where the state of perfection is unknown and will never be reached since at that state there will be complete commonness and no differentiation which will render the activities in the space of actual-potential dualities irrelevant as well as neuro-decision-choice actions, input-output processes, information-knowledge processes and other such living fundamentals in the space of imagination-reality dualities completely irrelevant with stoppage of all variety mater-energy, relational and information transformations.

In this theory of epistemic field in support of the theory of problem-solution dualities and polarities, the concept of possible world is a derivative from the concept of possibility space which is a derivative from the concept of possibility contained in the space of imaginations as a derivative from the potential space. By dualistic-polar conditional extension the concepts of possibility and possibility space have corresponding concepts of impossibility and impossibility space with corresponding characteristic dispositions. In the same logical system, the concept of possible world has the concept of impossible world. To the extent to which the concept of possibility is not clearly defined and explicated, the whole concepts of possibility space and possible world, irrespective what paradigm of thought is used, would reside in cognitive confusion, delusions, and illusions. Similarly, there are analytical confusions and linguistic unsureness and zone of semantic conflicts over the space of source-destination dualities if the concepts of imagination and reality and corresponding spaces are not clearly defined and explicated. Given the definitions, explications, and sources of the concepts of possibility and probability, it is useful to deal with those of imaginations and realities with alternative definitions in fuzzy logical continuity, and transformations with inclusions of the space of process-technological dualities including functors, categorial convertors and expectation operators. The nature of varieties in the space of process-technological dualities are important to understand the varieties in the space of identification-transformation dualities as they relate to the elements in the spaces of problem-solution dualities, input-output dualities, ignorance-knowledge dualities, and epistemic transversality conditions.

3.4.1 Processes and Technology over the Space of Knowing: A Reflective Note

It is useful to note the nature of the concepts and phenomena of process and technology in the space of identification-transformation dualities and how they relates to themselves and the varieties over the space of nature-social dualities within the space of static-dynamic dualities, where every variety is composed of static-dynamic dualistic-polar dispositions under input-output relationality with special social rationality in the state of conflicting social preferences defined over the spaces of cost-benefit dualities and individual-community dualities, changed-unchanged dualities, progress-nonprogress dualities and systems of varieties. I have discussed some aspects of the concepts and phenomena of the progress and technology in the monograph info-dynamics [R17.17] [R17.18] as well as the concepts and phenomena of variety identity and categorial identity in [R17,19]. Let us connect the concepts and phenomena with the essential elements that will allow us to understand the concept and phenomena of imagination, reality and imagination-reality dualities and the manner in which a concept of process is related to the concept of technology and process and technology giving rise to the space of process-technological duality to affect the elemental identities in the space of varieties viewed relative to the elements in the space of static-dynamic dualities. We shall then relate the process-technological conditions to elements in the space of identification-transformation dualities relative to the rise of the elements in the space of varieties and the space of parent-offspring dualities seen from the space of problem-solution dualities by fist defining the concepts of process and technology, where continual transformations define the paths of both natural and social telescopic history over the Sankofa-anoma system of the spaces of time-information dualities composed of the three dualistic-polar spaces of past-present dualities, present-future dualities and past-future dualities, where we identify natural telescopic history and social telescopic history.

The natural telescopic history in the Sankofa-anoma three dualistic-polar spaces is an enveloping path by natural dynamics of forces within the space of matter-energy dualities which then produce transformations in the space of variety property-relation dualities independent of social existence but in response to socio-natural outcomes of transformational dynamics, where continual recycling is within the space of natural input-output dualities. The social telescopic history in the Sankofa-anoma three dualistic-polar spaces is an enveloping path by social dynamics of forces within the space of matter-energy dualities which then produce transformations in the space of social variety property-relation dualities, the outcomes of which depend on social existence in response to socio-natural outcomes of transformation dynamics, where continual recycling is within the space of social input-output dualities of social existence and its conditions of know-how under the mercy of natural dynamics in the space of matter-energy dualities. The social telescopic history is generated by social neuro-decision-choice dynamics of social forces from the space of matter-energy dualities which together produce qualitative-quantitative motions for transformations in the space of variety property-relation dualities. In this respect, the dynamic set of neuro-decision-choice actions over the spaces of problem-solution and individual-community dualities constitutes the locomotive of social history, where revolutions are tipping points in historical shifts within the space of neuro-decision-choice actions with violent transformations as engineered by information-decision-choice-interactive processes.

Social revolutions are not locomotive of history that one may read from Karl Max's works. Social revolutions are trigged by conflicting conditions in the space of justice-injustice dualities relative to the conditions of the space of cost-benefit dualities translated to the space of social incentive-disincentive dualities as reflected to in the social systems over the space of individual-community dualities, where the conditions of the social distribution in space of incentive-disincentive dualities translate into the conditions over the space of cost-benefit dualities to set the directions on the telescopic path of history to connect the past to the present and the present to the future for continuity and unity. In these discussions, attention must be paid to the concepts and phenomena of the space of input-output dualities and the concept and phenomena of recycling as generating variety fuzzy optimality and maintaining variety equilibria and categorial balances through the dualistic-polar relational dynamics in the space of socio-nature dualities. The concept of recycling must be understood in terms of elements in the space of cost-benefit dualities where variety costs are turned into variety benefits and vice-versa to maintain the matter-energy equilibrium state. The current environmental difficulties in the space of socio-natural dualities may be explained by disequilibria of recycling know-how in the space of socio-natural dualities, where certain byproducts of human socioeconomic production cannot be fully dealt with in the space of input-output dualities and hence generating increasing cost externalities without variety social benefit. For the cost externalities to be accommodated the environment must lose some varieties which are benefits. It is this cost-benefit connectivity to the environment that the method of cost-benefit analysis is extensively used in environmental policy studies. The suggestion of carbon-catcher must be understood in terms of recycling within the space of input-output dualities. It is, however, unknown whether this satisfies the cost-benefit conditions in the space of input-output dualities. The conditions of variety transformations may be linked to the concepts and phenomena of process, technology, and the space of process-technology dualities. The spaces of processes and technologies may be linked to the space of variety transformations which may be linked to the spaces of functors and categorial convertors, all of which are derived from the space of variety imaginations which contains the spaces of possibility and probability, vision, goals, and objectives. Let us define them.

DEFINITION 3.4.1: Process, Technology, and the Space of Process-Technology dualities

Process is a series of interdependent actions providing instructions for the path to either transform or maintain a variety characteristic disposition to affect the variety identity in the space of positive-negative dualities. *Technology* is know-hows from the accumulated information-knowledge stocks containing applicational skills, and methods, to carry on processes that will allow transformation of variety identity in the space of input-output dualities contained in the space of variety potentials, where technology is embodied all *functors* and *categorial convertors*. The collection of all available process-technology varieties is the space of process-technological dualities, where *technique* is a specific process-skill within the variety transformation dynamics and the collection of all techniques in a process constitutes a *variety technology* that is design to act on the variety characteristic disposition to alter or maintain its identity. The collections of all variety technologies at any generation timepoint constitutes the space of generational know-hows of production-consumption dualities contained in the space of input-output dualities. The collection of all the proven and unproven information-knowledge stocks, techniques, skills, and methods of know-hows of production-consumption dualities, beliefs, propaganda, misinformation, disinformation and other such structures at any generational timepoint constitutes the *social philosophical consciencism*, while the space of its dynamics, use and improvements in the organization of intergenerational dynamics is the space of loss-gain dualities as derived from the space of cost-benefit dualities.

NOTE 3.4.1: Definitions of Process, Technology, and the Space of Process-Technology dualities

The system of defined concepts and phenomena in definition 3.4.1 is essential in the study and understanding variety qualitative and quantitative motions as discussed in the theory info-statics that establishes *anticipations, necessity* and *initial conditions* of change by establishing variety identities and corresponding characteristic dispositions [R17.17] and the theory of info-dynamics that establishes *expectations*, *freedom* and *sufficiency* of change by establishing the conditions of the space of process-technology dualities from the space of variety imaginations contained the space of potentials [R17.18]. From the space of imaginations, the possibility measures are developed to provide quantitative values specifying the degrees of anticipation powered by the internal necessity and external initial conditions on the path of transformations, while probability measures are developed to provide quantitative values to specify the degrees of expectations powered by the internal freedom and external sufficient conditions for crossing the dualistic-polar transversality line between duals and poles in the space of opposites, where the crossing may be gradually destructive, called, a change or violently destructive, called a revolution in the space of socio-natural dualities relative to the behaviors of varieties in the space of functors and categorial convertors. The concepts of functor, categorial converter and the corresponding spaces have been defined and explicated in the previous chapters. In the theory of knowing and information-knowledge production, the functor and categorial convertor are neuro-decision-choice dependent, where all neuro-decision-choice actions depend on the social philosophical consciencism to which we provide a workable definition.

DEFINITION 3.4.2: Social Philosophical Consciencism

The social philosophical consciencism is dynamic composite accumulations of variety information-knowledge stocks of all proven and unproven cognitions of individual and collective knowing from the past to the present intra-generationally and intergenerationally developed that are contained in the space of variety acquaintance-description dualities which also contains accepted constitution, cultural norms, illusions, misinformation, disinformation, beliefs, propaganda, religious ideas, laws, rights, and other such social norms in the space of individual-community dualities with enhanced elements about the conditions of the universal existence and its dynamics that guide neuro-decision-choice actions and conceptions of social relations and existence in the space of individual-collective dualities.

NOTE 3.4.2: Social Philosophical Consciencism and Organizational Analytics

It is very important to understand the concept and phenomenon of social philosophical consciencism and its relation to organizational analytics in social setup within generations and between generations. As has been defined and explicated in the theory of philosophical consciencism [R8.16a] with further extensions in the theory of epistemic field [R8.16b], the social philosophical consciencism is indispensable in all social systems and generations as one relates individual and collective behaviors to neuro-decision-choice actions. To understand the preponderating role of social philosophical consciencism in the space of social actions within the space of individual-collective dualities, one must understand socio-production-consumption-actions, fuzzy decomposition, and segmentation of social population at work over the spaces of acquaintance-description dualities, fundamental-applied dualities, cost-benefit dualities, and their distributions over the space of cost-benefit dualities conformally mapped onto the spaces of incentive-disincentive dualities, reward-punishment dualities, and pleasure-pain dualities. There is the segment of knowledge producers in the space of thinking, there is the segment of knowledge users through reasoning, innovations and applications for variety creation and there is the segment of consumers within the space of knowledge-reasoning dualities who simply follow the thinkers and reasoners by using some elements that are easily understood in the social philosophical consciencism and believed to produce distributive advantage over the space of social cost-benefit dualities without questioning their validities in the space of true-false dualities.,

The social philosophical consciencism is social system dependent on, as well as generates *national exceptionalism* over the space of general decision-choice space in such a way that if the contents of social philosophical consciencism emphasize collective violence in social actions, this will also lead to individuals using violence to settle conflicts and disputes in the space of dualities and polarities. In this sense, social conceptions and practices of philosophy and information-knowledge structures organized around imperialism, militarism, slavery and war within a nation create a social philosophical consciencism that contains thought of violence, continual destruction, oppression, racism and exploitation over nonviolence, construction, freedom indirect dictatorship in the space of war-peace dualities that provides conceptual support for the pillars of violent societies, where the individuals seek, following the precepts of the existing social philosophical consciencism, conflict resolutions through the techniques, and methods of imperialism, violence, militarism, oppression, racialism, and exploitation over the space of domestic-international dualities with a continual demonization of nation and races.

In this respect, one must conceptualize and understand the roles that concepts of war on poverty, war on drugs, war on crime, war on dictatorship, war on terror and more wars on other dualistic-polar structures contribute to the mental formation of the population over the social space, where such emphases on the concept of war are appeals and emphases on variety violence characteristic dispositions in the space of violence-nonviolence dualities where variety violence becomes characteristics of social behavior with or without guns. The alternative with emphasis on peace and nonviolence is also true. The development of conditions of racial superiority, slavery, and other, other discriminatory concepts as divine will in the space of propaganda for cost-benefit advantage enshrines in the social philosophical consciencism the elements of racism and exploitation which become internalized by some individuals as legitimate practice in the space of individual-collective duality contained in the social space and space of religious believes and practices. Here, it must be understood that it is within the social philosophical consciencism that *conceptual zombies* arise in the space of practice of thinking and reasoning; cruelty and non-human right practices arise in the space of practices of information and knowledge and racism; and racial demonization become accepted in national exceptionalism.

Similarly, the effectiveness of propaganda, misinformation, disinformation, and other similar structures is realized through the existing social philosophical consciencism at generational points. Intergenerational differences in neuro-decision-choice actions over the spaces of preferences, problem-solution dualities, cost-benefit dualities, incentive-disincentive dualities and many such dualistic-polar conditions are explainable through the dynamics of social philosophical consciencism as self-evolving and self-correcting system over the space of domestic-international dualities. It is within the space of individual-community dualities under the guidance of the ruling social philosophical consciencism that conflicts, injustice, oppression, and brutality arise with the purpose of obtaining individual or collective distributive advantage over the space of cost-benefit dualities and criminalities are produced by unjust laws from the legal structure and enforced by the political structure to distribute cost-benefit advantages in the economic structure. When zombie ideas arise in the social philosophical consciencism, they become internalized and had to go away as they are connected to cost-benefit distribution and hence linger on as zombie ideas that are hard to die away. This is also true over the space of imperialism-colonialism dualities from which the space of imperialism-independence dualities arises to contain neocolonialism, where the ills, atrocities and brutalities of imperialism linger on as zombie ideas that are held on in the analytical space of thought and reason under the *principle of fight-fight* (war) approach, where strength is considered as justice over the space of problem-dualities in relation to death-destruction promotion rather than on the *principle of talk-talk* (diplomacy) approach where right is justice over the space of problem-solution dualities in relation to life-maintenance promotion for development in the global space of relational-information dualities relative to human existence and ecological stability.

The social philosophical consciencism, as defined, forms the information-knowledge foundation of all human actions in the space of input-output dualities and preference formations affecting curiosity, creativity, courage, hope, despair, fear and many such elements over the space of quality-quantity dualities defined over the space of individual-collective dualities, neuro-decision-choice processes and in the conceptions of past-present-future dynamics within the space of beginning-end duality of ontological-epistemological-time dualities. In any nation and social setup, the social philosophical consciencism contains different subjectively acceptable information-knowledge blocks that give rise to dynamic distributions and changing distributions of different socio-political competing groups for social power in direction of control and managements of social systems over the space of nonviolence-violence dualities. The subjectively accepted information-knowledge blocks are constructed by *cognitive filtering* from the contents of the social philosophical consciencism, where similar cognitive filtering define groups of cognitive similarities. Corresponding to the social groups are subcultures of knowledge and practices over the space of conflict-cooperative dualities. The social philosophical consciencism defines foundations for positive-sum games, where everyone wins through a cooperative organizing system; negative-sum games, where everyone loses through an antagonistic organizing system; and zero-sum games, where somebody gains, and somebody loses through exploitative organizing system. In this respect, we have the problem of explaining and understanding the concept and phenomenon of cognitive filtering from the social philosophical consciencism and group formations which requires intensive research and the development of the *theory of cognitive filtering* from the social philosophical consciencism and group formation within a social set-up.

The social philosophical consciencism is a powerful analytical conceptual framework to understand the outcomes in social systems, where such outcomes involve input-output and self-correction processes defined over dualistic-polar spaces of destruction-construction, ignorance-knowledge, cooperative-noncooperative, good-evil, violence-nonviolence dualities encompassing other dualities in social systems. In the social philosophical consciencism, are the seeds of destruction, conflicts, violence, construction, visions, fear, propaganda, God, country, and other elements in the space of cost-benefit dualities. The powerful role of social philosophical consciencism must be clearly understood relative to the development of national and individual personalities with a particular direction within the general space of problem-solution dualities. When a child is born, he or she has a brain containing the mind as luggage. The luggage is then continually filled by the individual with filtered elements from the contents of the social philosophical consciencism without the knowing of other decision-choice agents. The act of filling the mental luggage is done by *cognitive filtering algorithms* as guides to choose elements from the social philosophical consciencism and its development in support of personality developments over the spaces of neuro-decision-choice activities and human actions. The explanations of the process come under the *theory of cognitive filtering* in support of personalities.

Let us now connect the *possibility space* containing *anticipation*, *necessity* and *initial conditions*, and *probability space* containing *expectation*, *freedom* and *sufficient conditions* that are generated from the space of imaginations contained in the space of potentials. Let us keep in mind that anticipation precedes expectation and hence we speak of expectation of anticipation and there is no expectation without anticipation. Similarly, we speak of possible world, impossible world, possible-world-impossible-world dualities, probable world, improbable world, probable-world-improbable-world dualities, and possible-probable dualities and how they relate to the space of imaginations to project the space of anticipation-expectation dualities. Let us provide the linkage through a system of definitions and explications.

DEFINITION 3.4.3A: Imagination and the Space of Imaginations (Alternative Verbal Disposition)

A variety from the space of varieties is said to be an *imagination* if its variety characteristic disposition is formed in terms of an abstract thinking power to construct a variety cognitive image from the potential space contained in the space of actual-potential dualities that exists as an identity space in the cognitive space of the imagining agent, where the variety imagination is equipped with a possibility-impossibility dualities containing variety necessity, initial conditions and corresponding *anticipation functors*, and probability-improbability dualities containing freedom, sufficient conditions with corresponding *expectation functors* and while the variety imagination is unknown to exist in the space of actuals in terms of perceived reality and not encountered in the space of acquaintance, and where the variety imagination with defined characteristic disposition resides only in the mind of the imagining agents. The variety characteristic disposition may be shared over the communication space of source-destination dualities as a signal disposition with functor or categorial characteristic converter, which may or may not be brought to the space of current realities or future realities, and where the space of realities is a sub-space of the space of actuals within the space of actual-potential dualities. The collection of all variety imaginations constitutes the *space of imagination* which houses the spaces of possibility-impossibility dualities from which the possibility and impossibility spaces are abstracted, the probability-improbability dualities from which the probability and improbability spaces are abstracted and possible-world-impossible-world dualities from which the spaces of possible world and impossible world may be abstracted and probable-world-improbable-world dualities formed from the spaces probable and improbable worlds are abstracted with dualistic-polar conditions and restricted by fuzzy conditionality through fuzzy selections with fuzzy indicator functions.

DEFINITION:3.4.3B: Imagination and the Space of Imaginations (Alternative Categorial Characteristic Disposition)

The variety $(\nu \in \mathbb{V})$ is a variety imagination if it may be expressed as $\left(\nu_{\mathfrak{g}} = (\nu, \mathbf{f}, \mathbf{e}, \mathbf{c}, \mathbf{a}) \in (\mathbb{V} \otimes \mathbf{F}_{\mathfrak{G}} \otimes \mathbf{E}_{\mathfrak{G}} \otimes \mathbf{C}_{\mathfrak{G}} \otimes \mathbf{A}_{\mathfrak{G}}) \mid \nu \in \mathbb{V} \subseteq (\mathfrak{U} \cup \mathfrak{A}), \mathfrak{g} \in \mathfrak{G} \subset \mathfrak{U}, \mathbf{f} \in \mathbf{F}_{\mathfrak{G}}, \mathbf{e} \in \mathbf{E}_{\mathfrak{G}}, \mathbf{c} \in \mathbf{C}_{\mathfrak{G}}, \mathbf{a} \in \mathbf{A}_{\mathfrak{G}}\right)$. This representation is supported by information $\left\{\mathbb{Z}_{\nu_{\mathfrak{g}}} = \mathbb{Z}_{(\nu, \mathbf{f}, \mathbf{e}, \mathbf{a})} \subset \mathbb{Z}_{(\mathbb{V} \otimes \mathbf{F}_{\mathfrak{G}} \otimes \mathbf{E}_{\mathfrak{G}} \otimes \mathbf{A}_{\mathfrak{G}})} \mid \mathbb{Z}_{\mathbb{V}} \subseteq \mathbb{Z}_{(\mathfrak{U} \cup \mathfrak{A})}, \mathbb{Z}_{\mathfrak{G}} \subset \mathbb{Z}_{\mathfrak{U}}\right\}$, where every information is defined as characteristic-signal disposition in the form $(\mathbb{Z} = \mathbb{X} \otimes \mathbb{S})$, where $(\mathbb{X})$ is a set of characteristic dispositions and $(\mathbb{S})$ is a set of signal dispositions. The varieties in the space of imaginations are epistemologically derived identities from the space of potentials $(\mathfrak{U})$ which are ontological identities in the sense that $(\mathfrak{G} \subset \mathfrak{U} \subset (\mathfrak{U} \cup \mathfrak{A}) = \Omega)$ such that $\mathfrak{G} = (\mathfrak{G} \cap \mathfrak{U}) \neq \varnothing$ where $(\omega \in \Omega)$ is the universal object space and $(\nu \in \mathbb{V})$ is the space of all varieties such that $(\mathbb{Q} = (\mathfrak{H} \otimes \mathfrak{G}) = (\mathfrak{H} \cup \mathfrak{G}) \subset ((\mathfrak{U} \cup \mathfrak{A}) = \Omega))$ is the space of imagination-reality dualities, where the set of the characteristics of every variety reality has some characteristics of variety imagination and similarly, the set of characteristics of every variety imagination has some characteristics of variety reality with dualistic-polar information conditions of the form $\mathbb{Z}_{\mathbb{Q}} = (\mathbb{Z}_{\mathfrak{H}} \cup \mathbb{Z}_{\mathfrak{G}}) = (\mathbb{Z}_{\mathfrak{H}} \otimes \mathbb{Z}_{\mathfrak{G}})$ such that $(\mathbb{Z}_{\mathfrak{H}} \cap \mathbb{Z}_{\mathfrak{G}}) \neq \varnothing$ for the fuzzy conditionality under the principle of opposites. The collection of all the variety imaginations constitutes the space of imaginations, $(\mathfrak{G})$ equipped with a functor space $(\mathbf{F}_{\mathfrak{G}})$ (space of categorial convertors $(\mathfrak{C}_{\mathfrak{G}}, \mathbf{F}_{\mathfrak{G}})$, the space of anticipation operators $(\mathbf{A}_{\mathfrak{G}})$ and the space of expectation operators $(\mathbf{E}_{\mathfrak{G}})$.

DEFINITION 3.4.4A: Reality and the Space of Realities (Alternative verbal Disposition)

A variety is a reality, the characteristic disposition of which has been abstracted by cognition from an ontological variety signal dispositions at the space of acquaintance to identify its identity, in difference and similarity, that is claimed to exist in the general ontological space with social agreement regarding approvals for intellectual acceptance of the abstracted characteristic disposition as information-knowledge of varieties in the space of actual-potential dualities, where every variety reality is a social construct from the space of actual-potential dualities through the knowing-transformation processes that has also pass through the space of anticipation-expectation dualities relative to the space of necessity-freedom dualities. The collection of all such variety realities constitutes the space of realities.

DEFINITION 3.4.4B: Reality and the Space of Realities (Alternative Categorial Characteristic Disposition)

The variety $(\nu \in \mathbb{V})$ is a variety reality if it may be expressed as $\left(\nu_{\mathfrak{h}} = (\nu, \mathbf{f}, \mathbf{e}, \mathbf{c}, \mathbf{a}) \in (\mathbb{V} \otimes \mathbf{F}_{\mathfrak{H}} \otimes \mathbf{E}_{\mathfrak{H}} \otimes \mathbf{C}_{\mathfrak{H}}, \mathbf{A}_{\mathfrak{H}}) \mid \nu \in \mathbb{V} \subseteq (\mathfrak{U} \cup \mathfrak{A}), \mathfrak{h} \in \mathfrak{H} \subset \mathfrak{A}, \mathbf{f} \in \mathbf{F}_{\mathfrak{H}}, \mathbf{e} \in \mathbf{E}_{\mathfrak{H}}, \mathbf{c} \in \mathbf{C}_{\mathfrak{H}} \mathbf{a} \in \mathbf{A}_{\mathfrak{H}}\right)$,and this representation is supported by information structure of the form

$\left\{\mathbb{Z}_{\nu_{\mathfrak{h}}}=\mathbb{Z}_{(\nu,\mathbf{f},\mathbf{e},\mathbf{c},\mathbf{a})}\subset\mathbb{Z}_{(\mathbb{V}\otimes\mathbf{F}_{\mathfrak{H}}\otimes\mathbf{E}_{\mathfrak{H}},\mathbf{C}_{\mathfrak{H}},\mathbf{A}_{\mathfrak{H}})}\mid\mathbb{Z}_{\mathbb{V}}\subseteq\mathbb{Z}_{(\mathfrak{U}\cup\mathfrak{A})},\mathbb{Z}_{\mathfrak{H}}\subset\mathbb{Z}_{\mathfrak{A}}\right\}$, where every information is defined as characteristic-signal disposition in the form $(\mathbb{Z}=\mathbb{X}\otimes\mathbb{S})$, where $(\mathbb{X})$ is the set of characteristic dispositions and $(\mathbb{S})$ is the set of signal dispositions such that $\mathbb{X}_{\nu_{\mathfrak{h}}}=\mathfrak{I}(\mathbb{S}_{\nu_{a}})$ or $\mathbb{X}_{\nu_{\mathfrak{h}}}=\mathfrak{I}(\mathbb{S}_{\nu_{g}})$ is an abstraction from a variety signal disposition of the space of actual in terms of knowing $(\mathbb{S}_{\nu_{a}})$ or an abstraction from a variety signal disposition of the space of imaginations in terms of transformation $(\mathbb{S}_{\nu_{a}})$, where $\mathfrak{I}(\bullet)=(\mathbf{F}_{\mathfrak{A}}(\bullet)\otimes\mathbf{E}_{\mathfrak{A}}(\bullet))$ is a *signal decipher function* from the space of actuals through the space of acquaintance or $\mathfrak{I}(\bullet)=(\mathbf{F}_{\mathfrak{U}}(\bullet)\otimes\mathbf{E}_{\mathfrak{U}}(\bullet))$ from the space of the potentials through the space of imaginations . The varieties in the space of imaginations are epistemologically derived identities from the space of potentials $(\mathfrak{U})$ which are ontological identities in the sense that $(\mathfrak{G}\subset\mathfrak{U}\subset(\mathfrak{U}\cup\mathfrak{A})=\Omega)$ such that $\mathfrak{G}=(\mathfrak{G}\cap\mathfrak{U})\neq\varnothing\mathfrak{A}$ where $(\omega\in\Omega)$ is the universal object space and $(\nu\in\mathbb{V})$ is the space of all varieties such that $(\mathbb{Q}=(\mathfrak{H}\otimes\mathfrak{G})=(\mathfrak{H}\cup\mathfrak{G})\subset((\mathfrak{U}\cup\mathfrak{A})=\Omega))$ is the space of imagination-reality dualities. The reality is said to be false if $\mathbb{Z}_{\nu_{\mathfrak{h}}}=\mathbb{Z}_{(\nu,\mathbf{f},\mathbf{e},\mathbf{c},\mathbf{a})}\subset\mathbb{Z}_{\mathfrak{H}}\not\subset\mathbb{Z}_{\mathfrak{A}}$ but $\mathbb{Z}_{\mathfrak{H}}\subset\mathbb{Z}_{\mathfrak{U}}$ and $\mathbb{Z}_{\mathbb{V}}\subset\mathbb{Z}_{(\mathfrak{U})}$ otherwise the variety reality is said to be justified, where the justification finds expression with the guidance in the space of true-false dualities with entropic conditionality.

DEFINITION 3.4.4C: Space of Imagination-Reality Dualities (Alternative Categorial Characteristic Disposition)

The space of imagination $(\mathfrak{G})$ with a generic element $(\mathfrak{g}\in\mathfrak{G})$ is a small collection of varieties, constrained by cognitive capacity limitations, from the infinite space of potentials $(\mathfrak{U})$

with generic element $(\mathfrak{u} \in \mathfrak{U})$ and $(\mathfrak{g} \in \mathfrak{G} \Subset \mathfrak{U})$ as well as containing the spaces of possibility-impossibility dualities $((\mathfrak{P} \cup \mathfrak{M}) \subset \mathfrak{G})$, probability-improbability dualities $((\mathfrak{B} \cup \mathfrak{H}) \subset \mathfrak{G})$, possibility-probability dualities $((\mathfrak{P} \cup \mathfrak{B}) \subset \mathfrak{G})$ and possible-world-impossible-world dualities $(((\mathfrak{W} \cup \mathfrak{V}) \subset \mathfrak{G} \Subset \mathfrak{U} \subset \Omega))$, where the set of the imagination characteristic dispositions $\left(\mathbb{X}_{\mathfrak{G}} \subset \mathbb{X}_{\mathfrak{U}} \subset \mathbb{X}_{(\mathfrak{U} \cup \mathfrak{A})}\right)$ is contained in the space of potential $(\mathfrak{U})$ as the imagination dual, the characteristic dispositions of which, dominate the set of reality characteristic dispositions $\left(\mathbb{X}_{\mathfrak{H}} \subset \mathbb{X}_{\mathfrak{A}} \subset \mathbb{X}_{(\mathfrak{U} \cup \mathfrak{A})}\right)$ contained in the space of the actuals $(\mathfrak{A})$ as the reality dual, and where every variety imagination $(\mathfrak{g} \in \mathfrak{G})$ in the space of possible-world-impossible-world dualities $((\mathfrak{W} \cup \mathfrak{V}) = (\mathfrak{W} \otimes \mathfrak{V}) \subset \mathfrak{G})$ resides as a variety imagination-reality dualities $\mathbb{Q} = (\mathfrak{H} \otimes \mathfrak{G}) = (\mathfrak{H} \cup \mathfrak{G})$ with dualistic-bipolar conditions in the space of imagination-reality dualities. Similarly, every variety reality in the space of actual-potential dualities $(\mathfrak{U} \otimes \mathfrak{A})$ resides as a variety imagination-reality duality with dualistic-bipolar condition in the space of imagination-reality dualities $\mathbb{Q} = (\mathfrak{H} \otimes \mathfrak{G}) = (\mathfrak{H} \cup \mathfrak{G})$ with fuzzy conditionality $(\mathfrak{H} \cap \mathfrak{G}) \neq \varnothing$ where $(\mathfrak{H} \otimes \mathfrak{G})$ represents dualistic-polar structure, $(\mathfrak{H} \cup \mathfrak{G})$ represents relational unity and $(\mathfrak{H} \cap \mathfrak{G}) \neq \varnothing$ represents relational dualistic-polar continuum. A variety imagination (reality) is one with a dominating dual characteristic disposition over the dual characteristic disposition of reality (imagination) in a duality as dualistic-bipolar condition and fuzzy relationality in a simultaneity of existence of opposites with a relational continuum, give-and-take sharing mode in unity within the same variety.

NOTE 3.4.4: Imagination, Reality and the Corresponding Spaces and Alternative Definitions

As a derived element from the space of the actual-potential dualities, a variety reality is information-knowledge supported and justified by cognitive agents from the space of acquaintance-description dualities that may or may not have successfully passed through the spaces of possibility-probability dualities, necessity-freedom dualities, doubt-surety dualities, certainty-uncertainty dualities and other such elements in the space of true-false dualities, where the space of realities is always in a disequilibrium process following the information-knowledge disequilibrium dynamics. In the knowing process, there are true reality and false reality. True reality is information-knowledge justified as an epistemic in the space of actuals with ontic characteristic dispositions, where the internal characteristic disposition (dual) dominates the internal imagination characteristic disposition (dual) in the imagination-reality dualities. False reality is either intentionally or unintentionally dis-information-knowledge justified as an epistemic variety in the space of actuals with no ontic characteristic disposition support, where the reality internal characteristic disposition (dual) does not exist but manipulated through disinformation-misinformation processes to dominate the internal imagination characteristic disposition (dual) in the imagination-reality dualities.

The twin concepts of imagination and reality are extremely important in theory of knowing and the practice of knowing as an intellectual-investment process and information-knowledge production as an intellectual capital accumulation process over the space of fundamental-applied information-knowledge dualities as well as decision-choice actions over the spaces of production-consumption dualities, problem-solution dualities and question-answer dualities, where variety imaginations are abstracted to feed into the production of variety realities and variety realities nourish the production of variety imaginations in the space of imagination-reality dualities contained in the space of actual-potential dualities. The dynamic processes of the space of imagination-reality dualities are under the power of *epistemic creativity* affecting the nature and discovery of variety functors and categorial convertors that form bridges between the space of variety imaginations and the space of variety realities. All human actions over the space of production-consumption dualities , all human conceptions with visions, goals, objectives, designs propaganda, deceptions relative to the space of progress-retrogress dualities, joy-sorrow dualities, happiness-sadness dualities, all human relations of war-peace dualities, love-hate dualities, friend-enemy dualities and all human creations and transformations are abstract images formed within the space of imaginations relative to the space of visible-invisible dualities and then transmitted by conformal mapping onto the space of realities over the space of risk-benefit dualities, where some of the varieties may be illusions and deceptions to effect the desired outcomes over the space of cost-benefit dualities. All ontological characteristic dispositions are ontic characteristic dispositions with information-knowledge equality perfection and certainty, while all epistemological characteristic dispositions are epistemic characteristic dispositions with information-knowledge inequality imperfection and uncertainty.

The ontological categories with ontological characteristics reside in ontic spaces while the epistemological categories with epistemological characteristics reside in epistemic spaces. The comparative analyses are among key building blocks of ontological-epistemological duality, ontic-epistemic dualities, actual-potential dualities, imagination-reality dualities in support of life, human existence. The conditions of ontic-epistemic dualities give rise to imagination-reality dualities as abstracted from the space of actual-potential dualities .The conflicts in the space of ontic-epistemic dualities give rise to the conditions of the space of imaginations abstracted from the space of potentials in response to the conditions of certainty-uncertainty dualities and linked to the space of reality through the spaces of possibility-impossibility and probability-improbability dualities equipped with respective spaces of functors or categorial converters that implicitly contain creativity, curiosity, courage, fear and other such qualitative structures which find residence in the space of imaginations.

To examine the relational complexities among imagination, reality, creativity, curiosity problem-solution dualities, question-answer dualities, information-knowledge production and social change, let $\mathbb{Q} = (\mathfrak{G} \otimes \mathfrak{H}) = (\mathfrak{G} \cup \mathfrak{H})$ defines the space of imagination-reality dualities with characteristic-signal condition $\mathbb{Z}_{\mathbb{Q}} = \mathbb{Z}_{(\mathfrak{G}\otimes\mathfrak{H})} = \left(\mathbb{X}_{\mathfrak{G}\otimes\mathfrak{H}} \otimes \mathbb{S}_{(\mathfrak{G}\otimes\mathfrak{H})}\right)$ as information support under dualistic-bipolar conditions with a generic element $(\varphi \in \mathbb{Q})$ that allows the definitions of the spaces of imaginations and realities to follow from conditions of dualistic-bipolar dominance with relational continuum and unity such that $(\mathfrak{G} \cap \mathfrak{H}) \neq \emptyset$ with the corresponding characteristic dispositions. Variety Imagination is also connected to the goals, objectives, and visions in the space of individual-collective dualities, where there are *individual visions* and *collective visions* that may reside in the space of conflict-cooperation dualities in relation to the elements in universal space of goals objectives and visions. At the space of goals, objectives and visions in the knowing process, there are the *individual imaginations* in relation to individual *visions , goals* and *objectives* and there are the social or collective imaginations in relation to social or *collective visions*, *goals* and *objectives* as constrained by social philosophical consciencism that has taken hold of the members of the social set-up at various generational time points, where individual and collective creativities and curiosities play important and interdependent roles in knowing as the production of intellectual investment flows and information-knowledge accumulation as the production of intellectual capital stocks within the space of fundamental-applied dualities relative to the spaces of variety production-consumption dualities and variety input-output dualities in the space of identification-transformation dualities in dualistic-polar relations of past-present-future telescopic conditions of the Sankofa-anoma information-time trinity.

In the epistemic processes of knowing, information-knowledge production, input-out processes and the dynamics of problem-solution dualities and polarities over the space of fundamental-applied dualities, *creativity* and *innovation* play extremely important roles supported by curiosity and audacity in connecting the space of imaginations to the space of realities through the spaces of possibility-probability and necessity-freedom dualities and to the space of potentials, and indirectly to connect the space of the potentials to the space of imaginations to possibility space, and possibility space to probability space over the epistemological space. These interrelated conditions are characteristics of social behavior generated by cognitive capacity limitations to effect variety uncertainties. The question is: what is creativity and what role does it play over the epistemological space in spaces of knowing, information-knowledge production, problem-solution dynamics, question-answer dynamics, and variety transformations? This question will be answered with a definition, an explication and an analytical exposition that will establish individual and collective creativity, over the space of individual-community dualities and its relationships to knowing, information-knowledge accumulation in the space of diversity-unity dualities for individual progress within nation-building, social progress, and management of social dynamics on the path of the historical process under the Sankofa-anoma traditions. The knowing is always in reference to the production of intellectual investment flows, while the information-knowledge accumulation is always in reference to the production of intellectual capital stocks over the space of fundamental-applied dualities in relational feedback system to maintain the information-knowledge system in disequilibrium dynamics as well as variety transformation disequilibrium dynamics, where the space of fundamental stock-flow information-knowledge system relates to logical foundations and thinking algorithms for transformative actions and the applied stock-flow information-knowledge system relates to logical foundations for transforming the thinking algorithms to reasoning for the applications of the fundamental information-knowledge stock-flow system to variety transformations over the space of input-output dualities.

CHAPTER 4

ONTICITY, EPISTEMICITY, CREATIVITY, CURIOSITY AND HOPE-DISPAIR DUALITY

ABSTRACT

Chapter 4 of this monograph discusses the roles of creativity and curiosity in knowing as the production of intellectual investment flows and information-knowledge accumulation as production of intellectual capital stocks in the in the space of problem-solution dualities and polarities within the space of imagination-reality dualities as contained in the space of actual-potential dualities relative to the space of ignorance-knowledge dualities within the space of ontic-epistemic dualities relative to the space of ontological-epistemological duality. The understanding of the roles of creativity and curiosity are related to the spaces of courage-fear dualities, hope-despair dualities, and success-failure dualities. Definitions of these concepts are presented, and their contents explicated with corresponding operator functions. Comparative analytics are undertaken to show their interdependencies and their relative role and degrees of efficiencies in the space of ignorance-knowledge dualities and problem-solution dualities showing the strengths and limits of organization in neuro-decision-choice actions over the space of human behaviors under the principle of opposites. Through comparative organizational analytics in varieties and categories, we show the relational connectivity in the space of ontological-epistemological dualities in knowing, differences, similarities ontological information-time processes and epistemological information-time processes in reference to the ontological beginning-end duality in relation to *Bennu-Anoma-Gyenyame* conditions and epistemological past-present duality of Sankofa conditions in the space of ontological and epistemological dualities where the essentiality

K. K. Dompere, *Theories of Knowing and Knowledge Production*, Studies in Systems, Decision and Control 501, https://doi.org/10.1007/978-3-031-42863-0_4

of symmetry and supersymmetry are discussed under the principle of opposites relative to the space of relativity-absoluteness dualities with illustration by epistemic geometric representation with the corresponding mathematical spaces in terms axiological structures of varieties in the space of real-nominal dualities relative to the space of cost-benefit dualities.

4.1 SIMPLE REFLECTIONS ON THE IMMAGINATION-REALITY PROCESS

In chapter three, we discussed some aspects of imagination and its connectivity to potential as well as the space of imagination and the space of the potential. We also discussed some aspects of the space of reality and its connectivity to the space of actuals as well as the connectivity of the space of reality to the space of imagination. The actual-potential dualities are defined in the ontological space, where the characteristics of natural varieties find expressions in the *variety ontic space* from which all the natural variety transformations are induced in an infinite beginning-end time-information conditions, where there is no past, present and future under the laws of nature the past is in the present, the present which in the future where the past-present-future varieties reside in the beginning-end varieties in continual processes over the ontological identification-transformation dualities. All transformations are about changes in natural variety properties and relations defined in characteristic dispositions over the space of natural identification-transformation dualities, where the natural dynamics are constrained by information-knowledge equality conditions in continual maintenance of universal permanency as variety properties and relations alter to also alter the ontological environment as may be defined.

Similarly, the imagination-reality dualities are defined in the epistemological space, where the characteristic dispositions of social varieties find expressions in the *epistemic space* from which all the social variety transformations including from ignorance to knowledge are induced in an infinite past-present-future time-information conditions under the laws of social cost-benefit process, where all social transformations are about changes in social-natural variety properties and relations over the space of social identification-transformation dualities, where all social dynamics are constrained by information-knowledge inequality conditions defining the space of certainty-uncertainty dualities in the space of categorial convertors and functor space. The conditions of imagination-reality dualities give rise in the social space to the development of conditions of possibility-impossibility dualities, possible-impossible-world dualities and probability-improbability duality, probable-improbable-world dualities, possibility-probability dualities, necessity-freedom-dualities, to deal with the conditions in the space of certainty-uncertainty dualities generated by the conditions of cognitive capacity limitations, which are linked to the social space of goals, objectives and visions over certainty-uncertainty dualities, real cost-benefit dualities in the space of incentive-disincentive dualities. The input-output behaviors of social varieties depend on neuro-decision choice actions over all areas of cognitive activities, the input-output are influenced by several cognitive characteristics of many other elements of which creativity, curiosity, hope, failure, success, and other important cognitive qualities are preponderating factors to which we tend our attention.

4.2 CREATIVITY, CURIOSITY AND THE SPACE OF IMAGINATION-RELITY DUALITIES

It has been stated that creativity supported by curiosity plays an important connection between variety imaginations and variety realities. Analytically, we may speak of the *creativity connector* as a *creativity function* and the supporting curiosity as *curiosity function*. The role of creativity function $\left(\varsigma(\cdot)\in\mathscr{G}\right)$ as a cognitive enhancer of categorial convertor $\left(\mathbf{c}\in\mathbf{C}\right)$ or functor $\left(\mathfrak{f}\in\mathbf{F}\right)$ with supporting curiosity function $\left(\vartheta(\cdot)\in\mathscr{K}\right)$ as a booster within the neuro-decision-choice module or as a natural cognitive machine of decision-choice agents is to enhance the actualization of varieties in the space of imagination $(\mathfrak{G})$ through external technological processes in order to bring them into the space of reality $(\mathfrak{H})$. Importantly, the creative function and the curiosity function may be combined to construct a hybrid function of the form $\xi(\cdot)=\left(\vartheta(\cdot)\varsigma(\cdot)\right)\in\mathscr{Y}=\left(\mathscr{G}\otimes\mathscr{K}\right)\left(\mathscr{G}\cup\mathscr{K}\right)$ where $\xi(\cdot)$ is the hybrid function on the basis of which *hybrid index* as a logical construct of creativity index and curiosity index may be developed for the set of degrees of successes in the space of ignorance-knowledge dualities, problem-solution dualities in relation to intellectual investment flows and intellectual capital stocks over the space of risk-benefit dualities. The symbols $\mathscr{G},\mathscr{K}$ and $\mathscr{Y}$ represent sets of *creativity functions, curiosity functions and hybrid of creativity and curiosity functions* respectively from which creativity index $\left(\mathbf{g}\in\mathbf{G}\right)$, where $\left(\mathbf{G}\right)$ is a set of *creativity indexes*, curiosity index $\left(\mathbf{k}\in\mathbf{K}\right)$ where $\left(\mathbf{K}\right)$ is a set of *curiosity indexes* and hybrid index of analytically combined functions of creativity as enhancer and curiosity as a booster $\left(\mathbf{h}\in\mathbf{H}\right)$, where $\left(\mathbf{H}\right)$ is a set of *hybrid indexes of curiosity and creativity* all of which are defined over the space of individual-collective dualities under the social philosophical consciencism that provides a guidance to knowing-knowledge and neuro-decision-choice frameworks over the space of general production-consumption dualities interpreted as input-output dualities, destruction-construction dualities and parent-offspring dualities to understand socio-natural variety dynamics.

The space of reality is a sub-space of the space of the actuals $(\mathfrak{H} \subset \mathfrak{A})$ under the epistemological variety identification-transformation conditions, where the space of imagination-reality dualities is an epistemological sub-space of the space of ontological actual-potential dualities, such that, $\mathbb{Q} = (\mathfrak{G} \cup \mathfrak{H}) \subset ((\mathfrak{U} \cup \mathfrak{A}) = \Omega)$. All the spaces, varieties and categories in dualistic-polar conditions are linked together by information-knowledge structures as defined by their characteristic-signal dispositions in the form $\left(\mathbb{Z}_{(\cdot)} = \left(\mathbb{X}_{(\cdot)} \otimes \mathbb{S}_{(\cdot)}\right)\right)$ to establish their identities, similarities and distinctions over the space of varieties $(\mathbb{V})$, where the subscripts identify the corresponding phenomenon of interest. The symbols $\mathbb{X}_{(\cdot)}$ is the set of characteristics and $\mathbb{S}_{(\cdot)}$ is the corresponding set of signals over the space of ontic-epistemic dualities, where ontic relates to the ontological varieties and epistemic relates to epistemological varieties.

Between the space of imagination $(\mathfrak{G})$ and the space of reality $(\mathfrak{H})$ are the spaces of possibility $(\mathfrak{P})$ probability $(\mathfrak{B})$ in dualistic-polar conditions for the test of degrees of neuro-decision-choice efficiencies and quality control over the space of necessity-freedom dualities relative to the space of uncertainty-certainty dualities abstracted from the space of ontic-epistemic dualities. The spaces of possible world, impossible world and possible-world-impossible-world dualities are sub-spaces of the space of imaginations, where the space of realities is a collection of neuro-decision-choice outcomes from the space of possibility-probability dualities which gives rise to the space of anomalies. The spaces of fantasies, illusions, alternative realities, fake news, delusions, dreams and many such visually unidentifiable structures are contained in the space of possible-world-impossible-world dualities as derived from the space of imaginations contained in the space of potentials and provide analytical spaces for cognitive conditions of virtual reality, augmented reality, perceptive reality giving rise to distorted information-knowledge inputs into the space of neuro-decision-choice actions which may lead to unintended and unwanted outcomes. We shall now turn to the concept and phenomenon of creativity.

4.2.1: Creativity in Knowing and Science over the Spaces of Ignorance-Knowledge and Problem-Solution Dualities

The concept and phenomenon of creativity in cognitive agents are very complex that affect the behaviors of cognitive agents in all aspects of existence relevant to the principle of non-satiation over the space of actions contained in the space of problem-solution dualities and ignorance-knowledge dualities with conformal mapping onto the space of cost-benefit dualities to minimize cost dispositions and maximize benefit dispositions over the spaces of success-failure and question-answer dualities. Irrespective of how one sees cognitive intents over the space of imagination-reality dualities, the objectives of the phenomenon of creativity are balancing actions through thinking and reasoning between duals of dualities and poles of polarities such as between the problem dual and poles, and the solution dual and poles, setting one against the other, where the results of such actions are translated onto the spaces of cost and benefit dispositions. In this respect, what is the structure, meaning and the phenomenon of creativity in space epistemological characteristic dispositions and how does creativity relate to ontological variety characteristic dispositions in neural systems?

In the space of ontological-epistemological dualities where neuro-decision-choice agents operate over the space of identification dualities for knowing as the production of intellectual investment flows and information-knowledge accumulation as the production of intellectual capital stocks over the space of problem-solution dualities, creativity is an indispensable qualitative asset over the space of thinking-reasoning dualities in the game of life and reproduction of real life. Creativity is essential in the space of theory-applied dualities where creativity must be enhanced by courage and curiosity. A question arises as to how creativity and curiosity enter the spaces of human actions, knowing, technology over the space of imagination-reality dualities. The first step to this question is to provide definitions as to what creativity and curiosity are and relate them to human behavior. Two definitions will be given in terms of verbal and categorial disposition.

DEFINITION 4.2.1.1 A: CREATIVITY (Verbal)

Creativity is a variety disposition with cognitive enhancing power over the space of thinking-reasoning dualities in the neural system that acts on social functors or categorial convertors of transformation actions on varieties in the space of imagination-reality dualities and transform them to new varieties with new compositions of characteristic dispositions in the space of reality or from the space of imaginations to the space of realities or from the space of realities to the space of potentials over the space of quality-quantity dualities in the sense that creativity is an implicit cognitive technological disposition (a system of cognitive categorial enhancers of categorial convertors or functors) that enhances transformation actions of varieties in the spaces of imagination-reality dualities, problem-solution dualities and ignorance-knowledge dualities under dualistic-bipolar dynamics within the spaces of input-output and production-consumption dualities. When creativity involves the improvement or alterations of the characteristics of varieties of existing functors or categorial convertors in the space of realities, it is then called *innovation*. When creativity involves in the discovery of new functors or categorial convertor in the space problem-solution dualities and ignorance-knowledge dualities, it is called discovery. The collection of all different powers of creative dispositions from different cognitive transformations constitutes the *spaces of creativity* and categorial-conversion enhancers with inter-supportive system of generational dynamics of social philosophical consciencism which opens the doors of houses that imprison the creative ideas, curiosity, audacity, variety possibilities and visions in relation to elements in the space of imagination-reality dualities that may facilitate or hinder individual and social advancements over the spaces of production-consumption dualities, input-output dualities, problem-solution dualities and parent-offspring (successor-predecessor) dualities. The transformation function that connects creativity power dispositions to variety characteristic dispositions is called *creativity function*.

DEFINITION 4.2.1.1 B: Creativity and Creativity Function (Categorial Characteristic)

Let the space of creativity functions be $(\mathscr{G})$ with a generic element $(\varsigma \in \mathscr{G})$ as a creative categorial convertor-enhancement-operator function $(\varsigma(\bullet) \in \mathscr{G})$ with supporting-booster curiosity function $(\vartheta(\bullet) \in \mathscr{K})$ that acts on varieties in the space of variety imaginations $(\nu_{\mathfrak{g}} \in \mathfrak{G})$ transforming them as variety actuals into the space of reality $(\mathfrak{h} \in \mathfrak{H} \Subset \mathfrak{A})$ with a generic element

$(\mathfrak{h} \in \mathfrak{H})$ and a corresponding variety $(\nu_{\mathfrak{h}} \in \mathbb{V})$ and space of imagination $(\mathfrak{G})$ with a generic element $(\mathfrak{g} \in \mathfrak{G} \Subset \mathfrak{U})$ and corresponding variety $(\nu_{\mathfrak{g}} \in \mathbb{V})$. The variety $(\nu_{\varsigma} \in \mathbb{V})$ is said to be a creative transformation from the space of imaginations to the space of realities if $(\nu_{\mathfrak{g}} \in \mathbb{V}, \mathfrak{g} \in \mathfrak{G} \subset (\mathfrak{W} \cup \mathfrak{V}) \subset (\mathfrak{U} \cup \mathfrak{A}) \ni \nu_{\varsigma} = \nu_{\mathrm{a}} = \varsigma(\nu_{\mathfrak{g}}))$. Similarly, the variety $(\nu_{\varsigma} \in \mathbb{V})$ is said to be an innovation within the space of realities if $(\nu_{\mathfrak{h}_1} \in \mathbb{V}, \mathfrak{h}_1 \in \mathfrak{H} \subset (\mathfrak{G} \otimes \mathfrak{H}) \subset (\mathfrak{U} \cup \mathfrak{A}) \ni \nu_{\mathfrak{h}_2} = \nu_{\mathrm{a}_2} = \varsigma(\nu_{\mathfrak{h}_1}))$, where all varieties in the spaces of realities $(\mathfrak{H})$ and imaginations $(\mathfrak{G})$ are epistemologically determined, while the varieties in the space of actuals $(\mathfrak{A})$ and the space of potentials $(\mathfrak{U})$ are ontologically determined in the sense that $(\mathfrak{H} \subset \mathfrak{A} \subset (\mathfrak{U} \cup \mathfrak{A}) = \Omega)$ and $(\mathfrak{G} \subset \mathfrak{U} \subset (\mathfrak{U} \cup \mathfrak{A}) = \Omega)$, where $(\omega \in \Omega)$ is the universal object space and $(\nu \in \mathbb{V})$ is the space of all varieties such that $(\mathbb{Q} = (\mathfrak{G} \cup \mathfrak{H}) \subset (\mathfrak{U} \cup \mathfrak{A}) \subseteq \Omega)$ is the space of imagination-reality dualities contained in the space of actual-potential dualities as complete universal existence under mater-energy stock-flow equilibrium dynamics ,where every variety in knowing has a reality dual with a supporting imagination dual and with an information condition of $\mathbb{Z}_{\mathbb{Q}} = (\mathbb{Z}_{\mathfrak{H}} \cup \mathbb{Z}_{\mathfrak{G}}) \subset \mathbb{Z}_{(\mathfrak{U} \cup \mathfrak{A})} \subseteq \mathbb{Z}_{\Omega}$ with dualistic-polar conditions and constrained by fuzzy conditionality $(\mathfrak{H} \cap \mathfrak{G} \neq \varnothing)$ and $(\mathfrak{U} \cap \mathfrak{A}) \neq \varnothing$. Within categories, transformations are called creativity in terms of innovation, while between categories transformations are called creativity in terms of discovery.

4.2.2. Curiosity in Knowing and Science over the Space Ignorance-Knowledge and Problem-Solution Dualities

The phenomenon of creativity resides in the neural system with supporting system of phenomena that make it useful and efficient in knowing and information-knowledge production over the space of problem-solution dualities as contained in the epistemic field [R8.16b]. Among the phenomena supporting creativity is the phenomenon of curiosity. Curiosity is an important characteristic disposition in the neural system that has an important function in assisting neuro-decision-choice process in identifying and ranking important varieties in the space of problem-solution dualities creating a sub-space of the space of negative-positive dualities with relational continuum and unity in time and over time. It relates to hope, despair, courage, fear encouragement, and creativity in relation to neuro-decision-choice activities over the action space defined over the spaces of problem-solution dualities, production-consumption dualities, input-output dualities and destruction-replacement dualities relative to socio-psychological principle of non-satiation with conformal mapping onto the space of cost-benefit dualities, where the variety identifications and rankings include the elements of knowing and information-knowledge production over the space of identification-transformation dualities which include the space of imaginations. Curiosity is a characteristic disposition of the neural system that allows cognitive concentration to reduce the areas of interest of the neural systems. Let us define and explicate the concept and phenomenon of curiosity to separate them from their use in the family of ordinary languages (FOL). How do we conceive of curiosity within the space of neuro-decision-choice actions on knowing as intellectual investment flows and information-knowledge accumulations as intellectual capital stocks? How does curiosity connect to the space of problem-solution dualities in relation to the space of identification-transformation dualities within the epistemic field and the corresponding relational and information fields? Let us offer a workable definition.

DEFINITION 4.2.2.1A: Curiosity and Curiosity Function (Verbal)

Curiosity is a cognitive characteristic disposition in the neural system of cognition and decision-choice actions that has a booster thinking-algorithmic power in the space of creativity to direct the thinking to isolate essential varieties for knowing and information-knowing production to form categories of essential interests for creative actions in the space of imagination-creativity dualities, the outcomes of which go to update information-knowledge stocks, where the formed categories of curiosity varieties are to be subjected to neuro-decision-choice actions over the space of identification-transformation dualities within the space of imagination-reality dualities as a sub-space of actual-potential dualities with curiosity as an essential part in the activities of the construction of the epistemic field supported by the relational and information fields.

DEFINITION 4.2.2.1 B: Curiosity and Curiosity Function (Categorial Characteristic)

Let the space of curiosity functions be $(\mathscr{K})$ with a generic element $(\vartheta \in \mathscr{K})$ as a curiosity categorial indicator function $(\vartheta(\cdot) \in \mathscr{K})$ as a booster function in support of creative function $(\varsigma(\cdot) \in \mathscr{G})$, where the categorial indicator function performs active selection process in placing variety elements in a category of curiosity $(\mathscr{K}_{(\mathfrak{U} \cup \mathfrak{A})})$ in degrees of belonging for knowing and information-knowledge production over the space of identification-transformation dualities from the space of actual-potential dualities to the sub-space of imagination-reality dualities $(\kappa \in \mathscr{K}_{(\mathfrak{U} \cup \mathfrak{A})}) \subset \{(\mathfrak{G} \otimes \mathfrak{H}) = (\mathfrak{G} \cup \mathfrak{H})\}$ as categorial priorities for critical examination by creativity actions in the categorial conversions under the general principles of opposites containing dualities and opposites, where $\left(\kappa_{u} \in \mathscr{K}_{(\mathfrak{U})} \subset \mathfrak{G}\right)$ is a curiosity set from the space of potentials and $\left(\kappa_{a} \in \mathscr{K}_{(\mathfrak{A})} \subset \mathfrak{H}\right)$ is a curiosity set from the space of actuals such that $\left(\kappa_{u}, \kappa_{a}\right) \in \mathscr{K}_{(\mathfrak{U} \cup \mathfrak{A})} = \left(\mathscr{K}_{(\mathfrak{U})} \otimes \mathscr{K}_{(\mathfrak{A})}\right) \subset \mathbb{Q} = (\mathfrak{G} \otimes \mathfrak{H}) = (\mathfrak{G} \cup \mathfrak{H})$ for neuro-decision-choice actions, where every variety has a reality dual with a supporting imaginative dual and with an information condition of $\mathbb{Z}_{\mathbb{Q}} = \left(\mathbb{Z}_{\mathfrak{H}} \cup \mathbb{Z}_{\mathfrak{G}}\right) \subset \mathbb{Z}_{(\mathfrak{U} \cup \mathfrak{A})} \subseteq \mathbb{Z}_{\Omega}$ with dualistic-polar conditions and constrained by a fuzzy conditionality. The curiosity function is a special indicator function for construction of category of varieties in the space of problem-solution dualities and ignorance-knowledge dualities. It is a booster function to the creative function in the sense that it helps to establish the space of interest and inquisitiveness from the space of acquaintance.

NOTE 4.2.2.1 Definition of Curiosity

As a cognitive characteristic disposition with a thinking-algorithmic boosting power, curiosity provides high-drive conditions in knowing and information-knowledge production for designing a selection framework in the space of ignorance-knowledge dualities that provides necessary and sufficient conditions to form the essential categories in the space of imagination-reality dualities to establish categories of variety imaginations and variety realities to be acted on in the space of identification-transformation dualities in the sense of what unknown actual must be known, what variety potentials must be brought to imaginations, what problems must be dealt with, what solutions must be considered. A curiosity, therefore, is a form of an *indicator function* in the search of or as a framework for all category formations and hence provides an inquisitive framework for the construct of indicator functions in all sub-spaces contained in the space of imagination-reality dualities as a derivative of the space of actual-potential dualities. The collection of all different categorial curiosities, as seen in degrees, from the neural system for categorial formation of interest constitutes the *spaces of curiosities* and possible categorial conversions over all spaces with inter-supportive system of generational dynamics of social philosophical consciencism which locks out the doors of the houses of fear, risk, inaction-curiosity inhibitors and opens the doors of houses of *courage, hope* and *freedom* under *variety curiosity, variety possibilities* and variety visions within the walls of ignorance in relation to elements in the space of imagination-reality dualities which may facilitate or hinder individual and social advancements over the space of social progress in knowing and information-knowledge accumulation over the space of problem-solution dualities.

4.2.3 Comparative Analytics of Creativity, Curiosity, and the corresponding functions in Knowing and Science.

Both curiosity and creativity are characteristic dispositions of the neural system existing in relational give-and-take sharing mode over the space of negative-positive dualities, where their inter-supporting relations may generate and enhance courage disposition that may be influenced by motivation, inspiration, belief and hope over the space of problem-solution dualities with conformal mapping onto the space of cost-benefit dualities retaining the relational structure with input-output dualities. The space of negative-positive dualities indicates the idea every variety is composed of negative, and positive characteristic dispositions for its identity. In terms of operational relationality, the curiosity constitutes the primary disposition while creativity constitutes the derived disposition in the mutual existence within the space of neuro-decision-choice activities over the space of fundamental-applied dualities in the space of production-consumption-dualities. It may also be the case that creativity constitutes the primary category and curiosity is a derived category in categorial formation, where creativity gives rise to curiosity which then influences and enhances creativity The curiosity indicates the relevant varieties in the neural system in the space of imagination-reality dualities for knowing as intellectual investment flows and then creativity takes over within the variety transformation processes. Over the space of parent-offspring (successor-predecessor) processes, creativity and curiosity exist in both primary-derived structure under the principle of opposites in the mind contained in the brain, where both the brain and the mind find residence in the space of visible-invisible dualities with continuum and unity. Under the principle of opposite, every visible characteristic disposition contains invisible characteristic disposition and vice versa and similarly every matter has antimatter and vice versa and so other similar dualistic-polar structures.

The *curiosity disposition* defines the conditions for minimizing the number of elements in category that may be subjected to investigation for knowing and information-knowledge production at any time point while the *curiosity function* operating under these conditions selects the varieties that will belong to the curiosity category, where these varieties will be subjected to knowing and transformation actions. Given the formed curiosity category, the *creativity disposition* is activated within the space of identification-transformation dualities for information knowledge conditions on variety identities and variety transformations. The *creativity disposition* indicates the conditions required to select the best path for variety knowing and variety transformation while the *creativity function* uses these conditions to maximize the net-benefit path of variety identifications and transformations through the relevant functors under the prevailing social philosophical consciencism within which epistemic bubbles may arise over the space of individual-collective dualities. The presence of epistemic bubbles is always an important constraint on the dynamics of knowing and information-knowledge production.

Analytically, curiosity and creativity are linguistic variables capturing qualitative dispositions in human behavior and hence may be seen as fuzzy sets with degrees of belonging. The curiosity operator function expresses the degrees of belonging to the space of efficiency of identifying the elements of categorial interest with desire to know for building intellectual information-capital stocks, where the degrees of belonging express the strength of the desire. Similarly, the creativity operator function expresses the degrees of belonging to the space of efficiency in the space of identification-transformation dualities in knowing and information-knowledge production containing the spaces of problem-solution and input-output dualities where the degrees of belonging express the strength of efficiencies in functors and categorial convertors. The curiosity operator functions for creating categorial varieties of interest involves the active pursuit of knowing of information-knowledge about new varieties and the widen and deepening of experiential information from the acquaintance space. The creativity operator function for activating the elements in the curiosity categories involves in the efficient functors and categorial convertors in transforming ignorance to knowledge by discovering new knowledge, ideas, and varieties to expand the information-knowledge system in support of the space of neuro-decision-choice system.

At this point, it is useful to distinguish the effects of *innovation creativity* from *discovery creativity* in either functors or categorial convertors. Innovation creativity brings new methods and techniques for improving the efficiencies of using a functor or categorial convertor from within the same functor or categorial convertor to increase the benefits over costs of variety the transformation. Innovative creativity is cognitive action which moves within the same functor operator function to usage efficiency. Discovery creativity involves finding new and more efficient functors or categorial converters containing new methods and techniques from without to increase the benefits over costs of variety transformation. The discovery creativity is cognitive action to abstract new functors and categorial convertor within ignorance-knowledge dualities and problem-solution dualities to increase usage efficiency of variety conversions of characteristic disposition within and without categories. The discovery creativity is cognitive action which moves shifts the path of functor operator function or categorial operator converter function to better usage efficiency. In this respect we can modify the functor operator functor and the categorial conversion operator into forms that contain both creativity and curiosity operator functions.

4.2.3.1 Innovation Creativity, Discovery Creativity, and the Spaces of Functors and Categorial Converter.

It is analytically to combine the phenomena of innovative creativity operator function and discovery creativity operator function and the Spaces of Functors and Categorial Converters over the space of imagination-reality dualities, where the space of imagination containing spaces of possibility and probability are equipped with a functor space $(\mathfrak{f} \in \mathbf{F})$, where $(\mathfrak{f})$ is a generic element or space of categorial convertors $(\mathfrak{c} \in \mathfrak{C})$ where $(\mathfrak{c})$ a generic element of the space of categorial convertors. The functor space and the space of categorial converters may be modified to include innovation creativity operator function and discovery creativity operator function such that we have $(\mathbf{F})$ may be written as $(\mathbf{F} \otimes \mathscr{G})$ with generic element $(\mathbf{t}(\cdot) = (\mathfrak{f}(\cdot), \varsigma(\cdot)) \in \mathbf{T} = (\mathbf{F} \otimes \mathscr{G}))$. In this respect, possibility space may be written as $(\mathfrak{P} \otimes \mathbf{T} \otimes \mathbf{A})$ with a generic element $(\mathfrak{p}, \mathbf{t}, \mathfrak{a}) \in (\mathfrak{P} \otimes \mathbf{T} \otimes \mathbf{A})$, where $(\mathfrak{a} \in \mathbf{A})$ is the space anticipation operator and $(\mathfrak{p},\mathbf{t},\mathfrak{a}) \in (\mathfrak{P} \otimes \mathbf{T} \otimes \mathfrak{A})$ is the modified possibility space with functor space or space of categorial converter containing the space of creativity where $(\mathbf{T})$ with generic element $(\mathbf{t} \in \mathbf{T})$ is either space of new functors or categorial convertors associated with possibility reasoning with cognitive capacity limitations. Similarly, the probability space with functors and categorial converters containing creativity may be modified as $(\mathfrak{B} \otimes \mathbf{T} \otimes \mathbf{E})$ with a generic element $(\mathfrak{b},\mathbf{t},\mathfrak{e}) \in (\mathfrak{B} \otimes \mathbf{T} \otimes \mathbf{E})$ with a generic element $(\mathfrak{b},\mathbf{t},\mathfrak{e}) \in (\mathfrak{B} \otimes \mathbf{T} \otimes \mathbf{E})$ associated with probabilistic neuro-decision-choice reasoning for social transformation with resource capacity constraints which are not associated with ontological transformations.

The space of functors, anticipations, and creativity in the space of imagination is associated with necessity and initial conditions for transformation, while the space of functors expectation and creativity is associated with freedoms and sufficiency that turn anticipations into realities as have been discussed in a previous monograph [R8.16b]]. The concepts of functors, functor spaces, space of categorial converters, anticipation operators, expectation operators, creativity and the relevant spaces will be further explicated in relation to the spaces of courage-fear dualities, hope-despair dualities and success-failure dualities and incentive-disincentive dualities all of which are traceable through the space of epistemic distances to the space of ontic-epistemic dualities which generates the space of certainty-uncertainty dualities.

The space of creativity and the creative functions $(\varsigma(\bullet) \in \mathscr{G})$ may be decomposed into innovation creativity $\mathscr{G} = (\mathscr{H} \otimes \mathscr{Q})$, where $(\mathscr{H})$ represents the space of innovation creativity operator functions $h(\bullet)$ and $(\mathscr{Q})$ represents the space of discovery creativity operator function $q(\bullet)$

such that $\{\varsigma(\bullet)=(\mathscr{h}(\bullet),\mathscr{g}(\bullet)) \mid \mathscr{h}(\bullet)\in\mathscr{H},\mathscr{g}(\bullet)\in\mathscr{Q}\}$ and every functor operator function or categorial operator converter function is in the form $\{(\mathbf{t}(\bullet)=(\mathbf{f}(\bullet),\varsigma(\bullet))\in\mathbf{T}=(\mathbf{F}\otimes\mathscr{G}))\mid\varsigma(\bullet)=(\mathscr{h}(\bullet),\mathscr{g}(\bullet))\}$, where the efficiency of functor $\mathbf{t}(\bullet)$ resides in the space of cost-benefit duality $(\mathbf{C}\otimes\mathbf{B})=\mathbf{C}\bigcup\mathbf{B}$ defined in characteristics, with membership characteristic functions of $\left\{\mu_{\mathbf{C}}(x,\mathbf{t}),\frac{d\mu_{\mathbf{C}}}{dx}\leq 0\right\}$ for cost disposition and $\left\{\mu_{\mathbf{B}}(x,\mathbf{t}),\frac{d\mu_{\mathbf{B}}}{dx}\geq 0\right\}$ for benefit disposition such that $\mu_{\mathbf{C}}(x,\mathbf{t})+\mu_{\mathbf{B}}(x,\mathbf{t})=1$ indicating the degree of belonging of each characteristic disposition given the cognitive creativity such that as the membership function of the cost characteristic disposition falls the membership function of the benefit characteristic disposition increases in relation to increasing creativity maintaining the logical unity of cost-benefit duality under the principle of opposites, where $\mathbf{C}\bigcap\mathbf{B}\neq\varnothing$ represents cost-benefit fuzzy continuity . In the case of *innovation creativity*, we move along the membership characteristic functions speeding up increasing benefit and speeding up decreasing cost. In the case of *discovery creativity*, the benefit characteristic membership function shifts $\mu_{\mathbf{B}}(x,\mathbf{t})$ upward to the left increasing the degrees of benefit at all levels while the cost characteristic membership function $\mu_{\mathbf{C}}(x,\mathbf{t})$ shifts downward to the left reducing the degrees of cost at all levels, where x is a cost-benefit characteristic.

The interplay of creativity and non-creativity in functor space may be examined in the space of cost-benefit dualities where the creativity dual is constrained by non-creativity dual and vice versa that may be represented as cost dual of functor is constrained by benefit dual and vice versa and the neuro-decision-choice action in the space of non-creativity-creativity duality relative to the functor space may be seen as cost-benefit interactions in term of fuzzy decision such that

$(\Delta = \mathbf{B} \cap \mathbf{C})$ with membership characteristic function $\mu_{\Delta}(x, \mathbf{t})$ to reconcile conflicts in the space of ignorance-knowledge duality under the relational connectivity among creativity functor, cost, and benefit, where a fuzzy optimal neuro-decision-choice action is obtained through the membership characteristic functions of $\{\mu_{\Delta}(x, \mathbf{t}) = (\mu_{\mathbf{B}}(x, \mathbf{t}) \wedge \mu_{\mathbf{C}}(x, \mathbf{t}))\}$. The fuzzy decision points to the conditions of duality and polarity, where benefit disposition and cost disposition are mutually constraining. The solution to this fuzzy decision problem in the space of non-creativity and creativity may be formulated as a fuzzy mathematical programing problem [R5.43], [R6.3] [R6.8] [R6.10] in the form:

$$\underset{x \in (\mathbf{B} \cap \mathbf{C})}{\text{opt}} \mu_{(\mathbf{B} \cap \mathbf{C})}(x) = \underset{x \in \Delta}{\text{opt}} \mu_{\Delta}(x) = \begin{cases} \max\limits_{x \in \mathbb{Q}} \mu_{(\mathbf{B})}(x), \ \frac{d\mu}{dx} \geq 0 \\ \text{s.t} \ \ \mathbb{Q} = (\mu_{\mathbf{C}}(x) - \mu_{(\mathbf{B})}(x)) \geq 0 \\ \qquad \text{OR} \\ \min\limits_{x \in \mathbb{Q}} \mu_{(\mathbf{C})}(x), \ \frac{d\mu}{dx} \leq 0 \\ \text{s.t.} \ \mathbb{Q} = (\mu_{\mathbf{C}}(x) - \mu_{(\mathbf{B})}(x)) \geq 0 \end{cases} \tag{4.2.3.1.1}$$

The solution is such that at the point of optimality $\{\mu_{\Delta}(\alpha^*) = \mu_{(\mathbf{B} \cap \mathbf{C})}(\alpha^*) = \mu_{\mathbf{B}}(\alpha^*) = \mu_{\mathbf{C}}(\beta^*)\}$ with $\alpha^* = \mu_{\mathbf{B}}(x^*) = \mu_{\mathbf{C}}(x^*) = \beta^*$ where $(\alpha^* + \beta^*) = 1$ with the corresponding solution specified as:

$$\text{solution} = \begin{cases} \mu_{\mathbf{B}}(\alpha) \in (\alpha^* + \varepsilon, 1) \Rightarrow \text{a zone of creativity with greater than } \alpha^*\text{-degree of creativity} \\ \mu_{\mathbf{C}}(\alpha) \in (0, \alpha^* - \varepsilon) \Rightarrow \text{a zone of uncreativity with less than } \alpha^*\text{- degree of creativity} \\ \mu_{\mathbf{B}}(\alpha^*) = \mu_{\mathbf{C}}(\alpha^*) \Rightarrow \text{a point of idifference} \end{cases}, \ \alpha^* \in [0,1] \tag{4.2.3.1.2}$$

The problem of eqn. 4.2.3.1.1 and the solution in the eqn. 4.2.3.1.2 may be conformally mapped onto the functor space or the space of categorial converters to locate the zone of efficient and in efficient functors or categorial converters to improve the conditions in the neuro-decision-choice actions where the effects of creativity may be examined over the space of identification-transformation dualities relative to the conditions over the spaces of problem-solution and ignorance-knowledge dualities. In this respect, we have:

$$\text{Functor solution} = \begin{cases} \mu_{\mathbf{B}}(\alpha) \in (\alpha^* + \varepsilon, 1) \Rightarrow \text{zone of efficient functors with creativity dominance } (\mathbf{Z}^{\rho}) \\ \mu_{\mathbf{C}}(\alpha) \in (0, \alpha^* - \varepsilon) \Rightarrow \text{zone of inefficient functors with uncreativity dominance } (\mathbf{Z}^{\lambda}) \end{cases}, \alpha^* \in [0,1] \quad (4.2.3.1.3)$$

We may the decomposed the functor space, $\mathbf{T} = \{\mathbf{Z}^{\rho} \cup \mathbf{Z}^{\lambda} \mid \mathbf{Z}^{\rho} \cap \mathbf{Z}^{\lambda} \neq \varnothing\}$ such that,

$$\begin{aligned} \mathbf{Z}^{\rho} &= \{(\mathbf{t}, \mu_{\mathbf{B}}(x^*)) \in (\mathbf{T}, \mu_{\Delta}(x)) \mid \mu_{\mathbf{B}}(x) \in (\alpha^* + \varepsilon, 1)\} \\ \mathbf{Z}^{\lambda} &= \{(\mathbf{t}, \mu_{\mathbf{C}}(x^*)) \in (\mathbf{T}, \mu_{\Delta}(x)) \mid \mu_{\mathbf{C}}(x) \in (0, \alpha^* - \varepsilon,)\} \end{aligned} \quad (4.2.3.1.4)$$

for examining the dynamics of creativity in the space of identification-transformation dualities relative to the functor space. The analytical solution to the fuzzy optimal problem may be found in [R5.43], [R6.3][R6.10].The fuzzy optimal neuro-decision-choice problem may be illustrated in an epistemic geometry as in figure 4.2.3.1, where there is dualistic-polar conflict resolution that may be applied to fuzzy games over space of conflict-resolution under the general principle of opposites in continuum and unity containing contradictions with non-excluded middle and negation of negation in continual process of variety transformations.

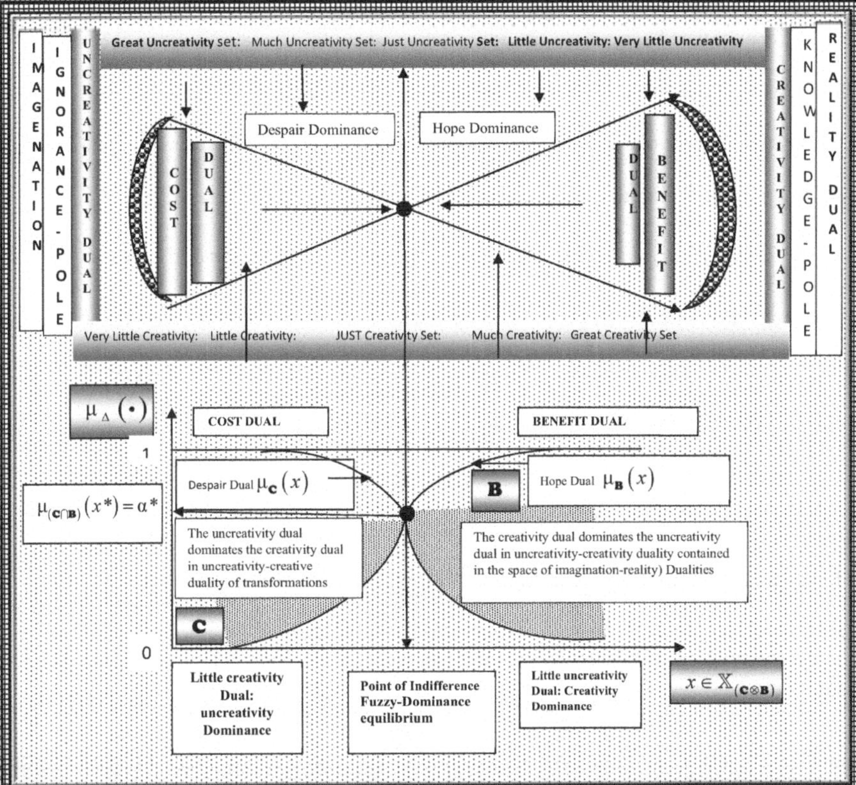

Figure 4.2.3.1: A Cognitive Geometry of Representation of Uncreativity-Creativity Duality in Continuum and Unity in Relation to the Fuzzy Paradigm, Approximate Reasoning on Variety Functors in Identification-Transformation Dualities of Knowing, Problem-Solution Dynamics, and Ignorance-Knowledge Dualities.

It is useful to note the nature and structure of fuzzy optimization and how they work in abstracting conditions of optimality. The optimization takes place by restricting the choice variable within the zone of positive differences between then negative characteristic membership function and the positive characteristic membership function which may be interpreted as the zone of excess negative duality or the zone of positive dualistic deficiency. In the case of eqn. 4.2.3.1.1 the constraint $\mathbb{Q} = \left(\mu_{\mathbf{C}}(x) - \mu_{(\mathbf{B})}(x)\right) \geq 0$ has two interpretations of the zone of excess costs when the benefit characteristic function is being maximized, and *zone of benefit deficiency* when the cost characteristic membership function is being minimized, where the optimization is considered from the origin of the axis.

4.3. COURAGE AND COURAGE OPERATOR FUNCTION IN KNOWING AND SCIENCE

Let us now connect courage, fear and courage-fear dualities to creativity, curiosity, and neuro-decision-choice actions over the space of identification-transformation dualities. Let us keep in mind that every neuro-decision-choice action is defined over the space of identification-transformation dualities, where the neuro-decision-action is either about identification of variety property or variety relation or both, or about transformation of variety property or variety relation or both or simultaneity about both identification and transformation over the space of qualitative-quantitative dualities. Variety problem dispositions are identified, variety solution dispositions are identified, and their relational structural dispositions also identified and established and transformed, where the space of problem-solution dualities are decomposed into fuzzy categories, the correction of which constitutes the epistemic field $(\mathscr{C})$. Over the epistemic field $(\mathscr{C})$ with relational field $(\mathscr{R})$ and information field $(\mathscr{I})$, curiosity and creativity are important inputs into variety courage disposition and variety hope disposition to activate variety motivation and variety inspiration dispositions within the complexity of the neural system in inter-supporting relational give-and-take modes that help *thinking* as a complex system of relational diversities in continua and unity to reconcile conflicts over the space of internal-external dualities under the principle of opposites defined over the space of fear-courage dualities.

The natures of curiosity and creativity, and their relationships will vary in intra-categories and inter-categories over the space of static-dynamic dualities relative to the space of problem-solution dualities as a sub-space of imagination-reality dualities which also contains the space of possibility-impossibility dualities linked to the space of ignorance-knowledge dualities in the epistemic field with the relational and information fields as well as over the space of individual-community dualities, where the degrees of creativity and curiosity will vary over individuals as well as between individuals and the collective. In all the spaces of human action, the possibility space presents necessity and initial conditions for categorial conversions which require curiosity with anticipation to overcome the constraint of necessity and initial conditions. The probability space presents freedom and sufficiency for functor actions or categorial conversions which require creativity with expectation to overcome the constraints of freedom and necessity. The overcoming of the constraints of necessity, initial conditions, freedom and sufficient conditions in variety identifications and transformations requires courage of actions with hope, motivation and inspiration and perspirations over the space of cost-benefit dualities, where benefits are maximized over costs and costs are minimized over benefits under the socio-economic principle of non-satiation in knowing as intellectual investment flows and information-knowledge accumulation as intellectual capital stocks. The difficulties with which these constraints may be overcome will vary from category to category over the epistemic field, accumulated social information-knowledge stocks and other cultural dimensions contained in the social philosophical consciencism which will then affect courage dispositions and the derived variety dispositions over the dynamic space of problem-solution dualities and parent-offspring problem-solution dynamics.

The background about courage-fear duality offers us a framework to define courage, relative to fear, that is required in dealing with the elements in the curiosity category in order to exercise the creative disposition over the space of imagination-reality dualities, where every neuro-decision-choice action involves a change of variety within the spaces of identification-transformation dualities and relation-property dualities by destructing a variety which is cognitively assessed to have cost characteristic disposition greater than an assessed benefit characteristic disposition in the space of realities containing the problems space and replacing it with a variety which is cognitively assessed to have benefit characteristic disposition greater than cognitively assessed cost characteristic disposition from the space of imaginations containing the solution space to the varieties in the space of problem-solution dualities and polarities. Let us keep in mind, as have been explained in the previous chapters, that the variety ranking over the space of problem-solution dualities is on the basis of relativity of cost-benefit conditions and that every neuro-decision-choice action in the space imagination-reality dualities is a change of property or a change in relation or both from ignorance to knowledge or from one variety to another and that variety problems are defined where the internal cost dispositions that dominate the internal benefit dispositions relative to a vision or a goal-objective element, and solutions are defined as varieties where the internal benefit dispositions dominate the internal cost dispositions. To expand the analytical work on the connectivity of courage with curiosity and creativity in terms of fuzzy decomposition, let us define the variety courage and corresponding complements in the space of problem-solution dualities that provides a foundation for knowing and information-knowledge stocks as viewed from the spaces of production-consumption dualities, input-output dualities, and social dynamics over the space of progress-retrogress dualities containing the space of success-failure dualities.

DEFINITION 4.3.1A: Courage and Courage Function (verbal)

Courage is a cognitive characteristic disposition within the neural system of cognition and decision-choice actions that has an enabling ability to generate motivation, perspiration and hope over despair to exercise creativity with elements of curiosity over very high uncertainty in the space of certainty-uncertainty dualities relative to the space of problem-solution dualities with vicissitudes of variety cost-benefit conditions to generate increasing degrees of inspiration, motivation and perspiration over the space of hope-despair dualities relative to the space of risk-benefit dualities to empower the thinking-algorithm to execute increasing effort over the space of problem-solution dualities containing problem varieties with increasing degrees of difficulty by disregarding conditions of variety failure and accepting success over the elements in the space of success-failure dualities relative to the element of interest in the variety curiosity category contained in the space of imagination-creativity dualities. The dependent relationship of courage to, curiosity and creativity in the space of neuro-decision-choice activities is the *courage function*.

DEFINITION 4.3.1B: Courage and Courage Function (Algebraic)

Let the space of courage functions be $(\mathbf{K})$ with a generic element $(\mathbf{k}(\cdot) \in \mathbf{K})$ as a courage categorial indicator function $(\mathbf{k}(\mathbf{g}, \mathbf{c}) \in \mathbf{K})$ where $(\mathbf{g} \in \mathscr{K})$ is curiosity disposition with a corresponding curiosity indicator function $\left(\kappa \in \mathscr{K}_{(\mathfrak{H} \otimes \mathfrak{G})}\right)$ and $(\mathbf{c} \in \mathscr{G})$ is creative disposition with a corresponding creativity function $\left(\varsigma(\bullet)\in\mathscr{G}\right)$ within the social philosophical consciencism containing social exceptionalism relative to the corresponding social personality at intragenerational and intergenerational time points, where the identification and transformation of its identity is revealed by information condition of $\mathbb{Z}_{\mathbf{K}}=\left(\mathbb{Z}_{\mathbf{c}}\cup\mathbb{Z}_{\mathbf{g}}\right)\subset\mathbb{Z}_{(\mathfrak{S}\otimes\mathfrak{H})}\subseteq\mathbb{Z}_{\Omega}$. The courage disposition, required to overcome increasing difficulties in knowing and science over the space of variety problem-solution dualities, depends on curiosity disposition and creativity disposition and may be written as an implicit functional dependence. The implicit courage operator function may be written as multiplicatively separable functional operator or additively separable functional operator which will depend on the type of category in the epistemic field. The courage functional operator in knowing and science may be written as:

$$
\begin{aligned}
&\left(\mathbf{k}(\mathbf{g},\mathbf{c})\in\mathbf{K}\right)\ldots\ldots\ldots\ldots\text{inplicit function cognitive operator}\\
&\mathbf{k}(\mathbf{g},\mathbf{c})=\left(\mathbf{g}(\bullet)^{\alpha}\mathbf{c}(\bullet)^{\beta}\right)\Rightarrow\log\mathbf{k}(\bullet)=\alpha\log\mathbf{g}(\bullet)+\beta\log\mathbf{c}(\bullet),(\alpha+\beta\leq 1)\ldots\text{multiplicative seperability cognitive operator}\\
&\mathbf{k}(\mathbf{g},\mathbf{c})=\sigma_1\mathbf{g}(\bullet)+\sigma_2\mathbf{c}(\bullet)\ldots\ldots\ldots\ldots(\sigma_1,\ \sigma_2>0)\ldots\text{additive separability cognitive operator}
\end{aligned}
\tag{4.3.1}
$$

Where (α) is the share of curiosity in courage and (β) is the share of creativity in the development of courage to overcome fear over the action space, and (σ_1) is the additive proportion of curiosity to courage and (σ_2) is additive proportion of curiosity to courage.

NOTE 4.3.1: Definition of Courage

The values of (α) and (β) depend on the elements from the social philosophical consciencism, its contents, corresponding national personality, social exceptionalism and their dynamics at any generational time point relative to the degrees of goal-objective importance as revealed by the conditions of social philosophical consciencism with individual and collective preferences over the space of reward-punishment dualities. The space of reward-punishment dualities always has conformal mapping onto the space of incentive-disincentive dualities linked to the space of cost-benefit dualities, where the curiosity parameter generates motivation and increasing confidence to excise increasing efforts to overcome conditions in the space of acceptance-rejection dualities within the social constraints, while the creativity parameter generates hope to exercise perspirations over the space of individual-collective dualities. The courage operator operates over the space of risk-reward dualities with efforts constrained by social philosophical consciencism in terms of action conditions over the space of social admissibility-inadmissibility dualities based on cost-benefit balances through the social value system. The social value system is contained in the social philosophical consciencism to overcome fear within the space of failure-success dualities. The space of success-failure dualities is contained in the space of the imagination-reality dualities that houses the space of problem-solution dualities, where the problem space is contained in the space of realities and the solution space is contained in the space of imagination at the behest of the space of certainty-uncertainty duality generated from the space of ontic-epistemic dualities.

4.3.1 Courage in Knowing, Science and Information-Knowledge Production

A brief reflection on the role of the phenomenon of courage in human behavior, especially in neuro-decision-choice actions over the general space of problem-solution dualities, will be useful for the understanding of success-failure interactions with problem-solution varieties and movements from the space of imaginations to the space of realities to overcome ignorance and create knowledge, or set knowledge against ignorance in the space of ignorance-knowledge dualities, to overcome nothingness and create somethingness or set somethingness against nothingness in the space of nothingness-somethingness dualities and in general to overcome costs and create benefits, where benefits are set against costs under the principle of opposites with relational continuum and unity. Courage is a characteristic disposition with multi-dependent factors that motivate the internal forces of neuro-decision-choice disposition to effect transformations of a variety imaginations by using combinations of curiosity and creativity to asset the variety possibilities and probabilities against the variety realities as variety actuals in the space of problem-solution dualities by overcoming direct and indirect constraints in the space of easiness-difficulty dualities, where every variety knowing or variety transformation has some easiness characteristics and difficulty characteristics in varying degrees of relativity. Over the space of imaginations, there is no courage without curiosity which gives rise to *hope* where hope activates creative and innovative actions over the space of production-consumption dualities over the space of goals, objectives and visions defined within the space of imagination-reality dualities. The degrees of courage in the space of courage-fear dualities depend on hope in the space of hope-despair dualities to which we turn our attention.

4.4. HOPE-DESPAIR DUALITIES AND THE CORRESPONDING OPERATOR FUNCTIONS IN KNOWING AND SCIENCE

Over the space of imaginations, cognitive activities over the space of imagination-reality dualities are based on *hope* as confident anticipation under the initial conditions of necessity in the possibility space and then as confident expectation under the sufficient conditions of freedom over the probability space, where the confidence is defined in terms of subjective degrees between zero and one with a specified membership characteristic function in terms of what variety of knowing and transformation enters into the fuzzy *set of hope*, $(\mathbf{H})$ with membership characteristic function defined as $\{\mu_{\mathbf{H}}(x) \in [0,1]\}$ which is an indicator function that selects varieties from the space of imagination-reality dualities $\{\mathbb{Q} = (\mathfrak{G} \otimes \mathfrak{H}) = (\mathfrak{G} \cup \mathfrak{H})\}$ contained in the space of actual-potential dualities $(\mathfrak{A} \otimes \mathfrak{U}) = (\mathfrak{A} \cup \mathfrak{U})$ and places them in the set of hopes $(\mathbf{H})$, the identities of which are to be known and transformation from the space of imaginations to the space of realities through thinking-reasoning processes over the spaces of problem-solution dualities and ignorance-knowledge dualities. Hope is a linguistic variable or fuzzy variable projected as fuzzy set, where the degrees of belonging are hope indicators with a membership characteristic function specifying the degrees that may be represented as:

$$\mu_{\mathbf{H}}(x) = \begin{cases} 1 \text{ ....complete hope with variety identification and transformation} \\ (0,1)\text{........dualistic polar condition with a combination of incomplete hope} \\ 0 \text{ ....complete despair with variety identification and transformation,} \\ \text{where } x \text{ is a variable of hope.} \end{cases}, \frac{d\mu_{\mathbf{H}}}{dx} \geq 0 \quad (4.4.1)$$

Cognitive activities over the space of imagination-reality dualities are based on *hope* as confident anticipation under the initial conditions of necessity in the possibility space and then as confident expectation under the sufficient conditions of freedom over the probability space, where the confidence is defined in terms of subjective degrees between zero and one with a specified membership characteristic function in terms of what variety of knowing and transformation enters the fuzzy *set of hope*, $(\mathbf{H})$ with membership characteristic function defined as $\{\mu_{\mathbf{H}}(x) \in [0,1]\}$ which is an indicator function that selects varieties from the space of imagination-reality duality.

Under the principle of opposites, there is corresponding opposite of despair as a fuzzy set of despair $(\mathbf{D})$ with membership characteristic function $\{\mu_{\mathbf{D}}(x) \in [0,1]\}$ that specify the degrees of despair hesitant anticipation of the same varieties under the initial conditions of necessity in the possibility space and then as hesitant expectation under the sufficient conditions of freedom over the probability space, where the hesitancy is defined in terms of subjective degrees between zero and one with a specified membership characteristic function in terms of what variety of knowing and transformation enters into the *fuzzy set of despair*, $(\mathbf{D})$ with membership characteristic function defined as $\{\mu_{\mathbf{D}}(x) \in [0,1]\}$ which is an indicative function that forms a constraint on the hope such that the despair indicator function selects the same varieties from the space of curiosity within the imagination-reality dualities $\mathbb{Q} = (\mathfrak{G} \otimes \mathfrak{H}) = (\mathfrak{G} \cup \mathfrak{H})$ contained in the space of actual-potential dualities $(\mathfrak{A} \otimes \mathfrak{U}) = (\mathfrak{A} \cup \mathfrak{U})$ and places them in the set of despair $(\mathbf{D})$ to show the degree of despair associated with the identities which are to be known and transformed from the space of imaginations to the space of realities or from the space of problems to the space of the space of solutions as contained in the space of ignorance-knowledge dualities. The hope indicator is a membership characteristic function that may be represented as:

$$\mu_{\mathbf{D}}(x) = \begin{cases} 1 \text{ ....complete dispair with variety identification and transformation} \\ (0,1) \text{........dualistic-polar condition with a combination of incomplete dispair} \\ 0 \text{ ....complete non-dispair with variety identification and transformation} \\ \text{where } x \text{ is variable of hope and despair as a characteristic} \end{cases}, \frac{d\mu}{dx} \leq 0 \quad (4.4.2)$$

The despair as fuzzy set is to form a constraint on hope and the hope as a fuzzy set is to form a constraint on despair in the neuro-decision-action space where the anticipated and perceived outcomes within the space of necessity-freedom dualities is mapped onto the space of real cost-benefit dualities $(\mathbf{B} \bigcup \mathbf{C}) = (\mathbf{B} \otimes \mathbf{C})$ for ranking and action. With respect to knowing as intellectual investment flows and information-knowledge accumulation as intellectual capital stocks, what are the meanings and roles of hope and despair in the space of hope-despair dualities in the general system of dualities and polarities under the principle of opposites over the *epistemic field* composed of categories of knowing over the space of problem-solution dualities? Let us offer some analytically workable definitions for hope and despair over the space of action-inaction dualities of cognitive activities under the general space of quality-quantity dualities.

DEFINITION 4.4.1: HOPE (DESPAIR)

Hope (despair) is a combined confident (hesitant) anticipation in the of possibility space and (hesitant) expectation in the probability space to push neuro-decision-choice agents for the application of consistent actions to overcome increasing difficulties over the success-failure dualities, where such anticipation and expectation reside in a justified zone of belief on the basis of information-knowledge support which is derived from the social philosophical consciencism prevailing at the generation points, and where the nature of the fuzzy characteristic functions of belonging depends on availability, interpretations and easiness to access the contents in the social philosophical consciencism at generation time points.

4.4.1 Problem-Solution Analytics over the Space of Hope-Despair Dualities

From the fuzzy sets of hope and despair, we can derive conditions of optimal hope and optimal despair in terms of subjective assessments over the space of hope-despair dualities by first defining a fuzzy decision problem as the fuzzy intersection of hope and despair in the form

$\Delta = (\mathbf{H} \cap \mathbf{D}) \neq \varnothing$ with a membership characteristic function of the form $\{\mu_\Delta(x) = \mu_{(\mathbf{H}\cap\mathbf{D})}(x) = \mu_\mathbf{H}(x) \wedge \mu_\mathbf{D}(x)\}$ such that $(\mathbf{H} \otimes \mathbf{D}) = (\mathbf{H} \cup \mathbf{D})$ with $(\mathbf{H} \cap \mathbf{D}) \neq \varnothing$ as the fuzzy conditionality under the general principle of opposites. To obtain either maximum hope or minimum despair as an optimal hope-despair combination within the space of hope-despair dualities, we define a fuzzy optimal problem by maximizing hope subject to despair or minimizing despair subject hope. This may be done with the membership characteristic functions in a useful form by translating the decision-choice problem into a fuzzy mathematical programing problem of the form:

$$\mu_\Delta(x) = \mu_{(\mathbf{H}\cap\mathbf{D})}(x) \operatorname*{opt}_{\nu\in} \mu_\Delta(\bullet) = \begin{cases} \max\limits_{\nu\in\mathbb{Q}} \mu_{(\mathbf{H})}(x), \; \frac{d\mu}{dx} \geq 0 \\ \text{s.t} \;\; \left(\mu_\mathbf{D}(x) - \mu_{(\mathbf{H})}(x)\right) \geq 0 \\ \qquad \text{OR} \\ \min\limits_{\nu\in\mathbb{Q}} \mu_{(\mathbf{D})}(x), \; \frac{d\mu}{dx} \leq 0 \\ \text{s.t.} \;\; \left(\mu_\mathbf{D}(x) - \mu_{(\mathbf{H})}(x)\right) \geq 0 \end{cases} \quad - \qquad (4.4.1\,.1)$$

The solution to this fuzzy optimization problem is such that at the point of optimality $\mu_\Delta(x^*) = \mu_{(\mathbf{H}\cap\mathbf{D})}(x^*) = \mu_\mathbf{H}(x) = \mu_\mathbf{D}(x^*)$ with $\alpha^* = \beta^*$ where $\mu_\mathbf{H}(x^*) + \mu_\mathbf{D}(x^*) = 1$ with the corresponding solution specified as:

$$\text{solution} = \begin{cases} \mu_\mathbf{H}(x) \in (\alpha^* + \varepsilon, 1) \Rightarrow \text{zone of hope with greater than } \alpha^*\text{-degree of hope} \\ \mu_\mathbf{D}(x) \in (0, \alpha^* - \varepsilon) \Rightarrow \text{zone of despair with less than } \alpha^*\text{- degree of hope} \\ \mu_\mathbf{H}(x) = \mu_\mathbf{D}(x) = \alpha^* \Rightarrow \text{condition of indifference and dominance equilibrium} \end{cases}, \alpha \in [0,1] \qquad (4.4.1.2)$$

The value $(\alpha^* = \beta^*)$ presents conditions of indifference supporting the principle of indifference over the space of certainty-uncertainty dualities for neuro-decision-choice actions , while the ε-value is a subjective confidence measure or hesitation measure and the confidenc (hesitation) interval $(-\varepsilon, \varepsilon)$ presents conditions, where doubts of hope may occur to provide decision-choice cushion for errors of analysis to effect a zone of *justified true hope* (JTH) and the zone of *justified true despair* (JTD). All types of hope and despair are defined by their values in ascending or descending order of magnitude between zero and one specifying the strengths of hope and despair in opposite values based on fuzzy abstractions from the available information-knowledge environment as contained in the social philosophical consciencism, enhanced by the prevailing social personality and national exceptionalism. Over the space of neuro-decision-choice actions, justified true hope (JTH) must satisfy conditions of possibility and probability and must have understanding connectivity to the space of ontic-epistemic dualities as through fuzzy processes to *de-fuzzify* the vagueness. In other words, varieties with JTH or JTD must belong the space of possibility-probability dualities as well as belonging to the spaces of anticipation-expectation duality and necessity-freedom dualities. As specified, hope may be increased with increasing knowing which provides an increasing confidence and decreasing hesitation in further knowing which is also dependent on increasing hope and decreasing despair which are abstracted from the general conditions of the prevailing social philosophical consciencism.

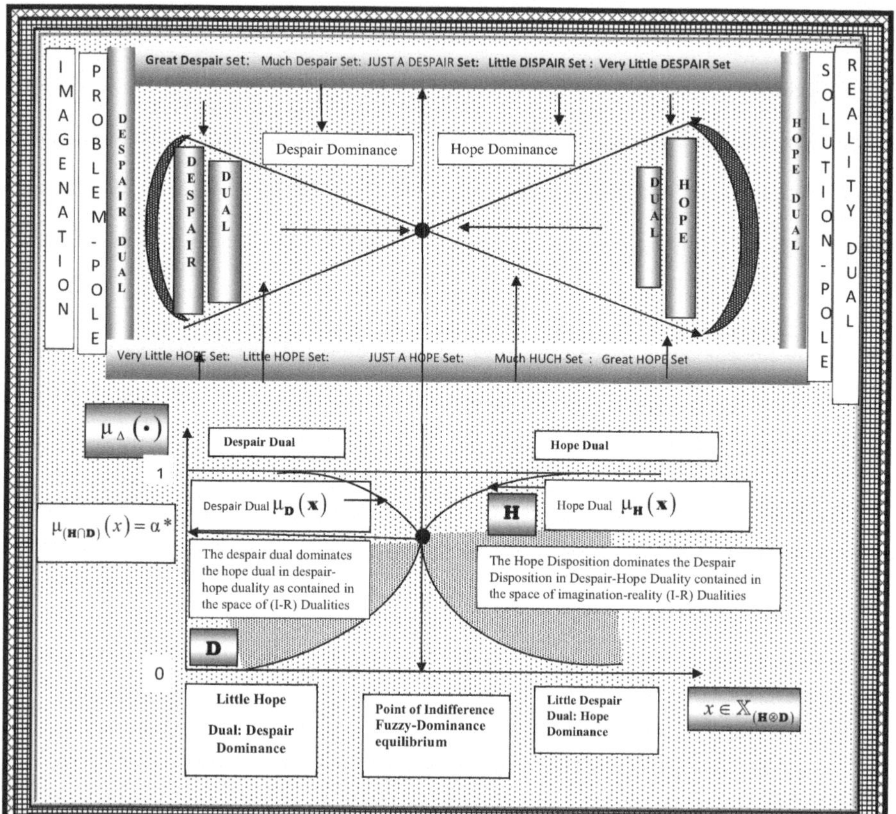

Figure 4.4.1.1: A Cognitive Geometry of Representation of Hope-Despair Duality in Continuum and Unity in Relation to the Fuzzy Paradigm, Approximate Reasoning on Varieties in Identification-Transformation Dualities of Knowing, Problem-Solution Dynamics and Hope-Despair Dualities.

The resolution to hope-despair conflict in any cognitive action is resolved through mutual constraining structure as shown in eqn. (4.4.1.1) and illustrated in Fig (4.4.1.1). The important analytical point in this approach is the justified framework of a general optimization under the principle of opposites where the opposites are of mutually constraining in give-and-take sharing modes over the neuro-decision-choice action space in the epistemic field, and where the opposites exist under conditions of supersymmetry in adinkra geometric representations in the process of knowing as the production of intellectual investment flows and information-capital accumulation as the production of intellectual capital stocks. The mutual constraints translate to goal-objective constrained interactions of human actions, where the optimal selection from the space of goals (benefits) is optimized subject to the varieties from the space of resource constraints (costs). In the neuro-decision-choice process under the understanding of hope-despair processes over the space of knowing in problem-solution dualities, it is such that there is the maximization of a hope (benefit) as a goal, restricted by a despair (cost) as a constraint, while there is a minimization of a despair (cost) as the goal, restricted by a hope (benefit) as the constraint under the *asantrofi-anoma rationality* and Sankofa-anoma info-temporal conditions. All types of variety hope, and variety despair fit into this analytical work and neuro-decision-choice actions, where hope may be related to optimism and despair may be related to pessimism. [R27.9] [27.10] [R27.11[R27.12] [R27.13] [R27.17].

With the conditions of optimal hope and optimal despair the variety curiosity set for knowing may be undertaken with indicator function $\mathcal{I}_{\mathcal{K}}(x_i)$:

$$\mathcal{I}_{\mathcal{K}}(x_i)=\begin{Bmatrix} 1,\ \text{if } x_i \in \mathbf{H}=\left\{\left(x_i,\mu_{\mathbf{H}}(x_i)\mid \mu_{\mathbf{H}}(x_i)\in\left[(\alpha^*+\varepsilon),1\right]\right)\right\},\ \mathbf{H}\subset\Omega\Rightarrow\text{Hopeful set} \\ \left(\left(\alpha^*+|\varepsilon|\right)\right),\ \text{if } x_i\in(\mathbf{D}\otimes\mathbf{H})=\left\{\left(x_i,\mu_{(\mathbf{H}\otimes\mathbf{D})}(x)\mid \mu_{(\mathbf{D}\otimes\mathbf{H})}(x_i)\in\left[(\alpha^*-\varepsilon),(\alpha^*+\varepsilon)\right]\right)\right\}\Rightarrow\text{Indifference} \\ 0,\ \text{if } x_i\in\mathbf{D}=\left\{\left(x,\mu_{\mathbf{D}}(x_i)\mid\mu_{\mathbf{D}}(x_i)\in\left[0,(\alpha^*-\varepsilon)\right]\right)\right\},\ \mathbf{D}\subset\Omega\Rightarrow\text{Despair set} \end{Bmatrix} \quad (4.4.1.3)$$

This set of distributions will indicate the constrained space for efficient resource allocations to the efforts of variety knowing as intellectual investment flows and variety information-capital accumulation as intellectual capital stocks, where variety hope resides in the space of possibility-probability dualities and variety despair resides in the space of impossibility-improbability dualities with hope and despair over the epistemic field are opposites that relate to the degrees of positive and negative confidence over the space of problem-solution dualities relative to the danger, and risk of uncertain outcomes in the space of cost-benefit dualities relative to the fear of losses of resources and time or courage for gains of resources and time. In optimal neuro-decision-choice actions over the space governed by the principles of opposite the internal value of variety is obtained by optimizing the positive (negative) characteristic disposition subject to the negative (positive) characteristic disposition in fuzzy space with the use of fuzzy optimization to obtain fuzzy-stochastic optimal values with fuzzy-stochastic conditionality. The fuzzy-stochastic optimal values are then used to rank the members in the set of interest for selection (for analytical techniques of fuzzy optimization see [R5.43] [R6,3] [R6.10].

It is at this point of human actions and behaviors within the epistemic field of the space of categories of problem-solution dualities, that hope and despair interact with internal courage, motivation, inspiration, creativity and perspirations over the space of knowing as the production of intellectual investment flows and information-knowledge accumulation as the production and consumption of intellectual capital stocks, where the drive to knowing depends on internal conditions of motivations and inspirations and may be influenced by the conditions of externalities. The external conditions of motivation, inspirations are unsustainable since they may not be coded in the social philosophical consciencism which generates asocial personality, and national exceptionalism, and hence may not generate continuity of actions, while the internal motivation and inspirations are sustainable drawing their intellectual energies from the prevailing social philosophical consciencism in support of courage containing curiosity and creativity of neuro-decision-choice actions over the epistemic field, where the conditions of curiosity motivate the will to know and the conditions of the will to know motivate the conditions of the search for the toolbox of knowledge, search for variety solutions to variety problems, and the search for variety knowledge to variety ignorance.

The curiosity function as the categorial indicator function performs active selection actions of varieties from the space of actual-potential dualities, and placing the variety elements in a category of curiosity $\left(\mathcal{K}_{(\mathfrak{U}\cup\mathfrak{A})}\right)$ in degrees of belonging for knowing action and information-knowledge production over the space of identification-transformation dualities from the space of actual-potential dualities to the sub-space of imagination-reality dualities such that $\left(\kappa \in \mathcal{K}_{(\mathfrak{U}\cup\mathfrak{A})}\right) \subset (\mathfrak{G} \otimes \mathfrak{H}) = (\mathfrak{G} \cup \mathfrak{H})$ as categorial priorities for critical examination by creative actions in the categorial conversions under the general principles of opposites containing dualities and polarities. where $\left(\kappa_{(\mathfrak{u}\in\mathfrak{U})} \in \mathcal{K}_{(\mathfrak{U})} \subset \mathfrak{G}\right)$ is curiosity set from the space of imaginations as a sub-set of the space of potentials and $\left(\kappa_{\mathfrak{a}} \in \mathcal{K}_{(\mathfrak{A})} \subset \mathfrak{H}\right)$ is a curiosity set from the space of realities as a sub-set of the space of actuals $(\mathfrak{A} \subset (\mathfrak{A} \otimes \mathfrak{U}))$ such that $(\kappa_{\mathfrak{u}}, \kappa_{\mathfrak{a}}) \in \mathcal{K}_{(\mathfrak{U}\cup\mathfrak{A})} = \left(\mathcal{K}_{\mathfrak{U}} \otimes \mathcal{K}_{(\mathfrak{A})}\right) \subset \mathbb{Q} = (\mathfrak{G} \otimes \mathfrak{H}) = (\mathfrak{G} \cup \mathfrak{H})$ and $(\mathfrak{G} \cap \mathfrak{H}) \neq \varnothing$ for neuro-decision choice actions, where every variety has a reality dual with a supporting imagination dual and with an information condition of $\mathbb{Z}_{\mathbb{Q}} = (\mathbb{Z}_{\mathfrak{H}} \cup \mathbb{Z}_{\mathfrak{G}}) \subset \mathbb{Z}_{(\mathfrak{U}\cup\mathfrak{A})} \subseteq \mathbb{Z}_{\Omega}$ with dualistic-polar condition and constrained by fuzzy conditionality. The curiosity function is a special indicator function for the construction of category of varieties in the spaces of problem-solution dualities and ignorance-knowledge dualities.

4.4.1.1: A Reflection on Hope and Despair in Human actions and Knowing.

Alternatively viewed, hope is a variety characteristic disposition in the neural system that assists the neuro-decision-choice agents to deal with the conditions of variety uncertainty in the space of certainty-uncertainty dualities to take actions over the space of problem-solution dualities. It helps to minimize anticipation of unpleasant outcome with greater cost over the present benefit with expectation of greater benefit over cost by taking risk and the corresponding benefit in Asantrofi-anoma tradition. Hope is composed of multiplicity of factors that are in relation to minimization of perceptive of uncertainty dual and maximization of the certainty dual within the space of uncertainty-certainty dualities relative to the space of human neuro-decision-choice actions, where such factors are abstracted from the content of social philosophical consciencism containing social personality. Similarly, despair is a variety characteristic disposition in the neural system that acts against hope by assisting the neuro-decision-choice agents to avoid conditions of variety uncertainty in the space of certainty-uncertainty dualities to not take actions over the space of problem-solution dualities. It helps to maximize anticipation of unpleasant outcomes with greater variety costs over the present variety benefits with expectation of greater variety benefits over variety costs by avoiding risk and the corresponding benefit in Asantrofi-anoma tradition in the sense that variety cost is substantially valued over variety benefit. It is composed of multiplicity of factors that are in relation to maximization of perceptive of uncertainty dual and minimization of the certainty dual within the space of uncertainty-certainty dualities relative to the space of human neuro-decision-choice actions, where such factors are abstracted from the content of social philosophical consciencism. In other words, hope encourages risk-taking and positive actions, while despair discourages risk-taking and encourages negative human behavior defined over the space of cost-benefit dualities relative to the space of identification-transformation dualities, imagination-reality dualities within the epistemological space relative to the ontological space.

4.4.1.2: Sources of Hope and Despair in Human actions and Knowing

The reflections on hope and despair lead us to the examination of sources of hope, despair and the space that contains the spaces of hope, despair, and hope-despair dualities. What are the factors for variety hope and despair dispositions defined? The spaces in which the phenomena are defined are essential to understand their roles and importance in human actions over the spaces of problem-solution and ignorance-knowledge dualities. In all reflections, neuro-decision-choice actions and problem-solution processes take place in the space of imagination-reality dualities, where variety problems are defined by greater internal cost dispositions over internal benefit dispositions and variety benefits are defined by greater internal benefit dispositions over the internal cost dispositions in varieties, producing paths to the space of goals, objectives, or visions. Problems reside in the space of realities while solutions reside in the space of imaginations. Neuro-decision-choice actions are about changes of variety realities with replacements of variety imaginations and hence neuro-decision-choice actions reside in the space of imaginations. Since curiosity, creativity, courage, hope, fear relate to neuro decision-choice actions, they are all defined and exist in the space imaginations in the sense that they exist in cognition. Hope and fear are defined and reside in the space of possibility-probability dualities and affect the paths of problem-solution dualities.

The variety hope and variety fear from the space of possibility-probability dualities are traceable through the space of imaginations to the space of certainty-uncertainty dualities generate by epistemic distance from the space of ontic-epistemic dualities. Hope is affected creativity which the affects curiosity and search for new varieties and new solutions as well as innovation of existing variety realities in innovation of processes and improvement of actions over the spaces of production-consumption and ignorance-knowledge dualities. The essential source of hope and fear is generated by conditions of degrees of knowledge and the degree of ignorance that are traced to actual-potential dualities abstracted from the ontological space. When hope and fear are related to the variety in the space of realities, they relate to the conditions of losses of variety benefits of variety realities relative to cost-benefit conditions of variety imaginations.

4.4.2: Creativity over the Space of Hope-Despair Dualities in Knowing and Science

Let us take a note on creativity in relation to the knowing process, information-knowledge production, and problem-solution process at the guidance of neuro-decision-choice actions in all areas of human activities that will be useful at this point for the examination of diversity and unity in the system of thoughts over the epistemic field as a collection of categories with the corresponding relational and information fields. Let us keep in mind that the whole process of knowing as production of intellectual investment flows and information-knowledge accumulation as the production of intellectual capital stocks are dynamics of neuro-decision-choice system containing the organic structure of thinking and reasoning over the space of problem-solution dualities and the conception of human progress under the space of preferences as governed by the conditions of social philosophical consciencism that has taken place in the society. The thinking-reasoning processes are instruments within the neuro-decision-choice system which relates the space of problem-solution dualities to the space of fundamental-applied dualities, where variety identities are identified and transformed to provide us with three motions of changes in properties, changes in relations and simultaneous changes in both producing motions over the space of quantity-quality dualities. It has been pointed out that the whole of knowing and information-knowledge accumulations is about balancing the dualistic-polar conditions over variety ignorance-knowledge dualities in relation to variety identification-transformation dualities, where the duals are mutually constraining in the space of neuro-decision-choice actions. The solution to the variety identification problem provides the initial conditions and necessity for the variety transformation. It is within the space of identification-transformation dualities that courage containing curiosity and creativity find meaning and expressions, and hope is weaponized to overcome ignorance, where the knowing process provides organizational conditions for basic knowledge and applied knowledge containing technical knowledge and engineering knowledge as inputs into the spaced of production-consumption dualities.

At level of particularity, creativity is an operator function that enhances categorial conversions of varieties from one category to another by changing the variety characteristics. At the level of generality, creativity constitutes the elements in the space of transformation operator functions, where each function has specific operator-conversion capacity for specific actions in relation to the varieties in the space of problem-solution dualities and polarities. The effectiveness of any of the transformation function depends on constraints exerted by the philosophical consciencism that have taken hold of the social setup, where the philosophical consciencism may be a facilitator or inhibitor over the spaces of positive-negative dualities and hope-despair duality within the epistemic field.

Let us keep in mind that over the space of imagination-reality dualities, the dynamics of the space of production-consumption dualities relative to the dynamics of the space of input-output dualities are nothing but neuro-decision-choice actions over the space of problem-solution dualities in a general system of parent-offspring processes or system of successor-predecessor dynamics over the solution space of continual disequilibrium dynamics under the powerful *nyansapo* (adinkra *wisdom knot*) principle showing the varieties in the space of past-present dualities contained in the space of *beginning-end dualities* and the space of emptiness-fullness dualities is contained in the space of *nothingness-somethingness dualities* with complex connections to dimensions of existence, where a solution to a problem generates a new problem or a set of problems and hence a new knowing and a new opportunity of learning within the environment of the space of production-consumption dualities with conformal mapping onto the space of input-output dualities and then into the space of cost-benefit dualities.

In this epistemic framework, if there is anything important in the space of challenges in human cognitive actions, relative to knowing as intellectual investment flows and information-capital accumulation as an intellectual capital stocks over the spaces of ignorance-knowledge dualities, peace-war dualities, imagination-reality dualities, poverty-prosperity dualities and other dualistic-polar structures under the general principles of opposites, it is *creativity* as exercised over the space of fundamental-applied knowledge dualities in diversity and unity of knowing, science and information-knowledge accumulation with hope exercised over the spaces of possibility-probability dualities, anticipation-expectation dualities and necessity-freedom dualities that give life its meaning and essence in human existence. It is here that the basic structure of general human organization, composed of the economic structure, political structure and the legal structure may foster the development of varieties in the space of hope-despair dualities contained in social philosophical consciencism and politico-legal culture to manifest disrespect, and even disdain for information-knowledge to evolve a popular culture that houses individual and social actions in the zones of irreducible ignorance, where neuro-decision-choice actions are not guided by thoughts, thoughts are not translated into actions, which then produce accidental events of human sufferings and unintended benefit achievements in the space of possibility-probability dualities, as one move from the space of imaginations to the space of realities. In the space of ignorance, the occurrences of these accidental cost events and undesirable conditions of life are explained by conditions of dark spiritual forces of nature in the zone of cognitive imbecility, where research in certain areas of human endeavors are discouraged, some literature on some subjects are banned and punishable as illegals contained in the social philosophical consciencism from which the notions are drawn.

4.5 COMPARATIVE ORGANIZATIONAL ANALYTICS OF THE ONTOLOGICAL AND EPISTEMOLOGICAL SPACES IN THE SPACE OF KNOWING

In chapter 2 of this monograph, the geometric structure of the relationships of the ontological and epistemological spaces were presented in different forms in Figures 2.1.2, 2.2.1 and 2.5.2.1 in Chapter 2 of this monograph. The development of our universal knowledge system is a continual organization of variety relations between the varieties in the ontological space with ontic characteristic dispositions and varieties in the epistemological space with epistemic characteristic dispositions, where the epistemic disposition may differ from the ontic disposition of the same variety giving rise to variety ontic-epistemic duality with variety epistemic distance defining variety certainty-uncertainty duality, with further establishment of the space of variety ontic-epistemic dualities with the space of variety ontic-epistemic distances for the definition of the space of certainty-uncertainty dualities as it has been discussed [R3.26][R17.17][R17.18]. The space of ontic characteristic dispositions is always governed by information-knowledge equality constraints, such that there are no uncertainties and risks in the ontological processes in the sense that *what there was,* is *what there was*, *what there is,* is, *what there is* and *what there would be* is *what there would be* as established in the ontic space as a conformal mapping from the ontological space. Similarly, the space of epistemic characteristic dispositions is always governed by information-knowledge inequality constraints, such that there are uncertainties and risks in the epistemological processes in the sense that the cognition of *what there was* may not be *what there was*, *what there is* may not be *what there is* and *what there would be* may not be *what there would be* in the epistemological space as established in the epistemic space abstracted from the ontological space through the space of acquaintance under cognitive capacity limitations. Both the ontic and epistemic varieties are seen in terms of the principle of opposites with relational continuum and unity as extensively discussed in [R8.16b]. The ontological-epistemological relativity structure is presented as a relational geometry in Figure 4.1.1 where the relational roles of *Bennu anaoma*, *Gyenyame* and *Sankofa anoma* in the development of knowing, information-knowledge stocks and human actions are placed in the analytical space of absoluteness-relativity dualities.

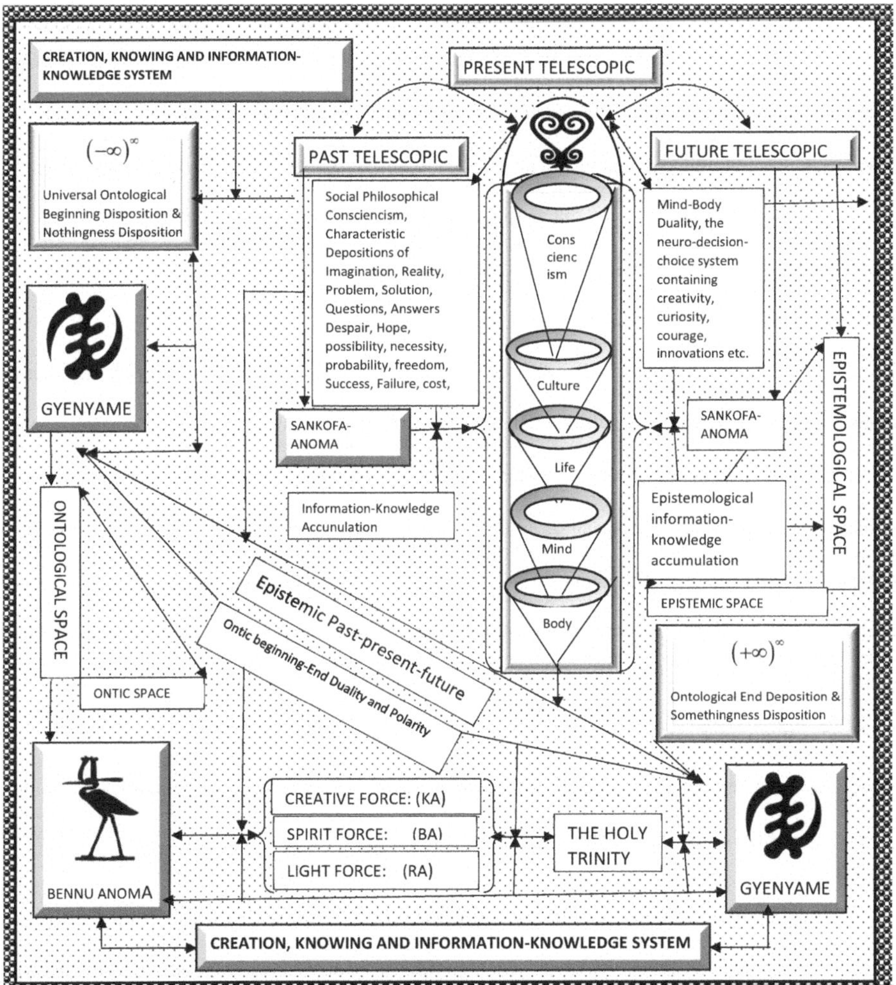

Figure 4.5.1: A Geometry of Organizational Paremiology of Relational Continuum and Unity in the Space of Ontological-Epistemological Duality for Knowing and Social Information-Knowledge Accumulation.

The ontic varieties and corresponding categories are established by natural laws of identity while the categorial conversions are established by natural laws of transformation through natural functors and categorial conversions. The epistemic varieties and corresponding categories are established by social laws of knowing as fuzzy decompositions through the use of fuzzy indicator functions that allow the establishment of interdependent neuro-categorial activities of the space of matter-mind dynamics through fix-level cuts of thinking-reasoning dualities that contain different degrees of qualitative activities of creativity, courage, innovation, hope, despair, risk and many other in the spaces of input-output dualities, production-consumption dualities and problem-solution dualities over the space of imagination-reality dualities contained in the space actual-potential dualities, where the space of actual-potential dualities is ontologically determined and the space of imagination-reality dualities is epistemologically determined, such that the ontic space is distinguished by information-knowledge equality constraints, while the epistemic space is distinguished by information-knowledge inequality constraints due to cognitive capacity limitations in all areas of knowing, information-knowledge production and human actions in the social existence and the space of ontic-epistemic dualities is distinguished by equality-inequality over the space of information-knowledge dualities.

It is important to keep in mind the distinguishing characteristics of the ontological space as universally creative identity and the epistemological space as constrained derived creative identity. The ontological space is composed of the space of actual-potential dualities as well as the space of ontic characteristic dispositions under natural dualistic-polar conditions as guided by the principle of opposites with relational continuum and unity which ensure ecological balances in intra-planetary, inter-planetary and exoplanetary systems under the principle of supersymmetry. The ecological balances in intra-planetary, inter-planetary and exoplanetary systems of the ontological space under the principle of supersymmetry are to ensure the polyrhythmic integrity of the universal relational continuum, unity and totality with continual creation and restructuring over the general spaces of destruction-construction and stability-instability dualities under the conflicts of opposites. The epistemological space is composed of the space of imagination-reality dualities as well as the space of epistemic characteristic dispositions under socio-natural dualistic-polar conditions as guided by the principle of opposites with relational continuum and unity which ensure social ecological structure in the space of imbalance-balance dualities within the space of social actions over the global system under the principle contained in the space of symmetry-asymmetry dualities. The social ecological structure in the space of imbalance-balance dualities within the space of social actions over our global system under the principle contained in the space of symmetry-asymmetry dualities in intra-planetary, inter-planetary and exoplanetary systems of the ontological space under the principle of supersymmetry leads to ensure the polyrhythmic integrity over the space of sustainability-unsustainability dualities of totality of human continual creation and restructuring over the general spaces of destruction-construction, input-output and stability-instability dualities and polarities.

Under the conditions of knowing as the production of intellectual investment flows and information-knowledge accumulation as the production of intellectual capital stocks, there is the *ontological time* defined by the space of beginning-end duality with beginning disposition as the beginning dual and the end disposition as the end dual under the principle of opposites with relational continuum and unity where, the beginning and end duals are defined by dualistic-polar dominance in the sense that the end is contained in the beginning and the beginning is contained in the end as the ontological existence with ontological time and space of time-information dualities defined within the beginning-end duality inclusively. The dualistic-polar conditions of the space of beginning-end dualities are conformally mapped onto the space of nothingness-somethingness dualities and vice versa with relational continuum and unity, where nothingness disposition is the nothingness dual and somethingness is somethingness dual. The space of nothingness-somethingness duality is the same as the space of beginning-end duality, where nothingness disposition (dual) corresponds to the beginning disposition (dual) and somethingness disposition (dual) correspond to the end disposition (dual) with natural conformal mapping. The nothingness-somethingness duality with nothingness-dual dominance correspond to beginning-end duality with beginning-dual dominance. Similarly, the nothingness-somethingness duality with somethingness dual dominance corresponds to beginning-end duality with the end dual dominance under dualistic-polar conditions with relational continuum and unity. The conditions of beginning dual disposition and end dual disposition are unavailable to the behavior at the epistemological space under the *Gyenyame* principle of universal existence.

This unavailability of conditions of beginning-end duality leads to conditions of the existence of knowable-unknowable duality and the space of knowable-unknowable dualities within the epistemological space relative to Sankofa-anoma conditions of information-time trinity of past-present-future telescopic structure. The conditions of knowability dual are related to varieties in the space of past-present dualities excluding the beginning dual and including the present dual in relational continuum and unity, while the conditions of unknowability dual is related to present-future duality including the end dual and excluding the present dual in relational continuum and unity. This does not mean that all elements in the space of knowability containing every present information-knowledge stock in relational continuum and unity within the space of static-dynamic dualities will be known and similarly, the elements in the space of unknowability containing every present information-knowledge stock in relational continuum and unity within the space of static-dynamic unity will not be known. The intersection of the space of knowability and the space of unknowability defines the present information-knowledge stocks over the space of static-dynamic dualities within the epistemological space.

From conditions of dualistic-polar structures, the relationships among the beginning-end duality, nothingness-somethingness duality and the ontological space must be clearly understood since these relationships establish the moving parameters of knowing as production of intellectual investment flows and information-knowledge accumulation as the production of intellectual capital stocks contained in the epistemological space. The construct of the epistemological space depends on the conceptions of the ontological space, while the integrity of the construct of the epistemological space depends on the conceptual conditions of the spaces of beginning-end dualities and nothingness-somethingness dualities in relational continuum and unity. The general understanding of the conceptual relationships of beginning, end, nothingness and somethingness within the ontological space will help us to minimize arrogance of knowledge claims and generation of paradoxes within the space of knowability and generations of phantom problems within the space of unknowability as well as maximize the conditions of cognitive effectiveness with the construct of the space of imagination-reality dualities contained in the epistemological space with consistent logical understanding of the phenomena of possibility, necessity, probability and freedom in the space of knowing.

There is also the *epistemological time* defined over the spaced of past-present dualities containing the present disposition. The rise of ill-posed problems, unsolvable problems, unsolvable problems, and conditions of incomputable problems in the space of abstract-nonabstract dualities is due to our poor conception of varieties in the ontological space as containing the conditions of the space of beginning-end duality, the nothingness-somethingness duality and their relational equality, continuum, and unity. The benefit of successful understanding of ontological order is to help an efficient construction of the space of acquaintance and its contents as abstracted from the ontological space and minimize the variety epistemic distances contained in the space of ontic-epistemic dualities, as well as help the understanding of the relational connectivity between the space of actual-potential dualities and the space of imagination-reality dualities and the epistemic journey over the space of ignorance-knowledge dualities by sharping our paradigms of thought from the conditions of the space of acquaintance and expanding the space of description.

By maximizing the space of acquaintance and minimizing the space of epistemic distance, phantom problems, and paradoxes, sharping our paradigms of thought over the space of imagination-reality dualities, our knowing, as production of intellectual investment flows, will be enhanced by shrinking the space of boundaries of limitations of the understanding of the conditions of the spaces of beginning-end and nothingness-somethingness dualities. The ontological space contains the spaces of beginning-end and nothingness-somethingness dualities under the dualistic-polar conditions with relational continuum and unity, where all ontological-epistemological varieties mimic this dualistic-polar stricture in the infinite universe, where the beginning is within the end and the end is within the beginning, the somethingness is within the nothingness and nothingness is within the somethingness in relational continuum and unity such that our linguist structures reflect the principle of opposites to establish the space of synonym-antonym dualities while our number system reflects the nothingness-somethingness duality to establish the form $[0,1]$ with extensions to $[-\infty,+\infty]$ and further extensions to $\left[(-\infty)^{\infty},(+\infty)^{\infty}\right]$ in the space of matter-energy dualities inclusive of the conditions of body-mind dualities over the spaces of symmetry-asymmetry and supersymmetry-super-asymmetry dualities that ensure the permanency of our universe in never-ending static-dynamic process.

4.5.1 Categorial Analytics of the Ontological Time, Epistemological Time, Beginning-End Dualities, Nothingness-Somethingness Dualities and Mathematical Spaces

Let us now turn our attention to the organization of categorial analytics of ontological and epistemological spaces, the construct of our number and algebraic systems. Let us note that if (∞) is an even number then $\left[(-\infty)^{\infty} = (+\infty)^{\infty}\right]$ and if (∞) is an odd number then $\left[\left|(-\infty)^{\infty}\right| = (+\infty)^{\infty}\right]$ in our mathematical systems. Similarly if $(\mathbf{N})$ is a non-empty set and $(\varnothing)$ is the empty set then the following conditions hold: 1) $(\mathbf{N}) \cup (\varnothing) = (\mathbf{N})$, 2) $(\mathbf{N}) \cap (\varnothing) = (\varnothing)$ which implies that $(\varnothing) \subset (\mathbf{N})$ .The relationships of these structures to ontological time, epistemological time, nothingness disposition, somethingness disposition, emptiness disposition and fullness disposition in the space of knowing are the production of intellectual investment flows and the space of information-knowledge accumulation as the production of intellectual capital stocks. At the level of ontological time and creational disposition with natural laws and processes, the structure represents the space of nothingness-somethingness dualities, where nothingness disposition is contained in somethingness disposition and vice versa. In this respect, $(0, \infty)$ presents ontological time structure with (0) as the ontological beginning containing nothingness-somethingness duality $(\mathbf{N}) \cup (\varnothing) = (\mathbf{N}) = (\varnothing) = (\mathbf{N})$ and the condition (∞) as the ontological end where

$(\mathbf{N}) \cup (\varnothing) = (\mathbf{N}) = (\varnothing) \subset (\mathbf{N}) \Rightarrow (\mathbf{N}) \cup (\mathbf{N}) \cap (\mathbf{N}) \cup (\varnothing) = (\mathbf{N})$ and $(-\infty, 0) \underset{\mathbf{T}^{-1}}{\overset{\mathbf{T}}{\rightleftarrows}} (0, \infty)$ represents the telescopic past-future duality with relational continuum and unity under parent-offspring process, where $(\mathbf{T})$ is natural categorial converter or functor and $(\mathbf{T}^{-1})$ is the negation functor.

At the level of epistemological time and production-consumption dispositions with social laws and processes, the structure represents the space of emptiness-fulness dualities $(\mathbf{E} \cup \mathbf{F})$ where emptiness disposition $(\mathbf{E})$ is contained in fullness disposition $(\mathbf{F})$ and fullness disposition is contained emptiness disposition $(\mathbf{E})$. In this respect, $(-\infty, 0)$ represents an epistemological time structure of past-present duality, where (0) shows the epistemic initial conditions containing ignorance, emptiness and necessity $(\mathbf{E})$ and $(-\infty)$ represents a telescopic past time in the space of knowability. Similarly, $(0, \infty)$ represents an epistemological time structure of present-future duality where (0) shows an epistemological initial disposition of fullness and freedom $(\mathbf{F})$ such that emptiness is contained in fullness and vice versa under the dualistic-polar relation and thus we have $(\mathbf{E} \subseteq \mathbf{F} \supseteq \mathbf{E})$ and (∞) represents telescopic future time in the space of unknowability and hence we have variety ignorance-knowledge process of the form $(-\infty, 0) \underset{\mathfrak{f}^{-1}}{\overset{\mathfrak{f}}{\rightleftarrows}} (0, \infty)$ where $(\mathfrak{f})$ is social ignorance-knowledge categorial converter or a functor and . $(\rightleftarrows)$ is parent-offspring process or negation of negation. There are similarities and differences between the ontological time and epistemological time. In a simple form, the ontological time relates to beginning-end and nothingness-somethingness dualities as establishing ontological information-time structure distinguished from epistemological time which relate to past-future and emptiness-fullness dualities as establishing epistemological information-time structure, where the epistemological information-time structure is contained in the ontological information-time structure. In this epistemic structure, it is important to observe two important principles in the space of ontological-epistemological duality of universal existence in terms of organic information-time process in knowing: 1) The union of all the spaces of emptiness-fullness dualities is an a subspace of (i.e. less than) the space of nothingness-somethingness duality; and 2) the union of the spaces of past-future dualities is also a subspace of (i.e. less than) the space of beginning-end duality. The first one is *matter-energy principle*, while the second is *information-time principle* of universal knowing.

4.5.2 Categorial Analytics of the Organization of Ontological, and Epistemological spaces

In the categorial analytics, let $(\mathscr{B})$ represent the space of beginning dispositions with a generic element $(\mathscr{b} \in \mathscr{B})$ while $(\mathfrak{E})$ represents the space of end dispositions with a generic element $(\mathfrak{e} \in \mathfrak{E})$ then the space beginning-end dualities is $(\mathfrak{U} = \mathscr{B} \cup \mathfrak{E})$ with relational continuum and unity and fuzzy conditionality of the form $(\mathscr{B} \cap \mathfrak{E} \neq \varnothing)$ with a fuzzy and dualistic-polar conditions $(\mathscr{B} \otimes \mathfrak{E}) = (\mathscr{B} \cup \mathfrak{E})$ where $\{(\mathfrak{u} = (\mathscr{b}, \mathfrak{e}) \in \mathscr{B} \otimes \mathfrak{E}) \mid \mathscr{b} \in \mathscr{B}, \mathfrak{e} \in \mathfrak{E}\}$. Similarly, let $(\mathbf{N})$ be the nothingness disposition and hence the nothingness dual with a generic element $(\mathbf{n} \in \mathbf{N})$ while $(\mathbf{S})$ represents the space of somethingness dispositions with a generic element $(\mathbf{s} \in \mathbf{S})$, then the space nothingness-somethingness duality is $(\mathbf{Q} = \mathbf{N} \cup \mathbf{S})$ with relational continuum and unity and fuzzy conditionality in the form $(\mathbf{N}\cap\mathbf{S}\neq\varnothing)$ and dualistic-polar condition $(\mathbf{N}\otimes\mathbf{S})=(\mathbf{N}\cup\mathbf{S})$ where $\{(\mathbf{q}=(\mathbf{n},\mathbf{s})\in(\mathbf{N}\otimes\mathbf{S}))\mid\mathbf{n}\in\mathbf{N},\mathbf{s}\in\mathbf{S}\}$.

At the ontological space, the beginning disposition $(\mathscr{B})$ contains nothingness disposition with equality such that $(\mathscr{B}=\mathbf{N}),(\mathscr{B}\cup\mathbf{N})=(\mathbf{N})=(\mathscr{B})=(\mathscr{B}\cup\mathbf{N})$ with $(\mathscr{B}\cap\mathbf{N})\neq\varnothing$. Similarly, the end disposition contains the somethingness disposition with equality such that $(\mathfrak{E}=\mathbf{S}),(\mathfrak{E}\cup\mathbf{S})=(\mathbf{S})=(\mathfrak{E})=(\mathfrak{E}\cap\mathbf{S})\neq\varnothing$ with all the dualistic-polar conditions and fuzzy conditionality holding over the space of actual-potential dualities $(\mathfrak{A}\cup\mathfrak{U})=(\mathfrak{A}\otimes\mathfrak{U})$. Similarly, the space of the end-beginning dualities $(\mathscr{B}\cup\mathfrak{E})$ and space of nothingness-somethingness dualities $(\mathbf{N}\cup\mathbf{S})$ have mutual containments such that $(\mathscr{B}\cup\mathfrak{E})=(\mathbf{N}\cup\mathbf{S})\Rightarrow(\mathbf{N}\otimes\mathbf{S})=(\mathscr{B}\otimes\mathfrak{E})\Rightarrow(\mathscr{B}\cup\mathfrak{E})\subseteq(\mathbf{N}\cup\mathbf{S})\subseteq(\mathscr{B}\cup\mathfrak{E})$. The space of actual-potential dualities $\Omega=(\mathfrak{A}\cup\mathfrak{U})=(\mathfrak{A}\otimes\mathfrak{U})$ is also contained in the spaces of beginning-end dualities $(\mathscr{B}\cup\mathfrak{E})$ and nothingness-somethingness dualities $(\mathbf{N}\cup\mathbf{S})$ inclusively such that $\Omega=(\mathfrak{A}\cup\mathfrak{U})\subseteq(\mathscr{B}\cup\mathfrak{E})=(\mathbf{N}\cup\mathbf{S})\subseteq(\mathfrak{A}\cup\mathfrak{U})$ and $\Omega=(\mathfrak{A}\otimes\mathfrak{U})\subseteq(\mathscr{B}\otimes\mathfrak{E})=(\mathbf{N}\otimes\mathbf{S})\subseteq(\mathfrak{A}\otimes\mathfrak{U})$, The beginning disposition $(\mathscr{B})$ with nothingness dispositions $(\mathbf{N})$ and the end disposition $(\mathfrak{E})$ with somethingness disposition $(\mathbf{S})$ and hence the space of beginning-end dualities $(\mathscr{B}\cup\mathfrak{E})$ and nothingness-somethingness dualities $(\mathbf{N}\cup\mathbf{S})$ are established by the *Gyenyame principle* called *Gyenyame creational structure*. The relational connectivity between the beginning-end duality and relational connectivity between nothingness-somethingness duality are established by *Bennu-anoma principle* called *Bennu-anoma connectivity structure.*

At the epistemological space, the beginning disposition and the nothingness disposition are unavailable for knowing under both Gyenyame principle and Bennu-anoma principle. What is available to knowing is always the past telescopic disposition $(\mathscr{M})$ and the present telescopic disposition $(\mathscr{N})$ that constitute *what there is* with either spaces of emptiness dispositions $(\mathbf{E})$ with the space of fullness dispositions $(\mathbf{F})$ such that there is the space of knowabilities $(\mathscr{E} = (\mathscr{M} \cup \mathscr{N}))$ defining the space of past-present dualities excluding the space of beginning disposition $(\mathscr{B})$ and nothingness disposition $(\mathbf{N})$ by the Gyenyame principle and creational structure such that the space of knowabilities is the symmetric difference (∇) between the ontological space and the space of the union of the beginning-nothingness dispositions $(\mathscr{B} \cup \mathbf{N})$ that may be specified as $(\mathscr{E} = (\mathscr{M} \cup \mathscr{N})) = \{(\Omega) \nabla (\mathscr{B} \cup \mathbf{N})\} = ((\mathfrak{A} \cup \mathfrak{U}) \nabla (\mathscr{B} \cup \mathbf{N})) = [(\mathfrak{A} \otimes \mathfrak{U}) \nabla (\mathscr{B} \otimes \mathbf{N})]$ such that $(\mathscr{M} \cap \mathscr{N}) \neq \varnothing$, $(\mathscr{B} \cap \mathbf{N}) \neq \varnothing$, $(\mathfrak{A} \cap \mathfrak{U}) \neq \varnothing$ as well as all types of the space intersectionality for fuzzy conditionality in relation to the conditions of the space of emptiness-fullness dualities $(\mathbf{E} \cup \mathbf{F})$ such that $(\mathbf{E} \otimes \mathbf{F}) = (\mathbf{E} \cup \mathbf{F})$ with $(\mathbf{E} \cap \mathbf{F}) \neq \varnothing$ and $(\mathscr{M} \cap \mathscr{N}) \subseteq (\mathbf{E} \cap \mathbf{F})$ with the space of past-present dualities $(\mathscr{M} \cup \mathscr{N})$ contained in the space of emptiness-fullness dualities $(\mathbf{E} \cup \mathbf{F})$ relative to the space of imagination-reality dualities $(\mathfrak{B} \cup \mathfrak{G}) \subset (\mathscr{M} \cup \mathscr{N})$. Let us observe that the space of beginning-end duality $(\mathfrak{U} = \mathscr{B} \cup \mathfrak{E})$ is partitioned into the spaces of past-future dualities in the form $\{\mathscr{H}_i = (\mathscr{M}_i \cup \mathscr{F}_i) \mid i \in \mathbb{I}^\infty\}$ such that $\left\{\bigcup_{i \in \mathbb{I}^\infty} \mathscr{H}_i\right\} \subset (\mathfrak{U} = \mathscr{B} \cup \mathfrak{E})$ for information-time principle of knowing with dualistic-polar condition of the form $\bigcap_{i\in\mathbb{I}^\infty} \mathscr{A}_i \neq \varnothing$ and $\{(\mathscr{M}_i \cap \mathscr{F}_j) \neq \varnothing \mid \forall i,j \in \mathbb{I}^\infty\}$. Additionally, there is the space of nothingness-somethingness duality in form $(\mathbf{Q} = \mathbf{N} \cup \mathbf{S})$ and the space emptiness-fullness dualities in the form $\mathbf{A}_i = \{(\mathbf{E}_i \cup \mathbf{F}_i) \mid i \in \mathbb{I}^\infty\}$ such that $\bigcup_{i\in\mathbb{I}^\infty} \mathbf{A}_i = \bigcup_{i\in\mathbb{I}^\infty} (\mathbf{E}_i \cup \mathbf{F}_i) \subset (\mathbf{Q} = \mathbf{N} \cup \mathbf{S})$ and fuzzy dualistic-polar conditions $\bigcap_{i\in\mathbb{I}^\infty} \mathbf{A}_i \neq \varnothing$, and $\{(\mathbf{E}_i \cap \mathbf{F}_j) \neq \varnothing, \forall i,j \in \mathbb{I}^\infty\}$ as the matter-energy principle of universal knowing with a summary of the form:

$$\left.\begin{cases} \mathbf{A}_i = \{(\mathbf{E}_i \cup \mathbf{F}_i) \mid i \in \mathbb{I}^\infty\} \ni \bigcup_{i\in\mathbb{I}^\infty} \mathbf{A}_i = \bigcup_{i\in\mathbb{I}^\infty} (\mathbf{E}_i \cup \mathbf{F}_i) \subset (\mathbf{Q} = \mathbf{N} \cup \mathbf{S}) \\ \text{with fuzzy conditionality} \bigcap_{i\in\mathbb{I}^\infty} \mathbf{A}_i \neq \varnothing, \text{ and} \{(\mathbf{E}_i \cap \mathbf{F}_j) \neq \varnothing, \forall i,j \in \mathbb{I}^\infty\} \end{cases}\right\} \text{Matter-Energy Principle}$$

$$(4.5.2.1)$$

$$\left.\begin{cases} \{\mathscr{A}_i = (\mathscr{M}_i \cup \mathscr{F}_i) \mid i \in \mathbb{I}^\infty\} \ni \left\{\bigcup_{i\in\mathbb{I}^\infty} \mathscr{A}_i\right\} = \bigcup_{i\in\mathbb{I}^\infty} (\mathscr{M}_i \cup \mathscr{F}_i) \subset (\mathfrak{U} = \mathscr{B} \cup \mathfrak{E}) \\ \text{with a fuzzy conditionality} \bigcap_{i\in\mathbb{I}^\infty} \mathscr{A}_i \neq \varnothing \text{ and } \{(\mathscr{M}_i \cap \mathscr{F}_j) \neq \varnothing \mid \forall i,j \in \mathbb{I}^\infty\} \end{cases}\right\} \text{Information-Time Principle}$$

The relational unity of the matter-energy and information-time principles of knowing provides a look of the space-time phenomenon over the spaces of problem-solution dualities and ignorance-knowledge dualities.

Similarly, the space of imagination dispositions $(\mathfrak{G})$ containing the space of emptiness dispositions $(\mathbf{E})$ in the space of imagination-reality dualities $(\mathfrak{B} \cup \mathfrak{G})$ containing the space of present-future dualities $(\mathscr{N} \cup \mathscr{F})$ excluding the present telescopic dispositions $(\mathscr{N})$ is not available to knowing and hence constitute the space of non-knowabilities $(\mathscr{U} = (\mathscr{N} \cup \mathscr{F}))$ with relational continuum and unity such that $(\mathscr{N} \cap \mathscr{F}) \neq \emptyset$ as the fuzzy conditionality which allows the decision-choice actions to examine the epistemic transversality conditions over the space of ignorance-knowledge dualities, where $(\mathscr{F})$ is future dispositions. In the space of knowing and information-knowledge production, the space of imagination-reality dualities $(\mathfrak{B} \cup \mathfrak{G})$ is contained in the space of knowability $(\mathscr{E} = (\mathscr{M} \cup \mathscr{N}))$ such that $(\mathfrak{B} \cup \mathfrak{G}) \subset (\mathscr{M} \cup \mathscr{N})$ and $(\mathfrak{B} \otimes \mathfrak{G}) \subset (\mathscr{M} \otimes \mathscr{N})$ with all the dualistic-polar conditions and fuzzy conditionality holding over the space of imagination-reality dualities. Similarly, the space of the reality dispositions and the space of the fullness dispositions have mutual containment $(\mathfrak{B} \subseteq \mathbf{F})$ where the space of nothingness $(\mathbf{N})$ contained in the ontological space (Ω) is different from the space emptiness $(\mathbf{E})$ contained in the epistemological space $(\mathfrak{E})$ and similarly, the space of somethingness $(\mathbf{S})$ in the ontological space $\Omega = (\mathfrak{A} \cup \mathfrak{U}) = (\mathfrak{A} \otimes \mathfrak{U})$ is different to the space fullness $(\mathbf{F})$ contained in the epistemological space $(\mathfrak{E}) = (\mathfrak{B} \cup \mathfrak{E})$. All the spaces and subspaces in the ontological space are primary identities and all spaces and subspaces in the epistemological space are derived identities such that the meaning of epistemic emptiness is in relation to ontological nothingness and the meaning of epistemic fullness is in relation to ontological somethingness and the meaning of the epistemological space of imagination-reality dualities is in relation to ontological space of actual-potential dualities.

The space of past dispositions with the space of emptiness dispositions and the future dispositions with fullness dispositions and hence the past-future dualities and emptiness-fullness dualities are established by the *cognitive principle* called *epistemic creational structure*. The relational connectivity between the past-future dualities, past-present dualities and present-future dualities and relational connectivity between emptiness-fullness dualities is established by Sankofa-anoma principle of information-time trinity. The elements of primary identities, spaces and subspaces and derived identities, spaces and subspaces present the comparative organizational essentialities of the ontological and epistemological spaces as well as the ontic and epistemic spaces. The Sankofa-anoma principle of time trinity is a characteristic of the epistemological space. It is not a characteristic of ontological space. Imagination, possibility, probability, problem, solution, success, failure, true, false, cost, benefit, vision, goal, objective, wellness, illness, progress, war, peace, love, hate, reality and many such varieties are cognitive constructs that belong to the epistemological space. They do not belong to the ontological space in which there are no meaning, essence and utility in the ontological existence and processes. Gyenyame and Bennu-anoma principles play important roles over the organizational analytics of the ontological space containing the space of actual-potential dualities, where the spaces of beginning-end and nothingness-somethingness dualities are defined with continual ontological variety transformations, where Sankofa-anoma principles play no role.

The Gyenyame and Bennu-anoma principles play no active roles over the space of the epistemological processes that connect cognition to the space of ontological processes. The Gyenyame and the Bennu-anoma principles establish the boundaries of the spaces of acquaintances, knowabilities and non-knowabilities for cognitive actions through knowing as intellectual investment flows, information-knowledge accumulation as production of intellectual capital stocks and variety epistemic distances resulting from the space of cognitive capacity limitations over the space of ontic-epistemic dualities through their impacts on hope disposition, fear disposition, belief disposition and other related variety dispositions. The cognitive capacity limitations which cannot be separated from the space of acquaintance-description dualities place boundaries on the space of scientific and mathematical knowabilities over the organizational space of fundamental-applied dualities from which technological dispositions and engineering dispositions are abstracted through creative understanding. The Sankofa-anoma principle of the info-time trinity which is composed of three info-time dualities of past-present dualities, present-future dualities and past-future dualities governs cognitive behavior with cognitive capacity limitations over the space of ignorance-knowledge dualities and all neuro-decision-choice actions over the space of problem-solution dualities relative to the space of human behaviors. The relative standing of the ontological and epistemological space is presented as cognitive geometry in Figure 4.5.1.

4.5.3 Categorial Analytics, Principle of Opposites, Supersymmetry in Ontological-Epistemological Dualities and Polarities

A careful analysis of the symbolic representations of Gyenyame and Bennu-anoma in the ontological space and symbolic representation of Sankofa anoma reveals conditions of dualistic-polar supersymmetries from which the *adinkramathics* and *adinkralogy* may be developed and connected to *anansimatics* and *anansiology* elements of which, we have discussed in the monograph [R8.16b]. The *adinkramatics* is a system of geometric-algebraic relational structures of the adinkra conceptual system of universal existence in a system of dualistic-polar symmetries. The *adinkralogy* is the philosophical system of adinkras that sees the universe as composed of varieties under the principle of opposites in relational continuum, unity, symmetries, and supersymmetries. The adinkramatics is complemented by *kentematics* while *the adinkralogy* is complemented by *kentelogy* in relation to the string geometry of Anansi (spider) web and its interpretations to observable universe as existing in a system of interconnected strings. The algebraic-geometric relational structure of the Anansi (spider) web is the *anansimatics*, while the corresponding philosophical system of its interpretation is the *anansiology* [R27.3], [R27.5] [R27.10] [R27.14]. The *kentematics*, is a system of geometric-algebraic relational structures of the strings in a conceptual system of relational network of fractals in a system of dualistic-polar varieties over the space of symmetry-asymmetry dualities. The *kentelogy* is the philosophical system of *kente* fractals in visual representations of conceptual varieties in paremiological structures under the principle of opposites in relational continuum, unity, symmetries, and supersymmetries. The paremiological structures are proverbial representations over the space of visual-verbal dualities. The conceptual family of the systems of *adinkramatics*, *adinkralogy*, *anansimatics, anansiology, kentematics* and *kentelogy* is concern with the geometries, mathematics, and paradigms of thought about variety identities and connecting relations that together constitute the universe existence in continuum and unity over the space of visible-invisible dualities.

The principle of opposites, with system of dualities and polarities in relational continuum, unity, symmetries, and supersymmetries, is the characteristics of Gyenyame, Bennu anoma and Sankofa anoma in the knowing and information-knowledge development. The theory of the principle of opposites is defined in the space of symmetry-asymmetry dualities in the space of dualistic-polar games of conflicts and changes with respect to negation and negation of negation of the duals of dualities and poles of polarity, where the space of visible-invisible dualities and polarities presents the condition, where the system of connecting strings is the system of network of relations which may also be viewed as a system of problem-solution processes connecting the relational strings. In the space of dualistic-polar games of dualistic-polar negation for identity dominance, we must always distinguish between ontological actions related to natural processes and epistemological actions related to social processes and their interactions, where negation and negation of negations over the ontological space are not open to cognitive agents. The epistemological negation and negation of negation over the dualistic-polar game space are through neuro-decision-choice actions through social contradictions. Over the space of dualistic-polar games, the neuro-decision-choice actions are the development of dualistic-polar balances in the choice of dualistic-polar identity dominance in the epistemological space contained in the space of ontological-epistemological dualities. The unified system of knowing and information-knowledge production and dualistic-polar system of problem-solution dynamics may also be seen in terms of systems of mathematics and philosophical interpretations of fractal analytics and visual analytics in the space of varieties under the principle of opposites [R27.14[[R28.1] [R28.6] [R28.7],[R28.24][R29.1][R.15].

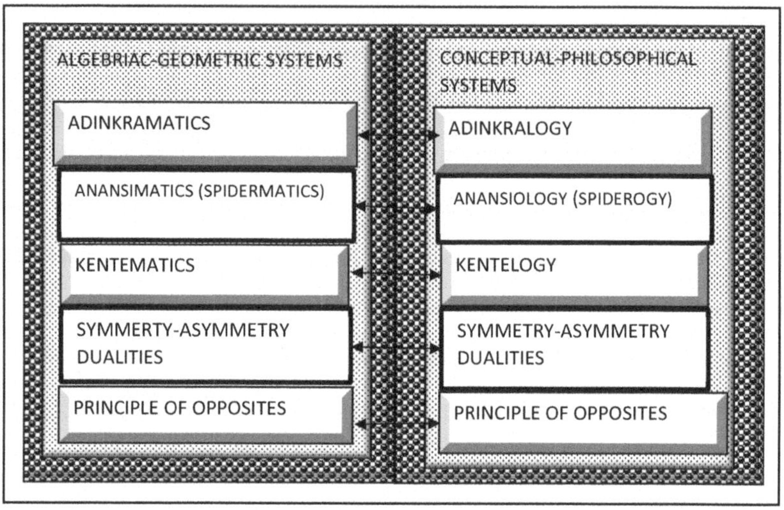

Figure 4.5.3.1 A Paremiological Relational Structure of the Principle of Opposites in Knowing and Information-knowledge Production.

From Figure 4.5.1, we can decompose the organization of the spaces of knowing processes and information-knowledge processes into conceptual structures of a natural organization of the ontological space and a social organization of the epistemological space with the understanding of their relational connectivity and unity through information defined as characteristic dispositions with relational dispositions also defined as characteristic dispositions for variety identities and similarities for categorial formations over the space ontological input-output dualities. The space of ontological input-output dualities is defined over the space of static-dynamic dualities for continual ontological actions in the maintenance and management of ecological and environmental balances in complexity and diversity with ontological variety transformations which are defined over the space of natural identification-transformation dualities. The space of natural identification-transformation dualities and the space of input-output dualities are mutually containing over the space of static-dynamic dualities for efficient recycling of the natural production, where the spaces of the natural outputs and natural inputs are mutually interdependent and reversable where inputs are outputs and outputs are inputs and existing in the space of internalities such that there are no externalities in the natural production processes.

Corresponding to the space of ontological input-output dualities is the space of epistemological input-output dualities also defined over the space of static-dynamic dualities for continual epistemological actions in the maintenance, control, and management of balances in social complexity and diversity with epistemological variety transformations which are defined over the space of social identification-transformation dualities. The space of social identification-transformation dualities and the space of input-output dualities are mutually containing over the space of static-dynamic dualities for inefficient recycling of the social production, where the spaces of the socio-natural outputs and socio-natural inputs are mutually interdependent and incompletely reversable where inputs are outputs and outputs are inputs and exist in the space of internality-externality duality such that the externalities are defined over the space of socially desirable, undesirable and naturally harmful dualities in the social production processes in relation to neuro-decision-choice actions over the space of problem-solution dualities that acquires meaning over the space of goals, objectives and visions relative to the space of cost-benefit dualities mapped onto the space of incentive-disincentive dualities, where theories of neuro-decision-choice processes acquire complex epistemic structures with give-and-take structures over the space of problem-solution dualities, where problems and solutions are defined in terms of cost-benefit information as facilitating or inhibiting the attainment of goals, objectives and visions over the space of individual-community dualities. The inefficient social input-output processes are foundations in explaining changing conditions of climatological and geomorphological changes.

4.5.3.1 A Geometric Representation of the Ontological Space

The ontological space contains all the variety actuals and potentials in terms of states, processes, identities, and transformations. The ontological space and all its contents constitute the universal primary identities. At the ontological space, there is the beginning-end duality in an indistinguishable existence, where the end resides in the beginning and the beginning resides in the end. Similarly, there is nothingness-somethingness duality in indistinguishable existence, where somethingness resides in nothingness and nothingness resides somethingness. This indistinguishability of mutual existence provides us with the understanding of the statement in mathematical construct that every set contains the null set which presents the difficulty for the classical paradigm of thought, where opposites cannot have simultaneous existence in a place and time under the principle of opposites and excluded middle. The geometric representation of the organizational analytics of the ontological space as abstracted from Figure 4.5.1 with the center of creation and natural variety transformations is shown in Figure 4.5.3.1.1.

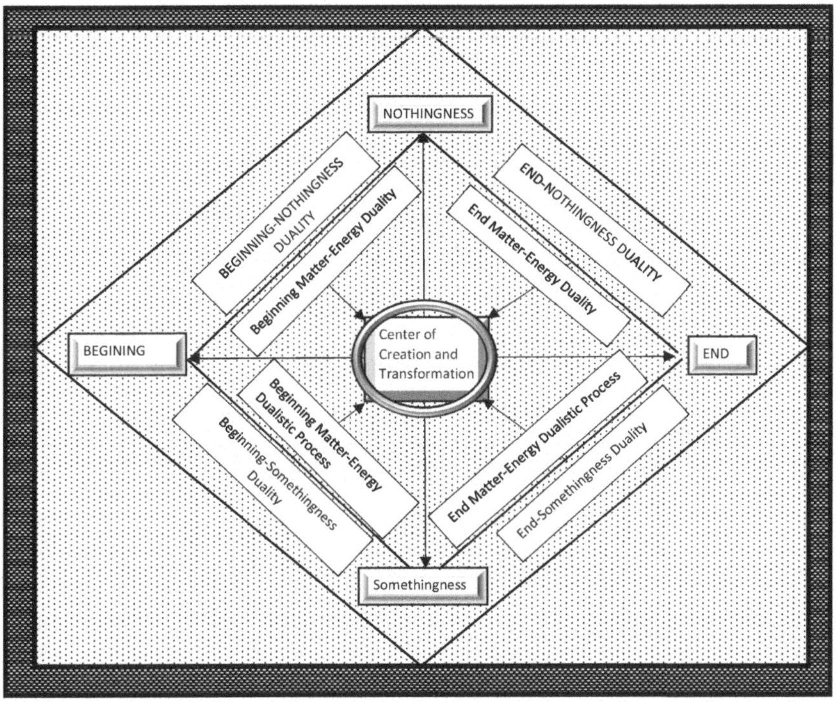

Figure 4.5.3.1.1 The Principle of Opposites in the Universal Creation and Ontological Processes with Ontological Information-Time in the Space of Ontological Input-Output Structure

4.5.3.2 A Geometric Representation of the Epistemological Space

At the epistemological space, there is the past-future duality in an indistinguishable existence, where the past that holds the initial conditions resides in the future and the future that holds goals, objectives and visions resides in the past. Similarly, there is the emptiness-fullness duality in indistinguishable existence, where emptiness resides in fullness and fullness resides in emptiness. The epistemological indistinguishability of mutual existence, just like ontological indistinguishability, provides us with the foundation of logical statement in mathematical construct that every set contains the null set [R3.6] [R3,15] [R3.57] [R3.88] [R17.15]. This mathematical construct presents a logical difficulty for the classical paradigm of thought, where opposites cannot have simultaneous existence in a space and time under the principle of opposites and excluded middle, where every statement is either true of false but not both and yet several proofs of theorems in mathematical uses the claim of contradiction as inconsistent in the classical paradigm of thought. The epistemological space contains all the variety imaginations and realities in terms of states, processes, identities, and neuro-decision-choice transformations. The epistemological space and all its contents constitute the universally derived identities with a geometric representation of the organizational analytics as a derived identity and abstracted from Figure 4.5.1 with the center of initial conditions and epistemological processes as shown in Figure 4.5.2.2 in a relative standing to the set of conditions of the ontological space. Let us notice that the space of past-future dualities is contained in the space of beginning-end dualities and the space of emptiness-fullness dualities is also contained in nothingness-somethingness dualities, that is. $(\mathbf{E} \cup \mathbf{F}) \subset (\mathbf{N} \cup \mathbf{S})$ and $(\mathscr{M} \cup \mathscr{N}) \subset (\mathscr{B} \cup \mathfrak{E})$.

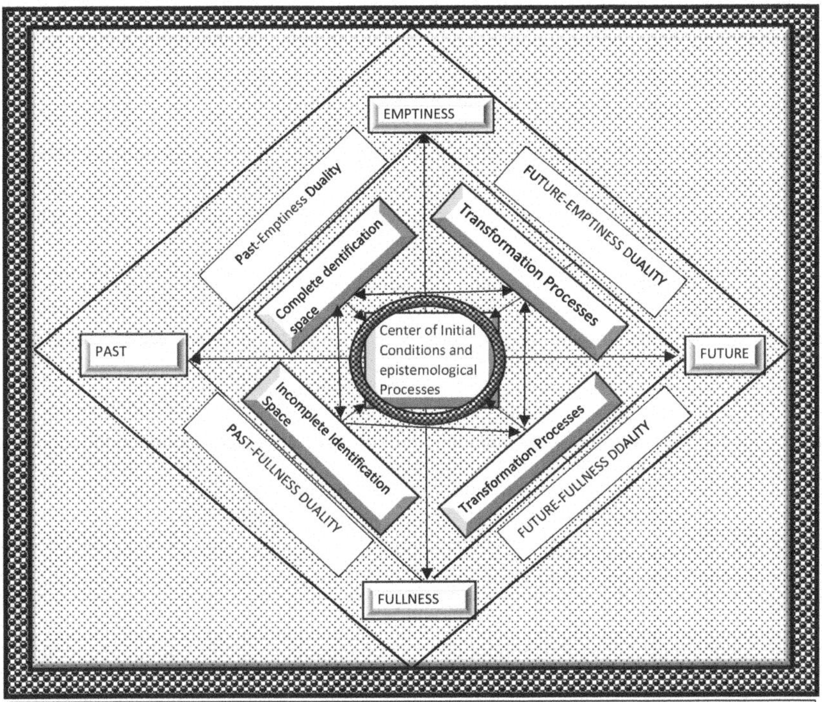

Figure 4.5.3.2.1 The Principle of Opposites in the Variety Identification and Epistemological Processes with Epistemological Information-Time Knowing Process

4.5.3.2.1 Ontological Space, Epistemological Space, and the Space of Absoluteness-Relativity Dualities

The varieties in the ontological space exist in the space of absoluteness in identity while their characteristic dispositions exist in the space of relativity under the principle of opposites with relational continuum and unity, where the space of relativity contains the space of absoluteness.

The space of relativity with excluded middle becomes the classical logical space of absoluteness, with separation and logical non-contradiction. The space of relativity with relational continuum and unity becomes the fuzzy logical space of relativity with acceptance of contradiction and simultaneous existence of opposites in the same variety just like every set contains fullness and emptiness.

The classical paradigm of thought with its mathematics and logics claims variety absoluteness. The fuzzy paradigm of thought with its mathematics and logics works towards absoluteness where every claim is a neuro-decision-choice action of dualistic-polar dominance in relative characteristic dominance. Both the classical and fuzzy spaces are under principle of opposites composed of dualities and polarities. The classical space of logical absoluteness is identified by conditions of opposites with excluded middle, disunity, and existence of non-contradiction such that, for example, there are conditions in the neuro-decision-choice space, where there is true or false but not both for the same statement in the same space and time [R3.13] [R3.86]. The fuzzy space of logical relativity is identified by conditions of relational continuum, unity, and existence of contradictions such that, for example, there are conditions in the neuro-decision-choice space where every truth contains falsity, and every falsity contains characteristics of truth for the same statement in the same space and time with the claim of truth or falsity determined by relative characteristic-disposition dominance. Similarly, ever variety is simultaneously cost and benefit relative to the elements in the space of goals, objectives, and visions, where cost-benefit relativity is the foundation of neuro-decision-choice actions in the space of acceptance-rejection dualities.

The understanding of the relational interrelated connectivity of the spaces of relativity, absoluteness, quality-quantity dualities, and invisible-visible dualities in the space of information-relational dualities is important to the understanding of the differences of classical paradigm of thought with its mathematics and logic and fuzzy paradigm of thought and its mathematics and logic. The classical information with its exactness and exclusion of vagueness, disregarding linguistic variables in the space of invisibles may be applied to varieties in the space of ontological quantities and situations where our technological processes provide conditions of transforming ontological qualities into quantitative characteristic by some form of measurements with reasonable degrees of exactness. The fuzzy information covering conditions of exactness and vagueness is applied to all elements of ontological varieties contained in the space of quantity-quality dualities relative to the space of visible-invisible dualities, where subjectivity is applicable to all varieties in the space of invisibles in such a way that neuro-decision-choice actions apply in all areas in the space of visible-invisible dualities in the processes of knowing and information-knowledge production. It is on the behavioral conditions of the classical paradigm of thought and the fuzzy paradigm of thought over the space of visible-invisible dualities that a claim is made that the analytical results of the classical paradigm is contained in the analytical results of the fuzzy paradigm of thought, where the fuzzy paradigm of thought under the principle of continuum and unity allows subjective assessments of invisible varieties, the classical paradigm of thought with an emphasize on objectivity, relational separation and disunity with the principle of excluded middle has epistemic difficulties over the space of invisibles that generates linguistic variables.

CHAPTER 5

THE THEORY OF EPISTEMIC ANOMALIES WITHIN THE PRINCIPLES OF DIVERSITY AND UNITY IN KNOWING AND SCIENCE: Organizational Algebraic Analytics

ABSTRACT

In chapter 4, the conceptual frame of anomalies in knowing and information-knowledge production was presented and related to some important elements in the space of quality-quantity dualities over the space of problem-solution dualities and polarities. Chapter 5 is a continuation with algebraic analytics of anomalies in knowing and information-knowledge systems and the roles that courage, curiosity, creativity, and innovation play in the space of imagination-reality dualities contained in the epistemological space and as a sub-space of the space of actual-potential dualities which is contained in the ontological space. The natures of the phenomena of thinking and reasoning relative to courage, curiosity and knowing are presented, defined, and explicated, where the differences and similarities among them are shown in relation to the neuro-decision-choice actions over the spaces of problem-solution dualities and certainty-uncertainty dualities. The concepts and phenomena of thinking and reasoning are related to possible worlds and the space of realities by the way of comparative analytics of relations among necessity, freedom and creativity which are then examined by way of analytical connectivity through definitions, explications, and analyses. From the analytical connectivity among possibility space and probability space in the space of imaginations to the space of realities as a subspace of the actuals, the concepts and phenomena of phantom problem, ill-posed problem, unsolvable problem, news, fake news, alternative facts and alternative reality are introduced, defined, explicated and

K. K. Dompere, *Theories of Knowing and Knowledge Production*, Studies in Systems, Decision and Control 501, https://doi.org/10.1007/978-3-031-42863-0_5

connected to impact analyses of neuro-decision-choice actions under the general social philosophical consciencism that generates systemic risk over the space of cost-benefit dualities at generational points, where the dynamics of the social philosophical consciencism help to explain the dynamics and differences in generational preferences and style of social management.

5.1 COURAGE, CURIOSITY, CREATIVITY AND CATEGORIAL CONVERSIONS

In the Chapter 4 of this monograph, devoted to the theory of intellectual anomalies, discussions were made regarding the give-and-take sharing relationships among courage, curiosity, creativity and their roles in knowing and information-knowledge production and their further complex relations with innovation, variety identifications and transformations over the spaces of possibility-impossibility dualities and probability-improbability dualities contained in the space of imagination-reality dualities as a sub-space of the space of actual-potential dualities, where knowing and information-knowledge production are viewed as neuro-decision-choice activities over the space of problem-solution dualities. The neuro-decision-choice activities are operated through categorial conversions by changing variety properties and relations and transforming varieties through destruction-construction processes, input-output processes, and destruction-replacement processes. It is here, over the space of problem-solution dualities, defined over the space of identification-transformation dualities, that the phenomenon of *thinking* (from the instrument of mind and brain) containing multiplicity of sub-phenomena arises, where thinking is an algorithmic guide to reasoning over all spaces of human actions. The multiplicity of the sub-phenomena, including curiosity, creativity and courage in relational give-and-take sharing modes with unity of work, occurs over the spaces of hope-despair dualities, success-failure dualities, and certainty-uncertainty dualities to create environments of thinking, reasoning, doubt, surety, fear, innovation, new ideas, variety discoveries, pain, pleasure, joy, sorrow, happiness, sadness within the social philosophical consciencism, the structure of which depends on the nature of the space of cognitive capacity limitations relative to the space of visible-invisible dualities.

The social philosophical consciencism unites the individual and collective consciousness with commonness of visions as well as separate the actions of the members of society into conditions of preferential conflicts and resolutions over the spaces of kindness-wickedness dualities, correct-wrong dualities and reward-punishment dualities with conformal mapping onto the space of cost-benefit dualities as variety categorial conversions occur within the epistemological space, where variety imaginations abstracted from variety potentials are transformed to variety realities relative to variety actuals, all under the neuro-decision-choice system through the development of functors and categorial converters.

In the thinking processes, over the space of fundamental-applied dualities, thinking is the organic processer that receives messages from the micro-units, interprets them, integrates them, and sends instructions to the essential components of the thought processes for execution of respective tasks by the elements of the set of micro-processers as forms of reasoning. The essential component of concern here is the micro-unit of the phenomenon of creativity under variety motivation and variety inspiration within the space of thinking-reasoning dualities. The phenomenal natures of thinking and reasoning and their differences have been discussed in chapter 4 of [R8.16b]. Here, definitions of thinking, reasoning and thinking-reasoning dualities are provided to allow for an added understanding of the phenomena of creativity in a system of categorial conversions, where the nature of creativity is constrained by the dynamics of the social philosophical consciencism.

DEFINITION 5.1.1A: Thinking (Verbal Disposition)

Thinking is a natural behavioral variety of a complex system of the mind as categorial converter from the brain and as an organic phenomenon containing sub-phenomena such as courage, curiosity, creativity, innovation, cost-benefit assessment capability, decision-choice modules in dealing with the complex multiplicity of interactions in relational continua and unity with give-and-take sharing modes over the space of imagination-reality dualities and other related phenomena with complex multiplicity to produce ideas, pictures, music, symbolic languages, non-symbolic languages, components of thought and theories over the space of problem-solution dualities processing the varieties in the space of ignorance contained in the space of imaginations as a sub-space of the space of the potentials and transforming them into the space of knowledge contained in the space of realities as a sub-space of the space of actuals which is contained in the space of actual-potential dualities defined over the spaces of matter-energy dualities in equilibrium dynamics and visible-invisible dualities in disequilibrium dynamics. Thinking is an invisible variety transformation operator as an algorithm and instrument of the mind to be used by neuro-decision-choice agents over the space of problem-solution dualities to accomplish outcomes in the space of visible-invisible dualities. Thinking as an algorithm is dependent on the mind of the developer and independent of the users.

DEFINITION: 5.1.1B: Thinking (Categorial Characteristic Disposition)

Let the space of thinking as categorial convertors be $(\mathbf{T})$ with a generic element $(\mathfrak{t} \in \mathbf{T})$, defined characteristic disposition $(\mathbb{X}_{\mathbf{T}})$ and a corresponding variety $(\nu_{\mathfrak{t}} \in \mathbb{V})$ and the space of reality $(\mathfrak{H})$ with a generic element $(\mathfrak{h} \in \mathfrak{H})$, characteristic disposition $(\mathbb{X}_{\mathfrak{H}})$ and a corresponding variety $(\nu_{\mathfrak{h}} \in \mathbb{V})$. The variety $(\nu_{\mathfrak{t}} \in \mathbb{V})$ is said to be a thinking categorial convertor if $(\nu_{\mathfrak{t}} \in \mathbb{V} \ni \mathfrak{t} \in \mathbf{T} \subset \mathfrak{G} \subset (\mathfrak{G} \otimes \mathfrak{H}) \subset (\mathfrak{U} \cup \mathfrak{A}))$ with corresponding characteristic dispositions $(\mathbb{X}_{(\mathfrak{G} \otimes \mathfrak{H})}$ and $\mathbb{X}_{(\mathfrak{U} \cup \mathfrak{A})})$ such that the varieties in the spaces of realities and imaginations are epistemologically determined while the varieties in the space of actuals $(\mathfrak{A})$ and the space of potentials $(\mathfrak{U})$ are ontologically determined in the sense that $(\mathfrak{H} \subset \mathfrak{A} \subset (\mathfrak{U} \cup \mathfrak{A}))$ such that $\mathfrak{H} = (\mathfrak{H} \cap \mathfrak{A})$ and $(\mathfrak{G} \subset \mathfrak{U} \subset (\mathfrak{U} \cup \mathfrak{A}) = \Omega)$, such that $\mathfrak{G} = (\mathfrak{G} \cap \mathfrak{U})$ where $(\omega \in \Omega)$ is the universal object space and $(\nu \in \mathbb{V})$ is the space of all varieties such that $(\mathbb{Q} = (\mathfrak{H} \otimes \mathfrak{G}) = (\mathfrak{H} \cup \mathfrak{G}))$ is the space of imagination-reality dualities with $(\mathfrak{H} \otimes \mathfrak{G})$ presenting the conditions of the principle of opposites, $(\mathfrak{H} \cup \mathfrak{G})$ representing relational unity, $(\mathfrak{H} \cap \mathfrak{G}) \neq \varnothing$ representing conditions of relational continuum with give-and-take sharing modes, where the characteristics of every variety reality have some characteristics of variety imagination with information condition $\mathbb{Z}_{\mathbb{Q}} = (\mathbb{Z}_{\mathfrak{H}} \cup \mathbb{Z}_{\mathfrak{G}}) = (\mathbb{Z}_{\mathfrak{H}} \otimes \mathbb{Z}_{\mathfrak{G}})$ such that $(\mathbb{Z}_{\mathfrak{H}} \cap \mathbb{Z}_{\mathfrak{G}}) \neq \varnothing$ for the fuzzy conditionality under the principle of opposites $(\mathbb{Z}_{\mathfrak{H}} \otimes \mathbb{Z}_{\mathfrak{G}})$ and $(\mathbb{Z}_{\mathfrak{H}} \cup \mathbb{Z}_{\mathfrak{G}})$ is relational unity.

DEFINITION 5.1.2A: Reasoning (Verbal disposition)

Reasoning is an individually self-induced activity of the mind as a thought process in neuro-decision-choice actions using the variety characteristic dispositions as inputs to follow the epistemic algorithms developed by thinking to arrive at the variety selection over the spaces of variety problem-solution dualities and ignorance-knowledge dualities in the sense that thinking is the product of the mind as an epistemic guide to reasoning as a cognitive process to exercise neuro-decision-choice actions over the spaces of problem-solution and question-answer dualities, where thinking is the *primary category* of mental action and reasoning is a *derived category* of mental action of neuro-decision-choice activities over the spaces of problem-solution and question-answer dualities which are contained in the space of imagination-reality dualities as derived from the space of actual-potential dualities . Reasoning, just like thinking, is an invisible variety of the category of the mind contained in the brain as a category of matter.

DEFINITION: 5.1.2B: REASONING (Categorial Characteristic Disposition)

Let the space of reasoning as the space of applicators of thinking over the space of problem-solution dualities be $(\mathfrak{R})$ with a generic element $(\mathfrak{r} \in \mathfrak{R})$, defined characteristic disposition $(\mathbb{X}_{\mathfrak{R}})$ and a corresponding variety $(\nu_{\mathfrak{r}} \in \mathbb{V})$ and the space of reality $(\mathfrak{H})$ with a generic element $(\mathfrak{h} \in \mathfrak{H})$, characteristic disposition $(\mathbb{X}_{\mathfrak{H}})$ and a corresponding variety $(\nu_{\mathfrak{h}} \in \mathbb{V})$. The variety $(\nu_{\mathfrak{r}} \in \mathbb{V})$ is said to be a reasoning applicator if $(\nu_{\mathfrak{r}} \in \mathbb{V} \ni \mathfrak{r} \in \mathfrak{R} \subset \mathfrak{G} \subset (\mathfrak{G} \otimes \mathfrak{H}) \subset (\mathfrak{U} \cup \mathfrak{A}))$ with corresponding characteristic dispositions $\left(\mathbb{X}_{(\mathfrak{G} \otimes \mathfrak{H})} \text{ and } \mathbb{X}_{(\mathfrak{U} \cup \mathfrak{A})}\right)$ such that the varieties in the spaces of realities and imaginations are epistemologically determined while the varieties in the space of actuals $(\mathfrak{A})$ and the space of potentials $(\mathfrak{U})$ are ontologically determined in the sense that

$(\mathfrak{H} \subset \mathfrak{A} \subset (\mathfrak{U} \cup \mathfrak{A}))$ such that $\mathfrak{H} = (\mathfrak{H} \cap \mathfrak{A})$ and $(\mathfrak{G} \subset \mathfrak{U} \subset (\mathfrak{U} \cup \mathfrak{A}) = \Omega)$, such that $\mathfrak{G} = (\mathfrak{G} \cap \mathfrak{U})$ where $(\omega \in \Omega)$ is the universal object space and $(\nu \in \mathbb{V})$ is the space of all varieties such that $(\mathbb{Q} = (\mathfrak{H} \otimes \mathfrak{G}) = (\mathfrak{H} \cup \mathfrak{G}))$ is the space of imagination-reality dualities with $(\mathfrak{H} \otimes \mathfrak{G})$ representing the conditions of the principle of opposites, $(\mathfrak{H} \cup \mathfrak{G})$ representing relational unity, $(\mathfrak{H} \cap \mathfrak{G}) \neq \varnothing$ representing a relational continuum with give-and-take sharing modes, where the characteristics of every variety reality have some characteristics of variety imagination with information condition $\mathbb{Z}_{\mathbb{Q}} = (\mathbb{Z}_{\mathfrak{H}} \cup \mathbb{Z}_{\mathfrak{G}}) = (\mathbb{Z}_{\mathfrak{H}} \otimes \mathbb{Z}_{\mathfrak{G}})$ such that $(\mathbb{Z}_{\mathfrak{H}} \cap \mathbb{Z}_{\mathfrak{G}}) \neq \varnothing$ for the fuzzy conditionality under the principle of opposites $(\mathbb{Z}_{\mathfrak{H}} \otimes \mathbb{Z}_{\mathfrak{G}})$ and $(\mathbb{Z}_{\mathfrak{H}} \cup \mathbb{Z}_{\mathfrak{G}})$ is relational unity.

DEFINITION 5.1.3: Thinking-Reasoning Duality and The Space of Thinking-Reasoning Dualities (Combined Verbal and Characteristic Dispositions)

Thinking-reasoning duality is a condition where the cognitive acti9on contain simultaneously thinking and reasoning dispositions in relative proportions in the same space, time and place. Given the spaces of thinking $(\mathbb{T})$ and reasoning $(\mathbb{R})$ defined in the space of quality-quantity dualities then the space of simultaneous existence of thinking and reasoning is the space of thinking-reasoning dualities $(\mathbb{D})$ defined by Cartesian product of the form $\{\mathbb{D} = (\mathbb{T} \otimes \mathbb{R}) \mid \mathbf{d} = (\mathbf{t}, \mathbf{r}) \in (\mathbb{T} \otimes \mathbb{R}), (\mathbf{t} \in \mathbb{T})(\mathbf{r} \in \mathbb{R})\}$ with fuzzy continuum $(\mathbb{T} \cap \mathbb{R}) \neq \varnothing$ and relational unity $(\mathbb{T} \cup \mathbb{R})$, principle of opposites $(\mathbb{T} \otimes \mathbb{R})$ and logical containment in the form $(\mathbb{R} \subseteq \mathbb{T} \subseteq \mathbb{R})$, where thinking and reasoning are mutually interdependent and mutual creating with information support $\mathbb{Z}_{\mathbb{D}} = (\mathbb{Z}_{\mathbb{T}} \cup \mathbb{Z}_{\mathbb{R}}) = (\mathbb{Z}_{\mathbb{T}} \otimes \mathbb{Z}_{\mathbb{R}})$ such that $(\mathbb{Z}_{\mathbb{T}} \cap \mathbb{Z}_{\mathbb{R}}) \neq \varnothing$.

NOTE 5.1.1: Thinking, Reasoning and Thinking-Reasoning Duality

Thinking and reasoning are mutually creating and interdependent and housed in the same matter, however, it is analytically useful to place then in a fuzzy decision space, where a decomposition may be made with fuzzy indicator function, where thinking and reasoning are linguistic variable residing in the space of invisible contained in the space of visible-invisible dualities in the space of relativity that allows infinite expressions of relative combination and comparisons within the set of $(0,1)$. The fuzzy decomposition is such that thinking dispositions are places in the space of algorithmic developments, where the development of such algorithms is individual-specific or collective-specific depending on the general usage by the public. The decomposition is such that the thinking algorithms become inputs into the space of reasoning actions as guides to neuro-decision-choice actions over the space of problem-solution dualities, where the reasoning processes are undertaken. The thinking-reasoning dualities are processors of information as inputs in the categories contained in the epistemic fields in the processes of knowing as the production of intellectual investment flows and information-knowledge accumulations as production of intellectual capital stocks. In this respect, all types of individual and collective thinking are seen as production of algorithms to be used over the general space of problem-solution dualities. Similarly, all types of individual and collective reasoning are seen as the uses of thinking algorithms over the space of problem-solution dualities and decision-choice activities over the social space of identification-transformation dualities.

The thinking-reasoning dualities are input-output products and consumptions of the neuro-decision-choice actions through the space of body-mind dualities in interactive complexities leading to the space of neuro-decision-choice actions for variety transformation over the space of imagination-reality dualities. It is within the complexities of the brain, mind, thinking and reasoning that the meaning of category formation and categorial conversion find meaning and serve as powerful analytical tools in understanding transformation dynamics containing visions over the space of quality-quantity dualities, where visibility becomes invisibility, matter becomes nonmatter, nothingness becomes somethingness and vice versa under the general principle of opposites. It is within this framework under the principle of opposites, categorial transformation, functor behavior, that one can understand process-technological dynamics relative to mind-body, primary-derived, parent-offspring, input-output, production-consumption dualities, and related phenomena in the universal space of socio-natural identification-transformation dualities. In the space of knowing and information-knowledge production, the important element is the understanding of the semantic foundation of lexical elasticity of language, the degrees of information-knowledge contents of experiential information, knowledge by description and cognitive filtering of information-knowledge contents from the social philosophical consciencism before becoming inputs into the space of neuro-decision-choice actions.

The space of thinking-reasoning dualities is intimately related to the functor space and the space of categorial converters (convertor) under the guidance of social philosophical consciencism in the sense that our neuro-decision-choice actions through thinking-reasoning processes are always transformative setting variety solutions against variety problems as well as reconciling conditions in the space of input-output dualities, incentive-disincentive dualities and cost-benefit dualities over the space of goals, objectives and visions that generates the space of problem-solution dualities connected to the spaces of ignorance-knowledge dualities and perception-actual dualities in the dynamics of imagination-reality dualities contained in the space of actual-potential dualities.

5.2 CREATIVITY AS ENHANCER OF CATEGORIAL CONVERTERS AND FUNCTORS OVER THE SPACE OF THINKING-REASONING DUALITIES

From the space of thinking we may isolate creativity acting as categorial converter enhancer represented as $\left(\mathbf{C}_{\varsigma(\cdot)}\right)$ which acts on an variety in one category $\left(\nu_{\mathbb{C}_i} \in \mathbb{C}_i\right)$ and convert it to another category $\left(\nu_{\mathbb{C}_j} \in \mathbb{C}_j\right)$ in the epistemological space, such as from the space of imagination $\left(\nu_{\mathfrak{g}} \in \mathfrak{G} \subset \mathfrak{U}\right)$ to the space of reality $\left(\nu_{\mathfrak{h}} \in \mathfrak{H} \subset \mathfrak{A}\right)$ performing a transferring function as $\left\{\mathbf{C}_{\varsigma(\cdot)}\left(\nu_{\mathfrak{g}} \rightarrow \nu_{\mathfrak{h}}\right) \Rightarrow \left(\nu_{\mathfrak{g}} \rightarrow \nu_{\mathfrak{h}}\right) \mid \nu_{\mathfrak{g}} \in \mathfrak{G},\ \nu_{\mathfrak{h}} \in \mathfrak{H},\ \varsigma(\cdot) \in \mathscr{G}\right\}$ and the space of all creative converters $(\mathscr{G})$. The variety cognitive creative converters change the inter-variety and inter-categorial relationalities in the input-output and knowing processes under dualistic-polar conditions of the principle of opposites relative to the space of production-consumption dualities. Corresponding to the space of variety creative convertors $(\mathscr{G})$ is the set of *categorial creative conversion moments* $\left(\mathscr{G}^{\mho}\right)$ of the form:

$$\left(\mathscr{G}^{\mho}\right) = \left\{\mathbf{C}_{\varsigma\nu}^{\mathrm{M}} = \mathbf{C}_{\varsigma}\left(\nu_{\mathfrak{g}} \rightarrow \nu_{\mathfrak{h}}\right) \Rightarrow \left(\nu_{\mathfrak{g}} \rightarrow \nu_{\mathfrak{h}}\right) \mid \mathfrak{g} \in \mathfrak{G} \subset \mathfrak{U},\ \mathfrak{h} \in \mathfrak{H} \subset \mathfrak{A},\ \nu \in \mathbb{V},\ \varsigma \in \mathscr{G}\right\} \quad (5.2.1)$$

The categorial conversion process is also functor process where variety imagination $\left(\nu_{\mathfrak{g}}\right)$ is transformed into variety reality $\left(\nu_{\mathfrak{h}}\right)$. Each categorial conversion moment shows the degree of efficiency of the creative action by thinking and reasoning over the dynamics of varieties. The creativity conversion function, like any process of transformation, resides in dualistic-polar conditions over the space of destruction-construction dualities in the sense that all creativities reside in destruction-construction dualities within the action space of cost-benefit dualities to transform real cost by destruction to real benefit by construction and vice versa in the space of positive-negative dualities such that in general creative dynamics with $\left(\mathbb{I}^{\infty}\right)$ as infinite index set of categories $\mathbb{C}_{(\bullet)}$ contained in the space of epistemic field $(\mathscr{E})$ we can write:

$$\mathbf{C}_{\varsigma}=\left\{\begin{array}{l}\left\{\left(\mathbf{C}_i,\mathbf{C}_j\right) \mid \left(v_{\mathfrak{g}} \rightleftarrows v_{\mathfrak{h}}\right) \& \left(v_{\mathfrak{h}} \rightleftarrows v_{\mathfrak{g}}\right), \mid v_{\mathfrak{g}} \in \mathbb{C}_i,\ v_{\mathfrak{h}} \in \mathbb{C}_j,\ \varsigma \in \mathscr{G}\ i,j \in \mathbb{I}^{\infty}\right\}, \text{the converters} \\ \left\{\mathscr{G}^{\mho}_{v_{\mathfrak{g}} \rightleftarrows v_{\mathfrak{h}}} = \left\{\left(\mathbf{C}_i^{\mathrm{M}},\mathbf{C}_j^{\mathrm{M}}\right) \mid \left(v_{\mathfrak{g}} \rightleftarrows v_{\mathfrak{h}}\right) \& \left(v_{\mathfrak{h}} \rightleftarrows v_{\mathfrak{g}}\right) \mid v_{\mathfrak{g}} \in \mathbb{C}_i,\ v_{\mathfrak{h}} \in \mathbb{C}_j,\ \varsigma \in \mathscr{G}\right\}\right\}, \text{conversion moments} \\ \text{where} (\rightleftarrows) \text{ represent mutual transformation in the space of imagination-reality dualities.}\end{array}\right\} \quad (5.2.2.)$$

The categorial *creative process* is also an enhancer of problem-solution process for the efficient destruction of greater real cost variety and construction of acceptable real benefit variety over the space of problem-solution dualities, where the solution-variety converter (transformation) is a creative problem solver, and problem-variety converter (transformation) is also a creative variety problem creator; all of which are always under the principle of opposites and conditions of cost-benefit dualities, where varieties of greater costs are viewed as problems and transformed and replaced with varieties of greater benefits viewed as variety solutions over the space of problem-solution-dualities generating variety knowing and variety information-knowledge accumulation . The categorial creative conversions over the space of problem-solution dualities and polarities are the works of dualistic-polar internal and external conflicts to induce cognitive creativity for problem-solution identifications and transformations as internal and external dualistic-polar games, where tactics and strategies are creatively designed. The categorial creative converters and categorial creative conversion moments are the guidance to human neuro-decision-choice balancing actions over the space of real cost-benefit dualities in knowing and information-knowledge accumulation, where benefits and costs are broadly defined in terms of the varieties in the spaces of input-output dualities and destruction-construction dualities defined over the space of easiness-difficulty dualities. They are used to shape the trajectory of human existence over the space of necessity-freedom dualities and anticipation-expectation dualities relative to the spaces of doubt-surety and hope-despair dualities with dualistic-polar conditions over all areas of cognitive endeavors generating knowing and information-knowledge over the spaces of success-failure dualities and input-output dualities.

5.3 REFLECTIONS ON INTERCREATIVE RELATION BETWEEN COURAGE, AND CURIOSITY IN KNOWING OVER THE SPACE OF ADENTIFICATION-TRANSFORMATION DUALITIES

There are some important reflections that must be made before we epistemically specify many dimensional interrelations in the knowing process as the production of intellectual investment flows and information-knowledge accumulation as the production of intellectual capital stocks, and how these interrelations constitute the foundations of all human activities over the space of problem-solution dualities and polarities, where knowing and information-knowledge production are defined over the spaces of input-output dualities and production-consumption dualities that houses the space of success-failure dualities as well as the space of cost-benefit dualities in the space of relativity. Over the general action space, the human progress is unbreakably dependent on the accumulated intellectual capital stocks contained in the social philosophical consciencism at generation times, where courage, curiosity and creativity find expressions over the spaces of possibility-impossibility and probability-improbability dualities. Difficulties, risks, failures, and successes arise over the space of problem-solution dualities at the presence of ignorance where difficulties, risks and failures increase with increasing ignorance and decrease with decreasing ignorance and successes and benefit increase with increasing knowledge and decrease with decreasing knowledge over the space of ignorance-knowledge dualities under the defined epistemic transversality conditions as produced by optimal fuzzy conditionality under the general principle of opposites. In this respect, the whole of knowing as the production of intellectual investment flows and information-knowledge accumulation as the production of intellectual capital stocks constitute the knowledge economy as the general dynamics of the space of input-output dualities and the space of production-consumption dualities. It is at this point of conditions over problem-solution dualities that a question arises as whether information or knowledge is the input into neuro-decision-choice processes. The position in this analytical work is that knowledge is the input for optimal neuro-decision-choice action and information is the output of the optimal neuro-decision-choice process where information is the primary category of knowing and knowledge is a derived category of knowing in the space of ignorance-knowledge dualities.

The possibility-impossibility dualities and the probability-improbability dualities, possibility-probability dualities with corresponding spaces and possible world arise in knowing and neuro-decision-choice actions due to cognitive capacity limitations that generate imperfect information conditions to create the space of ontic-epistemic differences to give rise to the space certainty-uncertainty dualities in the space of human actions and behaviors. The possibility-impossibility dualities are in relation to qualitative uncertainties which in turn are in relation to necessity and anticipation, while the probability-improbability dualities are in relation to quantitative uncertainties which in turn are in relation to expectations and freedom, where expectations and anticipations constitute dualities, and necessities and freedoms constitute dualities, possibilities and probabilities constitute hybrids of quality-quantity dualities with hybrid of anticipations and expectations in addition to the corresponding hybrids of necessities and freedoms in the space of human progress for cognitive activities over the action space. It must be kept in mind that the act of thinking, as a mega-phenomenon, complemented by a set of internal micro-phenomena including courage, curiosity, creativity and hope among other such qualitative dispositions, is to negotiate balancing activities between opposites in negation of negation over the space of varieties with conformal mapping onto the space of imagination-reality dualities which houses the spaces of possibility-impossibility, probability-improbability, possibility-probability and necessity-freedom dualities as generated by the conditions of the space of ontic-epistemic dualities relative to the space of ontological-epistemological dualities, where the space of ontic-epistemic dualities houses the space of information-knowledge systems in the space of perfection-imperfection dualities.

The space of perfection-imperfection dualities is always relative to the spaces of absoluteness-relativity dualities and certainty-uncertainty dualities. The concept of uncertainty must be distinguished and separated from the concepts of possibility and probability. Similarly, the theory of uncertainty must be distinguished from the theory of probability which must also be distinguished and separated from the theory of possibility even though all these concepts and phenomena are interdependent and constitute a sequence over the space of primary and derived definitional categories in the space of neuro-decision-choice actions. In every literature on decision and choice, the concept, phenomenon, and sources of uncertainty are taken as known and understood. Sometimes uncertainties are related to belief where belief functions are derived. Sometimes the uncertainties are related to repetitiveness of an event, where frequency functions are developed without providing the conceptual meaning of uncertainty. In response to these concerns, an introductory structure of *the theory of certainty-uncertainty dualities* is provided in chapter two of the monograph that helps to answer the question of the meaning phenomena and sources of uncertainty. This theory of certainty-uncertainty dualities provides a framework for the theory imagination, possibility, probability and their connectivity to necessity and freedom, anticipation and expectation, initial conditions, and sufficient conditions over the space of neuro-decision-choice actions.

5.3.1 Reflections on Possibility, Impossibility, Probability, Improbability and Different Conceptual Worlds in Possibility-Impossibility Dualities

In general discussions on theory and practice, the concepts and phenomena of possibility, probability, necessity and freedom, anticipation and expectation are assumed, poorly defined, explicated, and analyzed and imperfectly included in the space of neuro-decision-choice space without recognizing that these concepts are neuro-decision-choice generated and inseparable from the space of thinking-reasoning dualities. The sources of these concepts are taken for granted and never discussed. The analytical problems of these concepts and phenomena are also part of the general analytical problem of the concept of information, its definition and explication with sequential connectedness to data, knowledge, evidence, and evidential things. A failure to give a general definition of information in the space of variety-commonness dualities provides confusions in the space of knowing and knowledge production over the space of certainty-uncertainty dualities. Similarly, the sources of certainty, uncertainty and certainty-uncertainty duality are not indicated in such a way that one wonders whether possibility-probability duality generates the certainty-uncertainty duality or certainty-uncertainty duality generates possibility-probability duality.

All these concepts and corresponding dualities generate analytical confusions, the sources of which may be traced to lack of explicit definition of information that will allow the understanding of varieties over the space of identification-transformation dualities. The confusions are cleared when information is defined in terms of variety characteristic disposition with perfection in the ontological space containing the ontic space, where the space of acquaintance connects the epistemological space to the ontological space such that the epistemological variety characteristics are abstracted from ontological variety signal dispositions under cognitive capacity limitations in such a way that the epistemic varieties are incompletely identified and hence may differ from the ontic varieties in such a way as to generate the space of *ontic-epistemic dualities* which generates the spaces of completeness-incompleteness dualities and epistemic distances creating certainty-uncertainty dualities that connect neuro-decision choice behaviors to the space of actual-potential dualities from which the space of imagination-reality dualities is constructed by neuro-decision-choice actions for the epistemological space to be reconnected with the ontological space through the space of acquaintance.

In the its logical sequence, the space of ontic-epistemic dualities from the space of ontological-epistemological duality generates the space of certainty-uncertainty dualities to provide the space of imagination-reality dualities contained in the space of actual-potential dualities, where the space of imaginations is abstracted from the space of potentials and the space of realities is abstracted from the space of actuals under cognitive capacity limitations to give rise to spaces of possibility, impossibility, and the corresponding necessity, initial conditions, anticipation and possible world, possible-world-impossible-world duality; in addition to probability, improbability and corresponding freedom, sufficient conditions and expectation of final conditions, probable world, and improbable-world-probable-world dualities, all of which reside in the space of dualistic-polar conditions contained in the space of static-dynamic dualities. Thus, the sources of uncertainties are established as differences in terms of vagueness and volume incompleteness of variety characteristics within the space of ontic-epistemic dualities as a *primary category* of the knowing and information-knowledge production processes. The rise of the concepts and phenomena of possibility, probability and the sequential phenomena including the space of success-failure dualities are *derived categories* from the space of certainty-uncertainty dualities as contained in the epistemological space under neuro-decision-choice actions, where the epistemological space is contained in the space of ontological-epistemological dualities as discussed in the previous chapters and in [R8.16b]. [R17.17] [R17.18].

Following the definitional process under the principle of opposites, where the opposite duals mutually define themselves, their identities, and existence, the possibility space contains all varieties where the variety possibility duals dominating the variety impossibility duals, while the impossibility space is the collection all varieties with impossibility duals dominating the possibility duals. The possibility space is linked to the impossibility space by some socio-natural process to induce intra-categorial and inter-categorial transformations, where the concepts of possibility and impossibility are social in the sense of being constrained by human cognitive capacity limitations in observations and thought regarding variety outcomes of socio-natural processes in the space of variety identification-transformation dualities relative to the space of ontic-epistemic dualities. In other words, the variety impossibility in human cognition with cognitive capacity limitations, information-knowledge imperfection and disparity over the epistemological space is not necessarily a variety impossibility in the natural processes over the ontological space with conditions of information-knowledge equality and perfection where there are no uncertainties, successes, failure, and risks. The logical process over the epistemological space is such that the probability of a variety event in the space of imaginations is said to be zero if it belongs to the impossibility set and possibility of the event is zero under subjectively defined conditions. In this situation, there is no necessity and hence the freedom of actualizing the variety over the space of risk-benefit dualities is denied at the probability space.

The theory of probability has become central in the development of the theory stochastic processes, the theory of uncertain decision, the theory of risk, data science and many others in all areas of problem-solution dualities without explicit discussions of the source of uncertainty, possibility, and probability in knowing and neuro-decision-choice actions over the space of imagination containing the space of possibility. The contents of this monograph and the discussions in the previous ones connect us to how these concepts and phenomena arise in knowing as the production of intellectual investment flows and information-knowledge accumulation as the production of intellectual capital stocks by presenting a general model of production-consumption dualities that connects experiential information (knowledge by acquaintance) to knowledge by description (paradigmatic knowledge) through functor and categorial actions with quality control processes under the principle of opposites with relational continua and unity. The general model contains the space of theory constructions and the space of theory applications over the spaces of identification-transformation dualities and production-consumption dualities.

The dualistic-polar logical points to the conditions, where the probability of a variety event is positive belongs to the possibility set and the possibility of the event is positive under the defined conditions. It is here that the development of *the theory of possibility* acquires its relevance [R5.17]. In this situation, the necessity is defined at the possibility space and hence freedom of actualizing the variety with risk is available at the probability space as conceptually defined by neuro-decision-choice selections with fuzzy indicator function. It is also here, that the need to develop hybrid numbers of qualitative and quantitative uncertainties presents a useful approach to neuro-decision-choice theory that combines qualitative uncertainties that project belief system and quantitative uncertainties that project frequency system in understanding and numerical characterization [R5.34]. Every variety event with positive probability belongs to the space of freedoms and expectations as well as the possibility space and the space of necessities and anticipations. Every variety with zero probability belongs to the impossibility space and hence does not belong to the spaces of necessity-freedom dualities and anticipation-expectation dualities. The structure of relational continuum and unity under dualistic-polar conditions is shown as a paremiological geometry in Figure 5.3.1.1 which may be related to Figures (2.1.1), (2.2.1) and (2.5.2.1) of chapter 2 of this monograph. Let us keep in mind that the concepts and phenomena of imagination and imagination space arise from epistemic distance contained in the space of ontic-epistemic dualities which generate the phenomenon of uncertainty contained in the space of certainty-uncertainty dualities, where the epistemic defective varieties are generated from the space of acquaintance under cognitive capacity limitations.

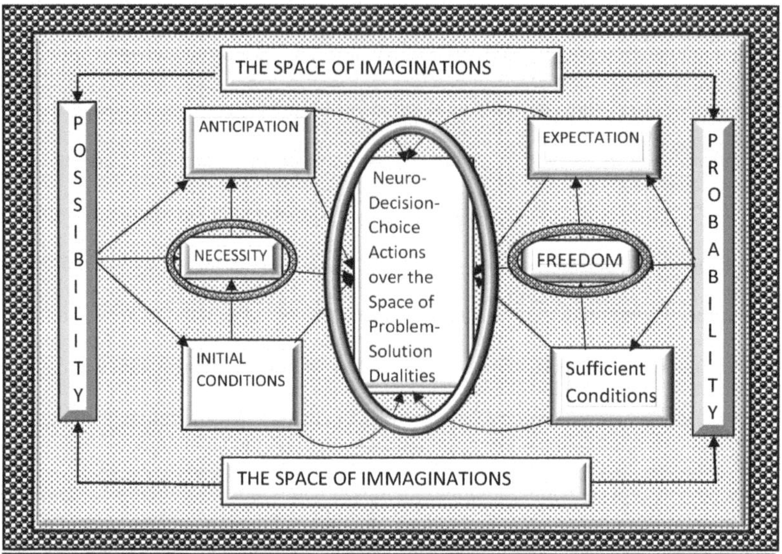

Figure 5.3.1.1. **Paremiological Geometry of Relationality and unity over the Space of Possibility-Probability Dualities with Initial Conditions, Necessity, Anticipations, Freedom, and Expectations of the Sufficient and Final Conditions.**

In the space of neuro-decision-choice actions, the possibility generates anticipation, necessity, and initial conditions of variety identities for transformation decision-choice actions and the probability generates expectations, freedom, sufficient conditions of action structures associated with variety transformations which then generate risk-benefit conditions over the spaces of problem-solution and ignorance-benefit dualities relative to the space of success-failure dualities, the outcomes of which generate new information in knowing and information-knowledge production to expand intellectual capital stocks. These conditions in the space of possibility-probability dualities relate to the principles of dividedness, diversity, unity, categorial conversions, functor processes and philosophical consciencism containing national personality and exceptionalism which are then related to conditions over the space of imagination-reality dualities under the principle of courage, creativity and hope within the thinking and reasoning operators relating variety costs to variety benefits, where courage, hope and success are find motivation from the ruling philosophical consciencism.

In this framework of knowing and information-knowledge production as foundation of theory of knowledge in the epistemological space, we can analytically divide the phenomenon of possible worlds and the space of possible worlds into *possible world semantics* with the *space of possible world semantics*, and the *possible world transformations* with the *space of possible world transformations.* In terms of knowing, information-knowledge production and the theory of knowledge, the possible world transformations relate to real varieties of goals, objectives, and visions, where scenarios may be designed for examinations of relative possible cost-benefit outcomes over the space of imaginations before implementation actions. The space of possible world transformations relates to the collections of the scenarios of nominal varieties of goals, objectives and visions that belong to the space of possibility-impossibility dualities as contained in the space of imagination-reality dualities which is also a derived sub-space of actual-potential dualities as a conformal mapping from the space of matter-energy dualities, the theory of knowledge is composed of conditions of knowing as investment flows and information knowledge accumulation as the production of intellectual capital accumulation. The theory of knowing as intellectual investment flows in the space of ignorance-knowledge dualities is the counterpart to the theory of investment in physical systems in the space of commodity production-consumption dualities. The theory of intellectual capital stocks through information-knowledge accumulation in the space of ignorance-knowledge production is the counterpart of the theory of capital in physical systems.

5.3.2 Reflections on Possible World and Possible Universe

The conditions of epistemic possibility and the derived possible worlds in relation to the varieties in the space of imagination-reality dualities are different from the conditions of epistemic possibility and derived *possible universes* in relation to the space of actual-potential dualities seen in terms of either semantics or transformations. The differences ae explainable by means of information-knowledge disparity conditions in the possession of cognitive agents over the ontological space, epistemological space, and the space of ontic-epistemic dualities. While we can speak of impossibility and improbability in the space of imagination-reality dualities contained in the epistemological space, these concepts have no analytical meaning over the space of actual-potential dualities contained in the ontological space from cognitive conceptions of universal existence in the past-present-future structure of Sankofa-anoma tradition within the space of beginning-end duality, where the space of actual-potential dualities relates to natural processes and find meaning and existence in the ontological space as the primary identity, and the space of imagination-reality dualities relates to social processes and find meaning and existence in the epistemological space as a derived identity from the primary identity.

The *possible world semantics* relates to nominal varieties of linguistic conditions that may be relate to logical claims about linguistic varieties relative to different sets of initial conditions that may be taken as necessity while the space of possible worlds semantics is the collection of nominal varieties from the space of imaginations that may be related to the space of claims on the basis of the set of the initial conditions of derived variety identities and necessities to project alternative possible worlds of varieties in the space of realities relative to the space of existing realities as has emerged from the space of imaginations where the variety initial conditions provide the information inputs into the space of thinking-reasoning dualities in order to derived possible outputs. Creating a set of possible worlds from the initial conditions, necessity and anticipation is also creating a set of alternative non-existing realities relative to, and in logical competition with the existing realities over the space of true-false dualities. Alternatively, changing the set of initial conditions is also changing the conditions of necessity and anticipations as may be seen to establish a possible world in the space of possible-world-impossible-world dualities which houses the space of conceptual domain of the ways that may be relative to what they are in the space of realities. The changes of the initial conditions, necessity and anticipation are used to specify epistemic spaces and logically possible combinations under the concept of possible world which is taken as a sub-space of the space of possible-world-impossible-world dualities, where a change of initial conditions creates a new necessity and a new possible world with a set of new logically possible outcomes which reside in the space of imaginations as a sub-space of the space of potentials contained in the space of actual-potential dualities.

The concepts of possible world semantics and the possible world transformations are not the use of epistemic spaces that have been developed in this monograph. In this monograph, epistemic spaces are not conceived in the space of imagination. They are epistemic decomposition analytics of space of problem-solution dualities contained in the space of varieties within the ontological space by partitioning it into super-categories and sub-categories of interactive spaces in knowing as production of intellectual flows and information-knowledge accumulation as production of intellectual capital stocks relative to socio-natural actions under similarity-difference processes. Here, we are not interested in how things might have been in the space of actual-potential dualities. We are interested in knowing the conditions of telescopic past, telescopic present and telescopic future, how these conditions may be used as information-knowledge inputs including forecasting from the telescopic past to the telescopic present and discounting from the telescopic future to the telescopic present in Sankofa-anoma tradition for the management and control of social existence and where these telescopic structures may be examined to create scenarios for neuro-decision-choice actions for actualizations of telescopic future in the space of alternative existences. The telescopic past excludes the beginning-nothingness dispositions and the end-somethingness dispositions. It is analytically useful to understand the epistemic differences of possible world semantic dispositions which relate to explanatory theory and explanatory science and possible world transformation dispositions which relate to *prescriptive theory* and *prescriptive science*.

The possible world semantics and its logical implications exist only in the known section of the space of knowabilities with initial conditions such that analytical comparability of possibilities is relatively meaningful. Comparative analytical possibilities of possible worlds are meaningless in the unknown section of the space of knowability. In the space of knowing, one may speculate on the existing possibilities of the varieties in the unknown section of the space of knowabilities as part of investigative actions, but the results are only speculative phenomena that may be subjected to competitive speculative claims over the space of past-present dualities, where the past-present varieties may be compared to possible world varieties as the initial conditions are altered to change necessity and anticipation. The construct of possible world is speculative but in comparison with some varieties in the space of existing realities as revealed by varieties in the space of past-present dualities. The construct of varieties in semantic possible world of unknown section of the space of knowability is also speculative but cannot be compared to any known variety since the initial conditions are unknown.

The *possible world transformation* relates to the space of social categorial converters or social functor space and exist in telescopic future, the varieties of which meet certain initial conditions and necessity that allow them to be anticipated to enter the space of realities under probabilistic constraints and conditions of freedom, sufficiency and expectation in the space of imaginations such that analytical comparability of possibilities and realities is cost-benefit meaningful over the space of categorial conversions. Comparative analytical possibilities of possible world transformations are always meaningful in telescopic future from telescopic present through the conditions of continual functor actions over the space of destruction-construction dualities relative to the space of present-future dualities. In the space of knowing regarding present-future dualities, one may speculate on the existing possibilities of the varieties in the unknown section of the space of knowabilities as part of investigative actions, where the results are speculative phenomena that may be subjected to functor actions for actualization of varieties into the space of present-future dualities. The construct of possible world transformation is speculative but in cost-benefit comparison with some varieties in the space of realities, where greater variety benefits are set against variety costs in comparison. The possible world semantics is distinguished from the possible word transformation by cost-benefit categorial conversions The possible word semantic is a theory of logical game, while the possible world transformation is the foundation of theory of planning.

The concept of possible universes is not analytically available to cognitive agents operating in epistemological space. This is due to the fact that the beginning disposition is conceptually absent and so there is no initial conditions to conceptualize from and hence there in no initial information inputs into the space of thinking-reasoning duality to create scenarios that may be compared to the space of actual-potential dualities, the completeness and totality of its varieties are not available to neuro-decision-choice agents to deliberate on alternative possible outcomes. Similarly, the conditions of the end-somethingness duality is not available to have comparative analytical possibilities of the types of the universe which is infinite in scope, size and varieties. It is in this sense that one may conceive of the concept of *the end disposition is in the beginning disposition, and the beginning disposition is in the end dispositions*. *Somethingness disposition is in nothingness disposition.* The *beginning disposition is in nothingness disposition* and the *somethingness disposition* is *in the end disposition,* where nothingness-somethingness duality and beginning-end duality have unbreakable mutual residence in dualistic-polar structures in relational continuum and unity in infinite existence.

5.3.3 Possible Worlds and the Space of Realities: Comparative Analytics

A question concerning the space of knowing and information-knowledge production in the space of neuro-decision-choice actions is how the concepts of possible world and the space of possible world relate to information, knowledge, data, and evidence and hence in the space of certainty-uncertainty dualities as seen over the space of possibility-probability dualities with a conformal mapping onto the space of necessity-freedom dualities. Are the concepts of possible world and impossible world related to the concepts of the spaces of decidability-undecidability dualities, doubt-surety dualities, uncertainty-certainty dualities, quality-quantity dualities, subjective-objective dualities, identification-transformation dualities, and other dualistic-polar extensions, and how do they find meanings and understandings under conditions of information and knowledge production in the general space of human actions. Critical difficulties arise over the space of true-false dualities when any possible world is taken to be an alternative space of realities. Technically, the space of realities has characteristic-signal dispositions which are not only supported by information, knowledge, data, facts, and evidence, but they must meet the existential conditions of the space of the actuals as a sub-space of the space of actual-potential dualities. Any possible world has an imagined characteristic-signal disposition which is not supported by information, knowledge, data, facts, and evidence but it is contained in the space of potentials through the space of imagination. It must be clear that the possibility space is a primary category in the space of imaginations and serves as a parent, through the probability space, to the space of realities which may serve as a comparative structure to the space of possible world semantics and possible world transformation. By disinformation, misinformation and propaganda any possible world semantics may be weaponized as a reality in the game of social cost-benefit distribution over the spaces of production-consumption and input-output dualities. It is in this analytical framework that one must understandings, injustice, slave economy, capitalism and socialism and other social organizations of production-consumption dualities where human body sovereignties are stripped by force or from the conditions in the unjust legal structure.

The epistemic conceptualizations of the dynamics of the theories of knowing and knowledge are such that the space of imagination-reality is contained in the space of actual-potential dualities, where the space of potentials is the primary category to the space of imaginations which constitutes the primary category to the possibility space which is the primary category to the probability space. The space of actuals is the primary category to the space of realities as offspring derived through indirect complex journeys over multiple paths from the space of imaginations that links the space of actuals to the space of potentials in the space of ontological-epistemological duality. In other words, the potential space is the parent to the space of imaginations as an offspring which becomes a parent to the possibility and probability spaces as offspring connecting the space of imaginations to the space of realities as a distant ancestor of the space of actuals through thinking-reasoning process, where thinking provides the space of mega-meta dualities of paradigms of thought and reasoning utilizes its mega and meta paradigms of thought as road maps in the complex system of the space of problem-solution dualities constrained by the space of information-knowledge dualities containing knowing as production of intellectual investment flows and information-knowledge accumulation as the production of intellectual capital stocks. The space of possibility-probability dualities is a derivative offspring of the space of imagination, where the possibility space serves as the primary category and a parent to the space of possible worlds as well as the space of impossible worlds which are their offspring. The space of possible worlds serves as the primary category of the space of incubators of ideas, goals, objectives, and visions that are to be transformed into the space of realities as a subspace of actuals.

There are cost-benefit conditions of possible world semantics and possible word transformations and the elements in the space of realities. Every possible world transformation considered in a full setting has a space containing variety potentials, imaginations, possibilities, and probabilities which contains the variety transformation capacity defined in terms of functor or categorial converter into the space of realities. The possible-world transformation also contains the solution sub-space, vision, goals and objectives while the space of reality contains variety problems defined by conditions of cost duals outweighing the benefit duals in the cost-benefit space which provide motivations within the space incentive-disincentive dualities to actualize varieties as solutions from the possibility space into the probability space and into the space of realities, where the actualization processes find expressions in *prescriptive theory* which provides a system of thinking algorithms to set the variety solutions as imaginations against variety problems as realities in the space of identification-transformation dualities.

The results of neuro-decision-choice actions over the space of identification problem-solution dualities present knowing characteristic-signal dispositions of variety existence in the space of realities which is contained in the space of the actuals. The results of neuro-decision-choice actions over the space of transformation problem-solutions dualities present knowing of changes in the characteristic-signal dispositions of variety negations between the space of possible world transformations and the space of realities. It must be understood that there are important differences between the space of possible world semantic and the space of possible world transformations as viewed relative to the space of realities as well as space of imaginations over the space of individual-collective dualities. The space of possible-world-impossible-world dualities is an offspring derivative of the space of possibility-impossibility duality which is its parent and primary category of existence.

The space of possible-worlds is made up of the sub-space of possible world semantics and the sub-space of possible world transformation in human action and analysis and is connected to the space of action-nonaction dualities. The space of *possible world semantics* belongs to the spaces of explanatory sciences and explanatory theories encompassing the space of telescopic past-present dualities as revealed by *what there was* and *what there is* in logical experiments, while the space of *possible world transformations* belongs to the spaces of prescriptive sciences and prescriptive theories encompassing the space of telescopic present-future dualities as revealed by *what there is* and *what there would be* in alternative cost-benefit scenarios for neuro-decision-choice actions over the spaces of problem-solution dualities and ignorance-knowledge dualities. The nature of explanatory science and the space of explanatory theories, and the nature of prescriptive science, and the space of prescriptive theories are discussed in [R4.10] [R13.7] [R13.9] [R13.15].

5.3.4 Reflections on Necessity, Freedom, Space of Problem-Solution Dualities and Polarities

Let us relate the possible-world-impossible-world dualities under transformation with the corresponding dualistic-polar conditions to possibility-probability dualities, necessity-freedom (accident-risk) dualities, initial-sufficient dualities and then to actual-potential dualities in the production and reproduction of real life with the dynamics of varieties in the variety space, where there are conditions of continual dividedness, oneness and relational changes for self-correction processes in nature and society, where such self-corrections in societies are defined over the space of problem-solution dualities and in nature are defined over the space of destruction-replacement dualities. Similarly, we consider possible-world-impossible-world dualities under semantic conditions over the space of source-destination dualities with all different logical implications. Let us keep in mind that natural variety transformations are defined over the ontological space through relational dualistic-polar conditions of the space of actual-potential dualities, where a variety actual is directly transformed into variety potential or vice versa through *natural technological processes* (natural categorial converter or functor) over the space of destruction-replacement dualities. The social transformations, on the other hand, proceed from the space of potentials to the space of imagination to the possibility space and then to the probability space, to space of realities and finally into the space of the actuals under fuzzy-stochastic conditionality as generated by cognitive capacity limitations from the space of acquaintance to the space descriptions under thinking algorithms, reasoning usage and cognitive creativity under a paradigm of thought.

In the transformation process in the space of the operators $\mathfrak{T}$, if $\nu \in \mathbb{V} \subseteq \mathfrak{U}$ and $\mathrm{P}(\nu) \in (0,1]$ as fuzzy-probability value with its corresponding entropic value then

$\nu \in (\mathfrak{P} \cup \mathfrak{B})$, the space of possibility-probability dualities that also contains the spaces of necessity-freedom dualities $(\mathfrak{F} \otimes \mathfrak{N})$ and risk-benefit dualities, where $(\mathfrak{F})$ is the space of freedoms and $(\mathfrak{N})$ is the space of necessities . If $\nu \in \mathbb{V}$ and $\mathrm{P}(\nu) = 0$ then $\nu \notin (\mathfrak{P} \cup \mathfrak{B})$, where $(\mathfrak{P} \cup \mathfrak{B}) \subset \mathfrak{G} \subset \mathfrak{U} \subset \Omega$ and $(\mathfrak{P} \cap \mathfrak{B}) \neq \varnothing$ .All categorial and variety relations must be see in terms of information relations such that to say $(\mathfrak{P} \cup \mathfrak{B}) \subset \mathfrak{P} \subset \mathfrak{U} \subset \Omega = (\mathfrak{A} \cup \mathfrak{U})$ is to say that $\mathbb{Z}_{(\mathfrak{P} \cup \mathfrak{B})} \subset \mathbb{Z}_{\mathfrak{G}} \subset \mathbb{Z}_{\mathfrak{U}} \subset \mathbb{Z}_{\Omega} = \mathbb{Z}_{(\mathfrak{A} \cup \mathfrak{U})}$. Over the epistemological space $\mathfrak{E}$, $\mathbb{Z}_{\mathfrak{E}} = (\mathbb{X}_{\mathfrak{E}} \otimes \mathbb{S}_{\mathfrak{E}}) \subset \mathbb{Z}_{(\mathfrak{U} \cup \mathfrak{A})} = (\mathbb{X}_{\mathfrak{U}} \otimes \mathbb{S}_{\mathfrak{U}}) \cup (\mathbb{X}_{\mathfrak{A}} \otimes \mathbb{S}_{\mathfrak{A}})$ where $\mathbb{X}_{\mathfrak{E}} \Subset \mathbb{X}_{(\mathfrak{U} \cup \mathfrak{A})}$ and that $\forall \nu \in \mathbb{V} \Rightarrow \mathbb{X}_{\mathfrak{E}_\nu} \subset \mathbb{X}_{(\mathfrak{U} \cup \mathfrak{A})_\nu} = \mathbb{X}_{\Omega_\nu}$, where $\mathbb{X}$, as the characteristic disposition, is the set of contents and $\mathbb{S}$ is the set of messages containing the contents from the source.

The implication for our understanding of cognitive behavior over the space of problem-solution dualities and the process of knowing is that every problem $(\mathfrak{p}_\nu \in \mathfrak{P})$ is such that $(\mathfrak{p}_\nu \in \mathfrak{A})$ where the solution is possible if $((\mathfrak{s}_\nu \in \mathfrak{S}) \subset (\mathfrak{P} \cup \mathfrak{B}) \subset \mathfrak{U})$ and the variety solution is said to be impossible if $(\mathfrak{s}_\nu \in \mathfrak{Y} \subset (\mathfrak{M} \cup \mathfrak{H}))$ where $(\mathfrak{M})$ is the impossibility space and $(\mathfrak{H})$ is the improbability space. Let us keep in mind that relational continuum and unity under the principle of opposite implies that $(\mathfrak{M} \subseteq \mathfrak{H} \supseteq \mathfrak{M})$ In this epistemic structure, if $\nu \in \mathbb{V}$ and $\nu \in \mathfrak{W} \subset \mathfrak{P}$ then the variety has a positive probability with freedom $(\mathfrak{f} \in \mathfrak{F})$ and hence it is said to be a possible world with a necessity $(\mathfrak{n} \in \mathfrak{N})$, where $\mathbb{Z}_\nu \subset \mathbb{Z}_{\mathfrak{U}}$. A variety problem $(\mathfrak{p}_\nu \in \mathfrak{P})$ is said to have no solution if all of its solution-characteristic dispositions belong to the impossibility space $(\mathfrak{M})$ and its initial conditions generate impossible worlds $(\mathfrak{Y})$ in dualistic-polar structures. Let us always keep in mind that all these varieties have corresponding characteristic-signal dispositions that provide the relevant information to establish their respective identities, similarities, differences, and categories of belonging with sub-categories, containments, unions, and intersections for the understanding of dividedness and oneness in the space of parent-offspring dualities. Information is defined in terms of characteristic-signal disposition, where the characteristic disposition is the contents, and the signal disposition is the messages. Every variety has a corresponding characteristic disposition that presents its identity and signal disposition that reveals its identity in the space of variety identifications, where the geometries and mathematics of connections reveal the relational give-and-take-sharing dynamics in continuum and unity within the space of linear-nonlinear dualities, and the geometries and mathematics of symmetry-asymmetry dualities reveal variety identities within linear-nonlinear relational structures in the space of dividedness-oneness dualities relative to the spaces of input-output dualities and production-consumption dualities.

It is analytically useful to note that the conceptual system that is being erected for the understanding of theory of knowing as the production of intellectual investment flows, information-knowledge accumulation as the production of intellectual capital stocks, and the theory of knowledge as the unity of the theory of intellectual investment and the theory of intellectual capital stock comparable to the theory of investment-capital process in physical production-consumption system [R13.10] [R13.11]. Every variety possibility and variety impossibility, belong to the space of imaginations as derived categories from the space of potentials and hence the epistemic possibility and impossibility spaces are contained in the ontological space of the potential space through the neuro-decision-choice abstraction from the acquaintance space by cognitive agents. The production of intellectual investment flows to update the intellectual capital stocks maintains the information-knowledge production system in continual disequilibrium states, where the intellectual investment flow is always non-zero and intellectual capital accumulation is always in an increasing mode. Every possible world belongs to the possibility space and hence the possible-world space is not only contained in the possibility space, but it is also contained in spaces of imaginations as abstracted from the space of potentials in order to establish necessity and initial conditions for variety knowing and transformation in relation to the varieties in the space of goals, objectives and visions under the psycho-economic principle of non-satiation which enhances increasing needs, individual greed, social progress and intensifies conflicts in the space of the game of life over the space of cost-benefit dualities.

Similarly, every impossible world belongs to the impossibility space, and hence the impossible-world space is not only contained in the impossibility space, but like the possible-world space, it is also contained in space of imaginations derived from the space of potentials by cognitive action under cognitive capacity limitations. The space of possible-world-impossible-world dualities is contained in the space of possibility-impossibility dualities which is contained in the space of imagination-reality dualities derived by cognitive actions as an offspring and a sub-space of the space of actual-potential dualities, where in the ontological space every actual is a potential and every potential is an actual by the dualistic-polar conditions over the space of static-dynamic dualities. Let us keep in mind that the ontological space is naturally defined and hence acts as the primary identity, while the epistemological space is a derivative and socially defined under cognitive capacity limitations in all areas of knowing to allow sub-spaces such as spaces of imaginations, possible worlds , possibilities and other relevant spaces to be abstracted in the process of knowing, information-knowledge development, input-output processes, and other spaces and sub-spaces within the guidance of principle of opposites with relational continuum and unity to provide the conditions of oneness and dividedness. All these spaces are cognitively defined with corresponding characteristic dispositions for their identities and communications over the space of source-destination dualities as well as provide cognitive pathways of cognitions and neuro-decision-choice actions.

5.4. Creativity, Transformation. Potentiality and Actualization of Varieties

In the previous section, we discussed the conditions of the phenomena of necessity, anticipation, freedom, expectation and the spaces of necessity-freedom dualities, and anticipation-expectation dualities, and their relationships to initial and sufficient conditions of knowing over the space of identification-transformation dualities through logical links to the space of problem-solution dualities contained in the space of imagination-reality dualities which also contains the space of possibility-probability dualities and its relation to the space of possible-world-impossible-world dualities. The possibilistic-probabilistic conditions of every possible-world transformation are such that not every possible world can be actualized from the space of imaginations within the epistemological space which alternatively implies that not every variety solution can be actualized in the replacement of a variety problem even though the variety solution exist in the space of possibility-probability dualities. The further implications are that there are variety problems in the space of realities contained in the space of the actuals that cognitive agents may not be able to solve by neuro-decision-choice activities due to cognitive capacity limitations on creativity in the space of imaginations over the space of actual-potential dualities contained in the space of ontological-epistemological dualities even though their variety solutions exist in the potential space. This leads us to examine the nature of variety problems in the problem space relative to variety solutions in solution space within the space of possibility-impossibility dualities and possible-world-impossible-world dualities. The neuro-decision-choice conditions over the space of variety problem-solution dualities must be such that variety problems must be identified over the space of the actuals before variety solutions can be searched over the space of potentials through the space of imaginations and then optimized over the space of possibility-probability dualities containing uncertainties generated by variety ontic-epistemic differences over the space of quality-quantity dualities for replacement selection at the space of realities derived by cognition as a sub-space of the space of the actuals under cognitive capacity limitations.

Let us keep in mind that a variety problem or a variety solution is defined in terms of cost-benefit balances relative to a goals, objectives or visions and that a variety problem is one with cost-dominance over benefit in the space of realities and as a hindrance to reach a goal, an objective or a vision in the space of imagination-reality dualities derived from the space of actual-potential dualities, where a variety solution is one with benefit-dominance to remove the variety with problem conditions and replace it with goal-facilitator varieties toward a vision. The goal-facilitator variety toward a vision rests in the space of imagination-reality dualities, where variety imagination is conceptualized and designed, and every variety reality is brought into being in the space of realities contained in the space of actuals $(\mathfrak{H} \subset \mathfrak{A})$ by creative processes in the space of creativities under the cost-benefit relationality over the space of destruction-construction dualities which is under the control and management of the neuro-decision-choice actions of cognitive agents. Over the space of problem-solution dualities, there is the problem identification of variety problem in the space of realities, and there is problem identification of variety solution in the space of imagination and there is problem identification of a functor or categorial converter in the space of imagination for destroying the variety problem and replacing it with optimal variety solution. The concept of optimization, here, implies the existence many variety solutions with differential degrees of variety cost-benefit replacement effectiveness for a variety problem and the neuro-decision-choice action is to rank the variety solutions in the degrees of cost-benefit effectiveness and select the best variety replacement. Similarly, it implies the existence of many variety problems with differential degrees of cost-benefit substitution effectiveness in ranking for which the variety solution will be good as the best cost-benefit variety. This is the simple foundation of the mini-max process over the space of problem-solution dualities, where solutions are obtained to problems through model representations of variety benefit (cost) disposition constrained by variety cost (benefit) disposition over the space of nominal-variety dualities such that variety benefit disposition is either the variety goal or variety constraint set relative to either variety cost disposition as either variety constraint set or the variety goal in the space of relativity where the best relative benefit-cost variety is selected.

Epistemically, it is useful to understand that the space of destruction-construction dualities is intimately connected to the space of cost-benefit dualities, where destruction resides in both cost and benefit dispositions with cost dominance and construction resides in both cost and benefit dispositions with benefit dominance, where balances are induced by thinking algorithms with curiosity, courage , creativity and innovation, and reasoning applications over the space of problem-solution dualities as the foundation of the space of meaning-essence dualities of life where the structure of the space of problem-solution dualities establishes the meaning of life and the manner of substitutions in the space of problem-solution dualities gives the essence of life over the space of individual-community dualities through neuro-decision-choice actions.

All human actions are neuro-decision-choice determined, not over the ontological space but over the epistemological space, where imagination, thinking, reasoning, fantasy, creativity, visions, dreams, goals, objectives, knowing and many others such qualities over the space of quality controls are conceived and actionized to enter the space of realities as determined by cognition as part of the space of actuals. The production of information-knowledge structures takes place over the space of fundamental-applied dualities, where every problem identification and every solution identification are the results of knowing and belonging to information-knowledge stocks. The important implication, here, is that the involved quality controls are undertaken over the space of mistake-correction dualities in the logical awareness within the space of cost-benefit dualities of identification-transformation processes over the space of ignorance-knowledge dualities. It is in this framework that the idea of mastering of nature finds meaning and essence in human life and efforts of utility in the control and management of social existence, where mastery of nature is the process of connecting the epistemological space to the ontological space as well as connecting the space of epistemic variety dispositions to the space of ontic variety dispositions regarding states and processes. The theory of mastering of nature is simply the theory of knowledge composed of the sub-theory of knowing as the production of intellectual investment flows and the sub-theory of information-knowledge accumulation as the production of intellectual stocks over the space of fundamental-applied dualities.

Based on the above definitions in this chapter, we now offer definitions of some anomalies that are common in the knowing as the production of intellectual investment flows and information-knowledge accumulation as production of intellectual capital stocks in terms of characteristic-signal dispositions for identities and differences in the process of mastering of nature. All anomalies are the results of thinking and reasoning with a paradigm of thought under cognitive capacity limitations and have nothing to do with conditions of ontological past-present-future variety existence in the space nothingness-somethingness dualities relative epistemological past-present-future variety existence, in the space of identification-transformation dualities between which ontic-epistemic differences arise to give rise to the space of certainty-uncertainty dualities which is the parent to the space of imaginations for the concepts and phenomena of possibility as a parent to anticipation and necessity to establish initial conditions, and probability as a parent to expectation and freedom to establish sufficient conditions under a social philosophical consciencism to give rise to the concept of reality and space of realities contained in the space of actuals on the path to the mastery of nature and its applications to relationships among varieties in the space of problem-solution dualities in the social space.

The anomalies are failures within the constructed telescopic past, present and future with linguistic representations and analysis from which conclusions are drawn over the space of fundamental-applied dualities contained in the epistemological space. It is these anomalies that constrain increasing knowing and place challenges on paradigms over the journeys in the space of ignorance-knowledge dualities providing motivations to thinking, innovations and creativity to develop new micro-paradigms and change of macro-paradigm in all areas of human actions in theories and applications toward mastery of nature. Within this framework over the epistemic field with information and relational fields, several anomalies may arise and may be examined by establishing definitions and explications using their characteristic dispositions as well as keeping in mind the definitions of the concepts of problem and solution in terms their cost-benefit relative existence.

DEFINITION 5.4.1A: Phantom Problem (Verbal Disposition)

A variety problem is said to be a *phantom problem* if its characteristic disposition belongs to the epistemic space and the intersections of the spaces of realities and imaginations contained in the epistemological space, however, there is no corresponding ontological signal disposition in its support, such that the epistemic variety disposition has no corresponding variety characteristic disposition from the ontological space that belongs to the epistemological space and, hence there is no ontological variety signal disposition to be decoded to find the corresponding ontic variety characteristic disposition and the corresponding variety, which implies that the variety problem belongs to the spaces of impossibility and impossible worlds as well as belonging to the space of imaginations with no corresponding variety solution which also implies that the variety problem is not real and does not belong to the space of realities. Every epistemological variety characteristic disposition must belong or has the potential to belong to the ontological space for one to claim it as existing in the space of actual-potential dualities as well as belonging to the ontological space, where every phantom problem is devoid of functor and categorial convertor with delusional expected operator (see also [R3.77] [R3.53]).

DEFINITION 5.4.1B: Phantom Problem (analytical characteristics Disposition)

A variety problem $(\mathfrak{p}_v \in \mathfrak{P})$ is said to be a *phantom problem* if its characteristic disposition$\left(\mathbb{X}_{\mathfrak{p}_v} \subset \mathbb{X}_{\mathfrak{E}}\right)$ belongs to the epistemological space and the space of imaginations $\left(\mathbb{X}_{\mathfrak{p}_v} \subset \mathbb{X}_{\mathfrak{G}}\right)$ such that$\left(\left(\mathbb{X}_{\mathfrak{p}_v} \subset \left(\mathbb{X}_{\mathfrak{G}} \cap \mathbb{X}_{\mathfrak{H}}\right) \subset \mathbb{X}_{\mathfrak{E}}\right)\right)$, however, there is no corresponding ontological signal disposition$\left(\mathbb{S}_{\mathfrak{p}_v} \not\subset \mathbb{S}_{\Omega}\right)$ in its support, such that the epistemological variety characteristic disposition$\left(\mathbb{X}_{\mathfrak{p}_v}\right)$ has no corresponding ontological variety characteristic disposition $\left(\mathbb{X}_{\mathfrak{p}_v} \subset \mathbb{X}_{\Omega}\right)$in the ontological space that belongs to the epistemological space and hence $\mathfrak{I}\left(\mathbb{S}_{\Omega_v}\right) = \varnothing \neq \left(\mathbb{X}_{\mathfrak{p}_v} \subset \mathbb{X}_{\mathfrak{H}}\right)$ with no corresponding ontological variety characteristic disposition and the corresponding variety, implying that the problem variety belongs to the spaces of impossibility and impossible worlds $(\nu \in \mathfrak{V})$ as well as belonging to the space of imaginations $(\nu \in \mathfrak{G})$ with corresponding solution $\left(\mathbb{X}_{\mathfrak{s}_\nu} \not\subset \mathbb{X}_{\mathfrak{W}}\right)$ which implies that $(\mathfrak{s}_\nu \in \mathfrak{V})$ where $(\mathfrak{p}_\nu \notin \mathfrak{H} \subset \mathfrak{A})$ and $\left(\mathbb{X}_{\mathfrak{p}_\nu} \not\subset \mathbb{X}_{\mathfrak{H}} \subset \mathbb{X}_{\mathfrak{A}}\right)$. Every epistemological variety characteristic disposition must belong or has the potential to belong to the ontological space with its characteristic disposition belonging to the space of ontic characteristic dispositions for one to claim it as existing in the space of actual-potential dualities as well as belonging to the epistemological space $\left(\mathbb{X}_{\mathfrak{E}_\nu} \subset \mathbb{X}_\Omega\right)$. If $\mathfrak{p}_\nu \notin \mathfrak{H} \subset \mathfrak{A} \Rightarrow \mathfrak{s}_\nu \notin \mathfrak{S} \subset \mathfrak{G}$. It may, however, be concluded that $\mathfrak{p}_\nu, \mathfrak{s}_\nu \in (\mathfrak{U} \cap \mathfrak{A})$ under dualistic-polar conditions and the principle of opposites with relational continuum and unity.

DEFINITION 5.4.1C: Phantom Problem (Categorial Characteristic Disposition)

Categorially, any problem $(\mathfrak{p}_\nu \in \mathfrak{P})$ is said to be a phantom problem if:

1) $\left(\mathbb{X}_{\mathfrak{p}_\nu} \subset \mathbb{X}_{\mathfrak{E}}\right)$ and $\nexists\, \mathbb{S}_{\Omega_\nu} \ni \mathbb{X}_{\mathfrak{p}_\nu} = \mathfrak{I}\left(\mathbb{S}_{\Omega_\nu}\right)$ and hence, there exists no corresponding variety solution.

2) $\left(\mathbb{X}_{\mathfrak{s}_\nu} \not\subset \mathbb{X}_\Omega\right) \ni \nexists\left(\mathbb{S}_{\mathfrak{s}_\nu} \subset \mathbb{S}_\Omega\right) \ni \mathbb{X}_{\mathfrak{s}_\nu} = \mathfrak{I}\left(\mathbb{S}_{\Omega_\nu}\right) \Rightarrow \mathfrak{s}_\nu \notin \mathfrak{S} \subset \mathfrak{G}$ as a replacement of $(\mathfrak{p}_\nu \in \mathfrak{P} \subset \mathfrak{A})$.

3) $\mathfrak{s}_\nu \in \mathfrak{V} \subset \mathfrak{G}$ and $\mathfrak{p}_\nu \notin \mathfrak{H} \ni \mathbb{X}_{\mathfrak{s}_\nu} \not\subset \mathbb{X}_\Omega \Rightarrow \nexists\, \mathbb{S}_{\mathfrak{p}_\nu} \ni \mathbb{X}_{\mathfrak{s}_\nu} = \mathfrak{I}\left(\mathbb{S}_{\Omega_\nu}\right)$

4) $\mathfrak{p}_v \notin \mathfrak{A} \Rightarrow \mathfrak{p}_v \in \mathfrak{G} \ni \nexists \mathfrak{s}_v \in \mathfrak{G} \subset \mathfrak{U}, \mathfrak{p}_v \notin \mathfrak{H} \subset \mathfrak{A} \Rightarrow \mathfrak{s}_v \notin \mathfrak{S}$

A phantom problem is a desert apparition, where the problem is in the space of impossible world semantics with no solution rather than the space of possible world transformation.

DEFINITION 5.4.2A: Ill-Posed Problem (Verbal Disposition)

A variety problem is said to be an *ill-posed problem* if its characteristic disposition belongs to the intersection of the space of imaginations and the space of potentials but not in the space of the actual in that part of its characteristics belongs to the space of reality while part belongs to the impossibility space in such a way that the part of solution disposition belongs to the possible world while a part of its characteristic disposition belongs to the space of impossible worlds such that its characteristic disposition conceptually belongs to the union of impossible world characteristic disposition and the characteristic disposition of reality, where the intersection of its characteristic disposition and the characteristic disposition of reality is conceptually non-empty and similarly, the intersection its characteristic disposition and that of impossible world is also conceptually nonempty making it difficult to find the variety solution which may exist or appear more than one thus a restructuring of the variety problem characteristic disposition is required in order to ascertain the corresponding disposition of the variety solution as a replacement to the variety problem.

DEFINITION 5.4.2B: Ill-Posed Problem (Analytical Characteristic Disposition)

A variety problem $(\mathfrak{p}_v \in \mathfrak{P})$ is said to be an *ill-posed problem* if its characteristic disposition $(\mathbb{X}_{\mathfrak{p}_v} \subset \mathbb{X}_{\mathfrak{C}})$ belongs to the space of imaginations $\mathfrak{G}$ and the space of potential $(\mathfrak{U})$

$\left(\mathbb{X}_{\mathfrak{p}_v} \subset \left(\mathbb{X}_{\mathfrak{G}} \cap \mathbb{X}_{\mathfrak{U}}\right)\right)$ but not in the space of the actual $(\mathfrak{A})$ in that, $\mathfrak{p}_v \in \mathfrak{G} \ni \mathbb{X}_{\mathfrak{p}_v} \not\subset \mathbb{X}_{\mathfrak{H}} \subset \mathbb{X}_{\mathfrak{A}}$ from which a part of its characteristic disposition belongs to the characteristic disposition of the space of realities $\mathbb{X}_{\mathfrak{H}}$ while part belongs to the characteristic disposition of impossibility space $\mathbb{X}_{\mathfrak{M}}$ in such a way that the part of solution disposition belongs to the possible world $(\mathfrak{s}_v \in \mathfrak{W})$ while a part of its characteristic disposition belongs to the space of impossible worlds $(\mathfrak{s}_v \in \mathfrak{Y})$ such that $\mathbb{X}_{\mathfrak{p}_v} \subset \left(\mathbb{X}_{\mathfrak{y}} \cup \mathbb{X}_{\mathfrak{H}}\right) \Rightarrow \left(\mathbb{X}_{\mathfrak{p}_v} \cap \mathbb{X}_{\mathfrak{H}}\right) \neq \varnothing$ and $\left(\mathbb{X}_{\mathfrak{p}_v} \cap \mathbb{X}_{\mathfrak{y}}\right) \neq \varnothing$ making it difficult to find the variety solution $(\mathfrak{s}_v \in \mathfrak{S} \subset \mathfrak{P})$ thus a restructuring of the variety problem characteristic disposition is required in order to ascertain the corresponding characteristic disposition of the variety solution $(\mathfrak{s}_v \in \mathfrak{S} \subset \mathfrak{W} \subset \mathfrak{P})$ as a replacement to the variety problem $(\mathfrak{p}_v \in \mathfrak{P} \subset \mathfrak{H} \subset \mathfrak{A} \mid v \in \mathbb{V} \subseteq \Omega)$.

DEFINITION 5.4.2C: Ill-Posed Problem (Categorial Characteristic Disposition)

Categorially, any problem $(\mathfrak{p}_v \in \mathfrak{P})$ is said to be an ill-posed problem if:

1) $\mathbb{X}_{\mathfrak{p}_v} \subset \left(\mathbb{X}_{\mathfrak{y}} \cup \mathbb{X}_{\mathfrak{H}}\right) \ni \left(\mathbb{X}_{\mathfrak{p}_v} \cap \mathbb{X}_{\mathfrak{y}}\right) \neq \varnothing$ and hence, there exists no corresponding variety solution $(\mathfrak{s}_v \in \mathfrak{S} \subset \mathfrak{U})$ such that $\mathbb{X}_{\mathfrak{s}_v} \subset \left(\mathbb{X}_{\mathfrak{y}} \cup \mathbb{X}_{\mathfrak{H}}\right) \ni \left(\mathbb{X}_{\mathfrak{s}_v} \cap \mathbb{X}_{\mathfrak{y}}\right) \neq \varnothing$ and $\left(\mathbb{X}_{\mathfrak{s}_v} \cap \mathbb{X}_{\mathbb{R}}\right) \neq \varnothing$ while $\left(\mathbb{X}_{\mathfrak{H}} \cap \mathbb{X}_{\mathfrak{y}}\right) = \varnothing$

2) $\left(\mathbb{X}_{\mathfrak{s}_v} \not\subset \mathbb{X}_{\Omega}\right) \ni \nexists \left(\mathbb{S}_{\mathfrak{s}_v} \subset \mathbb{S}_{\Omega}\right) \ni \mathbb{X}_{\mathfrak{s}_v} = \mathfrak{I}\left(\mathbb{S}_{\Omega_v}\right)$ and $\mathfrak{s}_v \notin \mathfrak{S}$ as a replacement for $(\mathfrak{p}_v \in \mathfrak{P})$.

3) $\mathfrak{p}_v \in (\mathfrak{Y} \cap \mathfrak{G}) \subset \mathfrak{W}$ and $\mathbb{X}_{\mathfrak{p}_v} \subset \left(\mathbb{X}_{\mathfrak{G}} \cap \mathbb{X}_{\mathfrak{H}}\right) \ni \mathbb{X}_{\mathfrak{s}_v} \not\subset \mathbb{X}_{\Omega} \ni \nexists\, \mathbb{S}_{\mathfrak{p}_v} \ni \mathbb{X}_{\mathfrak{s}_v} = \mathfrak{I}\left(\mathbb{S}_{\Omega_v}\right)$.

4) $\mathfrak{p}_v \notin \mathfrak{H} \subset \mathfrak{A} \Rightarrow \mathfrak{p}_v \in \mathfrak{G} \ni \exists\, \mathfrak{s}_v \in \mathfrak{U}$ such that $\mathfrak{p}_v \notin \mathfrak{H} \subset \mathfrak{A} \Rightarrow \mathfrak{s}_v \notin \mathfrak{S} \subset \mathfrak{G} \subset \mathfrak{U}$

DEFINITION 5.4.3: Unsolvable Problem (Analytical Characteristics)

A variety problem $(\mathfrak{p}_v \in \mathfrak{P})$ with $(\mathbb{X}_{\mathfrak{p}_v} \subset \mathbb{X}_{\mathfrak{P}})$ is said to be an *unsolvable problem* if its characteristic disposition belongs to the epistemological space $\mathbb{X}_{\mathfrak{E}_v} \subset \mathbb{X}_{\mathfrak{E}}$ but does not belong to the ontological space, $(\mathbb{X}_{\mathfrak{E}_v} \not\subset \mathbb{X}_{\Omega})$ and hence, there exists no corresponding variety solution in that $(\mathfrak{s}_v \notin \mathfrak{S} \subset \mathfrak{U})$ as a replacement to the variety problem $(\mathfrak{p}_v \in \mathfrak{P})$. It is said to be *unresolved problem* if the problem variety $(\mathfrak{p}_v \in \mathfrak{P})$ with $(\mathbb{X}_{\mathfrak{p}_v} \subset \mathbb{X}_{\mathfrak{P}})$ is such that $\mathbb{X}_{\mathfrak{E}_v} \subset \mathbb{X}_{\mathfrak{E}}$ and hence the corresponding solution $(\mathfrak{s}_v \in \mathfrak{S} \subset \mathfrak{U})$ exists with $(\mathbb{X}_{\mathfrak{s}_v} \subset \mathbb{X}_{\mathfrak{S}} \subset \mathbb{X}_{\mathfrak{U}} \subset \mathbb{X}_{\Omega})$ and may be found with a continual search over the potential space $\mathfrak{U} \subset \Omega$ with $(\mathbb{X}_{\mathfrak{U}} \subset \mathbb{X}_{\Omega})$ through the space of solutions contained in the space of imaginations $(\mathfrak{S} \subset \mathfrak{G})$ with characteristic disposition $(\mathbb{X}_{\mathfrak{S}} \subset \mathbb{X}_{\mathfrak{G}} \subset \mathbb{X}_{\mathfrak{U}})$.

DEFINITION 5.4.4: News and Faked News (Analytical Characteristic Dispositions)

A variety problem-solution $v \in \mathbb{V} \subseteq (\mathfrak{U} \cup \mathfrak{A})$ is said to be a *news item* if its epistemological characteristic disposition $\mathbb{X}_{\mathfrak{E}_v}$ is previously unknown and is currently revealed by its corresponding epistemological signal disposition $\mathbb{S}_{\mathfrak{E}_v}$ such that $\{\mathbb{X}_{\mathfrak{E}_v} = \mathfrak{I}(\mathbb{S}_{\mathfrak{E}_v}) \subset \mathbb{X}_{\mathfrak{H}}\}$. It is said to be *fake news* if the signal disposition $\mathbb{S}_{\mathfrak{E}_v}$ of the variety $v \in \mathbb{V}_{\mathfrak{p}} \subseteq (\mathfrak{U} \cup \mathfrak{A})$ exist but the corresponding epistemological characteristic disposition $\mathbb{X}_{\mathfrak{E}_v}$ does not exist in the epistemological space$(\mathfrak{E})$ such that $\mathbb{X}_{\mathfrak{E}_\nu} \neq \mathfrak{I}(\mathbb{S}_{\mathfrak{E}_\nu})$ and hence $\nu \notin \mathbb{V}_{\mathfrak{H}} \subseteq (\mathfrak{U} \cup \mathfrak{A})$ in that fake news have signal dispositions that belong to the space of imaginations $(\mathfrak{G})$ contained in the space of the potentials $(\mathfrak{U})$ and have no corresponding characteristic dispositions that belong to the space of the realities $(\mathfrak{H})$ contained in the space of the actuals $(\mathfrak{A})$.

PROPOSITION 5.4.1: Phantom, Unresolvable and Unresolved Problems and Fake News

Every variety phantom problem is an *unsolvable variety problem* in the space of problem-solution dualities in the sense that if$(\mathfrak{p}_\nu \in \mathfrak{P})$is a phantom problem then there is no variety solution $(\mathfrak{s}_\nu \in \mathfrak{S} \subset \mathfrak{P})$ as a replacement and$(\mathfrak{s}_\nu, \mathfrak{p}_\nu \in \mathfrak{V} \subset (\mathfrak{W} \cup \mathfrak{V}))$. Every *unsolvable variety problem* in the space of knowing is not necessarily a phantom problem, while all phantom problems are unsolvable problems in space of knowing. Every fake news about a variety problem $(\mathfrak{p}_\nu \in \mathfrak{P})$in the epistemological space $(\mathfrak{E})$ is a phantom problem and hence has no solution $(\mathfrak{s}_\nu \in \mathfrak{S})$. Any fake news $\mathbb{S}_{\mathfrak{E}_\nu}$ about a variety $\nu \in \mathbb{V}$ is a propaganda which is either disinformation or misinformation or both.

NOTE 5.4.1 A Simple Analytics on Phantom, Ill-posed, Unsolvable and Unsolved Problems

An important question arises as to what the differences and similarities are among phantom, ill-posed, unsolvable, and unsolved problems, Gödel's theorems over the space of incompleteness-completeness dualities, initial conditions, and paradigms of thought; and how they may be related to fake news and propaganda over the space of neuro-decision-choice actions as it relates to the spaces of cost-benefit dualities, decidability-undecidability dualities and true-false dualities. This question and its epistemic structures are left for further discussions in the development of a theory of knowledge as a theory of input-output dynamics that is composed of the theory of knowing as the production of intellectual investment flows and the theory of information-knowledge accumulation as the production of intellectual capital stocks over the spaces of problem-solution dualities, input-output dualities, and ignorance-knowledge dualities.

5.5 Reflections on Anomalies in Communications Over Source-Destination Dualities

Let us reflect on the role of these anomalies in knowing as production of intellectual investment flows and information-knowledge accumulation as the production of intellectual capital stocks viewed as input-output processes relative to problem-solution dualities, where the services from the intellectual capital stocks become inputs into the space of neuro-decision-choice actions under a cost-benefit rationality relative to the socio-psychological principle of non-satiation that motivates distortions over the space of cost-benefit distribution within the spaces of, individual-community dualities, incentive-disincentive dualities, greed-altruistic dualities, justice-injustice dualities, peace-war dualities, love-hate dualities, joy-sorrow dualities, happiness-sadness dualities, sickness-wellness dualities, wickedness-kindness dualities and many such dualistic-polar conditions, where thinking and reasoning with information-knowledge inputs are to balance conditions of these dualistic polar structure to achieve greater benefit over cost within the space of cost-benefit dualities. It is the role played by services from the information-knowledge intellectual capital stock as inputs into neuro-decision-choice action over the space of individual-collective dualities that cognitive anomalies acquire the characters of preponderating roles in the history of socio-human progresses over the space of individual-collective dualities viewed in telescopic past-preset-future processes of the Sankofa-anoma traditions contained in the space of beginning-end dualities. By the definition of information as characteristic-signal disposition, all varieties of nominal-real dualities in the space of input-output dualities are conceptualized in terms of information such that buildings, cars and other physical and non-physical elements are reduced to varieties in the information space and hence, all varieties are seen in terms of sequential structural derivative of information, data, knowledge, and evidence where all anomalies in thinking and reasoning are seen in terms of information.

5.5.1 Effects of Information-Knowledge Anomalies and Decision-Choice Actions

All communications in the space of source-destination enterprise, whether they are about problems or solutions, involve messaging and sending-interpretation processes in the space of source-destination dualities which requiring technologies of encoding, storage, and delivery at the source, while requiring the technologies of receival, decoding, storage and use at the destination. The messaging from the source about a variety $(\nu \in \mathbb{V})$ is through the signal disposition $(\mathbb{S}_{\mathfrak{C}_\nu})$ of the corresponding characteristic disposition $(\mathbb{X}_{\mathfrak{C}_\nu})$ in defined codes and the interpretation at the destination of the signal disposition $(\mathbb{S}_{\mathfrak{C}_\nu})$ of the codes is to associate the signal dispositions (received codes) to the corresponding characteristic disposition $(\mathbb{X}_{\mathfrak{C}_\nu})$ in order to identify the underlying variety $(\nu \in \mathbb{V})$ which is in the message, where $\mathfrak{I}(\bullet)$ is the decoding function.

The messages create *sending problems* in terms of correct signal problems that must be solved at the source as well as *interpretational problems* that must be solved at the destination, where the signal dispositions are correctly associated with the characteristic dispositions. For example, a book is a collection of messages that must be carefully coded in terms of a language by the author while the coded language must be interpreted through decoding with decoding function by the reader at the destination. In other words, there is the *source problem* and the *destination problem* in all communications in all information-knowledge processes in all human-nonhuman dualities. The source-destination problems may be amplified by conditions of intentional and unintentional credibility to generate phantom, ill-posed, unsolvable, disinformation, misinformation and propaganda problems from the source to affect the quality of information-knowledge input in the space of neuro-decision-choice actions in relation to the space of decidability-undesirability dualities over the space of source-destination dualities in the space of input-output dualities, where the size of information-knowledge structure and the speeds of analytical processing provide advantages to machines over the space of artificial intelligence-unintelligence dualities.

In the case of communication, a book is an instrument of information-knowledge transmission, where the book plays the role as smart material for storages and transmissions over the space of intra-generational and intergeneration source-destination dualities. It is within this space of source-destination dualities that the development of artificial intelligence must be understood with the need for smart material that can act as holding storage of large information-knowledge stocks, complex thinking algorithms produced by neuro-decision-choice agents for processing, where the technology of machine learning and storage will improve the flexibility in machine thinking, reasoning, and decision-choice actions independent of human actions.

These source-destination conditions are information-knowledge processes that pass through the spaces of certainty-uncertainty dualities and doubt-surety dualities under the principles of credibility and entropy. The check of credibility fails when the knowing process and the information-knowledge process are politicized and propagandized to shape the neuro-decision-choice outcomes and the corresponding cost-benefit distributions over the space of problem-solution dualities by distorting the real opportunity cost of existing real variety problem and the real potential net benefit of the variety replacement, where information is weaponized to take advantage in the game of life. An example is the current discussion of conditions of coronavirus in terms of its scientific information and knowledge and their implications for the socioeconomic dynamics, social policies, management, and socio-economic decisions in some countries. In most cases, the conditions of misinformation, disinformation and propaganda take place in the basic social structure domestically and internationally to affect the perceptions and calculations of cost-benefit distributions over the institutions of the domestic and international super-structures to affect the sources' willed directions of neuro-decision-choice actions of the destinations over the space of production-consumption dualities in relation to the space of input-output dualities broadly defined and specified to achieve a cost-benefit distribution advantage.

These source-destination problems are general to all areas of knowing with differential difficulties and complexities and must be solved as one travels from the *zone of variety ignorance* to cross the epistemic transversality points or the epistemic transversality line into the *zone of variety knowledge*, where between the zone of ignorance and the zone of knowledge is the zone of ignorance-knowledge conflicting claims as a *zone of dualistic-polar game* in the duals of the opposites, where the reward of the game is the dominance of a dual as the dualistic-power dominance affects cost-benefit distribution in social set-ups. It is useful to note the analytical power of game theory in variety transformations within the dualistic-polar relations in the space of all problem-solution dualities under the general principle of opposites with relational continuum and unity. It is also important to note that there are significant differences between the concept of *epistemic transversality* as a thinking algorithm and *cognitive transversality* as a reasoning process using the thinking algorithm in decision-choice actions, where the thinking algorithms may be used by anybody through reasoning over the space of problem-solution dualities not connected or connected to the development of the thinking algorithms. In fact, this is how the education enterprises work, where the information-knowledge accumulation and intellectual capital stocks may be used by developing logical keys to unlock the doors to the information-knowledge house for use in the processes of reasoning by learning in the space of ignorance-knowledge dualities as revealed by the Adinkra dualistic symmetry of *ohnim-sua-ohu* (ignorance-knowledge process) while keeping in mind that all the dualistic-polar structures are in symmetries and supper symmetries as seen in theory of adinkra graphs and geometry [R27.10][R27.1][R27.3].

The epistemic transversality conditions for both variety identifications and variety transformations in the space of ignorance-knowledge dualities as well as in the space of problem-solution dualities require the absence of phantom, ill-posed, unsolvable problems as well as misinformation, intellectual Ponzi scheme, fake news, propaganda, and disinformation conditions. The absence of these anomalies are the necessary conditions for the minimizing ignorance dual and maximizing knowledge dual about a variety in the cognitive journey from the *epistemic zones of ignorance* to the *epistemic zones of knowledge* about the same varieties through the *game zones* of ignorance-knowledge conflicting claims and hence optimizing the preferred outcome conditions, where the benefit dual outweighs the cost dual.

The point of emphasis is simply that the spaces of ignorance-knowledge and problem-solution dualities are general to all areas of knowing whether they are classified as science or non-science and that they exist in relational continuum and unity that allow the progress of knowing and information-knowledge production in all temporary equilibrium states and permanent disequilibrium states. Furthermore, there exist thresholds for all areas of knowing, where one passes from the zone of ignorance to the zone of knowledge, where the zone of ignorance is separated from the zone of knowledge by the game zone of conflicting ignorance-knowledge claims. Additionally, every knowing process as the production of intellectual investment flows and every information-knowledge accumulation as the production of intellectual capital stocks, irrespective of the area of knowing, pass through the zone of degrees of doubt to the zone of degrees of surety to claim acceptance with epistemic conditionality to define the applicable zone of neuro-decision-choice certainty over the zones of risk and benefit in all human actions over the space of risk-riskless dualities that is analytically traceable to the space of ontic-epistemic dualities.

The game zone which is ignorance-knowledge zone of conflicting claims is generated by the elements in the space of certainty-uncertainty dualities embedded in the space of quality-quantity dualities. The space of certainty-uncertainty dualities is a derivative and offspring of the space of ontic-epistemic dualities that generates cognitive variety differences. To the ε-left of the dualistic game conflict zone of the transversality line, claims of ignorance are assured with minimal certainty. To the ε-right of the dualistic game conflict zone of the transversality line, claims of knowledge are assured with minimal certainty with reference to [R8.16b].

The nature of certainty-uncertainty conditions is amplified by cognitive deceptions to shape neuro-decision-choice inputs and corresponding outputs for politico-legal decision-choice actions as we see in social practices. The conditions of the epistemic structure over the space of ignorance-knowledge dualities may be extended to the space of problem-solution dualities, where it must always be kept in mind that ignorance is a problem even though ignorance may be a solution, and a problem is a cost or may be a benefit in dualistic-polar structure relative to a goal or a set of goals. Similarly, knowledge is a solution even though it may be a problem, and a solution is a benefit even though it may be a cost under the principle of opposites in relation to all elements in the space of neuro-decision-choice actions of cognitive agents in search of a fuzzy goal or a set of fuzzy goals under the principle of non-satiation. There are some cases over the space of neuro-decision-choice actions, that ignorance may be a benefit and hence a solution while knowledge may be a cost and hence a problem relative to the space of goals, objectives, and visions. The principle of opposites provides us with foundations of the analytical claims of dual usage over the space of good-evil dualities or the *asantrofi-anoma* conditions of *musuo-seradie* dualities or synonym-antonym dualities of all nominal-real varieties such that every variety disposition is defined over the space of positive-negative dualities, and every negative disposition and every positive disposition are complexly defined over the space of cost-benefit dualities over the space of neuro-decision-choice actions.

5.5.2 Reflections on the Subject matter of Information, Knowledge, and Neuro-Decision-Choice Actions

The subject matters of information and knowledge are on the conception that there are three interdependent structures of the information-knowledge conditions. The *first structure* is the definitions, meanings and explications of information and knowledge and their derivatives such as facts, data and evidence in theory and applications. The *second structure* is the set of the input-output conditions of knowing and information-knowledge as varieties in the space of production-consumption dualities that will provide us with the cognitions and views of past-present-future telescopic processes to administer the affairs of the socio-natural environment to the advantage of cost-benefit conditions of social existence. The *third structure* is the communication of information-knowledge conditions and their uses as input-output processes among cognitive agents over the space of source-destination dualities in the epistemological space, where the information-knowledge conditions may be weaponized by the source agents against the destination agents to take advantage over the distributional conditions of the space of cost-benefit dualities through human uses of the services from information-knowledge capital stocks in the space of neuro-decision-choice actions, where all behaviors are neuro-decision-choice driven in program designs and executions within the space of imagination-reality dualities transforming imaginations to realities from the space of goals, objectives and visions, where such goals, objectives and visions are conceptualized and designed with characteristic dispositions in the space of variety imaginations contained in the space of potentials.

Much of the current and traditional developments of information-knowledge structure in theory and applications are about the third structure in the development of technologies in terms of information and its communications to take advantages over the space of cost-benefit dualities, given that we know the meaning and phenomena of information, data and knowledge which are the outcomes of knowing and information-knowledge system over the space of production-consumption dualities. The first structure and its approach to theory and practice over the space of identification-transformation dualities has been discussed in my monograph *The Theory of Info-statics* [R17.17] and *The Theory of Info-dynamics* [R17.18]. The second structure has two components of production-consumption dualities and the space of quality-quantity controls, where the space of production-consumption dualities constitutes the primary category to the space of quality-quantity controls. The outcomes of knowing and information-knowledge productions are always subjected to quality control over the spaces of true-false dualities and acceptance-rejection dualities.

The traditional approach to the study of information and knowledge has very little to say if any on the structure one and two. This approach to the study of information and knowledge is mostly concentrated on a small aspect of structure one on the concepts of information and knowledge in terms of their identities which find expressions in as quality control in the space of quantity-quality dualities, where definition of knowledge is provided as *justified true belief* (JTB). Here, much attention is concentrated on the conflicting discussions of the concepts like verification, falsification, corroboration, and others which may include some aspects of entropy without providing variety distinctions among information, knowledge, fact, data, and evidence. What is the definition of the concept of information, and what is its phenomenon in the classical epistemic setting? What are the differences and similarities among information, data, knowledge, and evidence? The answer of which requires the definitions of these concepts and to examine their conceptual connectivity if any. Which of these varieties of information, knowledge, fact, data, and evidence enters as an input into the neuro-decision-choice system?

The criticism of (JTB) is not helpful because the identity of variety knowledge is still unknown [R8.16b]. The difficulty, here, is that the structure of knowledge as output of production with information input is completely neglected in the information-knowledge discourse, where many a time, information is taken as knowledge and vice versa with conceptual equality extended to interchangeability of fact, data, and evidence without specifying the conditions of this validity. Similarly, the concept and phenomenon of information are not defined and specified allowing them to take on anonymous characters with no shape over the spaces of definition-explication dualities and input-output dualities relative to neuro-decision-choice dualities within the epistemic fields with relational and information fields. The separation of knowledge into knowledge by acquaintance and knowledge by description as suggested by Russel [R22.18] [R22.19] is not connected to the separation of information and knowledge leading us to misunderstanding of risk as social phenomenon without ontological roots, and there is no risk in nature or the ontological space. Risk, failure, truth, false love, hate and other such qualities reside in the epistemological space as linguistic variables with cognitive interpretations over the space of subjective-objective dualities.

It is the uncertified conditions of the standard approach to the understanding of information-knowledge systems that the current monograph in support of the monograph [R17.17] is being developed. Here, the conditions of the third of information-knowledge structure are under discussion following *The Theory of Info-statics* [R17.17] and *The Theory of Info-dynamics* [R17.18] in relation to variety behaviors over the space of static-dynamic dualities. Information-knowledge quality control over the space of true-false dualities is handled in *The General Theory of Entropy* [R21.5] [R21.23]. *The Theory of Problem-Solution Duality and Polarity* over the space of identification-transformation dualities, relative to the space of visible-invisible dualities[R8.16a] [R816b] is devoted in dealing with knowing as the production of intellectual investment flows and information-knowledge accumulation as the production of intellectual capital stocks showing input-output outcomes of neuro-decision-choice activities over the space of problem-solution dualities and polarities to give contents to information-knowledge relationality under the spaces of production-consumption dualities, parent-offspring dualities, predecessor-successor dualities and construction-destruction dualities in discussing the general nature of socio-natural behaviors and the meaning of life as problem-solution dynamics guided by social philosophical consciencism containing error-correction processes, where the universe is closed under input-output processes.

The theory of epistemic field [R8,16b] is developed to deals with variety category formation over the space of internal-external dualities relative to the space of ontic-epistemic dualities as the primary category of the space of certainty-uncertainty dualities relative to the space of ignorance-knowledge dualities in knowing. The current monograph discusses the theory of anomalies in paradigms of thought and their relationship to creativity, curiosity, courage, hope, failure and their relationality to the space of neuro-decision-choice actions over the space of incentive-disincentive duality mapped from the space of cost-benefit dualities. The analytical approaches taken here give explanations to the nature of information-knowledge disequilibrium dynamics, source-destination communication disequilibrium dynamics under matter-energy equilibrium dynamics where information is defined as characteristics over the space of content-signal dualities. All neuro-decision-choice analyses relate to the concepts of information, knowledge, data, evidence as sequential derivatives over the space of characteristic-signal dualities, where the characteristic dispositions are the encoded contents of information as ontic characteristic disposition to establish variety identities from the source, and the signal dispositions are messages in the information to be decoded at the destinations as epistemic characteristic dispositions to construct the epistemic variety and cand claim epistemic variety identities to reveal identities over the space of similarity-indifference dualities. The theories are about neuro-decision-choice actions over the space of variety identification-transformation dualities.

5.5.3: Anomalies, Communications and Source-Destination Dualities over the Epistemic Field

All variety knowing in terms of identities and changes come as source-destination enterprise within the space of *ontological-epistemological duality,* in relation to problems and solutions, involve messaging-interpretation processes in the space of source-destination dualities, where the source is composed of ontological elements containing the ontic varieties, and the destination is composed of epistemological elements containing epistemic varieties in knowing as the production of intellectual investment flows and information-knowledge accumulation as the production of intellectual capital stocks. It is always useful to keep in a full view within the cognitive search process, the *analytical tools* of epistemic relationality, connectivity, diversity and unity that ensure the variety and categorial give-and-take sharing processes in cognitive global understanding of universal information-knowledge systems for the developments of variety identifications, transformation, languages and neuro-decision-choice actions in socio-natural existence over the space of encoding-decoding dualities, where an important space of source-destination dualities is the complete nervous system with input-output messages carried by veins connecting the central nervous system to other parts of the body in the space of coding-decoding dualities in the give-and-take sharing processes.

In the whole process of knowing over the epistemic fields, supported by relational and information fields, the categories are defined over the space of primary-derived dualities, where each category is equipped with socio-natural functor or categorial converter making it possible to explain the conditions over the spaces of parent-offspring dualities, problem-solution dualities and many such dualistic-polar structures. The *primary category* of universal understanding is the four dimensions of matter, energy, information, and time, where information is a set of general property of all organic varieties including itself. The general property is the system of characteristic dispositions as basic to all things and is taken as the *primitive axiom of universal existence* which allows variety identities to be formed, compared and placed in categories of similarity and difference; languages to be developed and transmitted through a system of signal dispositions over the spaces of source-destination dualities and synonym-antonym dualities under the principle of opposites, with relational continua and unity, where organization of knowing is central to information-knowledge accumulation intra-generationally and inter-generationally over the epistemological space. In the framework of knowing and knowledge production, every generation, with a corresponding epistemic field containing the relational and information fields, presents a socially organizing opportunity, where every individual-community duality is an important creative actor, and where every country presents an important challenge to change the path of the global heritage of the universal information-knowledge process, where the fundamental building blocks are the ontological space and epistemological space from which the space ontological-epistemic dualities emerge to produce the space of ontic-epistemic dualities which contains the space of certainty-uncertainty duality.

The *primary category* of universal existence with its *characteristic disposition* is the ontological space (Ω), with ontological varieties and ontic characteristic dispositions, that acts as the identity of knowing, where its existence is independent of all the elements that it contains within the space of beginning-end duality with a mutual containment of the nothingness-somethingness duality relative to *Gyenyame-Bennunu-anoma principle* of ontological existence (see Figure 4.5.1 of Chap. 4). From the primary category, we obtain the *derived category of existence* which contains the decoded characteristic dispositions from the ontological *signal dispositions* of matter, energy, information, and time. The derived category of existence is the epistemological space that contains the past-present duality, present-future dualities, and past-future duality relative to the Sankofa-anoma principle of epistemological existence relative to the principle of ontological existence. Acquisition of knowledge whether intentional or unintentional by cognitive agents is a system of complex epistemic processes to solve *identification and transformation problems* of matter-energy varieties in the ontological space. The solutions are carried on through the interactive processes of neuro-decision-choice activities and information defined by characteristic-signal dispositions and through thinking to generate thinking algorithms and reasoning to utilize the thinking algorithms over the space of the problem-solution dualities, where the characteristic dispositions are the contents, and the signal dispositions are the messages of the varieties which enter as input-output over the neuro-decision-choice space.

In the process of solving the identification-transformation problems, information-knowledge awareness is developed through the stock-flow disequilibrium dynamics toward a general cognitive understanding, where the contents of information are defined by characteristic dispositions and the messages of the characteristic dispositions are defined by the signal dispositions. All the variety characteristic dispositions are seen in terms of belonging to the space of ontological-epistemological duality with messages of signal dispositions from the ontological space to the epistemological space. The ontological variety signal dispositions from the space of actual-potential dualities pass through the *acquaintance space* for encounters with cognitive agents, then through the space of imagination-reality dualities containing *possibility space* to generate *possibilistic information* and through the *probability space* to generate *probabilistic information* and to the *space reality* to generate *information-knowledge accumulation* with fuzzy-stochastic certainty in the space of realities as a sub-space of the space of the actuals.

The ontological-epistemological duality is composed of the ontological space of nature which generates the space of variety ontic characteristic dispositions and the epistemological space which generates the space of variety epistemic characteristic dispositions of society creating the ontic-epistemic differences under cognitive capacity limitations, the conflicts of which give rise to knowing and information-knowledge accumulation system, where perfect identification is achieved when a variety ontic characteristic disposition is equal to variety epistemic characteristic disposition. The epistemic system is such that in the space of the potentials do we have seeds of imaginations that may be abstracted; in the space of imaginations do we have the seeds of possibility space containing necessities, variety initial of conditions and anticipation; in space of necessity, do we have the seeds of probability space containing freedoms and sufficiency; in the probability space, do we have the seeds of actualities of necessity and freedom; and in the space of the actuals do we have the seeds of potentialities of the necessities and freedoms in the space of cognitive actions over the space of problem-solution dualities under variety transformation games, where creativity, curiosity, courage, fear, hope, despair and other such qualities find residence as enhancers..

A reflective note on the relational conditions of fake new, propaganda, disinformation and misinformation and how they affect the path of neuro-decision-choice processes will be useful and needed at this point of the explanatory and prescriptive understanding of the conditions of the dualistic-polar gaming processes of variety transformations through forces generated by dualistic-polar conflicts over the space of problem-solution dualities under cost-benefit conditions and preference orders within the space of relativity and principle of non-satiation. It has been argued that the information-knowledge process is also an input-output process as well as production-consumption process and destruction-replacement process over the space of cognitive actions. The information space is divided into two interactive and interchangeable dynamic spaces of the space of benefit dispositions and the space of cost dispositions. Every information-knowledge item is interchangeable as either cost disposition or as benefit disposition in relational continuum and unity. Thus any variety is simultaneously cost and benefit containing cost disposition and benefit disposition in the same place and time where the benefit or cost is relative to an element in the space of goals, objectives and visions, The misinformation, disinformation, propaganda, and fake news are tools of deception to abstract an increasing net benefit in the game of life where the deceptive varieties from the space of imaginations are presented as belonging to the space of realities.

The current development of digital systems involves activities over the space of source-destination dualities, where the effectiveness depends on clarity of signal transmission and the correct signal interpretation to abstract the intended contents. It is here that the theory of coding, just as the theory language, becomes important over the space source-destination dualities of the messaging system whether it is inter-human, intra-human, inter-nonhuman, or intra-nonhuman, where the source-destination must be broadly defined to include all forms over the space of sender-receiver dualities including forms of verbal and non-verbal codes which will affect knowing and knowledge production. In this respect, cognitive agents are actively searching for artificial material such as intelligent (smart) materials that have capability of holding codes and sending codes with instruction from the source with protection and capacity of self-learning without supervision.

5.6: The Limit of Knowing, Information, Knowledge, and Variety Relations

In chapter 4 of this monograph, the concepts and phenomena of the spaces of knowability-unknowability dualities $\mathscr{D} = (\mathscr{E} \cup \mathscr{U})$, space of knowability, $(\mathscr{E} = (\mathscr{M} \cup \mathscr{N}))$, where $(\mathscr{M} \cup \mathscr{N})$ is the space of past-present dualities, and the space of unknowability $(\mathscr{U} = (\mathscr{N} \cup \mathscr{F}))$, where $(\mathscr{N} \cup \mathscr{F})$ is the space of present-future dualities, The symbols $(\mathscr{M})$ as the space of telescopic past, $(\mathscr{N})$ as the space of telescopic present and $(\mathscr{F})$ as space of telescopic future are introduced and discussed, where knowing over the space of knowability-unknowability dualities is substantially constrained by cognitive capacity limitations over the epistemological space. The effects of the conditions of cognitive capacity limitations in the space of knowability-unknowability dualities may be extended and examined over the spaces of problem-solution dualities and question-answer dualities. Within the space of unknowability are questions like the questions surrounding the reasons for the existence of different planets, multiplicity of galaxies and the positions they occupy in the universe, the shapes and contents of the planets and planetary systems, galaxies, exoplanetary systems and more in the complexity of the universal existence. Over our planet earth, why are there mountains, valleys, earthquakes, volcanoes, rivers, lakes oceans and many such varieties not in terms of their cost-benefit implications to conditions of existence of lives in different forms but in the essence of their being in the ontological existence? Several of the questions involve why, what, how, when, where, and who contained in the space of question-answer dualities defined over the space of ignorance-knowledge dualities.

In this situation, what meaning do we attach to the claim of the expansion of the universe or is it meaningful to ask a question regarding the expansion and non-expansion of the universe relative to the space of visible-invisible dualities? Is there a beginning of the universe and is it knowable? Similarly, is there an end to the universe and if there is, is the end of the universe knowable? Is the universe permanent or temporary in terms of its contents and existence? How permanent are the varieties in the universe or the variety transitions? Are the forces in the space of static-dynamic dualities over the planet earth the same or like other conceivable planets? Are parent-offspring processes under the principle of opposites as observed over the planet earth the same over other planets? Where do the matters of different planets come from and what are the similarities and differences, and how do they relate to conditions of attraction, repulsion, spaces of rotational stability-instability dualities, invisible-visible dualities, and matter-nonmatter dualities? What is the meaning of emptiness in ontological dualities and is this meaning different from or similar to the meaning of emptiness in epistemological dualities viewed from the space of emptiness-non-emptiness dualities relative to the space of visible-invisible dualities? Are there limits to cognition of the universe and the forces of universe relative to the space of identification-transformation dualities?

At the level of cognition, what varieties are knowable over the space of acquaintance as our technologies improve relative the space of visible-invisible dualities? Are the input-output processes over the planet earth also taking place over other planets? In other words, is the phenomena of input-output and production-consumption processes universally general over all planets contained in the universe? How much and what can we know from the space of imagination containing creativity, curiosity, hope, failure, and others of such human qualities? The examinations of these questions and needed answers require that cognitive agents must learn not *what there is known* alone, but *how to think* in the space of *knowability-unknowability dualities* contained in the space of identification-transformation dualities as it is connected to ignorance-knowledge dualities. In other worlds, effective learning is about learning how to think with information-knowledge stocks as input-output varieties into neuro-decision-choice processes to establish epistemic interdependence over the space of problem-solution dualities and question-answer dualities. Effective teaching is about teaching how to think with input-output elements and not what to think in the space of neuro-decision-choice actions to encourage the development of epistemic interdependence.

Teaching and learning what to think is a path to cognitive enslavement, epistemic dependency, and external mind control over the space of neuro-decision-choice actions such that who controls the contents of what to teach and learn also controls the behavioral actions of the learner over the space of doubt-credulity dualities since what is learned, absolved, and retained becomes part of the input-output process in the space of cognitive actions. Teaching and learning how to think with and about the existing information-knowledge stock is a path to cognitive freedom, epistemic independence and liberated internal self-control of mind over the space of neuro-decision-choice actions such that the learner controls the contents of what to learn, absolve, and to hold for self-control of behavioral actions of the learner over the space of doubt-credulity dualities since what is learned, absorbed, retained as information-knowledge stock becomes part of the input-output process in the space of cognitive actions.

It is within this context that the works of Cater G Woodson [R17.68], Du Bois [R17.19b.], Asante [R17.3a][R17.3b], Cress Welsing [R17,67.] on the educational conditions of African Americans; and the works of Nkrumah [R17.44] Diop [R17.12b], Ben-Jocannan [R17.4a]

[R17.4b], and Abraham[R17.1a] Dompere [R17.15] on the educational conditions find transformative characters in education and socioeconomic development to deal with colonialism, neocolonialism, imperialism, and oppression of the subjugated. The contents of these works must be related to the theory of categorial conversion [R17.15] within the space of imaginations in establishing variety identities, initial conditions, necessity and anticipations through the space of input-output dualities, as well as related to the theory of philosophical consciencism [R17.16] [R17.44] within the space of imaginations in establishing conditions of variety transformation, sufficiency, freedom and expectations within the spaces of input-output dualities and necessity-freedom dualities relative to the space of production-consumption dualities to deal with imperial incumbency, neocolonialism and exploitation defined over the spaces of pain-pleasure dualities, joy-sorrow dualities and happiness-sadness dualities under the space of individual-collective sovereignties over the space of neuro-decision-choice actions relative to the space of goals, objectives and visions which is under creativity over the space of construction-destruction dualities under a preference scheme contained in human-right abuses from the imperial projects under the doctrine of discovery for looting and evil deeds.

5.7: Conceptions of Cost-Benefit Conditions over the Space of Ontological-Epistemological Dualities

The questions in section 5.6 and the need for answers lead to the introduction of the concepts and phenomena of costs and benefits over the space of ontological-epistemological dualities relative to the space of human actions and its existence, where events within the space of ontological-epistemological dualities are examined in terms of real costs and benefits over the space of static-dynamic dualities. Here, we have natural cost-benefit conditions of variety actions that affect variety balances over the ontological space. Similarly, we have social cost-benefit conditions of variety actions that affect balances over the epistemological space. How meaningful is it to speak of ontological costs and benefits of ontological actions over ontological space? Similarly, how meaningful is it to speak of epistemological costs and benefits of epistemological actions over the epistemological space? What are the analytical similarities and differences between the ontological cost-benefit dualities and epistemological cost-benefit dualities? Here, we want to examine effects of natural input-output processes on varieties over the ontological space in terms cost-benefit conditions over the universe. Similarly, we want to examine the effects of social input-output processes on variety transformations over the epistemological space and their ontological impacts in terms cost-benefit conditions over the societies and their surrounding socio-natural environments. The definition of cost as an opportunity cost in neuro-decision-choice space must be kept in mind and related to alternative uses of inputs relative to the best output outcome among possible outcomes, where cost is an alternative output benefit from the use of an input in the supply-demand chain over the complex space of production-consumption dualities relative to the space of imagination-reality dualities.

All discussions on the space of positive-negative dualities relative to the space of input-output dualities in relation to the space of cost-benefit dualities have been undertaken over the epistemological space, where the conditions of variety cost-benefit balances are the incentives and disincentives for social actions over the space of neuro-decision-choice actions. In this respect, natural input-output are viewed in terms social effects of cost-benefit conditions by neuro-decision-choice agents, where the conditions of socio-natural events are mapped onto the space of incentive-disincentive dualities under cognitive capacity limitations. Such cost-benefit evaluations of socio-natural events do not take place over the ontological space, where transformations take place under natural laws for natural balances under the principle of limitless transformational capacity with information-knowledge equality. Here, the social input-output externalities from the space of social production-consumption dualities are used in nature to create ontological varieties that may generate social costs or benefits to social existence. In this respect, cost-benefit analysis as a general foundation of decision and choice, allows the inclusion of internal and external cost-benefit conditions in the variety decision-choice processes over the space of relativity. Let us also keep in mind that the general concept of optimization (in terms of minimum and maximum) is based on cost-benefit analysis where the objective is either cost or benefit and the constraints are either benefit or cost such that benefit(cost) is maximized(minimized) as the objective subject to cost(benefit) as the constraint.

From natural processes, there are no accidents in the ontological space and hence the composition of the earth matter and things buried under earth are not accidents. There are ontological principles behind them for continual natural actions over the spaces of stability-instability dualities, ecological balance-imbalance dualities, stability-instability dualities, identification-transformation dualities under *ontological time*. The input-output of epistemological externalities generating news forms of ontological varieties or constraints on ontological transformation efficiencies must reveal how cost-benefit analysis of the environment must be viewed in relation to the space of ontological-epistemological dualities. A complete analysis and justification of the theory of cost-benefit analysis will require the development of the epistemics of cost-benefit analysis and neuro-decision-choice actions over the space of ignorance-knowledge dualities at epistemological times. The essence of the stated questions leads us to the discussions on the theory of epistemic transversality conditions over the spaces of ignorance-knowledge dualities, problem-solution dualities, question-answer dualities, failure-success dualities, and defeat-conquer dualities in the next chapter to allow us to answer the questions when does ignorance become knowledge, problem becomes a solution and others like these dualistic-polar structures.

CHAPTER 6

THE THEORY OF EPISTEMIC TRANSVERSALITY CONDITIONS OVER THE SPACE OF IGNORANCE-KNOWLEDGE DUALITIES AND POLARITIES

ABSTRACT

Chapter 6 of this monograph presents discussions on the theory of epistemic transversality conditions which relate to dynamic optimality of epistemic journeys between duals of duality and poles of polarity to obtain stopping rules of the epistemic journeys. The epistemic transversality conditions are made up of *identification transversality* conditions and *transformation transversality* conditions. The identification transversality conditions define the stopping rule for the cognitive journey to obtain the variety identity that reveals the inherent necessity, conditions of being and the knowledge about the variety at static state, where the varieties are made up of states and processes with properties and relations in the space of identification-transformation dualities. The transformation transversality conditions define the terminal state of the journey between variety ignorance and variety knowledge, variety problem to variety solution, variety question to variety answer and, variety cost to variety benefit presenting the conditions when variety knowledge is found for the replacement of variety ignorance or a variety solution is found to replace variety problem or variety answer is found to replace variety question over the spaces of dualities and polarities. The transversality conditions are related to cost-benefit marginal conditions of knowing as the production of intellectual investment flows and information-knowledge accumulation as the production of intellectual capital stock, where thinking-reasoning

K. K. Dompere, *Theories of Knowing and Knowledge Production*, Studies in Systems, Decision and Control 501, https://doi.org/10.1007/978-3-031-42863-0_6

processes are the vehicles of the journeys within and between dualities and polarities in the space of dualistic-polar structures. The epistemic transversality conditions are then related to telescopic past, present and future in all neuro-decision-choice processes and used to partition the knowing space into zones of knowledge and ignorance or zones of problems and zones of solutions leading to the discussions of their utility in neuro decision-choice space. A definition is provided and its relationship to the principle of opposites with relational continuum and unity are discussed and linked to fuzzy paradigm of thought where ignorance and knowledge are mutually constraining. The chapter is concluded with discussions linked with *asantrofi-anoma* problem, conditions, and rationality to provide computational understandings of epistemic transversality conditions in cognitive dynamics within ignorance-knowledge and problem-solution situations. The definition and discussions of the asantrofi-anoma principle and rationality are provided.

6.1 REFLECTIONS ON EPISTEMIC TRANSVERSALITY CONDITIONS IN ORGANIZATION OF KNOWING

The concept of epistemic transversality conditions has been used in the previous chapters of this monograph and in [R8.16a]. It is associated with epistemic journeys within all the dualistic-polar spaces that contain duals of dualities and poles of polarities under the principle of opposites with relational continuum and unity. A journey within a duality is a journey from a dual disposition to another dual disposition in terms of internal dualistic games. Similarly, a journey within a polarity is a journey from a pole disposition to another pole disposition within the same polarity in terms internal polar games. These are intra-dualistic-polar journeys where the destinations of the journeys are the dominance to maintain or change the nature of the variety dominance which is the reward of the dualistic-polar games. A journey between dualities is a journey from a disposition of duality to another disposition of duality in terms of external dualistic games that constitute internal polar games, where the combinations of external dualistic games and internal polar games gives rise to external polar games for general dualistic-polar dominance which is the destination of the journey as well as the reward of the general dualistic-polar games. The question that arises is simple. How does one know that variety identity dominance has been achieved for any dualistic-polar journey?

To answer this question and related ones, we need to know the definitional nature and the critical variables of the analytics of the epistemic journey. The critical variables in the dynamic path are the state variables composed of dualistic-polar variable ignorance and variety knowledge in opposite accumulation in the sense that as the variety characteristic disposition is knowing, the knowledge stock increases and ignorance stock decreases. In this way, there is positive knowledge stock accumulation and negative ignorance stock accumulation in the epistemic journey as the *state variables*, where the increase in the positive and negative accumulation is through knowing as the production of intellectual investment flow which acts as the *control variable* through the instruments of thinking and reasoning. As structured, the whole of intellectual investment-capital accumulation system is placed in the space of fuzzy optimal control problems, where there is the ignorance set and knowledge set which are fuzzy sets with fuzzy membership characteristic set of belonging and set intersects in the space of fuzzy decision as a space of fuzzy processes. The ignorance-knowledge-accumulation process is dualistic-polar process under the principle of opposites with relational continuum and unity as input-output processes of information. The process is such that the variety knowledge is always constrained by variety ignorance in degrees of accumulations in the fuzzy topological space in intra-categorial and inter-categorial journeys. The epistemic neuro-decision-control process for the intellectual investment-capital control

a set characteristic signals that presents its presence in the communication space of source-destination dualities to establish the space of ontic-epistemic dualities to give rise to the space of certainty-uncertainty dualities in order .The space of certainty-uncertainty dualities produces conditions to create the space of imagination-reality dualities containing the spaces of possibility-impossibility dualities and probability-improbability dualities, possible-impossible-world dualities from which goals, objective and visions under the principle of non-satiations take place in the space of human behavior within the space of production-consumption dualities. In this respect, the phenomenon of information is materialistic in essence as it has been argued in [R17.17] [R17.18] [R8.16b]. By the materialist definition, Information represents a sequential arrangement of properties of varieties into characteristic dispositions, where the alteration of any arrangements in a characteristic disposition changes the variety and its identity which then presents a different information and identity to reveal a variety transformation and identification in the space of genetic codes or epigenic structure.

Those who disagree with this definition of information must provide one that is either materialistic or non-materialistic such that the phenomenon of information covers all areas within the space of ontological-epistemological duality representing varieties in distinction and commonness. Whatever definition of information that may be advanced, it must be consistent with our linguistic and mathematical constructs of categories, groups, and sets. Furthermore, we must answer the question as whether any input variable of cognitive processors and their output variables are information, knowledge, data, fact, or evidence if their essences are different. It must be noted that information is powerful because its concept and phenomenon are used in the space of descriptive-analytical dualities over all areas of human endeavors and acts as inputs and outputs in the space of neuro-decision-choice activities to obtain knowledge in the space of identification-transformation dualities. It is powerful because it is the general instrument that has dualistic connectivity in the space of ignorance-knowledge dualities and offers a means of input-output dualities, controllability, and management of social existence.

Information, however is not elusive in any shape or form as seen and discussed in different writings of philosophers and non-philosopher on philosophy of information, theories of computer information systems, theories of communications, theory of languages of the family of ordinary languages (FOL] and the family of abstract languages[FAL] and many others theories of socio-natural phenomena and existence, where the concept and phenomenon of information is taken as given, understood and known creating analytical sloppiness where everything goes . The possible elusiveness stems from poor conceptions without critically providing workable definitions that contains its powerful concept and phenomenon of universality in use over all areas of human endeavor [R12.19] [R12.21] [R12.23] [R20.2] [R20.6] [R20.12]. One must answer the question of what the similarities and differences among economic information, political information, medical information, chemical information, biological information and many such phenomena are, and how does one establish the conditions of their identities over the space of source-destination dualities. It is through the definition of information that we define epistemic transversality conditions and optimality conditions.

6.2: EPISTEMIC TRANSVERSALITY CONDITIONS OF DUALISTIC-POLAR JOURNEYS BETWEEN DUALS AND BETWEEN POLES OVER THE SPACE OF STATIC- DYNAMIC DUALITIES AND POLARITIES

The development of information-knowledge conditions about our universal system, made up of the space of ontological-epistemological dualities, finds meaning in the cognitive development of information-knowledge conditions about the identity, behavior and transformations of matter-energy varieties in the space of actual-potential dualities and imagination-reality dualities in terms of cognitive conditions over the spaces of essentiality-non-essentiality dualities and relevance-irrelevance dualities. By entering the concept of information-knowledge condition, an important point is being made that information is not necessarily knowledge and versa in the space of knowing and neuro-decision-choice actions, the distinction, and similarities of which are discussed in [R17.17]. The epistemic information-knowledge development is seen in terms of variety problem-solution transformation dynamics in the space of variety ignorance which provides the analytical foundations through thinking-reasoning processes for the development of the theory of knowing. The knowing process as a production of intellectual investment flows integrates two opposing forces of ignorance and knowledge of system of balances in all identification-transformation problems of varieties, where each variety presents itself in the epistemological space as an ignorance-knowledge proportion in relative degrees of distribution of characteristic dispositions within the space of ignorance-knowledge duality of the relevant varieties contained in the space of imagination-reality dualities. The variety transformation dynamics is also seen in terms of neuro-decision-choice activities over the space of problem-solution dualities, where the variety problems are transformed into variety solutions in continual variety problem-solution replacement processes in an infinite stock-flow disequilibrium dynamics of intellectual information-knowledge capital accumulation that also acts as input-output processes in the space of knowing. The meaning of relevant or irrelevant alternative is always in relation to utility and decision in the spaces of production-consumption dualities, input-output dualities and availability-nonavailability dualities. The identification problem-solution dualities relative to the space of ignorance-knowledge dualities involves the revelation to variety identities to provide the initial conditions, necessity and anticipations for action or inaction, while the transformation problem-solution dualities relative to the space of ignorance-knowledge dualities involves the revelation to variety characteristic changes to provide the sufficient conditions, freedom and expectation for identity transformation under social philosophical consciencism at generational points where the space of information-knowledge dualities is continually expanding over the space of ignorance-knowledge dualities such that the more we know, the less we know to allow the development of disinformation, misinformation and propaganda to influence neuro-decision-choice actions and all their contents.

The variety problems in the reality problem-solution dualities, as identified by characteristic dispositions in the space of imagination-reality dualities, are established from the observational results as constructed from the variety signal dispositions over the space of acquaintance with all instruments of acquaintances and observations. The variety solutions in the potential problem-solution dualities are abstracted by cognitive constructs on characteristic-signal dispositions from the space of imaginations contained in the potential space under the destruction-replacement processes. How does one know that a variety problem has been solved? Alternatively, how does one know that a variety problem has been transformed into a variety solution and a variety solution has been transformed into a variety problem? In general, how does one know that the negative duality has been transformed into a positive duality in the system of negations of negations under the principle of opposites? Similarly, how does one know that the dominance of the negative dual has given way to positive dual and similarly with poles and polarities over the space of positive-negative dualities?

Over the space of ontological-epistemological dualities, there are two types of sets of problem-solution dualities of concern in the theory of knowing as an intellectual investment flow. Both types involve quantitative and qualitative dispositions over the space of quality-quantity dualities. The question is: how does one know the point over which one can claim that one qualitative (quantitative) variety has been transformed into another qualitative (quantitative) variety? In a sense, what is the demarcation line of variety ignorance and variety knowledge, variety problem and variety solution and variety negative and variety positive? The search for the lines of demarcations of transitions among different quantitative (qualitative) variety transformations leads to the discussion and development of the *epistemic transversality conditions* that provide frameworks of assessments of movements from a variety ignorance to a variety knowledge and from a variety problem to a variety solution. The dynamics of the space of problem-solution dualities provide the pathways to neuro-decision-choice efficiencies, knowing and information-knowledge accumulation. Knowing is seen in terms of intellectual investment flows over the spaces of ignorance-knowledge dualities and problem-solution dualities. Knowledge is seen in terms of accumulation of intellectual capital stocks to provide service flows from input-output conditions to thinking-reasoning processes balancing dualistic-polar forces over the spaces of ignorance-knowledge and problem-solution dualities in terms of varieties in the space of production-consumption dualities, where the concepts and phenomena of information, knowledge and knowing have been discussed in [R17.17] [R17.18] [R12.19] [R12.20]. The discussions are such that distinctions are placed on the theories of information, knowing, knowledge and ontic-epistemic duality to establish foundation of the space of certainty-uncertainty dualities. The concepts of certainty, uncertainty and certainty-uncertainty are destroyed when we equate information to knowledge. Information-knowledge equality is the characteristic of the ontological space where there are no uncertainties and risks in ontological behaviors. Information-knowledge inequality is the characteristic of the epistemological space where there are uncertainties and risks in epistemological behaviors.

6.2.1 Static-dynamic dualities and Types of Epistemic Transversality Conditions

The theory of knowing shows two varieties in the space of problem-solution dualities. They are *static problem-solution dualities* in the knowing process which involves variety identification problem-solution dualities, and *dynamic problem-solution dualities* involving variety transformation problem-solution dualities. To design and analyze the points or lines of dualistic-polar transitions, we must compare ignorance-knowledge structures before and after the transformation process to abstract the variety differences over the space of quality-quantity dualities within the space of imagination-reality dualities, where the space of imaginations is contained in the space of potentials and the space of realities is contained in the space actuals, where all spaces are defined and constructed under dualistic-polar conditionality over the space of characteristics [R17.17]. The variety information-knowledge structures over the epistemological space are established by their abstracted characteristic dispositions under cognitive capacity limitations, while they are established by original characteristic-signal dispositions of *what there was, what there is* and *what there would be* at the level of ontological space in terms of varieties as defined and distinguished by their characteristic dispositions.

It is here that a critical attention must be paid to analytical understandings between ontic variety characteristic dispositions and epistemic variety characteristic dispositions to establish *variety epistemic distances* which give rise to outcomes in the space of certainty-uncertainty dualities, where the outcomes are generated by the differences of variety characteristic dispositions over the space of ontic-epistemic dualities to provide us with pathways to explain the rise of degrees of uncertainty and degrees of risk within the space of human actions under cognitive capacity limitations and the varieties in the space of success-failure dualities. In a simple analytical understanding, the uncertainties are generated by the ontic-epistemic conditions, where the epistemic characteristic dispositions are derivatives from the ontic signal dispositions while the degrees of uncertainty are the results of the sizes of variety ontic-epistemic distances which are explainable by the results of cognitive capacity limitations in relation to information-knowledge differential equality.

The important question is what set of information conditions will allow one to define the points or the lines of variety problem changes to variety solution replacements and vice versa or variety falsity to variety truth and vice versa and many other dualistic-polar varieties. The search for an answer to this information-need question to define *points of separations* requires some judicious neuro-decision-choice analytics over the space of variety problem-solution dualities to relate variety problems to variety solutions in the destruction-replacement game dynamics. Similarly, the search process for solutions to the variety identification-transformation problems, demands the establishment of *epistemic transversality conditions of knowing* which is made up of *identification transversality conditions* and *transformation transversality conditions* in the knowing process through critical neuro-decision-choice journeys over the space of problem-solution dualities with comparative variety analytics that will reveal the information-knowledge structures for defining the respective *epistemic transversality conditions* between variety ignorance and variety knowledge with a defined level of epistemic conditionality. The *epistemic-transversality conditions* may also be used to specify the conditions of *tipping points* in all transformations such as climate changes, pollutions, truth, acceptability, non-acceptability and many other such transformations over the space of quality-quantity dualities as one epistemically moves from one zone to another different zone in the dualistic-polar game space of categorial conversions of varieties in intra-zone and inter-zone dynamics. The transversality conditions may be related to the traditional mode of knowledge verification, falsification, and justification over the space of acceptance-rejection dualities.

In all the transformations between the duals of dualities and the poles of polarities, where there are movements between dualistic-polar structures, there exist opposite zones that are established with membership characteristic functions in the degrees of belonging to the zones, where one epistemically moves from one zone to another different zone in the dualistic-polar game conditions of categorial conversions of varieties in inter-zone dynamics. The position of each variety will require fuzzy-stochastic zonal analytics to locate the optimal dualistic-polar transversal lines or points which separate the zones into opposites. In the knowing processes as the production of intellectual investment flows and information-knowledge accumulation as the production of intellectual capital stocks, the required epistemic transversality conditions indirectly relate to the conditions of fuzzy-stochastic entropy as quantitative solutions to the information-knowledge certainty problems in the sense of specifying the terminal conditions of cognitive journeys from unknown varieties (ignorance) to the known varieties (knowledge) with fuzzy-stochastic conditionality, where the fuzzy component relates to conditions of vagueness and subjectivity and the stochastic component relates to conditions of volume incompleteness [R17.15][R17.16]

[R21.5]. As it will become clear, the epistemic transversality conditions and the fuzzy-stochastic entropic conditions relate to methodological reductionism of verification, and falsification of fuzzy statistical tests of certainty within the general space of uncertainty-certainty dualities and the degree of truth disposition within the space of constructionism-reductionism dualities to establish justification-trust conditions in the space of acceptance-rejection dualities into the information-knowledge system as input into neuro-decision-choice actions in the space of theory-practice dualities and polarities for mutual creations and existence over the space imagination-reality dualities as derived from the space of actual-potential dualities.

The first step in the understanding of the journeys over the space of variety problem-solution dualities is to define the concepts of *epistemic transversality conditions* of knowing and then the *epistemic identification transversality* and *epistemic transformation transversality* conditions which are inter-zone traversing conditions as a rationality in the space of decidability-undecidability dualities. The definitions will establish the *characteristic dispositions* of these essential analytical concepts in knowing and knowledge production over the spaces of problem-solution dualities and ignorance-knowledge dualities relative to the space of identification-transformation dualities, where the characteristic dispositions are the contents of information about the varieties. Under this analytical space of variety-characteristic relationality, every definition in the space of real-nominal dualities in the universal existence is an attempt to establish differences, similarities and distinctions in this sense that we can distinguish the planet Earth from the planet Sun and planet Pluto, distinguish walk from stand and many concepts and phenomena involve in the space human understandings without which communication is impossible in the space of verbal-nonverbal dualities

From the structure of the definitions of the concepts of cognitive transversality and types, we can then *design quantitative measures* to represent the needed conditions for their characteristic-signal dispositions of varieties to provide us with a framework of stopping rules for terminal states for accepting variety solutions or variety problems, and similarly accepting conditions of variety knowledge or variety ignorance with fuzzy-stochastic conditionality, where such determinations are neuro-decision-choice based on the basis of characteristic dispositions as the contents of information inputs. The relationships among characteristics, characteristic disposition and varieties are discussed in [R17.17]. It is, here, that the theory of statistics, encompassing sub-theories of data analytics, fact analytics and evidential analytics, finds neuro-operational utility in decision-choice actions over the space of problem-solution dualities in all areas of human operations within knowledge and utilization of knowledge in the general neuro-decision-choice systems within the space of incentive-disincentive dualities with conformal mapping into the space of cost-benefit dualities. A question arises as to what the relational continuity of theories of possibility, probability, statistics data analytics, fact analytic and evidence analytics is and in relations to the concepts and phenomena they represent.

Within the general statistical theory, one may ask the following questions. What are the goals and objectives of statistical theory, what are the goals and objectives of data analytics or fact analytics and what are the goals and objective of evidential analytics and how do they differ from one another in knowing, information-knowledge accumulation and decision-choice systems? Alternatively, what are the differences among categories of problem-solution dualities, such as what are the differences and similarities of categories of economic, political, legal problem-solution dualities and the spaces containing them, and how do cognitive agents know them and relate problems and solutions? Similarly, what are the differences and similarities between categories of social problem-solution dualities and categories of natural problem-solution dualities, such as wellness-sickness dualities, social progress problem-solution dualities, hunger problem-solution dualities biological-problem-solution dualities, peace-war dualities, and many of such categories of problem-solution dualities and how do cognitive agents come to know them? How do the goals, objectives and visions affect knowing, information-knowledge input-output structures in neuro-decision-choice systems relative to the spaces of decidability-undecidability dualities, doubt-surety dualities, acceptance-rejection dualities, information-knowledge system and neuro-decision-choice system as infinitely self-correction systems under the principle of non-satiation over the space of social production-consumption dualities as a subspace in the space of imagination-reality dualities contained in the space of actual-potential dualities?

The self-correction systems are made possible by information-knowledge identification storage, learning and input-output processes through continual feedback processes and telescopic backward-forward conditions over the spaces of thinking-reasoning and teaching-learning dualities and polarities under the conditions of variety incentive-disincentive dualities over the general space of cost-benefit dualities to produce self-learning systems. It is these properties of the mind with thinking and reasoning processes as self-correcting through learning and knowing over the space of error-correct dualities that give us the logical foundation to see how to construct a self-learning machines, deep thinking machines and related automatic self-control systems in the spaces, under water, underground and in the skies by incorporating the understanding of methods and techniques of human cognition and decision-choice behavior into machine cognition with learning and intelligence over the space of input-output dualities. The condition of human intelligence is embodied in the space of thinking-reasoning processes that have capacity of flexibility of neuro-decision-choice actions over different terrains of categories of problem-solution dualities with knowing, learning and intellectual capital accumulation. Now we can go to the space of imaginations containing the possibility space and the ask the question as to whether possibilities exist to construct machine that can mimic the human thinking-reasoning-storage-learning capacity to allow, where we can incorporate the human thinking-reasoning-learning-storage quality into machines to actionize human activities.

Let us keep in mind that all communications take place through signal dispositions as message caring the characteristic dispositions that must be carefully encoded at the source, and where the signal disposition must be carefully interpreted by receiver at the source of destination to reveal the contents of the messages. The whole of inter-machine (machine-machine) communications, machine-human communication and inter-human communication are based on this structure of characteristic-signal process over the space of source-destination dualities containing the space of sender-receiver dualities. The variety characteristic dispositions administered by senders from the source have possibility of containing, or being misinformation, disinformation, propaganda and fake new. It is in this respect that interpretations of signal dispositions by the receiver at the destination present learning and teachable moments for decision-choice actions.

As usual and as linguistic requirements for communication in the use of words and concepts, the development of the definitions the cognitive transversality conditions will keep in mind the theoretically definitional conditions of real definition, nominal definition, explication, and the principle of diversity and unity of knowing and the principle of diversity and unity of science. Definitions and explications are essential tools of identity establishments and communications within the family of ordinary languages (FOL) and the family of abstract languages (FAL). Let us keep in mind that the essential driving force of knowing and information-knowledge accumulation is cognitive activities with thinking and reasoning through neuro-decision-choice actions over the space of variety problem-solution dualities which unites all areas of knowing whether science or non-science, and whether natural science or social science, and whether exact science or inexact science, and whether explanatory science or prescriptive science (engineering and all forms of planning). It is the cognitive activities over the space of problem-solution dualities contained in the space of imagination-reality dualities that define the path of human progress, culture and civilization, conflict, war, peace over the space of qualitative-quantitative dualities guided by the ruling social philosophical consciencism that contains the degrees of non-satiations and its distributions over the spaces of individual-collective dualities and greed-altruistic dualities. For the epistemic nature and usefulness of definitions and explications in methodological nominalism-constructionism-reductionism trilogy see the discussions in [R3.24] [R3.33] [R3.82] [R4.10] [R10.11].

6.2.2: Epistemic Transversality Conditions of Dualistic-Polar Journeys Over the Space of Thinking-Reasoning Processes

Let us examine the meaning and types of the concepts of epistemic transversality conditions, and their relationships to cognitive journeys over the spaces of ignorance-knowledge dualities and problem-solution dualities and polarities, and then extend the relationships to knowing and information-knowledge production for the understanding of intra-variety and inter-variety transformations under dualistic-polar game. Let us define the concept of epistemic transversality condition in knowing to establish the linguistic identity and then the analytical usefulness in knowing and information-knowledge production over the spaces of dualities and polarities.

Definition: 6.2.2.1: Epistemic Transversality Conditions in Knowing

The epistemic transversality conditions of knowing are the necessary conditions for an optimal variety identifications in space of info-statics and variety transformation in the space of info-dynamics to provide optimal beliefs for the acceptance of the identities of varieties and thus transformations of variety from the zone of ignorance to a variety in the zone of knowledge or variety in the zone of problems to variety in the zone of solutions of *what there was, what there is* and *what would be.* A point of transition from a zone of variety ignorance to a zone of variety knowledge or from a zone of variety problem to a zone of variety solution in identification of *what there was* and *what there is*, is called *epistemic identification transversality conditions* and a point of transition from a variety ignorance to a variety knowledge or from a variety problem to a variety solution in variety transformation of *what there is* to *what would be* is called *epistemic-transformation transversality conditions.* The *epistemic identification transversality conditions* and the *epistemic transformation transversality conditions* together constitutes the *epistemic transversality conditions* for the establishment of the terminal points (states) in the neuro-decision-choice journey from the zone of variety ignorance to the zone of variety knowledge and from the zone of variety problems to the zone of variety solutions in the variety-knowing processes, variety problem-solution processes and the development of general information-knowledge accumulation of intellectual capital stock-flows in disequilibrium dynamics.

NOTE 6.2.2.1 Establishing the Transversality points and lines.

The transversality points and lines are obtained from optimal resolutions of dualistic-polar conflicts of duals in the polarities and poles of polarities under fuzzy paradigm of thought where all the duals are in mutual constraint in the variety and categories of their belonging in degrees and similarly, all the poles are also in mutual constraints in the variety and category of their belonging in degrees, where the belongings are fuzzy sets of characteristic disposition. To make things clear let us characterize varieties as negative or positive. Let the positivity be a fuzzy set (A) with degrees of membership $\mu_A(x)$, $\frac{d\mu_A}{dx} \geq 0$ and the negativity as fuzzy set of (B) with degrees of membership $\left\{\mu_B(x), \frac{d\mu_B}{dx} \leq 0\right\}$. The conflict resolution is a fuzzy decision problem of the form $\Delta = A \cap B$ with a membership characteristic function of the form $\left\{\mu_\Delta(x) = \mu_{(A\cap B)}(x)\right\}$. The transversality point is obtained by solving the optimal fuzzy decision problem $\left\{\underset{x\in\Delta}{\text{opt}}\,\mu_\Delta(x) = \underset{x\in(A\cap B)}{\text{opt}}\,\mu_{(A\cap B)}(x)\right\}$, where $\left\{\mu_\Delta(x) = \mu_A(x) \wedge \mu_A(x)\right\}$. The structure and solution to this kind of problem is provided in Chapter 4 of equation (4.2.3.1.1) and equation (4.4.1.1) of this monograph. The structure is a fuzzy mathematical programing problem with an optimal solution taking the form $\left\{\mu_\Delta(x^*) = \mu_{(A)}(x^*) = \mu_{(B)}(x^*) = \alpha^*\right\}$ as the *transversality point* and the vertical line passing through this point as the *transversality line*. The horizontal strait line connecting the optimal point to the vertical axis of the degrees of belonging is the *line of optimality of belonging*. In this respect, the transversality point is the intersection of the transversality line and the line of optimality of belonging where there is dualistic-polar marginal equality in the degrees of belonging.

6.2.3 The Utility of Epistemic Transversality Conditions in the Space of Neuro-Decision-Choice Process

The epistemic transversality conditions presented here, are specified in two interdependent sets over the space of static-dynamic dualities for the understanding of variety identifications and variety transformations relative to *changes in property and changes in relation* or both. At the static level of variety identity, these changes are descriptions of how one in his, her or collective cognitive journey knows the point of crossing from either a zone of variety ignorance or variety problems to a zone of variety knowledge or variety solutions at the info-static state to claim a variety identification in the telescopic past and present to provide variety information-knowledge structures for identities of varieties of *what there was* and *what there is*. They also apply to all dualistic-polar conditions in all neuro-decision-choice actions over the space of decidability-undecidability dualities in dualistic-polar games of variety identifications, transformations and changes by providing terminal conditions as stopping rule of a search for knowledge or solution and to establish the possibility, initial conditions, necessity and anticipations for variety transformations under conditions over the space of cost-benefit dualities and rationality, where there are equalities of dualistic-polar marginal changes.

At the info-dynamic level, however, they provide information-knowledge structures of how one in his or her or collective journey knows the point of crossing from one zone of a dual to another zone of a dual to claim variety identifications of *what there is,* to *what would be* in the future existence after transformation and *what there is* to *what there was* in the past existence in the telescopic past-present dualities for the understanding of what was but no longer is. In other words, the epistemic transversality conditions of knowing, in info-dynamics are about the understanding of variety transitions from a variety ignorance or a variety problem to a variety knowledge or a variety solution of *what there was* (past) to *what there is* (present), from *what there is* (present) to *what would be* (future) and *what there was* (past). The engine of these processes is the set of epistemological converters and functors that map the conditions of imagination to conditions of reality through the space of identification-transformation dualities. The epistemic transversality conditions simply define finishing points of epistemic journeys over the spaces of problem-solution dualities, ignorance-knowledge dualities, intra-variety dualities and inter-variety dualities to indicate the areas of epistemic claims but not the end. They may be viewed in terms marginal analysis of slow transformation, where the set of ignorance characteristics are slowly reduced by transformation, and the set of knowledge characteristics are slowly increased by transformation in compensatory process of zero-sum game through knowing processes until there is equality in the marginal difference as may be seen or may not be seen in the space of cost-benefit dualities, where marginal losses equal marginal gains to produce conditions to define and establish the transversality point and line that separate dualistic-polar zones. Let us keep in mind that the characteristic disposition of every variety is composed of negative subset of characteristics and positive subset of characteristics within a fixed characteristic disposition that defines the identity of the variety, where, as the size of positive subset of characteristics increases revealed by an increasing membership characteristic belonging function $\left\{\mu_A(x),\ \frac{d\mu_A}{dx} \geq 0\right\}$, the size of negative subset of characteristics falls to maintain the variety characteristic disposition as revealed by decreasing membership characteristic belonging function $\left\{\mu_B(x),\ \frac{d\mu_B}{dx} \leq 0\right\}$.

In the knowing journey, the aggregate value of the epistemic conditions defines the ending value of ignorance as well as the beginning value of knowledge. The set of epistemic transversality conditions defines a *stopping rule* in the journey from either variety ignorance to variety knowledge or from a variety problem to a variety solution to claim the journey's outcome relative to a goal-objective element or a vision in the space of imagination-reality dualities. Since the information conditions for the identities of variety problems and variety solutions are specified in the space of cost-benefit dualities, the transversality conditions may be specified in marginal cost relative to marginal benefit where the marginal cost and marginal benefit sum up to zero, where further reductions in variety ignorance increase the increasing belonging of variety knowledge with increasing marginal benefits over marginal costs and the zones of conflict claims are established by subjective assessments in the knowing process around the fuzzy optimal values.

The knowledge of *what there is* initializes the necessity of the journey to the knowledge of *what there was* and vice versa through methodological reductionism. The knowledge of *what there is,* also initializes the necessity of the journey to the knowledge of *what would be* through methodological constructionism. Each stage of knowing involves cognitive activities over the space of problem-solution dualities in all areas of knowing, while the information-knowledge conditions provide variety identities over the space of problem-solution dualities in the space of imagination-reality dualities regarding information-knowledge conditions of telescopic past and telescopic present with the path to the telescopic future. The analytical approach of methodological problem-solution dualities under the principle of opposites is general and unites all areas of knowing whether science or non-science and whether under classical paradigm of thought over the space of absoluteness or under fuzzy paradigm of thought over the space of dualistic-polar conditions in the space of relativity. Since the identities of problems, solutions, ignorance, and knowledge are structured also in the space of cost-benefit dualities, the transversality conditions can be structured in the space of cost-benefit dualities with fuzzy-stochastic conditionality to capture fuzziness, relational continuity, volume limitation and cognitive capacity limitations.

It is analytically useful to note the epistemic importance of information-knowledge conditions of telescopic past, present and future and their relationships to generate analytical dualities of past-present dualities, present-future dualities, and past-future dualities in the journeys of knowing and information-knowledge production relative to elements in the space of varieties as contained in the space of ontological-epistemological dualities that is linked to the space of ontic-epistemic dualities. The processes of knowing, understanding and information-knowledge accumulation are telescopic time continuous and inseparably interdependent under the Sankofa-anoma principle of time-information trinity. At the space of the telescopic past-present dualities, the past and present are mutually creating, where the knowledge of the past resides in the knowledge of the present and the understanding of the present resides in the information-knowledge conditions of the past that establishes the initial conditions and necessity to achieve the present dispositions.

At the space of telescopic present-future dualities, the present and the future are also mutually creating, where the transformation conditions of the present reside in imagination conditions of the future knowledge of the future information-knowledge conditions which reside in the space of the present varieties through the neuro-decision-choice actions over the space of imagination-reality dualities, where the future resides in the space of imagination and the present resides in the space of realities in dualistic-polar structures of varieties in the space of identification-transformation dualities in variety dynamics over the space of destruction-replacement dualities. In other words, the Sankofa-anoma principle over the epistemological space postulates the idea that the past creates the present that contains the past and the present creates the future that contains the present in relational continuum and unity such that the past dispositions are the parents of the present dispositions as the offspring which are the parents of the future dispositions as the offspring in a continuous epistemological time.

6.2.3.1 Creativity, Variety Transformations, Comparative Analytics and Epistemic Transversality Conditions

From the solution to the optimal fuzzy decision problem over the space of positive-negative dualities under individual dualistic-polar game of dominance through the transformation of characteristic dispositions, we have the conditions of fuzzy set of positivity with characteristic function of the degree of belonging $\left\{(\mathrm{A}, \mu_{\mathrm{A}}(x,\varsigma)) \mid \frac{d\mu_A}{dx} \geq 0, \varsigma \in \mathscr{G}\right\}$, fuzzy set of negativity with characteristic function of degrees of belonging $\left\{(\mathrm{B}, \mu_{\mathrm{B}}(x,\varsigma)) \mid \frac{d\mu_B}{dx} \leq 0, \varsigma \in \mathscr{G}\right\}$, the transversality line$\left(\mu_{\Delta}(x^*,\varsigma), x\right)$ (a vertical line) and the optimal line of the degree of belonging $\left(\mu_{\Delta}(x^*,\varsigma), \mu(x,\varsigma)\right)$, a horizontal line. The four information conditions may be used to examine effects of creativity in variety problem solution conditions, where $\left\{(\mathrm{A}, \mu_{\mathrm{A}}(x,\varsigma)) \mid \frac{d\mu_A}{dx} \geq 0, \varsigma \in \mathscr{G}\right\}$ fuzzy set of knowledge and $\left\{(\mathrm{B}, \mu_{\mathrm{B}}(x,\varsigma)) \mid \frac{d\mu_B}{dx} \leq 0, \varsigma \in \mathscr{G}\right\}$ is a fuzzy set of ignorance. In the process of knowing, information-knowledge accumulation over the space of identification-transformation dualities and problem-solution dualities, creativity enters as an important parameter with creativity function of the form $\left\{\varsigma(\bullet)=(\mathscr{h}(\bullet), \mathscr{q}(\bullet)) \mid \mathscr{h}(\bullet) \in \mathscr{H}, \mathscr{q}(\bullet) \in \mathscr{Q}\right\}$, where $\varsigma(\bullet)$ is creativity function composed of creativity innovation $\mathscr{h}(\bullet) \in \mathscr{H}$ and creativity discovery $\mathscr{q}(\bullet) \in \mathscr{Q}$ are internal sub-parameters, $(\mathscr{H})$ is the space of creative innovation and $(\mathscr{Q})$ is the space of creativity discovery. The creativity affects the behaviors of both ignorance and knowledge in opposite directions. The information required for comparative analytics of the effect of creativity over the spaces of information-knowledge and problem-solution dualities is shown in figure 6.2.3.1.

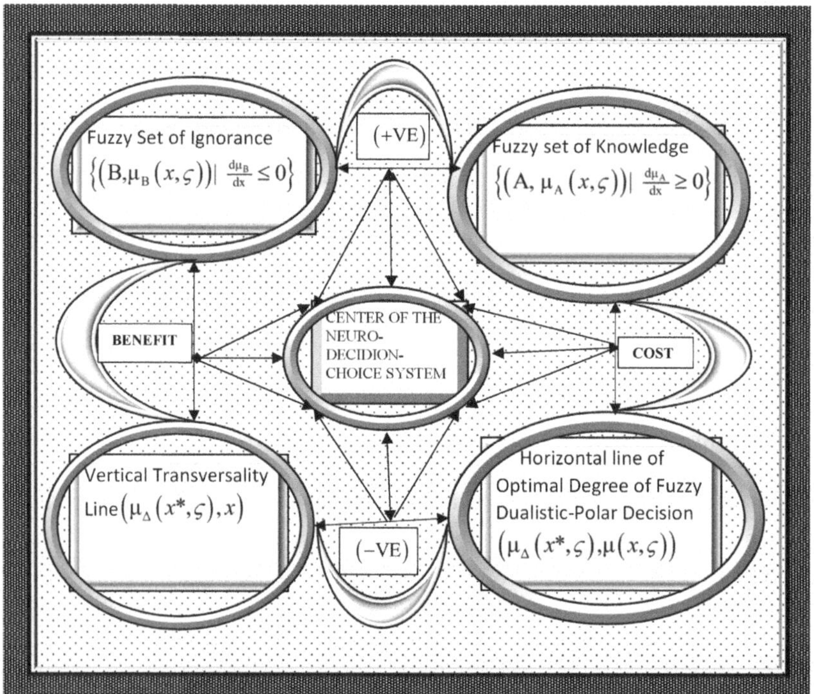

Figure 6.2.3.1 Geometry of Information for Comparative Analytics of the Effects of Creativity over the Spaces of Ignorance-Knowledge and Problem-Solution Dualities

The creative innovation moves the decision-choice process faster, traveling on the same membership functions by increasing the transformation rates of ignorance characteristics to knowledge characteristics and thus increasing the knowledge characteristic disposition while at the same time reducing the ignorance characteristic disposition as well as reducing the degrees of ignorance associated with the variety under the knowing process to cross the epistemic transversality line to reach the point of knowledge claims of an exit from the zone of ignorance to the zone of knowledge regarding the variety identity under knowing or from the zone of problems to the zone of solutions in the space of problem-solution dualities. The variety identity in the space of identification-transformation dualities always begins from a situation where the ignorance membership characteristic function is greater than the knowledge membership characteristic function providing the initial conditions and conditions of necessity. The creativity innovation is always in relation to the improvement of metaprocess in a metaparadigm or metaprocess in a mega paradigm.

The discovery innovation shifts the decision-choice process of knowledge upward to attain new parameters of the membership functions by discovering new path and rates for the transformation of ignorance characteristics to knowledge characteristics and thus increasing the size of the knowledge characteristic disposition while at the same time shifting the path of decision-choice process of ignorance downward to attain new parameters of ignorance membership function to reduce the characteristic disposition as well as reducing the degrees of ignorance associated with the variety under the knowing process to cross the epistemic transversality line to reach the point of knowledge claims of exists from the zone of ignorance to the zone of knowledge regarding the variety identity under knowing. The variety identity in the space of identification-transformation dualities always begins from a situation, where the ignorance membership characteristic function is greater than the knowledge membership characteristic function where the shifts may be such that the initial conditions may be the same or different. The creativity discovery is always in relation to the discovery of a mega process containing new metaparadigm or metaprocess.

The differences and similarities between the creative innovations and creative discoveries are reflected in varieties, time, effort, and energy over thew the space of identification-transformation dualities, where the effects of different types of creativity differently produce the same outcomes of varieties, time, energy, and effort. Creative innovation takes more characteristic transformations to cross the epistemic transversality line while creative discovery takes fewer characteristic transformations to cross the transversality line thus saving time, energy and effort with an increasing capacity for discovery of new varieties. In this respect, creative discovery has more impact than creative innovation over the spaces of ignorance-knowledge dualities and problem-solution dualities. The creative innovation does not affect the vertical transversality line, optimal fuzzy point, the horizontal line of optimal degree of fuzzy dualistic-dualistic decision and the sizes of the dualistic-polar zones. The creative discovery while not affecting the horizontal line of optimal degree of fuzzy dualistic-polar decision does affect the vertical transversality line, the positive and negative fuzzy membership functions, the optimal fuzzy point the zones of ignorance and knowledge by shifting the positive characteristic membership function upward to the left and similarly shifting the negative fuzzy characteristic membership function downward to the left to reducing the size of the zone of ignorance and increasing the size of the zone of knowledge for quicker decision on variety identity. The creative disposition is therefore a parametric shifter of all the essential functions to the left.

In all these analyses, one must always keep in mind the interactive relations among the principle of opposites with relational continuum and unity, fuzzy paradigm of thought and cost-benefit conditions in fuzzy optimization to abstract the optimal condition and their interpretations and the uses to which they may be used. it is always useful to keep in mind the nature and structure of fuzzy optimization and how they work in abstracting conditions of optimality. The fuzzy optimization over the space of dualities and polarities takes place by optimizing the behavior of a dual sech as positive dual subject to the constraint of the behavior of the corresponding dual such as negative dual and vice versa just as the framework of cost-benefit analysis and the variety dual usage relative the space of goals objectives and visions. The choice variable is restricted within the zone of positive differences between then negative characteristic membership function and the positive characteristic membership function which may be interpreted as the zone of excess negative duality or the zone of positive dualistic deficiency such as the zone of excess negative characteristics when the positive characteristic function is being maximizes, and *zone of positive deficiency* when the negative characteristic membership function is being minimized. where the optimization is considered from the origin of axis.

At the dualistic-polar identification-transformation games over the space of varieties, the points, and lines of demarcation between either the zones of variety ignorance and variety knowledge or the zones of variety problems and variety solutions in the knowing and decision process are either ignorance or problem terminal points or lines of demarcation that serve as the beginning points of either knowledge or solution claims from the zones of variety belonging. The solution to the problem of epistemic transversality conditions between either the zone of ignorance disposition and the zone of knowledge disposition or the zone of problem disposition and the zone of solution disposition of any variety may be seen in terms of characteristic disposition of either variety ignorance-knowledge duality or variety problem-solution duality in terms of degrees, where the degrees of ignorance and the degrees of knowledge are inter-supportive in dualistic-polar structures, and where every degree of knowledge has a corresponding degree of ignorance in dualistic-polar structures in terms of internal and external conditions of being in the space of relativity. The zones of ignorance, knowledge, problems, and solutions are fuzzy sets characterized by degrees of belonging expressed in terms of membership characteristic functions.

Every degree of ignorance has a corresponding degree of knowledge, and every degree of a problem has a corresponding degree of a solution expressed between zero and one such that their relative sum in each case is always equal to one also under the conditions of membership characteristic functions of belonging in the space of relativity. The space of relativity reflects the conditions of dualities containing duals in the same variety and conditions of polarities containing poles in the same variety under the general principle of opposites, where the same varirty is simultaneously a cost and a benefit relative to any goal-objective element or vision demanding organizational analytics of the opposites not with excluded middle and disunity but with relational continuum and unity over the space of ontological-epistemological duality. The epistemic identification transversality conditions involve variety info-statics while the transformation epistemic transversality conditions involve variety info-dynamics, where we must keep in mind that all relativities are structured in relativity characteristic dispositions. The epistemic transversality conditions show how the knowing path from a variety ignorance or a variety problem crosses the terminal line or a point of a zone of a variety ignorance or a variety problem to a zone of a variety knowledge or a variety solution without *epistemic Ponzi schemes* such as fake news, propaganda, misinformation, disinformation or intellectual deceptions to influence neuro-decision-choice actions or social policy, personal advantage or social advantage, all of which are expressed over the space of cost-benefit dualities over the space of quality-quantity dualities.

The epistemic transversality conditions involve inter-variety crossings in terms of information-knowledge conditions about cost-benefit, negative-positive and quality-quantity dispositions in relation to problem-solution dualities, ignorance-knowledge dualities, and neuro-decision-choice analytics, where variety identities over the space of real-nominal dualities are specified by characteristic dispositions. The epistemic transversality conditions are related to the space of general decidability-undecidability dualities, where every problem is a solution, and every solution is a problem in terms of neuro-decision-choice actions over the space of goals and objectives under the general preference schemes of individual and social collectivity. Similarly, the epistemic transversality conditions relate to the idea that every problem disposition presents a teachable moment in the production of intellectual investment flows, and every solution disposition presents a teachable moment in production of intellectual capital stocks through the decision-choice actions over the space of problem-solution dualities relative to the spaces of ignorance-knowledge dualities, certainty-uncertainty dualities, doubt-surety dualities and success-failure dualities, where a change of variety dominance in the game space is an information-knowledge condition presenting a teachable and learnable moments over the epistemological space. It is within this context that care must always be exercised on the choice of contents and methods of teaching and learning relative to philosophy of education, where questions arise regarding the concepts transformative education in relation to teaching what to think and teaching how to think and the goals and objective of learning and teaching and the effects of teaching, learning and content of knowledge under the principles of doubt and surety[R17.1a] [R17.16] [R17.19a] [R17.44] [R17.68].

The epistemic transversality conditions also define the intersection of negative and positive characteristic dispositions in dualistic-polar transformations in the knowing process, across which the negative or the positive dual dominates one another in the transformation game for variety identities concerning problems, solutions, ignorance, and knowledge in terms of succession of varieties and replacement of varieties. They are the terminal conditions for the change of dualistic-polar dominance to define the variety identity in categorial conversions over the spaces of input-output and quality-quantity dualities. Every zone of variety problems (solutions) contains both problem characteristics and solution characteristics, where the problem (solution) characteristics dominate the solution (problem) characteristics in the dualistic-polar conversion games in the inter-categorial structures and where problems and solutions are defined by their relative zonal cost-benefit dispositions in the space of imagination-reality dualities. The zonal conditions over the space of negative-positive dualities may always be translated to the zonal conditions of the space of cost-benefit dualities for neuro-decision-choice actions over the spaces of problem-solution dualities and input-output dualities. The relative dualistic-polar dominances in defining zones of duals and decision-choices over the spaces of decidability-undecidability, computability-non-computability dualities are very important way to understand knowing and information-knowledge accumulation, conflict and peace, truth, and false and other similar dualistic-polar structures in the space of human actions.

The concept of epistemic transversality conditions composed of epistemic identification transversality conditions in info-statics and epistemic transformation transversality conditions in info-dynamics that relate to quality-quantity-time processes in the journey over the space of ignorance-knowledge dualities through knowing must be clearly distinguished from the other concepts of transversality conditions in mathematics of optimal control that relate to quantitative time processes. The epistemic transversality conditions are about conditions of optimal cognitive travelling in the space of ignorance-knowledge dualities and dualistic-polar internal and external journeys through transformations of variety ignorance to variety knowing, and variety problem to variety solution. The point of essentiality is that the whole of human behavior is within the space of neuro-decision-choice actions as exercised over the space of variety identification-transformation dualities in the epistemological space contained by the space of ontological-epistemological dualities generating the space of ontic-epistemic dualities which provides the foundations of the space of certainty-uncertainty dualities. In this analytical setting, every variety is governed by the principle of opposites which exists as dualities and polarities in relational continuum and unity to generate internal and external conflicts and energy for variety transformations given their identities. Further discussions on this point and the theories behind it may be found in [R8.16a] [R8.16b] [[R17.17] [R17.18[[R21.5].

6.3 FUZZY-DECISION-CHOICE ANALYTICS OF THE EPISTEMIC TRANSVERSALITY CONDITIONS OF IGNORANCE-KNOWLEDGE AND PROBLEM-SOLUTION GAMES IN ALL AREAS OF KNOWING

The epistemic transversality conditions are in relation to variety ignorance-knowledge dynamics, where a variety ignorance may be a variety problem or a variety solution, and a variety knowledge may be a variety solution or a variety problem in all dualistic-polar dynamic games over the fuzzy-stochastic information space as connected to the space of goals, objectives and visions. The fuzzy-stochastic information space is an analytical concept that relates to information imperfections. The fuzzy component relates to the presence of vagueness in variety characteristic dispositions representing fuzziness in observations within the acquaintance space to create conditions over the space of doubt-surety dualities. The stochastic component relates to the presence of volume incompleteness of variety characteristic dispositions representing shortfalls in the contents of observations within the acquaintance space. The presence of vagueness and volume incompleteness over the acquaintance space creates fuzzy-stochastic information and fuzzy-stochastic information space leading to the space of ontic-epistemic dualities that produces the space of certainty-uncertainty dualities and is the product of cognitive capacity limitations of neuro-decision-choice agents. In this dualistic-polar dynamic ignorance-knowledge game and problem-solution game of life, every variety in the space of knowing is an ignorance-knowledge duality and every variety in the space of identification-transformation process is a problem-solution duality in degrees of combination defined by fuzzy-stochastic information under the principle of qualitative-quantitative constraints.

The variety information is defined by a characteristic-signal disposition, where the characteristic disposition is the set of the contents, and the signal disposition is the set of the signals. The variety $(\nu \in \mathbb{V})$ with characteristic disposition $(x \in \mathbb{X}_\nu)$ where the characteristic disposition is composed of dual (1) $(\mathbb{X}_\nu^1)$ and dual (2) $(\mathbb{X}_\nu^2)$ where dual (1) may be positive, benefit, output and others and dual (2) is the corresponding dual of say negative, cost, input and others such that $\mathbb{X}_\nu = (\mathbb{X}_\nu^1 \cup \mathbb{X}_\nu^2)$, defines relational unity, $(\mathbb{X}_\nu^1 \cap \mathbb{X}_\nu^2) \neq \varnothing$ defines relational continuum, $\mathbb{X}_\nu \neq \mathbb{X}_\nu^1 \subset \mathbb{X}_\nu \Rightarrow \mathbb{X}_\nu^2 \neq \varnothing$ and similarly $\mathbb{X}_\nu \neq \mathbb{X}_\nu^2 \subset \mathbb{X}_\nu \Rightarrow \mathbb{X}_\nu^1 \neq \varnothing$. In terms of membership characteristic functions, we have $\mu_{\mathbb{X}_\nu^1}(x)$, $\mu_{\mathbb{X}_\nu^2}(x)$, $\mu_{\mathbb{X}_\nu}(x)$ and $\mu_{(\mathbb{X}_\nu^1 \cap \mathbb{X}_\nu^2)}(x)$, where

$$\mathbb{X}_\nu = (\mathbb{X}_\nu^1 \cup \mathbb{X}_\nu^2) \Rightarrow \mu_{\mathbb{X}_\nu}(x) = \left(\mu_{\mathbb{X}_\nu^1}(x) \vee \mu_{\mathbb{X}_\nu^2}(x)\right), \text{and } (\mathbb{X}_\nu^1 \cap \mathbb{X}_\nu^2) \neq \varnothing \Rightarrow \mu_{\mathbb{X}_\nu^1 \cap \mathbb{X}_\nu^2}(x) = \left(\mu_{\mathbb{X}_\nu^1}(x) \wedge \mu_{\mathbb{X}_\nu^2}(x)\right)$$

such that $\left(\mu_{\mathbb{X}_\nu^1}(x) \lesseqgtr \mu_{\mathbb{X}_\nu^2}(x)\right)$, in the dualistic-polar game process, where neuro-decision-choice action on a dual is constrained by the conditions of the corresponding dual structured in the degrees of membership belonging where $x \in \mathbb{X}_\nu^1 \Rightarrow x \in \mathbb{X}_\nu$ and similarly $x \in \mathbb{X}_\nu^2 \Rightarrow x \in \mathbb{X}_\nu$ and $\left(\mu_{\mathbb{X}_\nu^1}(x) + \mu_{\mathbb{X}_\nu^2}(x)\right) = 1, \forall x \in \mathbb{X}_\nu$. All the characteristic dispositions have the corresponding signal dispositions with ontological characteristics presenting exact and full-volume dispositions for perfect variety identities in the ontic space and epistemological characteristics presenting inexact and less-than-volume dispositions for imperfect variety identities, where the ontic and epistemic variety dispositions combine to generate *epistemic distances* [R8.16b].

The concepts and phenomena of ignorance and knowledge are defined in the fuzzy space in relative degrees such that for every degree of ignorance, there is a corresponding degree of knowledge where the relative sum is always equals to a unity in opposite directions of change in the space of zero-sum games under dualistic-polar conditions of relational continuum and unity in the characteristics. As the degree of variety knowledge increases with increasing characteristic at the margin by transforming ignorance characteristics through a cognitive search, the corresponding degree of variety ignorance falls at the margin increasing characteristic as cognitive agents journey over the path of the dynamics of knowing and through the space of problem-solution dualities. It is under this ignorance-knowledge journey over the space of the problem-solution dualities that the methods of possibilistic and probability reasoning with statistical tests acquire an epistemic importance in the neuro-decision-choice activities over the space of decidability-undecidability dualities to help to abstract the degree of the uncertainty contents that is being accepted in abstracting the optimal decision-choice action in the space of dualistic-polar structures under linguistic variables. Other qualitative tests may be developed to support the available statistical tests to identify the epistemic transversality conditions. The need of qualitative tests for the position of qualitative motion and transformation dynamics over the space of quality-quantity dualities will depend on the nature of variety characteristic dispositions over the relativity space of quality-quantity dualities, where the degrees of belonging may be established by fuzzy information structured and analyzed by fuzzy paradigm of thought.

There are some discussions in the literature of knowing and knowledge in all areas of human activities involving the questions that some discussants refer to as the chicken-egg problem in the spaces of parent-offspring dualities and primary-derived dualities. It is not analytically and epistemically important to raise the questions of chicken-egg problem in the space of parent-offspring dualities or successor-predecessor dualities, where the parents are primary category of identities, and the offspring are the category of derived identities. Under structurally dualistic-polar conditions, where every offspring is a parent, and every parent is an offspring involving the initial conditions and variety outcomes in the sense that every problem is a solution, and every solution is also a problem in dualistic-polar structures in relational continuum and unity. The chicken-egg problem involve also the notion of universal beginning-end problem, where the beginning is unknown to any being ever lived and the end also will never be known to any living being in the complex system of knowing and knowledge. One thing that is clear is that the epistemological space is an offspring of the ontological space that functions as the parent of information-knowledge processes through the acquaintance space under the control of neuro-decision-choice actions. How do we see the relation of information and decision in the space of parent-offspring dualities in the sense of whether information is the parent of decision-choice actions, or the decision-choice actions are parents of information over the epistemological space? Information is a third dimension of ontological and epistemological existence and inseparable from matter-energy dualities with ontological time, as such it is the parent to all neuro-decision-choice actions without which variety identities are undefinable in the space of universal existence.

The differential distribution of variety information defined by variety characteristic dispositions relative to the space of ontological-epistemological duality is the parent of the space of ontic-epistemic dualities. As we have discussed, possibility space and probability space are the offspring of the space of imaginations which is an offspring of the space of ignorance-knowledge dualities as contained in the space of actual-potential dualities. The universe is conceived as defined in an inseparable trinity of matter, energy. and information over the identification space with neutrality of time, where all elements are connected in the space of relations, where additions of different varieties and transformation require the entry of time element in terms of ontological time and epistemological time. The space of matter, energy, relation and information contains the space of visible-invisible duality in relational continuum and unity and by the principle of opposites, extensions may be made to the space of antimatter, nonmatter, nonenergy and information, where information is an inseparable internal property of all things and the conditions of invisibility are relative to technology and instruments of observation in the space acquaintance and qualitative dispositions may be transformed to quantitative dispositions by surrogate measures.

The space of ignorance-knowledge dualities is the offspring of the space of ontic-epistemic dualities that generates the space of certainty-uncertainty dualities. In this respect, we can analytically say that the space of possibility, necessity, initial conditions, and anticipations is the parent serving as the primary category to the space of probability, freedom, sufficient conditions, and expectations as the derived category. The space of data from the space of acquaintance is the parent to the space of statistics, where the theory of probability from the space of imagination is the parent to the theory of statistics , where the theory of statistics is seen as a set of techniques, methods and collective instruments to process the variety stochastic information and insufficient conditions in the space of possibility-probability dualities within the space of imaginations as a sub-space of the space of potentials to fill in the observation imperfections to deal with information-knowledge inequality conditions in the epistemological space of epistemological transformations relative to the information-knowledge equality conditions in the ontological space of transformations.

The possibility space, on the other hand, is the parent to development of the space of fuzzy analytical techniques and methods for inexact and approximate reasoning, where the theory of possibility is the parent to the theory of fuzzy paradigm of thought. The fuzzy paradigm of thought is seen as a set of techniques and methods as a collection of instruments to process the variety vagueness information, inexactness and subjective conditions also in the space of possibility-probability dualities within the space of imaginations as a sub-space of the space of potentials to fill in the observation imperfections to deal with information-knowledge inequality conditions in the epistemological space of epistemological transformations to the space of realities relative to the information-knowledge exactness conditions in the ontological space of the ontic-epistemic dualities as the space parent-offspring dualities which are sequentially contained in the space of actual-potential dualities. A question may be asked as to what are the relationships of the space of possibility-probability duality, to the concepts and phenomena of fuzziness, randomness, stochasticity, random variable, stochastic variable, fuzzy variable and their mutual interdependence over the information-knowledge processes relative to the space of thinking-reasoning dualities, and how are they interrelated in the space of thought-practice dualities, where the power of thought demands practice and integrity of practice demands thought over the space of production-consumption dualities.

6.3.1 A Fuzzy-Decision-Choice Representation of the Epistemic Transversality Conditions of Ignorance-Knowledge and Problem-Solution Dualities

The variety transformation problem-solution dualities in relation to variety ignorance-knowledge processes of matter-energy varieties are of differently dualistic-polar structures depending on the areas of knowing which establish and define the differences in tools, techniques and methods as thinking algorithm of knowing within an organic paradigm of thought as well as establish the corresponding transversality conditions. Let us keep in mind that in the theory of knowing and the theory of knowledge, the paradigms of thought are established within the assumed information structures while methods and techniques for problem-solution dualities are established within the paradigms of thought. The solution to the problem of transversality conditions involves in initializing the idea that variety knowledge is seen as a *fuzzy set of knowledge* of the form of set $(\mathbf{A})$ with a generic element $(\mathbf{x} \in \mathbf{A})$ and a membership characteristic function of the form $(\mu_{\mathbf{A}}(\mathbf{x}) \in [0,1])$ where $(\mu_{\mathbf{A}}(\mathbf{x}) = 0)$ implies complete ignorance and $(\mu_{\mathbf{A}}(\mathbf{x}) = 1)$ implies complete knowledge in the space of absoluteness and $(\mu_{\mathbf{A}}(\mathbf{x}) \in (0,1))$ implies simultaneity of the presence of ignorance and knowledge in the distribution of relative structural degrees of knowing in the space of relativity. The set $(\mathbf{A})$ is a set of knowledge characteristics contained in the total characteristics disposition $(\mathbf{D})$ defining the identity of a variety and $(\mathbf{x})$ is the characteristic such that $(\mathbf{x} \in \mathbf{D})$ and $(\mathbf{A} \subset \mathbf{D})$. The fuzzy set $(\mathbf{A}, \mu_{\mathbf{A}}(\mathbf{x}) \in [0,1])$ may be viewed as the set of incomplete variety knowledge system, where the knowledge about each variety has a fuzzy number defined by $(\mu_{\mathbf{A}}(\mathbf{x}) \in [0,1])$ constructed from the fuzzy information structure containing vagueness and volume incompleteness showing the degree of knowledge attained. Similarly, in the same space of knowing and in line with the principle of opposites and dualistic-polar conditions, we can specify the opposite as a *fuzzy set of ignorance* about the same variety in the form of the fuzzy set $(\mathbf{B})$ with a generic element $(\mathbf{x} \in \mathbf{B})$ such that $(\mathbf{x} \in \mathbf{D})$ and $(\mathbf{B} \subset \mathbf{D})$ an membership characteristic function in the form $(\mu_{\mathbf{B}}(\mathbf{x}) \in [0,1])$ where $(\mu_{\mathbf{B}}(\mathbf{x}) = 0)$ implies a complete variety knowledge and $((\mu_{\mathbf{B}}(\mathbf{x}) = 1))$ implies a complete variety ignorance. The conditions $(\mu_{\mathbf{A}}(\mathbf{x}) \in (0,1))$ and $(\mu_{\mathbf{B}}(\mathbf{x}) \in (0,1))$ imply a simultaneity of the presence of either an ignorance-knowledge or a problem-solution combination in the distribution of relative structural degrees of knowing. Here, we conceive the total variety characteristics in relational unity as $(\mathbf{D} = \mathbf{A} \cup \mathbf{B})$ and in relational continuum that $(\mathbf{A} \cap \mathbf{B}) \neq \varnothing$ where $(\mathbf{A} = \mathbf{D} \Rightarrow \text{full knowledge and perfect variety identification})$ with surety and no uncertainty and doubt, and $(\mathbf{A} = \mathbf{D} \Rightarrow \text{full ignorance and no variety identification})$ with complete doubt and uncertainty.

In this ignorance-knowledge or problem-solution dualization, if $(\mu_{\mathbf{A}}(\mathbf{x}) = 0)$ then $(\mu_{\mathbf{B}}(\mathbf{x}) = 1)$ and if $(\mu_{\mathbf{A}}(\mathbf{x}) = 1)$ then $(\mu_{\mathbf{B}}(\mathbf{x}) = 0)$ to define the fuzzy logical extremes under the principle of opposites. These logical extremes define conditions of *classical true-false analytics* over the space of absoluteness with excluded middle and disunity of classical paradigm of thought, where a statement cannot be simultaneously true and false. This is the principle of opposites with relational dualistic-polar separation and excluded middle [R3.89]. Let us keep in mind that the fuzzy information corresponds to the use of fuzzy paradigm of thought in knowing and in the space of decidability-undecidability dualities within the space of relativity, where $(\mu_{\mathbf{B}}(\mathbf{x}) \in (0,1))$ implies a simultaneous existence of truth and falsity in distribution of relative proportions. The exact information corresponds to the classical paradigm of thought in knowing and in the space of decidability-undecidability dualities within the space of absoluteness constrained by information-volume limitations without explicit condition of cognitive capacity limitations. In general, the classical information structure exists by assumption by discounting the conditions of cognitive capacity limitations. Human decision-choice failures over the space of problem-solution dualities are the results of cognitive capacity limitations. If things are perfect in the acquaintance space implying perfect information, then there will be no ontic-epistemic dualities since all epistemic variety dispositions will be equal to the ontic variety dispositions and there will be no way to attribute and explain the rise of identity uncertainties in neuro-decision-choice systems. In this respect, the space of imaginations containing possibility and probability will not exist and in general mistakes in neuro-decision-choice actions will not occur over the space of problem-solution dualities, where problems and solutions are defined by cognitive agents relative to some varieties in the space of goals, objectives and visions which is mapped onto the space of cost-benefit dualities in such a way that problems and solutions are cost-benefit defined.

In the fuzzy paradigm of thought, under the principle of opposites with relational continuum and unity, the cognition about any variety resides simultaneously in ignorance and knowledge duality in relational continuum and unity in such a way that every proposition has a support of its opposite without which the proposition has no meaning in the space of relativity. This is the *fuzzy logical support principle* which is translated into an epistemic conditionality that helps to account for the general conditions of cognitive capacity limitations in the spaces of knowing, problem-solution dynamics, and ignorance-knowledge dualities *under fuzzy true-false analytics*. The phenomenon of cognitive capacity limitations of neuro-decision-choice agents implies that complete exactness, perfection, and knowledge in all areas of neuro-decision-choice activities are unattainable and complete vagueness, imperfection and ignorance have no existence over the epistemological space. What we have are the combinational outcomes in the spaces of completeness-incompleteness and perfection-imperfection dualities under dualistic-polar conditions and logical cutoffs determined by cognitive agents in the space of acceptance-rejection dualities through neuro-decision-choice actions.

The system of logical claims is such that in the fuzzy set $(\mathbf{B}, \mu_{\mathbf{B}}(\mathbf{x}) \in [0,1]) \subseteq (\mathbf{A}, \mu_{\mathbf{A}}(\mathbf{x}) \in [0,1]) \subseteq (\mathbf{B}, \mu_{\mathbf{B}}(\mathbf{x}) \in [0,1])$ such that the decision on the characteristic set $\mathbf{A}$ is constrained by the conditions of characteristic set $\mathbf{B}$ and vice versa, where one seeks an optimal value that separates either variety ignorance from variety knowledge or variety problem from variety solution in the fuzzy set $(\mathbf{A} \cap \mathbf{B}, \mu_{(\mathbf{A} \cap \mathbf{B})}(\mathbf{x}))$ as the common decomposition boundary of the duals, $(\mathbf{A} \cap \mathbf{B})$ of the duality or the poles of the polarity, where the fuzzy paradigm of thought deals with both the classical extremes and the classical excluded middle with the acceptance of contradictions in the thinking-reasoning processes and contradictions in the decision-choice space as the natural processes of neuro-decision-choice actions constrained by cognitive capacity limitations in human actions.

6.3.2 The Epistemic Transversality-Conditions, the Principle of Indifference and Fuzzy Conditionality

The classical logical contradictions are handled in the fuzzy paradigm of thought with fuzzification of the classical exact information into fuzzy information and the use of fuzzy paradigm of thought for defuzzification of the neuro-decision-choice actions with fuzzy conditionality, where the neuro-decision-choice actions over the space of certainty-uncertainty dualities are handled with fuzzy-stochastic entropy. It must be pointed out that the processes of fuzzification and defuzzification of the classical systems find meaning and analytical power in information space which may be empirical, axiomatic of both. The fuzzification of classical exact statements is to convert the classical exact and or complete information to fuzzy information and convert the space of absoluteness with excluded middle and disunity to the space of relativity with relational continuum and unity. Defuzzification does the opposite by converting the fuzzy optimal solution in the space of relativity to an exact solution with fuzzy or fuzzy-stochastic conditionality to provide the relative positions of the duals and poles of the optimal dualistic-polar values within the space of decidability-undecidability dualities. The fuzzification and defuzzification processes are through the thinking-reasoning tool of fuzzy membership analytics in representation, computation, and decision over the space of relativity, where all the duals of dualities and all the poles of polarities are in inter-supportiveness of relational diversities and unity under the mutual give-and-take-sharing principle. An important aspect of the elements over the space of fuzzification-defuzzification dualities is the integration of conditions of cognitive capacity limitations to achieve humanness over the space of error-correction dualities in all areas of neuro-decision-choice actions, where today's knowledge to ignorance or solution to a problem may be found to be false or inappropriate and then corrected.

The principles of duality and polarity suggest to us that in all neuro-decision-choice actions, the opposites exist as mutual constraints, where cost dispositions constrain benefit dispositions and vice versa, problem dispositions constrain solution dispositions and vice versa, truth dispositions constrain false dispositions and vice versa, and negative dispositions constrain positive dispositions, and vice versa in all cognitive actions through neuro-decision-choice activities in defining either knowledge to ignorance or solution to a problem in the spaces of ignorance-knowledge and problem-solution dualities with the help of the *principle of indifference* or *the principle of insufficient reasoning* which is affirmed by the conditions of fuzzy optimality under the principle of opposites in the space of relativity within the space of imagination-reality dualities, where the principle of indifference or insufficient reasoning are sequentially connected through the space of possibility-probability dualities contained in the space of imaginations traced to the space of certainty-uncertainty dualities as an offspring of the space of ontic-epistemic dualities contained in the space of actual-potential dualities relative to the ontological space. Under the general principle of opposites, the duals of duality and the poles of polarity are mutually constraining in give-and-take-sharing processes for their mutual existence and mutual destruction such that real cost, and real benefit are mutually constraining and mutually self-destructive and negative and positive are mutually constraining as well as mutually self-destructive, where mutually variety self-destruction may be seen in terms mutual categorial conversions through dualistic-polar game.

The *principle of indifference* or *the principle of insufficient reasoning* is affirmed in the principle of *dualistic-polar indispensability condition* in existence and neuro-decision-choice activities over the action space which is related to the spaces of problem-solution dualities and cost-benefit dualities, projected by the *asantrofi-anoma principle* as a universal transformation phenomenon and the *asantrofi-anoma rationality* as a universal neuro-decision-choice phenomenon, where the spaces of definability-indefinability dualities, decidability-undesirability dualities, predictability-unpredictability dualities are fuzzy information-knowledge constrained under the fuzzy paradigm of thought, where paradoxes in thinking, reasoning, computability, definability, explicability and many other such actions are resolved with fuzzy paradigm of thought in the space of relativity with multiplicity of fuzzy conditionality in relation to creativity, curiosity, courage , fear and many others in the space of thinking-reasoning dualities. At this point, let us provide analytical definitions of asantrofi-anoma problem, principle, and rationality, where the rationality is a solution principle.

6.4 THE ANALYTICS OF THE_ASANTROFI-ANOMA PROBLEM, PRINCIPLE AND RATIONALITY IN NEURO- DECISION-CHOICE PROCESSES OVER THE SPACES OF DUALITIES AND POLARITIES

The concept and phenomenon of *asantrofi-anoma* have been introduced as part of analytical tools to support the concepts of duality and polarity in relational continuum and unity under the general principle of opposites and fuzzy paradigm of thought as an alternative to the classical concepts of duality and polarity under excluded middle and disunity over the space of information-knowledge duality. The asantrofi-anoma concept and phenomenon were initially introduced in my paper of on rationality in the process of examining the relational connectivity between decision-choice rationality and epistemology [R17.13]. The nature of the asantrofi-anoma phenomenon and conditions were also revealed by March when discussing bounded rationality [R17.34]. The asantrofi-anoma concept is split into sub-concepts of asantrofi-anoma problem, asantrofi-anoma principle and asantrofi-anoma rationality, all of which are related to neuro-decision-choice action over the space of decidability-undecidability dualities relative to the conditions of the action space. Let us provide a definition and analytical notes for the concept in the understanding of its meaning, characteristic disposition, relevance, usage, and generality in all areas over the space of imagination-reality dualities. We must always keep in mind that the objective of definition is to establish identities, difference, commonness, and distinction of real and nominal varieties, where we have real definition in the system of variety existence and nominal definition in the system of languages.

DEFINITION 6.4.1: The Asantrofi-Anoma Principle and the Asantrofi-anoma Rtionality

The asantrofi-anoma principle specifies a universal phenomenon that every variety in the universal existence, in the space of actual-potential dualities is composed of negative-positive duality with negative-positive characteristic dispositions, such that the relative composition of the characteristic disposition defines variety identities in dualistic-polar existence, where both the negative and positive duals are individually composed of cost-benefit duality with different negative-positive dualities residing in different poles of a polarity to give the structure of dualistic-polar conditions called the *asantrofi-anoma conditions* for variety identification and transformations in the space of imagination-reality dualities contained in the space of actual-potential realities. The *asantrofi-anoma problem* finds expression in the space of problem-solution dualities, where every variety exists simultaneously with a problem dual and a solution dual in the space of cost-benefit dualities, and all variety problems exist simultaneously with cost and benefit duals, and all variety solutions also exist simultaneously with cost and benefit duals and hence the *asantofi-anoma problem* is how to reconcile the variety cost dispositions and variety benefit dispositions before neuro-decision-choice actions in the space of imaginations that holds opportunities over the space of positive-negative dualities. The *asantrofi-anoma rationality* specifies a universal neuro-decision-choice phenomenon that every decision-choice variety with alternatives over the universal existence in the space of reality as contained in the space of actuals, and the space of imagination as contained in the space of potentials is governed by cost-benefit balances for selection, where varieties are increasingly ranked in order of greater relative net benefit and the variety with maximum relative net cost-benefit structure is selected and similarly varieties are decreasingly ranked in order of least relative net cost and the variety with minimum relative net cost-benefit structure is selected providing a stopping rule, where the asantrofi-anoma rationality is defined in terms of maximum benefit over cost or minimum cost over benefit in the space of relativity for the conditions of cost-benefit optimization.

6.4.1 The Asantrofi-Anoma Conditions and Extensions in the Theory of Knowing

The asantrofi-anoma problem, conditions, principle and rationality find expressions in vagueness in thought, characteristic incompleteness, language deficiency, subjectivity, cognitive capacity limitations under the universal dualistic-polar conditions in the space of cost-benefit dualities that produces the space of attractive-repulsive dualities, where attractiveness and repulsiveness exist in the same variety and thus in the same space and time for neuro-decision-choice processes to require the need to use fuzzy paradigm of thought and not the classical paradigm of thought in dealing with neuro-decision-choice actions over penumbral and non-penumbral spaces of problem-solution dualities under fuzzy information and neuro-decision-choice activities within the space of imagination-reality dualities, in which the space of possibility-probability dualities is defined, where the space of imaginations houses the space of necessity-freedom dualities; and similarly, within the space of imagination-reality dualities in which the space of problem-solution dualities is defined, where the space of realities houses the spaces of static-dynamic dualities, anticipation-expectation dualities connected to the space of doubt-surety dualities that contains the space of success-failure dualities with the space of sufficiency-insufficiency dualities. The simultaneity of existence of opposites in the same space and time is incompatible with the use of the classical paradigm of thought under the conditions of excluded middle with its laws of thought, reasoning, and mathematics. The space of anticipation-expectation dualities in the space of realities is always in relation to the space of possibility-probability dualities within the space of imaginations, where anticipation and necessity relate to possibility and expectation and freedom relate to probability to generate the space of success-failure dualities.

The variety structures in all areas of knowing have one thing in common. They reveal the essence of dualistic-polar conditions of all varieties, where maximization (minimization) of variety internal benefit (cost) dispositions is constrained by the variety internal cost (benefit) dispositions and maximization (minimization) of external benefit (cost) dispositions is constrained by external cost (benefit) dispositions over the epistemological space. The usefulness of the fuzzy analytical approach is the emphasis placed on the notion that all claims over the epistemological space are decision-choice-determined with either fuzzy, stochastic, or fuzzy-stochastic conditionality under epistemic transversality conditions defined over the space of information cost-benefit dualities. The usefulness of the of the principle of opposites is the emphasis on the notion of dualistic-polar relationality of continuum and unity of the opposites that all claims over the epistemological space are not only neuro-decision-choice determined but that all logical values are in dualistically inseparable states within the space of fuzzy-stochastic uncertainty, where the duals of duality are mutually constraining and similarly the poles of polarity are also mutually constraining and similarly the dualities and polarities are mutually constrained over the space of internal-external dualities such that the role of thinking-reasoning in neuro-decision-choice processes is to balance opposing forces to achieve selected goal, objective and vision in the space of goals, objectives and visions under a preference scheme. The fuzzy analytical process also brings into focus the objective-constrained logico-mathematical reasoning where a selection is made from a non-fuzzy objective set $(\mathbf{A})$ or fuzzy objective set $(\mathbf{A}, \mu_{\mathbf{A}}(\mathbf{x}) \in (0,1))$ constrained by the elements in the non-fuzzy constrained set $(\mathbf{B})$ or fuzzy constraint set $(\mathbf{B}, \mu_{\mathbf{B}}(\mathbf{x}) \in (0,1))$ and vice versa in an interdependent mode of a search of knowledge about varieties in the dualistic-polar space or search for change of variety in the space of problem-solution dualities relative to varieties in the space of goals, objectives and visions.

The same variety negative-positive dispositions define the real cost-benefit dispositions in the variety transformations over the journey from variety ignorance to variety knowledge, where ignorance is either a real cost or benefit and knowledge is either a real benefit or cost as inputs into decision-choice behaviors in the management and control of socio-natural systems. The neuro-decision-choice system is a self-contained module that uses thinking to develop algorithms and use reasoning guided by thinking algorithms to process the information-knowledge input-output structures to relate the space of real cost-benefit dispositions to the space of varieties of all areas of knowing given individual or collective preferences. It is the view, here, that the foundations of constrained optimizations are the relative cost-benefit conditions and the foundations of unconstrained optimization is the net cost-benefit conditions, where the cost-benefit structures govern preference ordering over the decision-choice space in relation to the space of attractive-repulsive dualities as seen over the space of incentive-disincentive dualities in knowing and information-knowledge accumulation as intellectual capital stock-flow dynamics. As have been explained in the previous chapters, both variety negative disposition and positive dispositions are defined over the space of cost-benefit dualities as viewed over the space of goals, objectives, and visions.

6.4.2 The Asantrofi-Anoma Rationality and Neuro-Decision-Choice Analytics

In the neuro-decision-choice analytics of the principles of diversity and unity of knowing, and diversity and unity of science, every problem statement and solution statement reside first in a problem-solution duality and then in a true-false duality within relational balances with ignorance-knowledge dualities in a relational continuum and unity, and in mutually and existentially inter-supportive give-and-take-service modes. Translated into knowing, universalities of varieties and categorial varieties of ignorant-knowledge dualities over the space of problem-solution polarities, we have the space of awareness-unawareness dualities in a relational continuum and unity, where unawareness resides in ignorance dual as its modal supporting and awareness resides in knowledge dual as its modal support. Similarly, ignorance as an actual pole resides in knowledge as the potential pole for its support and knowledge, as a potential pole, resides in ignorance as an actual pole in a relational continuum and unity in such a way that both ignorance and knowledge complete the existential universality of the actual-potential polarity in the general process of variety knowing from *knowledge by acquaintance* to *knowledge by description* through the space of imagination-reality dualities, where the space of imaginations is connected to the space of realities through the neuro-decision-choice actions over the space of identification-transformation dualities. In this respect, the neuro-decision choice analytics is to examine the connecting pathways between variety ignorance and variety knowledge in variety identifications and transformations.

The meanings of the concepts of duality, negative-positive duals, polarity, actual-potential poles, variety and the knowledge about their states and processes are made available over the space of epistemic actions through the epistemological information as the third dimension of universal existence. The epistemological information, viewed either as the knowledge by acquaintance or an experiential information, is the primary category of knowing from which all other information-knowledge structures are derived categories as inputs into neuro-decision-choice activities. All these information-knowledge varieties find meanings and values over the space of problem-solution dualities and neuro-decision-choice activities within the space of ignorance-knowledge dualities, where these dualities establish the principles of diversity and unity in knowing, diversity and unity of science. The cognitive activities over any area of knowing are about finding correct variety knowledge solutions to variety ignorance problems to establish a transition from ignorance to knowledge over varieties under the principle of transversality over the space of dualistic-polar structures. The claim of superiority of science in the categories of knowing is its relative consistency of variety characteristic dispositions to allow for multiple verifications of its existence or occurrence at the spaces of acquaintance and quality-quantity dualities. The importance of social science lies in its conditions for organizing knowing as the production of intellectual investment flows and information-capital accumulation as the production of intellectual capital stocks.

Over the epistemological space, the variety information-knowledge strictures, as knowledge by acquaintance and knowledge by description, come to cognitive agents as fuzzy-stochastic information in the sense of vagueness and volume incompleteness under the principle of acquaintance with variety signal dispositions. From the knowledge by acquaintance under processing with a paradigm of thought we arrive at knowledge by description. The representation of vagueness in information was the central debate between Russell [R3.79] [3.81] [R16.58] and Brouwer [R3.14] [R3.15] [R3.22] that also formed the foundational debate between the intuitionist school and the formalist schools of mathematical development in terms of representation of information and the logic of analysis [R3.6] [R3.26] [R18.53]. The thinking-reasoning conflicts over the path of knowing do not involve information representation alone but also the paradigm of thought over the important spaces of certainty-uncertainty dualities, doubt-surety dualities, decidability-undecidability dualities and acceptance-rejection dualities in relation to neuro-decision-choice actions over the spaces of problem-solution dualities, ignorance-knowledge dualities, success-failure dualities, and epistemic transversality conditions where creativity, curiosity, and hope enter. Overcoming the vagueness and volume limitation of information, translated as qualitative and quantitative dispositions, is a common factor that unites all areas of knowing, irrespective of technique and methods used, to allow the application of measures of uncertainties in possibilistic and probabilistic reasoning in order to obtain fuzzy-stochastic conditionality on information-knowledge surety, and within the space of certainty-uncertainty dualities for all areas of knowing without exception.

The possibilistic and probabilistic reasonings are applications of possibilistic and probabilistic thinking dealing with uncertainties and unsureness by the use of the fuzzy-stochastic paradigm of thought, where the conditions of fuzzy-stochastic conditionality define the applicable areas of surety and doubt over the spaces of acceptance-rejection dualities and ignorance-knowledge dualities, and where anticipations based on necessity translate into expectations base on freedom within the space of imaginations, and expectations translate into beliefs within the space of realities contained in the space of actuals, and where necessity translates into freedom, freedom translates into decision-choice decidability and initial conditions become sufficient conditions of actions and actions become successes or failures and variety imaginations becomes variety realities, where necessity, with anticipations, and freedom with expectations constitute individual and collective sovereignties over the space of cognitive actions. In this organizational structure of knowing and the practice of knowing, restraints on sovereignty is a deny of either necessity, freedom, or both in the space of anticipation-expectation dualities over the space of imagination as a process of controllability. The representation of incompleteness finds expressions in the stochasticity or probability with either exact or inexact structures [R2.17] [R2.23] [R2.24] [R2.32] [R2.40] [R3.6]. The representation of vagueness finds expressions in fuzziness or possibility with inexact information conditions [R4.4] [R4.7] [R18.18]. The vagueness and stochasticity combine to define possibilistic-probabilistic information structures with hybrid units that help to trace the source of fuzzy-stochastic uncertainties and the development of fuzzy-stochastic entropy of various kinds [R21.2][R21.4][R21.5] [R21.15] [R21.29].

The concepts and the methods of exact and complete representations of epistemological or experiential information as the *classical information* are the position of the formalist with an artificial imposition in the universal understanding to allow the use of the classical paradigm with the principle of excluded middle to dismantle the simultaneous existence of opposites in the same space and time. These conditions of the thinking-reasoning approach of the classical paradigm of thought in knowing and information-knowledge production relate difficulties in the understanding of some important aspects of socio-natural gradual transformation processes of ignorance varieties in the space of opposites with human try-and-error and learning-by-doing processes, scientific experimentations, knowledge abstractions from data and witness, evidence in judicial decisions, the whole construct of varieties hat enters our food chain, medicine box from their natural conditions of existence, where the classical logic deals with a change or no change with complete excluded middle under the principle of opposites and where the acceptance of partial change is a logical contradiction and logically unacceptable. The use of excluded middle, where the duals are disconnected in dualities, the poles are disconnected in polarities and the dualities are disconnected from their respective poles means one cannot meaningfully understand internal process of change in terms characteristic transformations such as from either negative disposition to positive disposition, from liquid disposition to solid disposition, from cost disposition to benefit disposition, from seed to gemination and to a tree and other important socio-natural transformations with reverse transformations that are essential to the information-knowledge stock-flow disequilibrium dynamics, where every disposition is a set of characteristics defining the identity of the corresponding variety with a simultaneous existence of costs and benefits in the same variety and at the same time , where the negative and the positive duals are simultaneously benefit and cost in the same space of neuro-decision-choice actions in all areas of human endeavor relative to goals, objectives and visions. Let us also keep in mind that a variety characteristic disposition defining a variety identity is a fixed set of characteristics composed of positive and negative subsets, the changes of its distribution change the identities and hence the varieties from the old varieties to new varieties with the space of parent-offspring dualities. Changes in the natural variety dispositions take place in nature by natural functors and categorial converters in the space of input-output dualities, while changes in social variety dispositions are undertaken with social functors and categorial converters in the space of production-consumption dualities.

The inexact and incomplete epistemological information or knowledge by acquaintance is the natural factor of all characteristic-signal dispositions of varieties due to cognitive capacity limitations in all forms. This is the general *fuzzy information structure.* The representation of fuzzy information structure must find types of meaning in a paradigm of thought that allows conditions of subjectivity, vagueness, inexactness, and incompleteness of experiential information to be represented with a logical analysis, within a paradigm of thought that accepts opposites in the same space and time with relational continuum, unity, and contradictions. Such paradigm of thought must reject excluded middle and relational non-contradictions and disunity in thought and practice, where the experiential characteristic dispositions are abstractions from ontological signal dispositions with errors and error-correction process in the neuro-decision-choice actions in knowing and information-knowledge productions over the space of quality-quantity dualities. The variety experiential characteristic dispositions as abstracted from the ontological signal dispositions constituting epistemic space may be different from the variety ontological characteristic dispositions constituting the ontic space. The differences are the variety *epistemic distances* which have been defined and discussed in the previous chapters of this monograph such that the epistemic distances provide the foundation of analysis of the space of certainty-uncertainty dualities. The knowledge by acquaintance and the knowledge by description are developed in the space of imagination-reality dualities, where thoughts are developed in the space of imaginations and actions are developed in the space of realities such that behind every action there is a thought for action to avoid variety emptiness and behind every thought there is action to avoid variety blindness [R17.44].

To understanding the general and continual socio-natural transformations and information-knowledge stock-flow disequilibrium dynamics, and hence variety transformation disequilibrium dynamics with continual self-correction processes, one must understand the nature of variety forward-backward connectivity of transformations and technological processes of reverse engineering and reengineering, where solid becomes liquid and vice versa, hot becomes cold and vice versa, dry becomes wet and vice versa, and many other relational structures under the general principle of opposites and dualistic-polar transversality conditions with the understanding that acquisition of knowledge is a cognitive journey over the multiplicity of paths connecting ignorance to knowledge within the variety internal dualities and polarities, where a neuro-decision-choice action is to select the optimal path to cross either the point or the line of epistemic transversality. It is important to place the economic concept of opportunity cost within the conditions of the asantrofi-anoma problem where variety benefit is a variety opportunity cost and variety cost is variety opportunity benefit, variety input is variety output and variety output is a variety input over the space of supply-demand chain dualities in variety transformations over the space of imagination-reality dualities. In the system of dualities and polarities, the asantrofi-anoma problem, in the adinkra conceptual system, is a derivative from *nyansapo problem* (the wisdom knot) which presents conditions of general cognition under the principle of opposites as *"nfitiase-she-nawieyemu-nawieye-she-nefitiasemu"(* the beginning is in the end and the end is in the beginning-(*beginning-end-dualitiy)*) to present the system of beginning-end dualities over the space of primary-derived dualities as extended to the space of parent-offspring processes or predecessor-successor dynamics that give explanation and prescription to problem-solution dynamics over the space of theory-practice dualities, where in theory do we find practice and in practice do we find practice in unbreakable chain of information-knowledge stock-flow dynamics as illustrated in Figure 4.5.1 of chapter 4 of this monograph to present the knowledge conflict between the ontological beginning-end duality and epistemological past-future dualities.

6.5 EPISTEMIC REFLECTIONS ON THE PRINCIPLE OF OPPOSITES, THE VARIETY SPACE, AND KNOWING IN THE SPACE OF ONTOLOGICAL-EPISTEMOLOGICAL DUALITY.

The general cognitive battles and claims of information-knowledge over the socio-natural space must be placed in the space of relativity rather than in the space absoluteness if we are to make continual progress conditional on individual, collective and generational preference under the general principle of non-satiation and cognitive capacity limitations. The principle of non-satiation is only meaningful in the space of relativity, where continual work towards a space of absoluteness is meaningful within the space of developments and variety transformations widening the space of varieties under stock-flow disequilibrium in the matter-energy space. The space of absoluteness is also the space of complete perfection which is exclusive in in the space of realities but conceptually in existence in the space of imaginations contained in the potential space, where the space of imaginations contains individual and collective conflicting goals, objectives and visions expressed in needs and wants in the dualistic-polar space of social confusions over cost-benefit varieties defined over the social space of non-violence-violence dualities.

The central analytical concept in understanding the human conditions over the space of the principle of non-satiation, confusion, violence is through a conceptualization of the principle of opposites over the space of ontological-epistemological dualities. Since the ontological space with ontic space constitute the identities of knowing and understanding in the epistemological space with epistemic space, we will begin with the ontological space. The ontological space is identity to the epistemological space. The ontic space is the identity to the epistemic space. The principle of opposites projects the conceptual system where the universe is composed of varieties and these varieties are identified by characteristic dispositions in closed sets where the characteristic dispositions constitute the contents of what we refer to as information. Each variety characteristic disposition is made up simultaneous existence of positive and negative characteristics dispositions at the same time and place in the *closed set* of characteristics such that the relative positive-negative combinations provide the variety's identities in the space of ontological identifications. The positive-negative varieties become input-output relations in the ontological transformation under information-knowledge equality conditions to create varieties over the space of positive-negative dualities. The ontological dualistic-polar structure is provided in Figure 6.5.1.

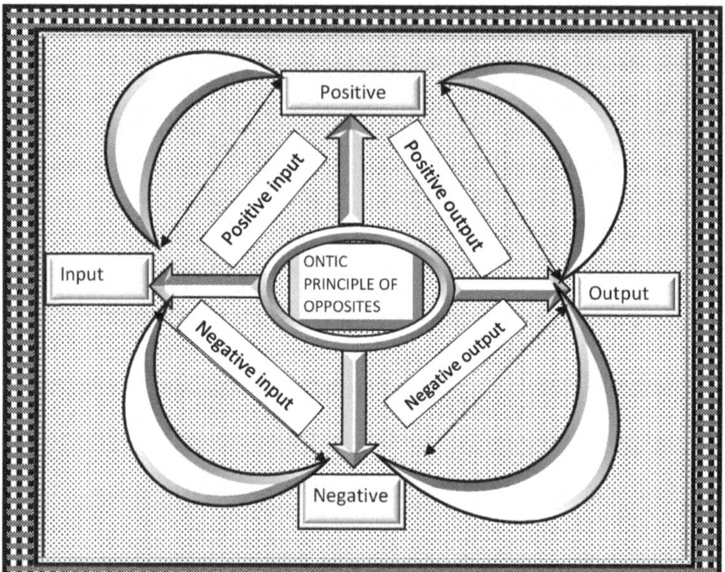

Figure 6.5.1 The Geometry of the Principle of Opposite over the Ontological Space

6.6 COST-POSITIVE, COST-NEGATIVE, BENEFIT-POSITIVE, BENEFIT NEGATIVE, THE VARIETY SPACE AND TRANSFORMATIONS IN THE SPACE OF ONTOLOGICAL-EPISTEMOLOGICAL DUALITY.

It is important to notice the complexity of the concept of the variety and the further complexity in its relation to the *principle of opposites*, paradigms of thought, and principles of diversity and unity in knowing, science and neuro-decision-choice actions. The principle of opposites is central to all conflicts, decision-choice actions, internal and external transformation, and generation of socio-natural forces, where every variety disposition exists as an opened set of characteristics. The understanding of the principle of opposite provides a framework to analyze conditions of variety problem, variety solution and their relationship, where the space of problem-solution dualities constitute the center of all neuro-decision-choice actions. The principle of opposites is composed of interdependent sub-principle of duality in relational continuum and unity, and sub-principle of polarity also in relational continuum and unity where the duality and polarity are in relational continuum and unity with each other. The continuum and unity are defined by give-and-take mode in terms of existence over the space of construction-destruction dualities. The dualistic-polar system is such that every pole of a polarity has residing duality where the nature of the duality defines the pole and the internal activities of the pole bring changes in the pole, all specified in characteristics, where in the space of variety transformations every variety disposition is an *open set* of characteristics composed of cost positive, cost negative, benefit positive, benefit negative for ranking and neuro-decision-choice actions for variety preferences.

Every variety in the universal system exists as a multiplicity of dualistic-polar objects. Every element in any given space is a variety such that the universe at any time is a complete collection of varieties under identification with variety dispositions as closed sets and transformation with variety disposition as open sets, and hence equals to the *space of varieties*, which is an infinite collection of actual and potential varieties. The variety dispositions as closed sets allow variety identifications and the variety dispositions as open sets allow transformation in the space of input-output dualities (for open and closed sets, categories, groups, fields see [R8.16a[R816b]). The multiplicity of dualities and polarities includes *negative-positive dualities* and *polarities* in deferential relative combinations at the level of the ontological space. At the level of epistemological space, where cognitive agents operate under neuro-decision-choice actions with preferences, each variety does not only appear as negative-positive duality and polarity, but the negative-positive duality and polarity are transformed into *cost-benefit dualities* and *polarities* which are then transformed into *problem-solution dualities* which in turn become identification-transformation dualities, all in a complex system of dualities and polarities for decision-choice activities over the epistemological space in relation to information-knowledge disequilibrium dynamics. The basic structure is shown as an epistemic geometry in figure (6.5.2). In this geometric representation, the conditions of the space of input-output is assumed, where the inputs may be positive or negative and the outputs may also be positive or negative. The negative variety and the positive varieties are defined over the space of cost-benefit dualities, where there are cost negative disposition, and cost positive disposition and similarly, there are benefit negative disposition and benefit positive disposition presenting information-decision structures for cost-benefit analytics over the space on internal-external dualities [R5.13] [R5.14]. As I have pointed out that the concept of cost-benefit dualities, incentive-disincentive dualities and many such dualistic-polar structures reside in the epistemological space where cognitive agents exercise neuro-decision-choice actions based on imperfect information structure. These dualistic-polar concept and phenomena are not in the ontological space, where natural transformations are executed under perfect information in the space of information-knowledge equality. The ontological principle of opposites involves a system of input-output dualities and positive-negative dualities in a closed system where every variety is simultaneously an input and an output in transformation without externalities. The epistemological principle of opposite involves a system of production-consumption dualities and cost-benefit dualities in an open system where every epistemological variety is simultaneous a cost and a benefit in transformation with externalities.

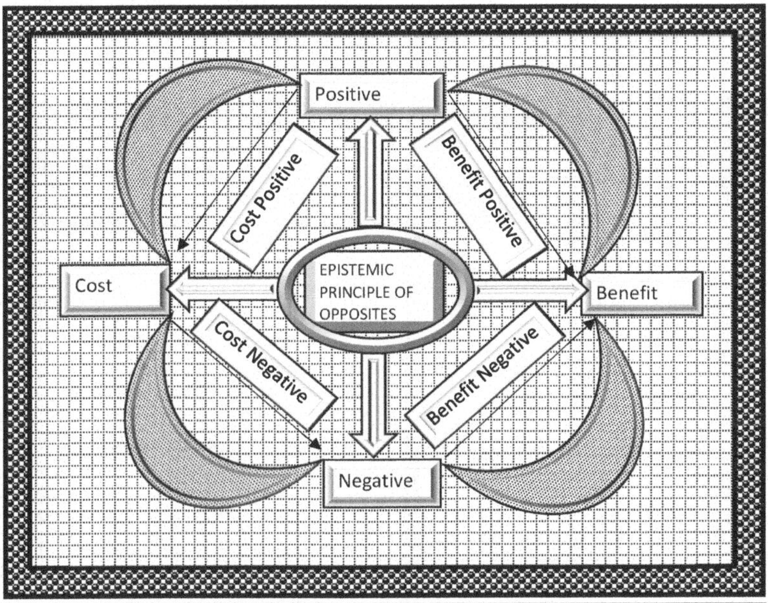

Figure 6.6.1: Geometry of the Principle of Opposites Over the Epistemological Space

It is useful to present some of the elements of comparative analytics of the ontological space and the epistemological space that together make up the space of ontological-epistemological dualities. There are differences and similarity of the elements of dualistic-polar conditions over the ontological space and epistemological space. The similarities connect the epistemological space to the ontological space in the contents for knowing and of the development of information-knowledge stocks for the maintenance of individual and collective lives. The differences provide cognitive agents with analytical methods and techniques executing the actions to effect the ontological-epistemological connectivity paths in knowing and information-knowledge stock-flow disequilibrium dynamics for individual and collective existence, stability and progress under the socially acceptable preferences contained in social philosophical consciencism. The comparative elements are displayed in Figure 6.5.3.

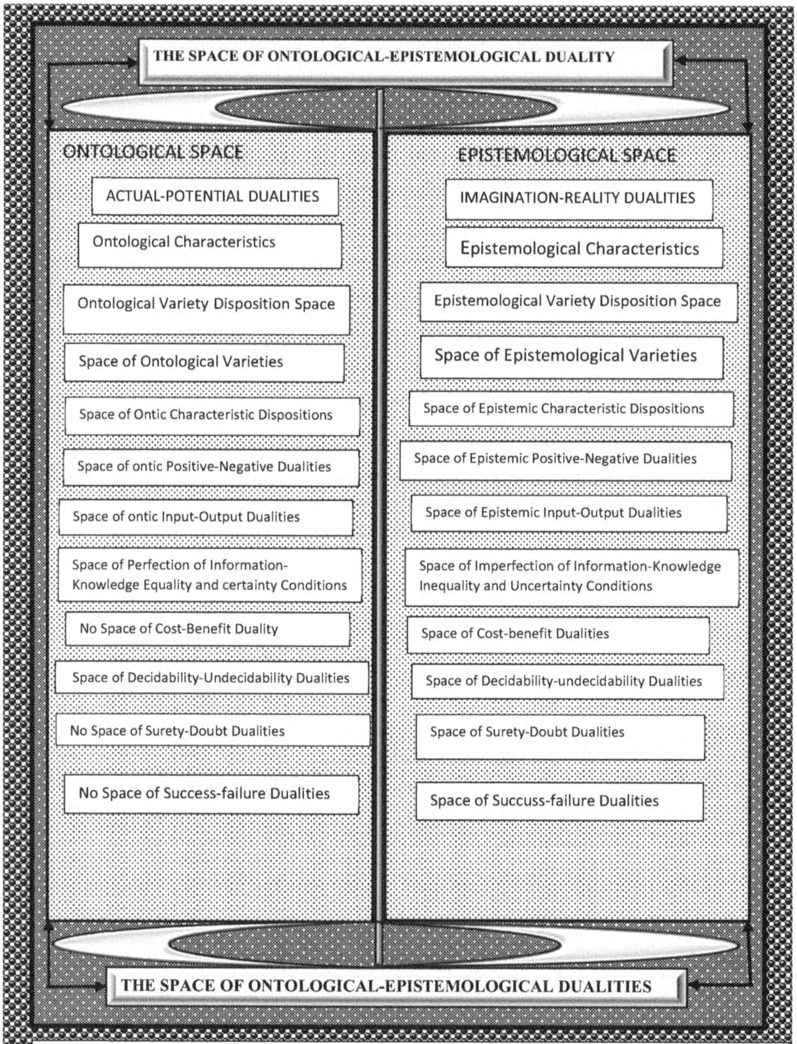

Figure 6.6.2 Comparative Elements of the Ontological and Epistemological Spaces

CHAPTER 7

FURTHER REFLECTIONS ON THE PRINCIPLE OF OPPOSITES, THE VARIETY SPACE AND KNOWING IN THE SPACE OF ONTOLOGICAL-EPISTEMOLOGICAL DUALITY

ABSTRACT

Chapter 7 of this monograph presents some important discussions and reflections on comparative analytics about the differences and similarities of the ontological principle of opposites in the ontological space and epistemological principle of opposites in epistemological space. It is argued that the principle of opposites is composed of a system of dualities and polarities. There is the system of *key opposites* of dualities and polarities, and a system of *supporting opposites* of dualities and polarities which are defined and specified. The key aggregate opposites which are common to both the ontological and epistemological spaces are positive-negative dualities, input-output dualities, and identification-transformation dualities, where the ontological space contains the space of actual-potential dualities and epistemological space contains the space of imagination-reality dualities. The supporting opposites are specific to the epistemological space. They are specified and explained to be generated by the requirements of neuro-decision-choice processes under individual and collective preferences which are admissible in the social philosophical consciencism over the epistemological space containing the space of imagination-reality dualities and constrained by all forms of cognitive capacity limitations. Among the supporting opposites are cost-benefit, minimum-maximum, acceptance-rejection, and other similar dualistic-polar structures which relate to the category of the key aggregate opposites.

K. K. Dompere, *Theories of Knowing and Knowledge Production*, Studies in Systems, Decision and Control 501, https://doi.org/10.1007/978-3-031-42863-0_7

Comparative analytics are undertaking to show the similarities and differences of the ontological and epistemological spaces, where organizational analytics allow for the differences and similarities to be connected to the space of knowing as intellectual investment flows and information-knowledge accumulation as the production of intellectual capital stocks.

7.1 A REFLECTION ON THE PRINCIPLE OF OPPOSITES AND THE VARIETY SPACE

The whole development of conceptual understanding of behaviors of cognitive agents over the phenomena of problem, solution, question, answer, ignorance, and knowledge as relate to human actions and activities over the space of goals, objectives and visions is motivated from the fundamental idea of the principle of opposites within the behaviors of natural and social variety events over the space of static-dynamic dualities. The opposites are composed of dualities, sub-dualities, polarities, and sub-polarities in primary and derived categories. Here, the concept of ontological space is used to represent a system of natural states, processes, and varieties, while the concept of the epistemological space is used to represent a system of social states, processes, and varieties. The connectivity of the ontological space and the epistemological space is established by a system of pathways of dualistic-polar structures of the space of ontological-epistemological duality, where such dualistic-polar structures are endowered with conflicts to generate energy for internal and external transformations and expansion of the space of varieties and information-knowledge system over the space of past-present dualities and potentially projected to the space of present-future dualities.

The space of the ontological-epistemological duality generates the space of social-natural duality, where the space of variety ignorance-knowledge dualities provides the information-knowledge conditions for neuro-decision-choice process in dealing with nature as it is, as well as the organization of control and management of all social set-ups with cognitive understanding of natural processes, functors, categorial converters, innovations, and uses coded in the social philosophical consciencism. Given the socio-natural existence of cognitive agents, the overriding elements of continuity of existence are information-knowledge elements that must be acquired processed and used. The information-knowledge elements are, here, seen as socio-economic production in a conflict space of input-output dualities through cognitive battles, where every variety is identified and established by its characteristic disposition and all categories are formed by similarities of variety characteristic dispositions. The concepts of characteristic, characteristic disposition and varieties are linguistic primitive and essential foundation in the development of theory of knowing and information-knowledge production.

The general cognitive battles and claims of information-knowledge elements over the socio-natural space must be placed in the space of relativity rather than in the space absoluteness if we are to make continual progress conditional on individual, collective and generational preferences under the general principle of non-satiation, where cognitive agents continually work toward greater preference state. The principle of non-satiation is only meaningful in the space of relativity, where continual work towards the space of absoluteness is connected to the space of developments and claims of progress. The space of absoluteness is also the space of complete perfection which is elusive in the space of reality but conceptually in existence in the space of imaginations contained in the potential space, where the space of imaginations contains individual and collective conflicting goals, objectives, and visions in the dualistic-polar space of social confusions over cost-benefit varieties in the social space of non-violence-violence dualities of social actions. The central analytical concept in understanding the human conditions over the space of the principle of non-satiation, confusion, violence and other similar phenomena are through the conceptualization of the principle of opposites over the space of ontological-epistemological dualities. Since the ontological space with ontic space constitute the identity of knowing and understanding in the epistemological space with epistemic space, we will begin with the ontological space and contrast it with the conditions of the epistemological space. The cognitive agents have no control over ontological events and transformations. In the space of ontological-epistemological duality, cognitive agents have only one toolbox set , and this toolbox set is the toolbox which is composed of compromises, adjustments and self-management to ontological process and outcome by knowing and learning, where the ontological responses to epistemological processes and outcomes are unknown and will not be known until they happen. It is this ontological-epistemological differences in response space that also provide the drive to know and development of information-knowledge stocks.

THEOREM: The Laws of Opposites

The principle of opposites is composed of the *laws of duality* and *laws of polarity* which together maintain the variety behavior over the space of identification-transformation processes.

THE LAW OF DUALITY: Every variety exists as an internal duality of negative dual and positive dual defined in qualitative-quantitative characteristic dispositions for its identity which constitutes an external dual which then respectively constitutes a negative and positive pole of a polarity for initial conditions and necessity for change.

THE LAW OF POLARITY: Every variety is a negative or a positive duality which resides in either a negative pole or positive pole of a polarity, where the external negative duality defines the negative pole and the external positive duality defines the positive pole of the duality.

THE LAW OF MOTION: The dualistic-polar variety transformations are induced by dualistic conflicts in the duality generating internal energy to induce qualitative-quantitative motions which affect the transformation behavior of the poles and polarity relative to variety identity, where every significant variety transformation is induced by the behavior of the variety identities.

7.2 ONTOLOGICAL POSITIVE-NEGATIVE AND INPUT-OUTPUT DUALITIES

The ontological space is the identity to the epistemological space. The ontic space is the identity to the epistemic space. The principle of opposites projects the conceptual system where the universe is composed of varieties and these varieties are identified by characteristic dispositions which constitute the contents of what we refer to as information in the space of varieties. Each variety characteristic disposition is made up simultaneous existence of a combination of positive and negative characteristic dispositions at the same time and place such that the relative positive-negative combinations provide the variety's identities in the space of ontological identifications. The positive-negative varieties become input-output relations in the ontological transformation under information-knowledge equality conditions to create varieties over the space of positive-negative dualities where variety potentials are internally transformed into variety actuals, and variety actuals are also internally transformed into the variety potentials. The ontological dualistic-polar structure is provided in a relational geometry as in Figure 7.2.1., where there are inter-relational interdependent spaces of positive-negative dualities, actual-potential dualities, and input-output dualities for ontological processes.

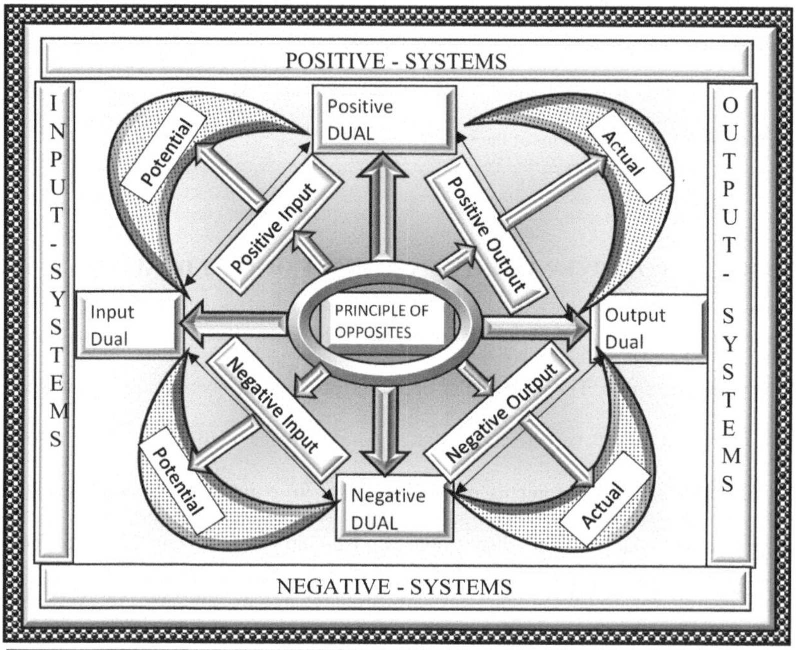

Figure 7.2.1 The Geometry of the Organizational Analytics of the Principle of Opposites over the Ontological Space with Actual-Potential Dualities as Foundation of Internal Variety Transformation through Dualistic-Polar Conflicts.

It is important to notice the complexity of the concept of the variety and the further complexity in relation to the *principle of opposites* over the ontological space which acts as the identity in the space of knowing and information-knowledge production under the conditions of neuro-decision-choice actions. The principle of opposites is central to all conflicts, internal and external transformations, and the generation of natural forces over the ontological space. The understanding of the principle of opposite provides an important framework to analyze conditions of ontological variety identities, differences, commonness. natural forces and behaviors in relationality of continuum and unity. The principle of opposites over the ontological space is composed of interdependent sub-principle of dualities in relational continuum and unity, and sub-principle of polarities also in relational continuum and unity, where the duality and polarity are in relational continuum and unity with each other establishing a system of variety dualistic-polar structures. The dualistic-polar structures in relational continuum and unity are defined by a system of give-and-take sharing modes in terms of existence over the space of construction-destruction dualities where the give-and-take sharing modes established complexities in mutual existences and destructions also in a system of input-output dynamics relative to the space of negative-positive dualities for mutual variety existence and destruction in a continual ontological variety creation and transformation that provides meaning to the concept of ontological food-chain for the maintenance of ontological lives, where the ontological input-output dynamics provide meaning to ontological variety supply-demand chain for ontological transformation on the based on information-knowledge equality.

Every variety in the universal system exists as a multiplicity of dualistic-polar objects and every variety disposition is a closed set that defines the identity of the variety. Every element in any given space is a variety such that the universe is a complete collection of varieties under identification and transformation acting simultaneously as inputs and outputs, and hence equals to the *space of varieties*, which is an infinite collection of actual and potential varieties for the existence of the ontological space composed of varieties contained in the space of matter-energy in stock-flow disequilibrium states. The multiplicity of dualities and polarities includes *negative-positive* and input-output *dualities* and *polarities* in deferential relative combinations at the level of the ontological space, the relational structure of which is shown in Figure 7.2.1.

The structure of the ontological principle of opposites is simple with a complex generating system of a complex family of negative inputs, positive inputs , negative outputs and positive output in the spaces of actual varieties and potential varieties, where every variety simultaneously exists as an input and output in the space of actual-potential dualities over the space of identification-transformation dualistic-polar structures establishing a complete control and management of varieties in the space of waste-non-waste dualities, where there is a continual ontological recycling without externalities over the space of ontological input-output dualities. The non-existence of externalities in the ontological space presents a situation, where every output is ontologically useful and every input is ontologically useful in the input-output transformations under the conditions, where unwanted outputs become are useful inputs and unwanted inputs become useful outputs to the ontological transformation processes. Built into the ontological variety identification-transformation processes is an efficient *waste management system* that preserves the integrity of the ontological space in environmental input-output equilibrium.

The organizational structure of the principle of opposites over the ontological space is simple over the space of identification-transformation dualities and over the space of construction-destruction dualities, where there is a matter-energy equilibrium in the sense that ontological matter-energy is taken as existing in the state of non-destructibility but in unknown dynamics over the space of creational-noncreational duality. The variety space is in continual disequilibrium dynamics over the space of construction-destruction dualities, where the existing varieties may be destroyed or transformed to create new varieties and or to transform existing varieties with the process of the disappearance of the old and the emergence of the new leading to a continual expansion of the space of ontological information from the *infinite ontological time* of the beginning to the present $\left((-\infty)^{\infty}, 0\right)$ and from the present to the infinite ontological future time of the end $\left(0,(+\infty)^{\infty}\right)$ where (0) initializes the present on the ontological telescopic path of the form $\left[(-\infty)^{\infty},0,(+\infty)^{\infty}\right]$.

The ontological telescopic time set may be related to telescopic histories of varieties in variety set representations as they have been done in section 4.5.2 of chapter 4 in terms of categorial analytics of the organization of ontological and epistemological spaces. It may be noted that $\left[\left((-\infty)^{\infty},0\right)\cup(0+\infty)^{\infty}\right]=\Omega$ in terms of ontological space of varieties where $\left((-\infty)^{\infty},0\right)=\mathfrak{A}$ is a representation of the space of actual varieties containing the beginning to the present varieties inclusively and $\left\{\left(0,(+\infty)^{\infty}\right)=\mathfrak{U}\right\}$ is the space of potential varieties containing the present varieties to the end variety. The ontological space Ω contains the space of knowable and unknowable varieties in open and closed sets established by the conditions of epistemic capacity limitations within the epistemological space where all cognitive agents operate. The space of past-future dualities is the symmetric difference between the space of actuals and the space of potentials in the form $\left\{\left((-\infty)^{\infty},0\right)\nabla\left(0,(+\infty)^{\infty}\right)\right\}$ and where the present is always specified as $\left[\left((-\infty)^{\infty},0\right)\cap(0+\infty)^{\infty}\right]=0$.

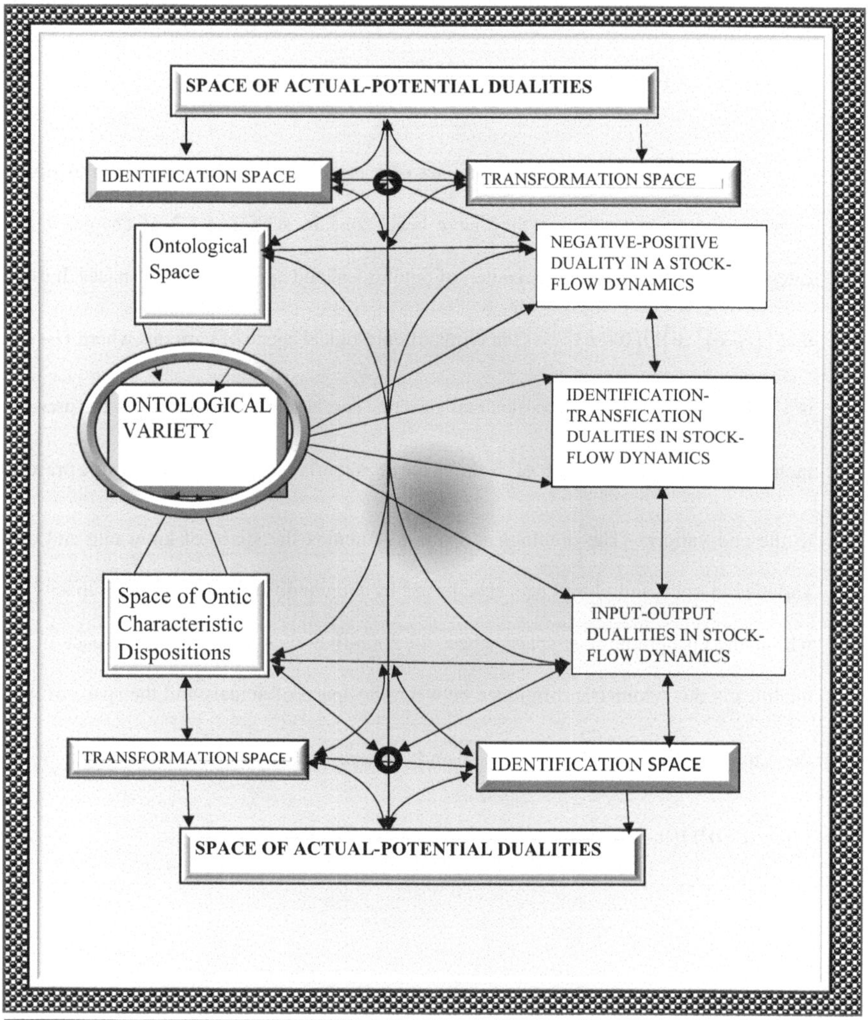

Figure 7.2.2 An Epistemic Geometry of the Relationships among the Principle of Opposites, Ontological Space, and the Space of Varieties over the Space of Identification-Transformation Dualities where is a system of Zones of Ontological Interchanges

The ontological principle of opposites of input-output and negative-positive dualities and polarities in the spaces of relationality of continuum and unity over the space of identification-transformation dualities and polarities with zones of ontological interchanges is shown in Figure 7.2.2. The space of actual-potential dualities and polarities constitutes the collections of positive-negative varieties as input-outputs in the space of identification-transformation dualities and polarities, where every pole of a polarity has a residing active duality with conditions of dualistic-polar dominance. The dualistic dominance of the duals determines the identities of the dualities which then determine the identities of the poles. The polar dominance determines the identity of the polarity in the system of dualistic-polar structures, where transformations of polarities are brought about by changes in the respective poles instigated by changes in the residing dualities, where such changes are internally brought about by the conflicts of the residing duals in the game of transformations within the zones of variety interchanges destroying existing variety identities by transforming the corresponding characteristic dispositions and creating new variety identities with new characteristic dispositions under the space of input-output and actual-potential processes without mater-energy destructions in the sense of matter-energy equilibrium, where the equilibrium is defined in terms of stock-flow conditions in the sense that zero matter-energy flow implies constant and unchanging matter-energy stocks.

In this respect, the ontological principle of opposites establishes ontological dualistic-polar game space for internal variety identifications and transformations. The space of ontological dualistic-polar games provides a framework to study and understand ontological variety dynamics such as volcanos, earthquakes, climatological changes, geomorphological changes, species extinctions, emergence of new species and how they may relate to ontological recycling in the space of input-output dualities for ecological balances over the space of stability-instability dualities in dualistic-polar ontological supersymmetry as depictured by adinkra geometry of symbolism supported by *adinkramatics* as the mathematical system of the adinkras and *adinkralogy* as the philosophical conceptual system for knowing constrained by cognitive capacity limitations. The principle of opposites provides us with a framework to *questions of ontological how* but not to the *questions of ontological why*. The space of ontological how allows us to investigate the mechanics of ontological transformations and how they occur under cognitive capacity limitations and find some meaningful answers to questions of how and possible utilization by ontological varieties. The space of ontological questions of why does not avail itself to answers for knowing in the sense that it is possible to know how earthquakes, volcanoes, and star rebirth occur but not why they occur in the ontological space. We may know how volcanoes, hurricanes, typhoon, earthquake are formed through internal motions based on ontological dualistic-polar conflicts which generate energies and forces for variety self-transformations, where nature is always viewed in terms of oneness. We shall, however, have the questions involving why and the purposes of the ontological transformations. Every approach to formulate the ontological why-problems leads to the production of phantom problems or unsolvable problems. The reasons may be found from the conditions of existence of cognitive capacity limitations that generate the space of unknowability contained in the space of beginning-end dualities where cognitive agents are completely blind to the required initial conditions of universal beginning which is unavailable to cognitive agents except speculative processes through the space of imaginations contained in the space of potentials.

From the system of ontological how-why questions, there are several troubling questions that must be brought into focus. What is the ontological system of decision-choice processes and how does it work? What is the nature of information as input into the ontological decision-choice processes over the space of identification-transformation processes and how is this information processed? Is there a system of incentive and disincentive conditions as motivation to ontological identification-transformation decisions? If there is an incentive-disincentive space, then what is it and can it be conceptualized in terms of ontological variety costs and variety benefits? If there is no incentive-disincentive space, then what is behind the ontological variety transformations in the space of destruction-construction processes? Alternatively, what are the reasonings for earthquakes, volcanoes, monsoons, hurricanes, stars, the rise of new star galaxies, exoplanetary system, and many undiscovered ontological elements in the ontological space-time $\left[(-\infty)^{\infty},0,(+\infty)^{\infty}\right]$? The analytical structure is such that the beginning dispositions constitute the primary category in the ontological creational dynamics and all other varieties are secondary and derived categories. In this structure, the beginning disposition serves as the primary parent which is the nothingness-somethingness duality, where by the principle of opposite, the beginning disposition is contained in the end disposition and the end disposition is contained in the beginning disposition, such that the beginning and the end are mutually creating and inter-supportive in the manner, where the beginning, in whatever disposition, is a parent to the end, in whatever disposition, as an offspring and the end is the parent to the beginning as offspring under the dualistic-polar conditions in relational continuum and unity in the sense of *Gyenyame-Bennu-anoma* existential conditions, where the beginning is conceived to be parent to multiple offspring and the end is conceived to be just one offspring just as the beginning. The beginning-end ontological interchange is different from past-present-future conditions as may be conceived by cognitive agents on the ontological space-time conditions.

7.3 EPISTEMOLOGICAL POSITIVE-NEGATIVE AND INPUT-OUTPUT DUALITIES

The ontological space with positive-negative, input-output, and actual-potential dualities constitutes the identity and primary category for knowing within the ontological time and space of beginning-end creational dynamics. The ontological space with its time structure is the parent to the epistemological space with its time structure as an offspring in knowing and information-knowledge production. The parent-offspring process is through the space of source-destination dualities, where the sources are the ontological varieties, and the destinations are the space of acquaintance of epistemological varieties based on the mind as the essential instrument of cognitive actions. The ontological-epistemological connectivity is such that all knowing, and information-knowledge productions are about *what there was* in the ontological space, (epistemological past connected to the ontological beginning), *what there is* in the ontological space (epistemological present connected to the ontological present) and *what there would be* in the ontological space, (the epistemological future connected to the ontological end). Available to us is the system of ontological beginning, ontological present and ontological end on one hand and a system of epistemological past, epistemological present and epistemological tomorrow, where the epistemological past-present-future time structure is contained in the ontological beginning-present-end time structure. At the level of epistemological space, where cognitive agents operate under neuro-decision-choice actions with preferences, each epistemological variety does not only appear as negative-positive duality and polarity, but the negative-positive duality and polarity are transformed into *cost-benefit dualities* and *polarities* which are then transformed into *problem-solution dualities* which in turn become identification-transformation dualities under preference scheme of cognitive agents, all in a complex system of dualities and polarities for neuro-decision-choice activities over the epistemological space in relation to information-knowledge disequilibrium dynamics and epistemological time, where the concept of disequilibrium is seen in terms of stock-flow conditions, where positive information-knowledge-flow conditions imply disequilibrium of information-stock disequilibrium. The basic structure of the epistemological space with identification-transformation processes under the principle of opposites is shown as an epistemic geometry in Figure 7.3.1.

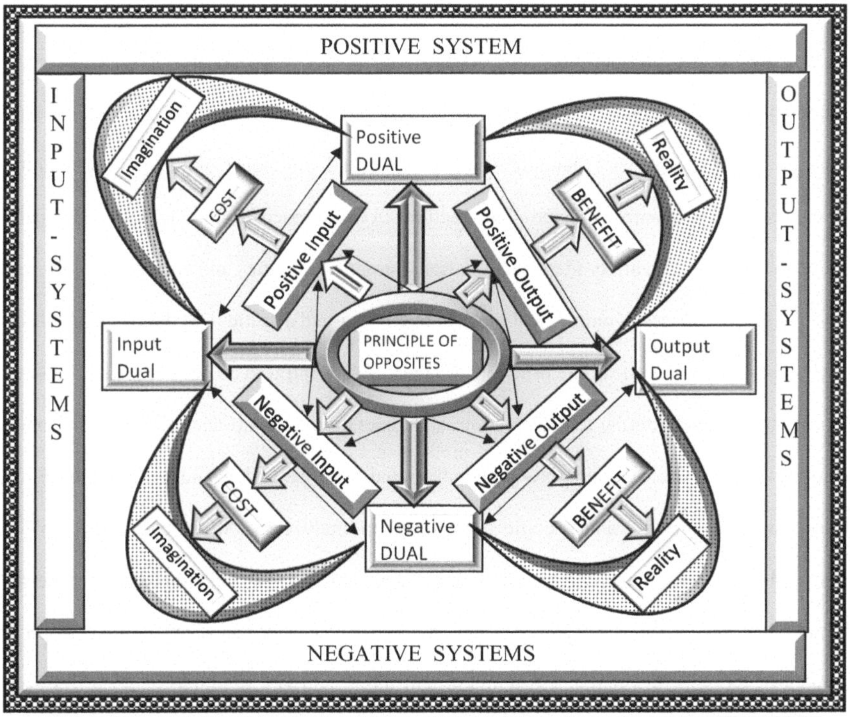

Figure 7.3.1 The Geometry of the Organizational Analytics of the Principle of Opposites over the Epistemological Space with input-Output and Cost-Benefit Dualities within the space of Imagination-Reality Dualities

7.3.1 Cost-Benefit, Input-Output, Positive-Negative Actual-Potential and Imagination-Reality Dualities.

Let us now examine the concepts of input, output, cost, benefit, negative, positive, actual, potential, imagination, reality under dualistic-polar conditions in relation to the principle of opposites over the ontological space and the epistemic space in terms of their similarities and differences in knowing as production of intellectual investment flows and information-knowledge accumulation as the production of intellectual capital stocks. It is important to notice the complexity of the concepts of information and variety and the further complexity in their relations to the *principle of opposites*, paradigms of thought, and principles of diversity and unity in knowing, science and neuro-decision-choice actions over the space of thinking-reasoning dualities relative to the space of variety identification-transformation dualities and socio-natural dynamics viewed within the spaces of ontological times of beginning-end dualities and the space of epistemic timed of past-future dualities. The principle of opposites is central to all conflicts, decision-choice actions, internal and external transformation, and generation of socio-natural forces for complex motions over the space of qualitative-quantitative dualities. The understanding of the principle of opposite provides general and specific frameworks to analyze conditions of variety problems, variety solutions, variety identities, variety transformations and their relationships over the space of static-dynamic dualities. The principle of opposites is composed of interdependent key principle of dualities in relational continuum and unity, and key principle of polarities also in relational continuum and unity, where the dualities and polarities are in relational continuum and unity with each other in such a way that each duality has simultaneous positive and negative dispositions in the same variety, and every pole of polarity has a residing duality, the identity of which determines the pole's identity, where the relative positive-negative dominance determines the identity of the polarity and the activities of the residing duality are connected to the space of destruction-construction dualities over the dynamic behavior of the space of identification-transformation dualities. The continuum and unity are universal and are defined by give-and-take sharing modes in terms of existence and non-existence over the spaces of socio-natural construction-destruction dualities and parent-offspring dualities.

The key principle of opposites is simply made up of input-output dualities and positive-negative dualities which are the commonness between the ontological space and the epistemological space containing the space of variety matter-energy dualities where the space of matter-energy space exists in an equilibrium state and the space of varieties exists in a disequilibrium state. The epistemological space contains the sub-principle of opposites which distinguishes it from the ontological space containing the space of variety actual-potential dualities within the space of destruction-construction dualities. This sub-principle of opposites is the results of behavior of cognitive agents operating under neuro-decision-choice actions with subjective preferences under the principle of non-satiation defined over the space of varieties within the general space of individual-collective dualities. The space of sub-principle of opposites includes the space of cost-benefit dualities which generates the space of incentive-disincentive dualities for variety ordering, decision, and choice over the space of imagination-reality dualities linked to the space of goals, objectives, and visions over the spaces of identification-transformation dualities and problem-solution dualities within the space of past-future dualities in the epistemological space. Our theories of neuro-decision-choice actions are limited to the space of imagination-reality dualities where it is always difficult to extend them to the space of actual-potential dualities.

Every variety in the universal system exists as a multiplicity of dualistic-polar objects. Every element in any given space is a variety such that the universe is a complete collection of varieties under identification and transformation, and hence equals to the dynamic *space of varieties*, which is an infinite collection of actual and potential varieties all ontological times constituting the ontological space. The multiplicity of dualities and polarities includes *negative-positive dualities* and *polarities* in deferential relative combinations at the level of the ontological space. At the level of epistemological space, where cognitive agents operate under neuro-decision-choice actions with preferences, conditions of non-satiation and cognitive capacity limitation, each variety does not only appear as negative-positive duality and polarity, but the negative-positive duality and polarity are transformed into *cost-benefit dualities* and *polarities* which are then transformed into *problem-solution dualities* defined over the space of identification-transformation dualities over the space of ignorance-knowledge dualities, all in a complex system of dualities and polarities for decision-choice activities over the space of imagination-reality dualities contained in the epistemological space in relation to information-knowledge disequilibrium dynamics. The disequilibrium dynamics are such that the variety imaginations are drawn from the space of potentials and then transformed into the variety realities with appropriate functors or categorial converters relative to the elements in the space of actuals. The set of complex relations of epistemological dualities with relational interchanges in the highways of information-knowledge self-correcting system provides an important framework for the driving forces for knowing and information-knowledge production over the space of fundamental-applied dualities containing variety identities of states and processes.

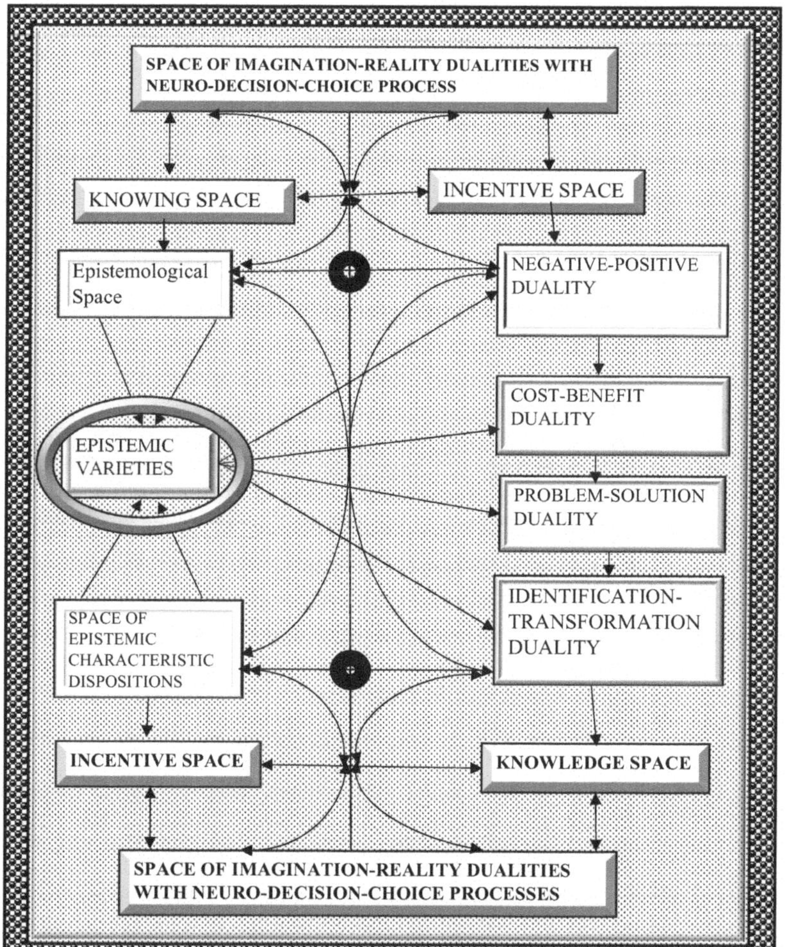

Figure 7.3.2 An Epistemic Geometry of the Relationships among the Principle of Opposites, Epistemological Space, the Spaces of Varieties, Negative-Positive, Cost-Benefit, Problem-Solution, and Identification-Transformation Dualities with the Zones of Epistemological Interchanges on the Highways of Information-knowledge Systems

The epistemological principle of opposites of input-output and negative-positive dualities and polarities in the spaces of relationality in continuum and unity over the space of identification-transformation dualities and polarities connected to the space of imagination-reality dualities containing the spaces of possibility-probability, necessity-freedom and cost-benefit dualities for variety ordering by neuro-decision-choice within the zones and highways of epistemological interchanges is shown in Figure 7.3.2. The space of imagination-reality dualities and polarities constitutes the collections of positive-negative epistemological varieties as input-output and cost-benefit dualities in the space of epistemological identification-transformation dualities and polarities, where just like the ontological space, every epistemological pole of a polarity has a residing active epistemological duality with conditions of dualistic-polar dominance. The dualistic dominance of the duals, just like the conditions of the ontological space, determines the identities of the dualities which then determine the identities of the poles.

The relative polar dominance determines the identity of the polarity in the system of dualistic-polar structures, where transformations of polarities are brought about by changes in the respective poles instigated by changes in the residing dualities, where such changes are internally brought about by neuro-decision-choice conflicts of cognitive agents about the relationally preference structures of cost-benefit conditions of the residing duals in the game of transformations within the zones of variety interchanges, destroying existing variety identities by transforming the corresponding characteristic dispositions and creating new variety identities with new characteristic dispositions under the space of input-output and imagination-reality-transformation processes without mater-energy destructions in the sense of ontological matter-energy equilibrium, where the equilibrium is defined in terms of stock-flow conditions in the sense that zero matter-energy flow that implies constant and unchanging matter-energy stock with continual constancy in level and volume.

At the level of human actions, knowing and information-knowledge production, the epistemological principle of opposites establishes epistemological dualistic-polar game space for variety identifications and transformations by cognitive agents through neuro-decision-choice actions. The space of epistemological dualistic-polar games provides a framework to understand the organic nature of information-knowledge process dynamics of epistemological varieties over the spaces of production-construction and production-consumption dualities relative to social progress and retrogress, peace and war complexities of elements, similarities, differences, categories and many such structures in the ontological space, regarding *what there was, what there is* and *what there would be* under cognitive capacity limitations. In the epistemological space, the space of question-answer dualities is mapped onto the space of problem-solution dualities where such questions as who, what, how, where and why about the epistemological varieties may have answers in that the corresponding problems belong to the space of knowability in the sense that in the sense that what there was, what there is in the epistemological telescopic past-present history belongs to the space of knowability to amplify the space of ontological knowability. There is no spaces of ontological question-answer dualities and problem-solution dualities. These spaces arise as cognitive agents seek to create the epistemological space and connect it to the ontological space. The questions of what, who, how, where, and why and the corresponding space of problem-solution dualities do not arise in the ontological processes that proceed with conditions of information-knowledge equality over the space of input-output dualities. The spaces of question-answer dualities and problem-solution dualities arise when ontological-epistemological duality is sought for variety knowing, control and management of social existence in the space of variety ignorance in the universe.

The epistemological conditions of the space of knowability remind us of conditions of ontological perfect recycling in the space of input-output dualities which acts as the space of production-consumption dualities for ecological balances over the space of stability-instability dualities in dualistic-polar ontological supersymmetry as depictured by adinkra geometry of symbolism supported by *adinkramatics* of mathematical system and *adinkralogy* of conceptual system where there are no externalities in ontological input-output processes over the space of matter-energy identification-transformation dualities. In an alternative view, the *adinkramatics* with the theory of adinkra geometry and graph corresponds to *anansimatics* representing the mathematics of Anansi (spider) string waving in symmetry; *adinkralogy* corresponds to *ansnsiology* revealing the philosophical contents of Anansi stories in the saying "to the wise, we speak in proverbs and codes containing the contents of the messages, and to the unwise, we speak in linguistic planeness containing the contents of the messages." [R27.11] [R27.12] [R27.18].

This perfect ontological recycling is not available to cognitive agents in the epistemological space, such that the epistemological input-output processes generate externalities that affect both ontological varieties and epistemological varieties over the spaces of negative-positive and quality-quantity dualities such that the effects of the externalities of epistemological varieties over the space of production-consumption dualities enter the ontological processes as unwanted inputs to generate harmful ontological outputs to the epistemological space. The epistemological principle of opposites provides us with a framework to *questions of ontological how, what, and where* but not to the *questions of ontological who and why*. The epistemological principle of opposites allows us to investigate the mechanics of ontological transformations of what, where and how they occur under cognitive capacity limitations and find some meaningful answers to questions of what, where and how to generate information-knowledge system for utilization of the epistemological answers to the ontological questions in the epistemological space relative to elements in the space of goals, objectives, and visions of cognitive agents that generate the epistemological space of problem-solution dualities which is conformally mapped onto the space of epistemological question-answer dualities, where answers to ontological whys are hard to obtain if they exist in the cognitive space.

From the space of epistemological questions, there are several troubling questions that must be brought into focus. What questions are generated in the epistemological space and what relevance are these questions and relevant in what sense? What is the nature of information as input into the ontological decision-choice processes over the space of identification-transformation processes and how is this information processed? Is there a system of incentive and disincentive conditions as motivation to epistemological identification-transformation decisions and does the nature of incentive system establish elements of similarity and different between conditions of ontological space and epistemological space? If there is a space of incentive-disincentive dualities contained in the epistemological space, then what is it, and can it be conceptualized in relation to variety costs and variety benefits relative to conditions of input-output processes? How do the elements in the space of incentive-disincentive dualities influence the epistemological variety identifications and transformations in the space of destruction-construction processes in the variety space of epistemological production-consumption dualities? Alternatively speaking, what are the reasonings for production and consumption of varieties, meaning of individual and social progress in the space of relativity-absolute dualities, and past-present and present future telescopic processes of the Sankofa-anoma principles of information-decision-choice interactive processes in the histologically epistemological space-time of the form $[(-t),0,(+t)]$, where $(-t)$ represents the epistemological past, (0) represents epistemological present and $(+t)$ represents epistemological future in the Sankofa-anoma information-time structure and principle?

The analytical structure is such that the past dispositions constitute the primary category in the epistemological transformation dynamics over the spaces of ignorance-knowledge and knowledge-wisdom dualities, where all other epistemological varieties are secondary and derived categories. In this logical structure, the epistemological past disposition serves as the primary parent of the present which is the past information-time duality, where by the principle of opposite, the past information-time disposition is contained in the present information-time disposition and the present disposition is contained in the past disposition, such that the past and the present are mutually creating and inter-supportive in the manner, where the past, in whatever disposition, is a parent to the future, in whatever disposition, as an offspring and the future is the parent to the past as offspring under the dualistic-polar conditions in relational continuum and unity in the sense of Sankofa-anoma existential conditions, where the present is conceived in the past to be parent to multiple offspring and the future is conceived to be just multiple offspring just as the past relative to the present. The epistemological past-future interchange is different from ontological beginning-end space-time conditions. The epistemological and ontological information-time structures are such that $[(-t),0,(+t)] \in (-\infty^{\infty},0,\infty^{\infty})$ where, $\{[(-t),0] \in (-\infty^{\infty},0,)\}$, as past-present information-time structure contained in beginning-present information-time structure , and $\{[0,(+t)] \in (0,\infty^{\infty})\}$ present-future information-time structure contained in the present-end information-time structure. The *space of knowability* is $[(-t),0,]$ and the space of unknowability is symmetric difference of the form $\{(-\infty^{\infty},0,\infty^{\infty})\nabla[(-t),0]\}$ where $\{(-\infty^{\infty},0^{+},\infty^{\infty})\nabla[(-t),0^{-}]\}=((-\infty^{\infty},-t)\cup(0^{+},\infty^{\infty}))$, such that $(-\infty^{\infty},-t)$ represents the beginning disposition and $(0^{+},\infty^{\infty})$ represents the space of future-end dualities.

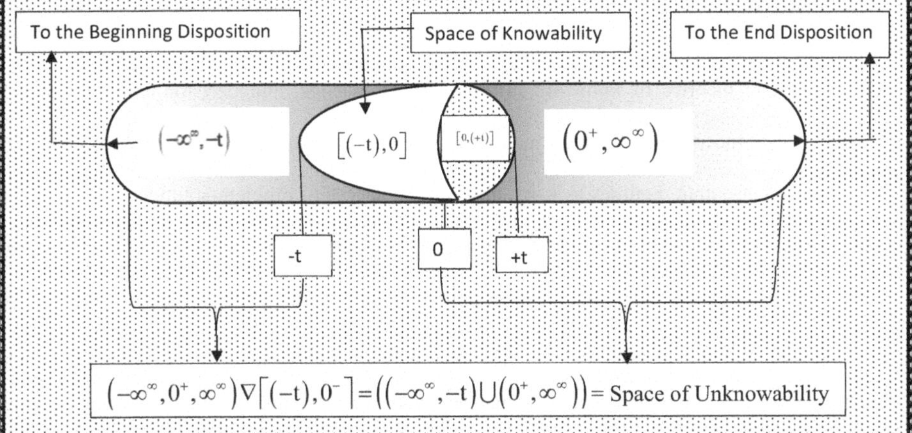

Figure 7.3.3 The Set Representation of Ontological-Epistemological Information-Time Spaces

Let us take a closer look at the time domain of the relationship between the ontological time and epistemological time of beginning-end dualities and past-future dualities. let the epistemological past-future be $[(-t_i),0,(+t_i)]$ and the ontological beginning-end be $(-\infty^{\infty},0,\infty^{\infty})$ in such a way that $\bigcup_{i\in\mathbb{I}^{\infty}}[(-t_i),0,(+t_i)]=(-\infty^{\infty},0,\infty^{\infty})$. The sum of all spaces of epistemological past-future dualities is an improper subspace of ontological space. This simply points to the statement the epistemological information is contained in the ontological space in such a way that the ontological space is composed of the system of knowability and unknowability

7.3.2 Further (Simple) Comparative analytics of the Ontological and Epistemological Spaces

In the theories of knowing, information, knowledge, and decision-choice system are collectively conceptualized as information-knowledge-decision-choice interactive system in the space of input-output dualities as it is related to the spaces of problem-solution and production-consumption dualities. In this respect, it is useful to know what are the objects of knowing, what is information, what is knowledge, what is decision, what is choice and what are their relationships that provide relational continua and unity in the space of human actions relative to the spaces of ontological-epistemological duality, actual-potential dualities, and imagination-reality dualities? The first important difference in nature and social systems in the theory of knowing is that the ontological space-time structure is characterized by beginning-end duality while the epistemological space-time structure is characterized by past-future duality where present disposition is neutral. In this framework, the beginning of ontological varieties and substantial sequence of ontological derivatives as offspring are illusive to neuro-decision-choice agents such that neuro-decision-choice agents do not even know the beginning of variety dispositions in the space actual including the beginning of existence of cognitive agents in terms of *what there was*. The ontological space is composed of the space of variety actual-potential dualities which constitutes variety identities and parents to the epistemological space which is composed of the space of imagination-reality dualities from which we have the evolving derived epistemological identity as offspring. There are no variety cost-benefit dispositions for the ontological varieties, in the ontological space and ontological identification-transformation dualities while every epistemological variety is dualistically defined over the space of cost-benefit dualities to serve as the space of incentive-disincentive dualities for the guidance of neuro-decision-choice actions over the space of variety identification-transformation dualities.

The ontological identification-transformation processes relative to the space of actual-potential dualities are not guided by epistemologically knowable incentive-disincentive elements such as cost-benefit elements in terms of why, and the nature of ontological information-decision-choice processes in terms of who is also not explicitly known to epistemological agents and may not be known, such that the creation of the complex planetary and exoplanetary system regarding *what the beginning dispositions were*, as contained in the space of actuals are not knowable to cognitive agents, and similarly *what would be* including the *end dispositions* in the space of potentials is unknowable to cognitive agents operating in the epistemological space where every claims are defined in the speculative of thinking within the space of speculative-unspeculative dualities . The epistemological identification-transformation processes are guided by epistemologically knowable incentive-disincentive elements that are drawn from the space of cost-benefit dualities in relation to the space of punishment-reward dualities. The nature of epistemological information-knowledge system is developed by cognitive agents for the management of preservation and management of individual and collective existence under cognitive capacity limitations and information-knowledge inequalities, where each epistemic characteristic disposition may differ from the corresponding ontic characteristic disposition. The ontological activities relative to neuro-decision-choice processes are also not explicitly known to epistemological agents such that the understanding of the creation of the complete planetary and exoplanetary system and all extensions are completely elusive over the space of knowing and information-knowledge development.

There are similarities between the ontological space and the epistemological space. Every variety in both ontological and epistemological space resides in the dualistic-polar spaces of input-output dualities and positive-negative dualities. The ontological space derives its space of actual-potential dualities from the spaces of positive-negative and input-output dualities as the primary category of identification-transformation dualities as identities of existence which constitutes the category of the primary principle of opposites without further dualistic-polar dividedness. Similarly, the epistemological space derives its space of imagination-reality dualities from the spaces of ontological positive-negative dualities and input-output dualities as the derived category of identification-transformation dualities as derived category of identities of existence through the space of actual-potential dualities from the acquaintance space with dualistic-polar dividedness. The ontological actual-potential space gives rise to *the space of ontic characteristic dispositions* which resides in the space of absoluteness while the space of epistemological imagination-reality dualities gives rise to *the space of epistemic characteristic dispositions* which resides in the space relativity. The analytical nature of differences and similarities of the spaces of relativity and absoluteness under the principles of opposites with dualistic-polar structures have been discussed in the previous chapters of this monograph and in [R8.16b].

7.3.3 The Organizational Analytics and the Principle of Opposites in Spaces of Knowing.

The neuro-decision-choice activities include information-knowledge constructs and institutional creations for policy creations and transmissions, as well as engineering of all forms including invention of new forms (creativity discovery), innovations in some existing forms (creativity innovation) and destruction of some old forms to define and establish the ontological and epistemological spaces, how the ontological space constitutes the primary identity and a parent to the epistemological space which constitutes a derived identity and an offspring. From the organizational analytics of knowing and information-knowledge production, the space of *ontological-epistemological duality* is defined and established to bring into focus the space of *acquaintance* that analytically connects the epistemological space to the ontological space. From the differences and similarities of the *ontological space* and the *epistemological space*, the spaces of *ontic characteristic dispositions* and *epistemic characteristic dispositions* are defined and established to also provide the space of *ontic-epistemic dualities* with variety characteristic differences called the epistemic distances. The ontic space establishes complete variety characteristic dispositions and *absolute variety identities* in the space of *actual-potential dualities* with information-knowledge equality conditions, certainties, and recklessness of processes over the ontological space of identification-transformation dualities. The *epistemic space* on the other hand establishes incomplete variety characteristic dispositions and *relative variety identities* in the space of *imagination-reality dualities* with information-knowledge inequality conditions, uncertainties, and risk of processes over the epistemological space of identification-transformation dualities.

The space of *ontic variety characteristic dispositions* establishes *ontological certainty* due to information-knowledge equalities and information perfection. The space of *epistemic variety dispositions* establishes *epistemological uncertainty* due to cognitive capacity limitations that generate incompleteness of variety characteristic dispositions, information-knowledge inequalities, and information deformities. The space of *ontic-epistemic dualities* generates variety epistemic distances which then establish the space of *certainty-uncertainty dualities* over the space of ontological-epistemological dualities with certainties as properties of ontological space and uncertainties as the properties of the epistemological space. The space of ontic-epistemic differences generates the space *epistemic distances* that establishes the degrees of uncertainty in cognitive actions within the epistemological space. The space of *certainty-uncertainty dualities* creates the space of doubt-surety dualities that generates the space of *imaginations* in the epistemological space as a derivative from the *potential space* contained in the ontological space. The space of imaginations generates the *space of possibilities* with corresponding spaces of *necessity, anticipations, possible worlds, impossible worlds, and possible-impossible world dualities*. Similarly, the same space of imagination sequentially generates the *spaces of probability* with corresponding *spaces of freedom, sufficiency, expectations, probable worlds, improbable worlds, and probable-improbable world dualities.*

The *space of possibility-probability dualities* is the result of the existence of the *space of certainty-uncertainty dualities* which is the result of the existence of the *space of ontic-epistemic dualities* of the behaviors of neuro-decision-choice agents operating under cognitive capacity limitations in the space of ontological-epistemological duality, where the epistemological space is connected to the ontological space by the space of acquaintance generating knowledge by acquaintance to be processed by some paradigm of thought to arrive at knowledge by description. This is the organizational structure of the theory of knowing as the production of intellectual investment flows and information-knowledge accumulation as the production of intellectual capital stocks. The theory of intellectual investment and the theory of intellectual capital stocks constitute the theory of knowledge as the general theory. This general theory of knowledge is the *theory of information-knowledge economy*. It is a critique of the classical theory of knowledge, the classical theory of economics the theory of information and the current theory of knowledge economy.

A behaviorally dualistic-polar complexity arises when we integrate preferences of cognitive agents with the space of neuro-decision-choice actions over the space of variety identification-transformation dualities as defined within the space of relativity-absoluteness dualities. The neuro-decision-choice behaviors of cognitive agents require variety ordering over the space of quality-quantity dualities for comparative and ordering analyses before action. The comparative-order analytics require the existence of the space of incentive-disincentive duality for variety comparative analytics and order for choice based on what is the best. The comparative and order analytics are defined under the secondary principle of opposites superimposed on the primary principle of opposites. The secondary principle of opposites includes cost-benefit dualities as representing incentive-disincentive dualities, minimum-maximum dualities, true-false duality and many orders in representation dualities, computation dualities such as big-small dualities , linguistic dualities, relational dualities, analytical dualities, definitional dualities such as antonyms and synonyms behavioral dualities such as doubt-surety dualities, love-hate dualities, medical dualities, such as wellness-sickness dualities, economic dualities such as production-consumption dualities, supply-demand dualities, political dualities such as imperialism-colonialism dualities, legal dualities such as justice-injustice dualities and many others to establish dualistic-polar structures for decisions and choices over the space of all human behavior.

The analytical geometry of the interactions of the spaces of ontological, epistemological, and ontological-epistemological dualistic-polar structure, with the ontic, epistemic, and ontic-epistemic dualities may be quickly obtained through relative impositions of Figure 7.3.1 on Figure 7.2.1 and similarly Figure 7.3.2 on Figure 7.2.2. These impositions in additions and relational

Figures 2.5.2 and 2.5.3 in chapter 2 of [R8.16b] present to us a *multiverse* in complexity of knowing and variety existence in information-knowledge details over the space of micro-macro dualities in paremiologically geometry of connectivity. It is through the complexity of multiverse extending to *megaverse*, to *gigaverse* to *petaverse* that curiosity, creativity, courage, innovation, hope, despair, disappointments and other such qualitative elements in the space of thinking-reasoning processes for the discovery and efficiencies of ontological functors and categorial convertors and the development of epistemological functors and categorial convertors find expressions over the space of successes-failure dualities in epistemic searches toward the space of variety absoluteness. The elements in the space of success-failure dualities relative to neuro-decision-choice actions over the space of identification-transformation dualities are induced by cognitive capacity limitations where the understanding of the cognitive capacity limitations may be helped by cognitive science which may also help in the understanding of the system of paremiological geometries that may be constructed for the relational connectivity under the principle of opposites, symmetries, and supersymmetries in the multiverse, where the paremiological geometries may be seen in terms of metaphorical structures in knowing.

It is the understanding of information-knowledge conditions of natural forces in the ontological space that ensures the continual efficient understanding of unity of knowing, information-knowledge accumulation, technological development and efficient socioeconomic transformations and control and management of societies. It is also the understanding of information-knowledge conditions of social forces that ensures the continual efficient understanding of diversity and unity of knowing, science, and the indivisible understanding of the interplay of socio-natural forces of explanatory and prescriptive processes of information-knowledge disequilibrium of knowledge stock-flow conditions, paradoxes, ill-posed problems and paradigms of thought. The knowing process and the information-knowledge accumulation, just like every variety, are also subject to the conditions of the principle of opposites over the space of cost-benefit dualities. It is also useful to note that the *epistemological time* is always contain in the *ontological time* such that $\left[(-t),0,(+t)\right]\subset\left[(-\infty)^{\infty},0,(+\infty)^{\infty}\right]$ as it is illustrated in figure 7.3.3 for all generational times. The epistemic framework is Such that *(what-there-was)-(what-there-was-not) dualities, (what-there-is)-(what-there-is-not) dualities* and *(what-would-be)-(what-would-be-not) dualities* are always contained in the space of beginning-end dualities, where the epistemological past and present dispositions belong to the space of knowability connected to the space of ignorance-knowledge dualities and the epistemological future dispositions, ontological beginning dispositions and ontological end dispositions belong to the space of unknowability.

EPILOGUE

ON THE PRINCIPLE OF OPPOSITES, EPISTEMIC FIELDS, AND FUZZY PARADIGM OF THOUGHT

ABSTRACT: Epilogue

The epilogue of this monograph provides a summary of analytical connectivity of the principle of opposites composed of dualities and polarities, epistemic field, and fuzzy paradigm of thought in the development of knowing as the production of intellectual investment flows and information-knowledge accumulation as the production of intellectual capital stocks. The principle of opposites places the system in the space of dualistic-polar games to cross the transversality conditions where the rewards of the game are variety identity and polar dominance. The conditions of functors and categorial converters are presented and linked to curiosity, creativity, hope and despair with suggestion of further research.

I: REFLECTIONS AND CONCLUSIONS

The development of this monograph is concluded by connecting it with a framework for analyzing the relationality of the principle of opposites, dualistic-polar games, conflicts, fuzzy paradigm of thought over the space identification-transformation dualities containing problem-

K. K. Dompere, *Theories of Knowing and Knowledge Production*, Studies in Systems, Decision and Control 501, https://doi.org/10.1007/978-3-031-42863-0

solution and ignorance-knowledge dualities in the space of imagination-reality dualities in the epistemological space relative to the space of actual-potential dualities containing the subs-pace of input-output dualities for continual ontological variety transformations in the space of life-death dualities and old-new dualities in the space of internal-external dualities in continuum and unity through the gradual processes of ontological time of beginning-end structure. The conclusions are based on interactive concepts and phenomena of principle of opposites, epistemic field and fuzzy paradigm of thought in the space of neuro-decision-choice actions, where the information-knowledge is infinitely self-correcting system linked to the space of problem-solution dualities and ignorance-knowledge dualities.

II: THE PRINCIPLE OF OPPOSITES

The principle of opposites states that every variety is made up of a characteristic disposition containing the opposites of sub-sets of positive and negative characteristic dispositions in relative proportions in continuum and unity, where the negative characteristic disposition is the internal negative dual and the positive characteristic disposition is the internal positive dual in relational continuum and unity. The natures of the internal relative proportions over the space of negative-positive dualities define the variety identities and existence under internal conflicts in the space of *dualistic games* for establishing variety identities in relative dominance. The dualities are said to be negative or positive to the extent to which the negative (positive) characteristic dispositions (duals) dominate the positive (negative) characteristic dispositions (duals) over the space of characteristic continuum-unity dualities in establishing variety differences, identities and commonness such that at the level of internality, the variety identities are organized as a system of dualities with internal conflicts and contradictions of the duals for mutual existence and destruction in relational continuum and unity under the principle of give-and-take sharing mode, where the give-and-take sharing mode may be cooperative, antagonistic or exploitive in the space of internal dualistic games, where the rewards of the games are identity controls over the space of identification-transformation dualities.

At the external level of existence, the space of varieties are organized as a system of polarities in mutual conflicts, where the polarities resides in the space of negative-positive polarities with continual conflicts and contradictions in the space of mutual existence-destruction *polar games* for establishing their existence, dominance and elimination of other variety polarities. The poles of each polarity have residing dualities. The identity of each pole is defined as negative or positive in terms of whether the residing duality is negative or positive, thus a pole is said to negative (positive) if the residing duality is negative (positive). The identity of polarity is negative (positive) to the extent to which the negative (positive) pole dominates the positive (negative) pole. A negative (positive) polarity is registered if the negative (positive) pole relationally dominates the positive (negative) pole. The combined spaces of dualistic and polar games constitute the space of dualistic-polar games in conflicts, contradictions, and characteristic transformations over the epistemic field under the principle of opposites. The polar games are negations of poles and polarities through transformations carried on from the results of variety internal dualistic games, where a change in duality produces a change in the residing pole in relative comparison of the other pole of the same polarity in the spaces of identification-transformation dualities and ignorance-knowledge dualities.

The activities over the identification-transformation processes over the epistemological space is under the neuro-decision-choice actions in the developments and constructions of functors and categorial converters that map variety dualities to other variety dualities. The functor and categorial converter mapping are said to be conformal in the space of varieties and epistemic field if they preserve the categorial dualistic-polar structures over the space of input-output dualities. The functor and categorial-converter actions take place in the dualities with conflicts of the opposites transforming the poles and hence the polarities in the dualistic-polar games under the general principle of opposites in continual and gradual processes over the space of identification-transformation dualities and ignorance-knowledge dualities. The conflicting actions are such that as the set of the negative (positive) characteristics are slowly transformed into the set of positive (negative) characteristics, the negative (positive) dual is reduced in size and role, while the set of the positive (negative) characteristics increases in size and role in the variety. The mutual opposite-transformation actions move the dualistic-polar system to transversality point dividing the zones of negative and positive characteristic dispositions as well as maintaining the dualistic structure of the opposites to transform the poles of the polarity and the polarity itself in relational continuum and unity. The natural functors and categorial converters are developed with information-knowledge equality, where the dualistic-polar games are played with information-knowledge equality. The social functor and categorial converter are developed with information-knowledge inequality under cognitive capacity limitations, where the dualistic-polar games are also played under cognitive capacity limitations and information-knowledge inequality which generate uncertainty on outcomes. The uncertainties of outcomes generate the space of imaginations containing error-collection space in all knowing processes, where the give-and-take sharing modes are such that the outputs of the neuro-decision-choice actions become inputs in the neuro-decision-choice actions over the space of the fundamental-applied dualities.

The space of dualistic-polar games presents three frameworks of the game space. The first one is defined in the space of cooperative give-and-take sharing space, where the game space is positive-sum and win-win games with shared benefits and mutual existence. The second one is defined in the space of antagonistic give-and-take-sharing space, where the game space is negative-sum and lose-lose games with increasing shared costs and mutual destruction. The third one is defined in the space of exploitive relational give-and-take-sharing space, where the game space is zero-sum and lose-win games with benefit to one for dominance and cost to the other with destruction and extinction. The conditions of the three segments of the game space relate to variety and categorial dynamics over the space of identification-transformation dualities with ontological information-knowledge equalities and epistemological information-knowledge inequalities in al processes. At the social setting these three game situations held to understand competitive imperialism, cooperative imperialism, competitive fringe, colonialism, neocolonialism, slavery, oppression, war, peace, and other behaviors over the global space containing the space of socio-natural dualities, where knowing and information-knowledge production are active instruments in the game space. More research and work are required on the abstract linguistic representations and analyses of relational conditions and the roles of functors, categorial converters and social philosophical consciencism in the general space of neuro-decision-choice actions. The game space is illustrated in figure II.1.

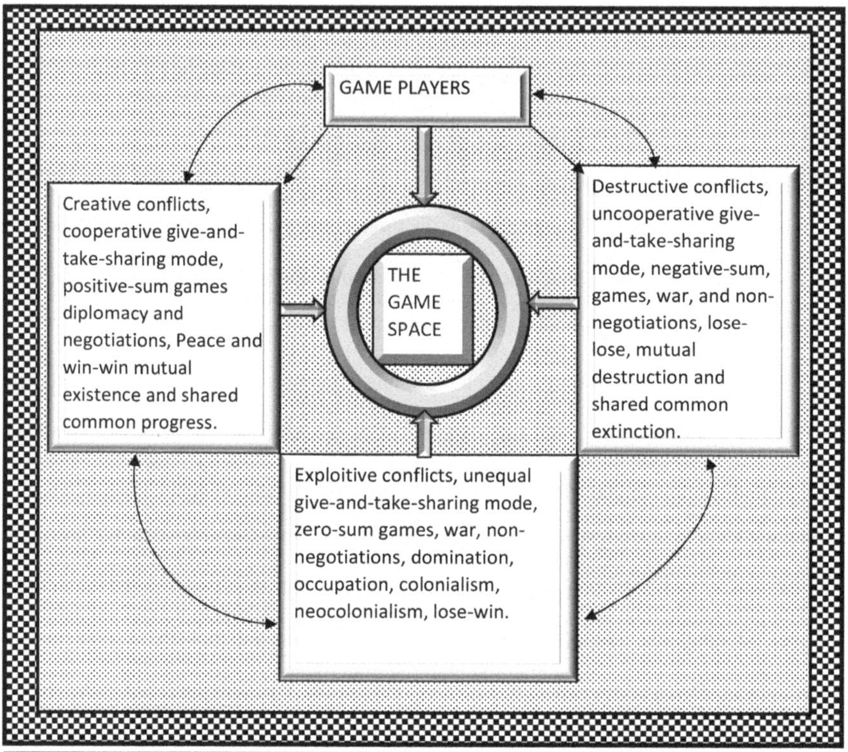

Figure II.1: An Epistemic Geometry of Dualistic-Polar Game Space of Positive Sum, Negative Sum and Zero-Sum Games

III: THE EPISTEMIC FIELDS

The concept of epistemic field is introduced into the development of the theory of knowing and information-knowledge accumulation of intellectual investment-capital processes to deal with dynamic categories of areas of knowing in the development of intellectual capital stock-flows. The epistemic field supported with corresponding relational and information fields contains categories of the space of problem-solution dualities in characteristic disposition for functor and categorial-converter actions on the space of characteristic dispositions and system of characteristic dispositions contained in the epistemic field that has continual field embedding. The categories in the epistemic field with relational and information fields are continually evolving relative to the evolving space of problem-solution dualities, where every functor or categorial-convertor mapping generates conditions of requirements for the development of new functor or categorial converter action over the space of thought-action dualities. The conditions of the development of the functors, categorial converters and the epistemic field with relational and information fields find their foundations in the social philosophical consciencism that contains proven and unproven accumulated intellectual capital stocks including believes, misconceptions, disinformation, misinformation, propaganda, misinterpretations and many such elements. The social philosophical consciencism gives rise to conditions of national personality and exceptionalism that affect the structures of neuro-decision-choice behavior over the space of individual-collective dualities in the dynamics of the categories contained in the epistemic field. Through the concept of the epistemic field, allowances are made to integrate conditions of curiosity, creativity, hope, despair, cognitive capacity limitations into the constructs of the functors and categorial converter over the space of imagination-reality dualities. Similarly, conditions are created to understand the relationships of intellectual stock-flow processes and input-output processes in the space of functor and categorial converter actions over the space of imagination-reality dualities with cognitive capacity limitations over the space of production-consumption dualities contained in the epistemological space.

The construct and the epistemic field with the relational and information fields may be seen in terms of industrialization of intellectual production-consumption system relative to all input-output systems containing categories of knowing and information-knowledge production. Each category of knowing represents an industry with a system of factories containing departments of specified knowing and information-knowledge production over the space of fundamental-applied dualities containing a system of specific space of variety ignorance-knowledge dualities relative to the space of ignorance-knowledge dualities relative to the space of variety identification-transformation dualities in establishing the varieties in the space of imagination-reality dualities, where variety transformations are undertaken with specific functors and categorial converters in the defined categories of knowing and information-knowledge production where the functor and categorial converter actions are seen in terms variety and categorial mapping of characteristic dispositions in conformal structures maintaining all relational continuum and unity as seen in the space of supply-demand chains linked to the space of input-output dualities. At the level of epistemology, the categories are relationally link in continuum and unity by a dynamic system of categories of institutions, the efficiencies of which are drawn from the effects of the ruling social national philosophical consciencism. The epistemic field contains socio-natural varieties and categories over the space of fundamental-applied dualities relative to the space of identification-transformation dualities, where the categories are created with fuzzy identification function. In this structure, the epistemic field is an organization of organic system of human behaviors composed of information-knowledge production and information-knowledge usage with cognitive capacity limitations over the space of ontological-epistemological dualities [R8.16a] [R8.16b]. The epistemic field contains categories of institutions of knowledge production, categories of institutions of knowledge use and institutions of system's organizing, managing. and controlling actions.

IV: ON THE FUZZY PARADIGM OF THOUGHT

Given the principle of opposites with relational continuum and unity, and epistemic field containing categories of knowing over the space of fundamental-applied dualities, the experiential information is defined by vagueness and volume limitation due to cognitive capacity limitations of all forms to create analyses are defined by vague and incomplete information defined in the space of characteristic-signal dualities from the acquaintance space. The vague and volume-incomplete information is called fuzzy information in the space of exact-inexact dualities. The fuzzy information is processed with fuzzy paradigm of thought containing fuzzy laws of thought and mathematics for representation and analysis in the space of thinking-reasoning dualities with fuzzy logic that allows contradictions as logical value, where opposites can exist in the same space and time. The fuzzy paradigm of thought contains fuzzy logic and fuzzy mathematics to deal with information representation and processing of fuzzy information containing vagueness and volume incompleteness where the vagueness and volume-incompleteness are generated by cognitive capacity limitations in the space of inexactness-exactness dualities relative the spaces of identification-transformation dualities, ignorance-knowledge dualities, certainty-uncertainty dualities in the space of absoluteness-relativity dualities. The conditions of graduality in variety transformations reside in the space of relativity but not in the space of absoluteness and cannot be well processed with the classical paradigm thought with the principle of excluded middle generating opposites with relational separation and disunity. The conceptual framework of the theory allows simultaneous treatment categories of industries in term of input-output system of knowing and information-knowledge production in the space of life-death dualities linked to the space of individual-collective dualities over the space diversity-commonness dualities.

The theory of knowing and information-knowledge production is seen as general theory of production and reproduction of real life in the epistemic field with relational and information fields. The principle of opposites presents to us variety identities, essences, and behavior over the space of identification-transformation dualities in the dualistic-polar game space where every variety resides in the space of both duality and polarity. The duality characterizes the internal conditions of existence, and the polarity characterizes the external. conditions of existence. The dualistic-polar game space is about identifications and transformations that form the foundations for the space of ignorance-knowledge dualities The vehicles for transformations are the functors and categorial converters to carry the duals and poles across the transversality conditions. The epistemic field presents the organization of categories of knowing and information-knowledge production in diversity and unity, where every category represents industry with corresponding system of functors and categorial converters which are defined in terms of processes and technologies. The fuzzy paradigm of thought with its mathematics and logic provides analytical instruments to process the vague and incomplete information (fuzzy information) in the space of absoluteness-relativity dualities. All these require further analytical work to improve the analytical modeling of curiosity, creativity, hope, despair, and others in the general space of production-consumption dualities.

APPENDIX

GLOSSARY OF NON-ENGLISH TERMS

PAREMIOLOGY OF SOME ADINKRA DUALISTIC-POLAR TERMS

This appendix is used to connect the reader to the understanding of certain terms from the Adinkra conceptual system that have been introduced in this monograph. These conceptual terms are used to capture the dualistic-polar conditions in the space of abstract-applied dualities under the principle of opposites with relational continuum and unity but not in separation and disunity. In the adinkra conceptual system, every variety exists as an internal duality with a relational continuum and unity of all the duals as well as exists as external duality and as poles of a polarity with relational continuum and unity. The adinkra conceptual system presents paremiological universal understanding in symbols where *adinkralogy* presents its epistemological foundations which is represented the conceptual system of *anansiology* as the theory of knowing in thoughts through the Anansi stories. The *adinkramatics* presents the mathematical foundations of adinkras as mathematical processes of multiplicity of relational connectivity of string geometry and transformations of the varieties in the space of imagination-reality dualities for each of the adinkras as polyrhythmic approaches to information-knowledge development based on the principle of opposites with relational continuum and unity as governed by the principle of give-and-take sharing mode over the space of linear-nonlinear dualities.

Both the *adinkralogy* and *adinkramatics* present a system of thinking and reasoning for the development of *anansiology* and *anansimatics while anansiology and anansimatics provide feedback inputs into further refinements of adinkralogy and idinkramatics* in a dualistic-polar logic in knowing, reasoning and neuro-decision-choice activities in organizational creation, development and social management within the spaces of life-death dualities and new-old dualities

K. K. Dompere, *Theories of Knowing and Knowledge Production*, Studies in Systems, Decision and Control 501, https://doi.org/10.1007/978-3-031-42863-0

of the same material in linear and nonlinear structures over the space of property-relational dualities relative to the space of visible-invisible dualities. The socio-natural varieties exist in the space complex dualities with internal symmetry and external symmetry to form supersymmetry under the laws of opposites and recycling in relation to parent-offspring processes. The system of dualistic-polar reasoning may also be referred to as the Dogon system of dualities. The dualistic-polar thinking and reasoning are exercised over the space of relativity and not the space of absoluteness.

The opposites in terms of duality and polarity involve all varieties to create tensions, energy, power, and force of internal and external variety transformations in diversity and unity. The opposites as dualities and polarities show themselves at all levels through which one can understand life-death, war-peace, love-hate, solid-liquid, and many others without which we cannot understand the information-knowledge conditions that shape the forces of socio-natural transformations within the space of relativity toward the space of absoluteness with neutrality of time. The opposites do not exist as diversities in in separation and disunity with excluded middle but rather in the space of relativity that allows for relational continuity and unity that maintain the give-and-take sharing mode for variety maintenance The understanding of the characteristics of opposites provides one the possibility containing the conditions of necessity and anticipation may create the initial conditions that may set the desirable opposite dual against the undesirable opposite dual in the zone of transformations.

The adinkra conceptual system, like the Dogon conceptual system and the Nile-valley conceptual system presents a system of paremiological geometries in metaphors in a system of logical relations of thoughts in similarities, indifference and differences to create a system of opposites in symmetric string orders to attribute the existence of the universe to dualistic-polar conditions with relation-string give-and-take-sharing threads to establish diversity and unity through connectivity, where from one arise diversities, and from diversities arises oneness, and from oneness arises dividedness in the wisdom construction of *anansi* (spider) tapestry in strings to maintain duality and unity, where the beginning and the end are relationally indistinguishable as well as nothingness and somethingness are in inseparably relational unity and visible varieties are linked and connected to one another by invisible varieties such that the concept and phenomenon of vacuum must be seen in the space invisibilities relative to degrees of variety weights, where variety invisibilities form the unseen complex relational connectivity to produce relational continua and unity in the universal existence that allow the explanation.

ADINKRA: A system of symbols representing a conceptual system that provides paremiological structure of thinking-reasoning system about socio-natural varieties in the space of static-dynamic dualities as well as neuro-decision-choice actions over the space of decidability-undecidability dualities under a system of incentive structures to maintain relationality in socio-natural continuum and unity over the space of varieties in static processes of variety knowing and dynamic processes of variety transformations, where each symbol is a paremiology of thought and the system of symbols constitutes a complete a philosophy. The elements in the Adinkra system are such that each symbol belongs to the space of principles of supersymmetry, opposites, continuum and unity as defining the integrated structure of universal existence.

***ADINKRAMATICS*:** is the mathematics of adinkra symbols in symmetry and supersymmetry.

ADINKRALOGY is the philosophical and conceptual system of the adinkras and representations of thought under the principle of opposites in relational continuum and unity relative to the space of visible-invisible dualities and polarities.

ANANSI is spider in the epistemic space of human information-knowledge system and its dynamics over the space of identification-transformation dualities and polarities.

ANANSIMATICS is the mathematics of the spider network system of fractals in the space of symmetry-antisymmetry dualities and polarities.

ANANSIOLOGY is the philosophical and conceptual system of Anansi stories (*anansisem*) relative to the different complex forms of Anansi fractal tapestries and string geometries as representations of concepts and information-knowledge structures over the space of variety identification-transformation dualities and polarities in learning over the space of social-nature dualities and polarities.

ANOMA-KOKONE-KONE is an individual-community duality in a relational continuum and unity within decision-choice systems projecting relational continuum, cost-benefit commonness, interdependencies and common vision over present-future dualities and polarities , where the individual is simultaneously a cost and a benefit to the community and the community is also simultaneously a cost and a benefit to the individual in the dualistic-polar conditions over the general preference space in conflict under the principle of *each-for-all-and-all-for-each* (*ubuntu* principle, or *dibi-ma-me-nibi* [give-and-take-sharing] principle) with the community preferences dominating in collective decisions for social transformations, and stability for common vision for telescopic tomorrow.

ANOMA-KOKONE-KONE PROBLEM is to reconcile the conflicts of individual preferences in the collective decision space in a social setup. It is here that people's democracy finds philosophical and analytical support in theory and practice. It is within the space of reconciliations between individual and collective preferences that different system of social organizations arise in the space of individual-community dualities with complete community preference and complete individual preference as extremes with various combinations within the individual-community duality under "each for all and all for each", relative to "each for himself and socio-natural survivability for all" characterized as *pesemekonyi* (selfishness)principle.

ASANTROFI-ANOMA is a variety choice that represents simultaneous existence of attractiveness-repulsiveness conditions in terms of cost (*musuo*) and benefit (*serade*) in the same variety where opposites reside in the same space and time in violation of the Western classical logic, where opposites cannot exist in the same space and time.

THE *ASANTROFI-ANOMA PROBLEM* is a decision-choice problem, where every choice variety simultaneously contains cost and benefit in a duality, and one cannot select the benefit dual without the cost dual in a give-and-take relation in the space of relativity.

ASANTROFI-ANOMA RATIONALITY is a cost-benefit rationality, where neuro-decision-choice actions are based on the principle of selecting either the maximum benefit over a cost or minimum cost over a benefit in the space of varieties.

ASANTROFI-ANOMA PRINCIPLE is the principle of using cost-benefit balances for decision-choice actions,

ASANTTOFI-ANOMA CONDITION is the combined conditions of asantrofi-anoma principle and rationality.

MUSUO is the set of conditions of cost characteristics in neuro-decision-choice actions.

SERADE is the set of conditions of benefit characteristic in neuro-decision-choice actions.

MUSUO-SERADE DUALITY is the cost-benefit duality.

BENNU-ANOMA is the relational connectivity between present and the beginning over the space of visible-invisible dualities and polarities.

BENNU-ANOMA-GYENYAME CONDITION is the condition of relational connectivity between the past-present duality and the beginning-end duality, where present-future dualities are contained in the beginning-end dualities.

FUNTUMIREKU-DENKYEMMEREKU is a two headed crocodile with one stomach representing duality with relational continuum and unity in a give-and-take zone of cost-benefit (*musuo-serade*) conflicts. It presents a general dualistic-polar system of varieties in internal and external dualities with internal relational continuum and unity.

FUNTUMMIREKU-DENKYEMMERIKU PROBLEM is how to reconcile duals and poles of the principle of opposites in continuum and unity projecting dualistic-polar conflicts with give-and-take sharing dynamics within the spaces of parent-offspring dualities and visible-invisible dualities of continual processes relative the disappearance of the old and the emergence of the new in the space of static-dynamic dualities contained in the spaces of nothingness-somethingness dualities and beginning-end duality.

GYENYAME is a conceptual representation that means nothing is more powerful than the Creative Force, the Spirit Force and the Light Force structured in relational continuum and unity in the space of oneness-dividedness dualities and polarities as the Holy Trinity, where the concept splits into Creative-Force-Spirit-Force Duality, Creative-Force-Light-Force Duality and the Spirit Force-Light Force duality from the beginning to the end where the beginning and the end will never be known through cognitive existence.

KENTE is a system of complex fractal in strings in details with relational symmetry and supersymmetry representing thoughts and information-knowledge dynamics.

KENTEMATICS is the mathematics of *kente* concepts and geometric representation of socio-natural forms and their relations.

KENTELOGY is the philosophy of *kente* concepts and geometric representation of socio-natural forms for their meanings and explanations of the *kente* fractals in the space of acquaintance-description dualities.

KENTE FRACTALS **are** representations of mathematical shapes that are complex pattern with limiting process in an infinitely repetitive processes over the space of finite-infinite duality. Each fractal has a representation of socio-natural phenomenon for identity and distinction.

NYANSAPO is a wisdom knot representing the beginning-end duality, where the beginning dual is in the end dual and end dual is in the beginning dual in an interlocking process of interchange in an inseparable continuum and unity.

It also represents nothingness-somethingness duality, where the somethingness dual is in the nothingness dual and nothingness dual is in the somethingness dual in relational continuum and unity such that the space of nothingness is the collection of all varieties where the nothingness characteristics dominate the somethingness characteristics constituting the space of nothingness pole and the space of somethingness varieties is the collections of all varieties, where the somethingness characteristics dominate the nothingness characteristics to constitute the somethingness pole. The nothingness duality is housed in the nothingness pole while the somethingness duality is housed in the somethingness pole for all polarities. It is this nothingness-somethingness duality that provides us the reason to claim that the null set is in every set, group and category and provides us the conditions for the universal set $\left[0, \frac{m}{n}, \frac{n}{m}, \infty\right]$ where $m,n \in \left[0, \infty\right]$.

NYANSAPO PROBLEM is how to distinguish between the duals in dualities and poles in polarities and the conditions of their continuum and unity such that all diversities proceed from unity to dividedness and return to unified state through the diversity. This dualistic-polar thinking and reasoning provide us a way to understand matter-energy stock-flow equilibrium and information-knowledge stock-flow disequilibrium relative to the space of invisible-visible dualities, where the matter-energy stock-flow equilibrium is maintained within the nothingness-somethingness duality, and information-knowledge disequilibrium is maintained within the visible-invisible duality. Under the principle of opposite, the nyansapo duality forces us to confront the knowing complexities of dualistic-polar interdependencies and relationalities of all levels of knowing within the *ohnim-sua-ohu* (ignorance-knowledge) duality to journey from ignorance to knowledge under the application of Sankofa-anoma (entropic)principle of quality control over the space of epistemic rationality of transversality conditions between duals of dualities and poles of polarities.

OHNIM-SUA-OHU is the concept of ignorance-knowledge duality where ignorance (ohnim) is connected through research, teaching and learning through (*sua)* to knowledge (*ohu*) .

OHNIM-SUA-OHU PROBLEM is how to implement the conditions of research, teaching and learning to connect the *ohnim dual* to *ohu dual* in any variety duality of ignorance-knowledge state.

OHNIM **is** a state of ignorance.

ONIM *is a state of knowing.*

OHU is a state of knowledge.

PESEMEKONYA selfishness in each for himself as opposed to each-for-all-and-all-for-each (*dibi-mami-nibi*) in the space of individual-collective duality.

SANKOFA-ANOMA is a time trinity of the past, present and future regarding information trinity of what there was (past), what there is (present) and what there would be (future) needed as inputs into present neuro-decision-choice actions. The Sankofa-anoma time trinity splits into three dualities of *past-present duality*, *Present-future duality,* and *past-future duality*.

SANKOFA-ANOMA PROBLEM is how to bring the past information as an input by forecasting and future information as an input by discounting into present to create a present organic information input into the neuro-decision-choice actions for variety transformations.

SUA is the act of learning with or without supervision.

BLURB

The monograph is devoted to exploring the organizational dimensions of the general space of problem-solution dualities within epistemic fields to examine the diversity and unity of knowing and information-knowledge production in science and non-science and how to account for and explain the forces of variety transformations, as induced by decision-choice actions in creating the functor space and the space of categorial converters as well as to integrate decision-choice qualities of curiosity, creativity, hope, despair, success, failure and their relationships with epistemic anomalies in the production of intellectual investment flows and capital stocks, and the rise and fall of institutions in respond to the dynamics of the problem-solution dualities under the principle of non-satiation. The conditions and sources of uncertainties in human actions are traced though the space of ontic-epistemic dualities as generated by cognitive capacity limitations leading to the concepts of imagination, possibility, and probability with corresponding worlds and then link to the discussions on thinking-reasoning processes with goals objectives and visions of education in social set-ups. The concepts and phenomena of anticipation, necessity, expectation, freedom, anomaly, curiosity, creativity, hope, despair, success, and failure contained in the space of imaginations are defined under the concepts of onticity and epistemicity, where their roles in the dualistic-polar journeys are discussed, analyzed, and connected to epistemic transversality conditions under the principle of opposites relative to the conditions of input-output and cost-benefit dualities. The monograph is to fill in some of the important phenomena needed in my investigation of theory of knowing as a production of intellectual investment flows and

K. K. Dompere, *Theories of Knowing and Knowledge Production*, Studies in Systems, Decision and Control 501, https://doi.org/10.1007/978-3-031-42863-0

information-knowledge accumulation as the production of intellectual capital stocks relative to the disequilibrium dynamics of human intellectual heritage based on fuzzy paradigm of thought.

INTERDISCIPLINARY AND MULTIDISCIPLINARY REFERENCES

R 1. CATEGORY AND FUNCTOR THEORIES IN MATHEMATICS, LOGIC, AND SCIENCES

[R1.1] Arzi-Gonczaworski, Z., "Perceive This as That — Analogies, Artificial Perception, and Category Theory", *Annals of Mathematics and Artificial Intelligence*, 26 (1), 1999, pp. 215–252.

[R1.2] Awodey, S., "Structure in Mathematics and Logic: A Categorical Perspective," *Philosophia Mathematica,* Vol. 3. 1996, pp. 209-237.

[R1.3] Baez, J. & Dolan, J., "Categorification", *Higher Category Theory* (Contemporary Mathematics, Volume 230), Ezra Getzler and Mikhail Kapranov (eds.), Providence: AMS, 1998, pp.1–36.

[R1.4] Baez, J. & May, P. J., *Towards Higher Category Theory*, Berlin: 2010,Springer.

[R1.5] Bain, J., "Category-theoretic Structure and Radical Ontic Structural Realism", *Synthese*, Vol. 190, 2013, pp.1621–1635.

[R1.6] Bell, J. L., "Category Theory and the Foundations of Mathematics," *British Journal of Science,* Vol. 32, 1981, pp. 349-358.

[R1.7] Bell, J. L., "Categories, Toposes and Sets," *Synthese,* Vol. 51, 1982, pp. 393-337

[R1.8a] Bell, J. L, 1995, "Infinitesimals and the Continuum", *Mathematical Intelligencer*, Vol, 17 (2), pp. 55–57.

K. K. Dompere, *Theories of Knowing and Knowledge Production*, Studies in Systems, Decision and Control 501, https://doi.org/10.1007/978-3-031-42863-0

[R1.8b] Bergman, S. *The Kernel Function and Conformal Mapping.* New York: Amer. Math. Soc., 1950

[R1.9] Black, M., *The Nature of Mathematics,* Totowa, N.J., Littlefield, Adams, and Co., 1965.

[R1.10] Blass, A., "The Interaction between Category and Set Theory," *Mathematical Applications of Category Theory*, Vol. 30, 1984, pp. 5-29.

[R1.11] Brown, B., and J Woods (eds.), *Logical Consequence; Rival Approaches and New Studies in exact Philosophy: Logic, Mathematics and Science*, Vol. II Oxford, Hermes, 2000.

[R1.12] Buchsbaum, D.A., "Exact Categories and Duality", *Transactions of the American Mathematical Society*, 80 (1), 1955, pp. 1–34.

[R1.13a] Carathéodory, C. *Conformal Representation.* New York: Dover, 1998.

[R1.13b] Domany, J. L.et al., *Models of Neural Networks III,* New York, Springer, 1996

[R1.14] Feferman, S., "Categorical Foundations and Foundations of Category Theory," in R. Butts (ed.), *Logic, Foundations of Mathematics and Computability*, Boston, Mass., Reidel, 1977, pp.149-169.

[R1.15] Freyd, P., *Abelian Categories. An Introduction to the Theory of Functors*, New York: Harper & Row. 1964.

[R1.16] Freyd, P., "The Theories of Functors and Models". *Theories of Models*, Amsterdam: North Holland, 1965, pp.107–120.A

[R1.17] Freyd, P., Friedman, H. & Scedrov, A., "Lindembaum Algebras of Intuitionistic Theories and Free Categories", *Annals of Pure and Applied Logic*, Vol. 35 (2), 1987, 167–172.

[R1.18] Glimcher, P. W., *Decisions, Uncertainty, and the Brain: The Science of Neoroeconomics,* Cambridge, Mass., MIT Press, 2004

[R1.19] Gray, J.W. (ed.) *Mathematical Applications of Category Theory* (American Mathematical Society Meeting 89th Denver Colo. 1983)., Providence, R.I., American Mathematical Society, 1984.

[R1.20] Healy, M. J., & Caudell, T. P., "Ontologies and Worlds in Category Theory: Implications for Neural Systems",*Axiomathes*, Vol. 16 (1–2), 2006, pp.165–214.

[R1.21a] Hellman, G., "Does Category Theory Provide a Framework for Mathematical Structuralism?", *Philosophia Mathematica*, Vol. 11 (2), 2003, pp.129–157.

[R1.21b] Hellman, G, "A Small Complete Category", *Annals of Pure and Applied Logic*, Vol. 40 (2), 1988, pp. 135–165.

[R.1.22] Johansson, Ingvar, *Ontological Investigations: An Inquiry into the Categories of Nature, Man, and Society,* New York, Routledge, 1989.

[R1.23] Kamps, K. H., D. Pumplun, and W. Tholen (eds.) *Category Theory: Proceedings of the International Conference*, Gummersbach, July 6-10, New York, Springer, 1982.

[R1.24] Kan, D. M., "Adjoint Functors", *Transactions of the American Mathematical Society*,Vol. 87, 1958, pp. 294–329.

[R1.25] Kythe, P. K. Computational Conformal Mapping. Boston, MA: Birkhäuser, 1998.

[R1.26] Lal, R. and Teh, N., "Categorical Generalization and Physical Structuralism", *British Journal for the Philosophy of Science*, Vol. 68 (1), 2017: pp. 213–251.

[R1.27a] Lambek, J. and Scott, P.J., "Intuitionistic Type Theory and Foundations", *Journal of Philosophical Logic*, Vol.10 (1), 1981: pp. 101–115.

[R1.27b] Lambek, J. and Scott, P.J, "New Proofs of Some Intuitionistic Principles", *Zeitschrift für Mathematische Logik und Grundlagen der Mathematik*, 29 (6), 1983, pp.493–504.

[R1.28] Landry, E., "Category Theory: The Language of Mathematics", *Philosophy of Science*, Vol.66 (3) (Supplement) 1999, pp. S14–S27.

[R1.29] Lawvere, F. W., "Functorial Semantics of Algebraic Theories", *Proceedings of the National Academy of Sciences U.S.A.*, Vol 50, 1963, pp. 869–872.

[R1.30] Lawvere, F. W., "Foundations and Applications: Axiomatization and Education. New

Programs and Open Problems in the Foundation of Mathematics", *Bulletin of Symbolic Logic*, 9 (2), 2003: pp. 213–224.

[R1.31] Landry, E., Category Theory: The Language of Mathematics," *Philosophy of Science,* Vol. 66, (Supplement), pp. S14-S27.

[R1.32] Landry E. and J.P Marquis, "Categories in Context: Historical, Foundational and Philosophical," *Philiosophia Mathematica*, Vol. 13, 2005, pp. 1-43.

[R1.33] Linnebo, O. & Pettigrew, R., "Category Theory as an Autonomous Foundation," *Philosophia Mathematica*, Vol. 19 (3), 2011, pp. 227–254.

[R1.34] Mac Lane, S., "Dualities for Groups", *Bulletin of the American Mathematical Society*, Vol. 56, 1950, pp, 485–516.

[R1.35] Makkai, M. & Paré, R., *Accessible Categories: The Foundations of Categorical Model Theory*, Contemporary Mathematics Vol.104, Providence: AMS1989, .

[R1.36] Makkai, M. & Reyes, G., *First-Order Categorical Logic*, Springer Lecture Notes in Mathematics Vol. 611, New York: Springer,1977.

[R1.37] Makkai, M. & Reyes, G, "Completeness Results for Intuitionistic and Modal Logic in a Categorical Setting", *Annals of Pure and Applied Logic*, Vol. 72 (1), 1995, pp. 25–101.

[R1.38] Marquis, J. –P., "Three Kinds of Universals in Mathematics," in B. Brown, and J. Woods (eds.), *Logical Consequence; Rival Approaches and New Studies in exact Philosophy: Logic, Mathematics and Science*, Vol. II Oxford, Hermes, 2000, pp. 191-212.

[R1.39] McLarty, C., "Category Theory in Real Time," *Philosophia Mathematica,* Vol. 2, 1994, pp. 36-44.

[R1.40] McLarty, C., "Learning from Questions on Categorical Foundations," *Philosophia Mathematica,* Vol.13, 2005, pp. 44-60.

[R1.41]Moon, P. and Spencer, D. E. "Conformal Transformations." in *Field Theory Handbook, Including Coordinate Systems, Differential Equations, and Their Solutions, 2nd ed.* New York: Springer-Verlag, pp. 49-76, 1988.

[R1.42] Nehari, Z., Conformal Mapping. New York: Dover, 1982.

[R1.43] Pareigis, B., *Categories and Functors*, New York: Academic Press, 1970 .

[R1.44] Peruzzi, A., "The Meaning of Category Theory for 21st Century Philosophy", *Axiomathes*, Vol. 16 (4), 2006 pp. 424–459.

[R1.45] Peruzzi, A., "Category Theory and the Search for Universals: A Very Short Guide for Philosophers", in *Modern Logic 1850–1950, East and West*, F. Abeles & M. Muller (eds.), Basel: Birkhauser, 2016, pp. 159–182.

[R1.46] Rodabaugh, S. et. al., (eds.), *Application of Category Theory to Fuzzy Subsets*, Boston, Mass., Kluwer, 1992.

[R1.47] Rodin, A., *Axiomatic Method and Category Theory* (Synthese Library: Volume 364), New York: Springer, 2014.

[R1.48] Spivak, D., *Category Theory for the Sciences*, Cambridge: MIT Press, 2014.

[R1.49] Taylor, P., "Intuitionistic sets and Ordinals", *Journal of Symbolic Logic*, Vol. 61 , 1996, pp.705–744.

[R1.50] **Ross**, Don, *Economic theory and Cognitive Science; Microexplanation*, Cambridge, Mass., MIT Press, 2005.

[R1.51] Sieradski, Allan, J., An Introduction to Topology and Homotopy, PWS-KENT Pub. Boston,1992.

[R1.52] Spivak, D., *Category Theory for the Sciences*, Cambridge: MIT Press, 2014.

[R1.53] Taylor, J.G. (ed.), *Mathematical Approaches to Neural Networks*, New York North-Holland, 1993.

[R1.54] Taylor, P., "Intuitionistic sets and Ordinals", *Journal of Symbolic Logic*,Vol. 61, 1996, pp. 705–744.

[R1.55] Van Benthem, J. et al.(eds.), *The Age of Alternative Logics: Assessing Philosophy of Logic and Mathematics Today,* New York, Springer, 2006.

R2. CONCEPTS OF INFORMATION, FUZZY PROBABILITY, FUZZY RANDOM VARIABLE AND RANDOM FUZZY VARIABLE

[R2.1] Bandemer, H., "From Fuzzy Data to Functional Relations," *Mathematical Modelling*, Vol. 6, 1987, pp. 419-426.

[R2.2] Bandemer, H. et. al., *Fuzzy Data Analysis*, Boston, Mass, Kluwer, 1992.

[R2.3] Kruse, R. et. al., *Statistics with Vague Data,* Dordrecht, D. Reidel Pub. Co., 1987.

[R2.4] El Rayes, A.B. et. al., "Generalized Possibility Measures," *Information Sciences,* Vol. 79, 1994, pp. 201-222.

[R2.5] Dumitrescu, D., "Entropy of a Fuzzy Process," *Fuzzy Sets and Systems*, Vol. 55, #2, 1993, pp. 169-177.

[R2.6] Delgado, M. et. al., "On the Concept of Possibility-Probability Consistency," *Fuzzy Sets and Systems*, Vol. 21, #3, 1987, pp. 311-318.

[R2.7] Devi, B.B. et. al., "Estimation of Fuzzy Memberships from Histograms," *Information Sciences*, Vol. 35, #1, 1985, pp. 43-59.

[R2.8] Dubois, D. et. al., "Fuzzy Sets, Probability and Measurement," *European Jour. of Operational Research*, Vol. 40, #2, 1989, pp. 135-154.

[R2.9] Fruhwirth-Schnatter, S., "On Statistical Inference for Fuzzy Data with Applications to Descriptive Statistics," *Fuzzy Sets and Systems*, Vol. 50, #2, 1992, pp. 143-165.

[R2.10] Gaines, B.R., "Fuzzy and Probability Uncertainty logics," *Information and Control*, Vol. 38, #2, 1978, pp. 154-169.

[R2.11] Geer, J.F. et. al., "Discord in Possibility Theory," *International Jour. of General Systems*, Vol. 19, 1991, pp. 119-132.

[R2.12] Geer, J.F. et. al., "A Mathematical Analysis of Information-Processing Transformation Between Probabilistic and Possibilistic Formulation of Uncertainty," *International Jour. of General Systems*, Vol. 20, #2, 1992, pp. 14-176.

[R2.13] Goodman, I.R. ct. al., *Uncertainty Models for Knowledge Based Systems*, New York, North-Holland, 1985.

[R2.14] Grabish, M. et. al., *Fundamentals of Uncertainty Calculi with Application to Fuzzy Systems*, Boston, Mass., Kluwer, 1994.

[R2.15] Guan, J.W. et. al., *Evidence Theory and Its Applications*, Vol. 1, New York, North-Holland, 1991.

[R2.16] Guan, J.W. et. al., *Evidence Theory and Its Applications*, Vol. 2, New York, North-Holland, 1992.

[R2.17] Hisdal, E., Are Grades of Membership Probabilities?" *Fuzzy Sets and Systems*, Vol. 25, #3, 1988, pp. 349-356.

[R2.18] Höhle Ulrich , "**A** Mathematical Theory of Uncertainty," in R.R. Yager (ed.) *Fuzzy Set and Possibility Theory: Recent Developments*, New York, Pergamon, 1982, pp. 344 – 355.

[R2.19] Kacprzyk , Janusz and Mario Fedrizzi (eds.) *Combining Fuzzy Imprecision with Probabilistic Uncertainty in Decision Making*, New York, Plenum Press, 1992.

[R2.20] Kacprzyk, J. et. al., *Combining Fuzzy Imprecision with Probabilistic Uncertainty in Decision Making*, New York, Springer-Verlag, 1988.

[R2.21] Klir, G.J., "Where Do We Stand on Measures of Uncertainty, Ambignity, Fuzziness and the like?" *Fuzzy Sets and Systems*, Vol. 24, #2, 1987, pp. 141-160.

[R2.22] Klir, G.J. et. al., *Fuzzy Sets, Uncertainty and Information*, Englewood Cliff, Prentice Hall, 1988.

[R2.23] Klir, G. J. et. al., "Probability-Possibility Transformations: A Comparison," *Intern. Jour. of General Systems*, Vol. 21, #3, 1992, pp. 291-310.

[R2.24] Kosko, B., "Fuzziness vs Probability," *Intern. Jour. of General Systems*, Vol. 17, # (1-3) 1990, pp. 211-240.

[R2.25] Manton, K.G. et. al., *Statistical Applications Using Fuzzy Sets*, New York, John Wiley, 1994.

[R2.26] Meier, W., et. al., "Fuzzy Data Analysis: Methods and Industrial Applications," *Fuzzy Sets and Systems*, Vol. 61, #1, 1994, pp. 19-28.

[R2.27] Nakamura, A., et. al., "A logic for Fuzzy Data Analysis," *Fuzzy Sets and Systems,* vol. 39, #2, 1991, pp. 127-132.

[R2.28] Negoita, C.V. et. al., *Simulation, Knowledge-Based Computing and Fuzzy Statistics*, New York, Van Nostrand Reinhold, 1987.

[R2.29] Nguyen, H.T., "Random Sets and Belief Functions," *Jour. of Math. Analysis and Applications*, Vol. 65, #3, 1978, pp. 531-542.

[R2.30] Prade, H. et. al., "Representation and Combination of Uncertainty with belief Functions and Possibility Measures," *Comput. Intell.* ,Vol. 4, 1988, pp. 244-264.

[R2.31] Puri, M.L. et. al., "Fuzzy Random Variables," *Jour. of Mathematical Analysis and Applications*, Vol. 114, #2, 1986, pp. 409-422.

[R2.32] Rao, N.B. and A. Rashed, "Some Comments on Fuzzy Random Variables," *Fuzzy Sets and Systems*, Vol. 6, # 3, 1981, pp.285-292.

[R2.33] Sakawa, M. et. al., "Multiobjective Fuzzy linear Regression Analysis for Fuzzy Input-Output Data," *Fuzzy Sets and Systems*, Vol. 47, #2, 1992, pp. 173-182.

[R2.34] Schneider, M. et. al., "Properties of the Fuzzy Expected Values and the Fuzzy Expected Interval," *Fuzzy Sets and Systems*, Vol. 26, #3, 1988, pp. 373-385.

[R2.35] Stein, N.E. and K Talaki, "Convex Fuzzy Random Variables," *Fuzzy Sets and Systems*, Vol. 6, #3, 1981, pp. 271-284.

[R2.36] Sudkamp, T., "On Probability-Possibility Transformations," *Fuzzy Sets and Systems*, Vol. 51, #1, 1992, pp. 73-82.

[R2.37] Walley, P., *Statistical Reasoning with Imprecise Probabilities*, London Chapman and Hall, 1991.

[R2.38] Wang, G.Y. et. al., "The Theory of Fuzzy Stochastic Processes," *Fuzzy Sets and Systems*, Vol. 51, #2 1992, pp. 161-178.

[R2.39] Zadeh, L. A., "Probability Measure of Fuzzy Event," *Jour. of Math Analysis and Applications*, Vol. 23, 1968, pp. 421 – 427.

R3. EXACT SCIENCE, INEXACT SCIENCES, INFORMATION AND KNOWLEDGE

[R3.1] Achinstein, P., "The Problem of Theoretical Terms," in Brody, Baruch A. (Ed.) *Reading in the Philosophy of Science*, Englewood Cliffs, NJ. Prentice Hall, 1970.

[R3.2] Amo Afer, A. G., *The Absence of Sensation and the Faculty of Sense in the Human Mind and Their Presence in our Organic and Living Body, Dissertation and Other essays 1727-1749,* Halle Wittenberg, Jena, Martin Luther University Translation, 1968.

[R3.3] Beeson, M. J., *Foundations of Constructive Mathematics*, Berlin/New York, Springer,1985.

[R3.4] Benacerraf, P., "God, the Devil and Gödel," *Monist,* Vol. 51, 1967, pp.9-32.

[R3.5] Benecerraf, P and H. Putnam (eds.), *Philosophy of Mathematics: Selected Readings* , Cambridge, Cambridge University Press, 1983.

[R3.6] Black, Max, *The Nature of Mathematics,* Totowa, Littlefield, Adams and Co. 1965.

[R3.7] Blanche, R., *Contemporary Science and Rationalism*, Edinburgh, Oliver and Boyd, 1968.

[R3.8] Blanshard, Brand, *The Nature of Thought*, London Allen and Unwin, 1939.

[R3.9] Blauberg, I. V., V.N. Sadovsky and E.G. Yudin, Systems Theory: Philosophical and Methodological Problems, Moscow, Progress Publishers, 1977.

[R3.10] Braithwaite, R. B., *Scientific Explanation*, Cambridge, Cambridge University Press. 1955.

[R3.11] Brewer, M.B., and B. E. Collins (eds.) Scientific Inquiry and the Social Sciences, Washington, Jossey-Bass Pub, 1981

[R3.12a] Brody, Baruch A. (ed.), *Reading in the Philosophy of Science*, Englewood Cliffs, N.J., Prentice Hall, 1970.

[R3.12b] Brody, Baruch A., "Confirmation and Explanation," in Brody, Baruch A. (ed.) *Reading in the Philosophy of Science*, Englewood Cliffs, N.J., Prentice-Hall,1970, pp. 410-426.

[R3.13] Brouwer, L.E.J., "Intuitionism and Formalism", *Bull of American Math. Soc.*, Vol. 20, 1913, pp81-96.; Also, in Benecerraf, P. and H. Putnam (eds.), *Philosophy of Mathematics: Selected Readings* , Cambridge, Cambridge University Press, 1983. pp. 77-89.

[R3.14] Brouwer, L.E.J., "Consciousness, Philosophy, and Mathematics," in Benecerraf, P. and H. Putnam (eds.), *Philosophy of Mathematics: Selected*

Readings , Cambridge, Cambridge University Press, 1983, pp.90-96.

[R3.15] Brouwer, L. E. J., *Collected Works, Vol. 1: Philosophy and Foundations of Mathematics* [A Heyting (ed.)], New York, Elsevier, 1975.

[R3.16] Campbell, Norman R., *What is Science?* New York, Dover, 1952.

[R3.17] Carnap, R., "Foundations of Logic and Mathematics," in *International Encyclopedia of Unified Science*, Chicago, Univ. of Chicago,1939, pp.143-211.

[R3.18] Carnap, Rudolf, "On Inductive Logic," *Philosophy of Science*, Vol. 12, 1945, pp. 72-97.

[R3.19] Carnap, Rudolf, "The Methodological Character of Theoretical Concepts," in Herbert Feigl and M. Scriven (eds.) *Minnesota Studies in the Philosophy of Science, Vol. I*, 1956, pp. 38-76.

[R3.20] Charles, David and Kathleen Lennon (eds.), *Reduction, Explanation, and Realism*, Oxford, Oxford Unive3rsity Press, 1992.

[R3.21] Cohen, Robert S. and Marx W. Wartofsky (eds.), *Methodological and Historical Essays in the Natural and Social Sciences*, Dordrecht, D. Reidel Publishing Co. 1974.

[R3.22] Dalen van, D. (ed.), *Brouwer's Cambridge Lectures on Intuitionism,* Cambridge, Cambridge University Press, 1981.

[R3.23] Davidson, Donald, *Truth and Meaning: Inquiries into Truth and Interpretation*, Oxford, Oxford University Press, 1984.

[R3.24] Davis, M., *Computability and Unsolvability*, New York, McGraw-Hill, 1958.

[R3.25] Denonn Lester E. (ed.), *The Wit and Wisdom of Bertrand Russell*, Boston, MA., The Beacon Press, 1951.

[R3.26] Dompere, Kofi K., *Fuzziness and Foundations of Exact and Inexact Sciences*, New York, Springer, 2013

[R3.27] Dummett, M., "The Philosophical Basis of Intuitionistic Logic," in Benecerraf, P. and H. Putnam (eds.), *Philosophy of Mathematics: Selected Readings* , Cambridge, Cambridge University Press, 1983, pp.97-129

[R3.28] Feigl, Herbert and M. Scriven (eds.), *Minnesota Studies in the Philosophy of Science,* Vol. I, 1956.

[R3.29] Garfinkel, Alan, *Forms of Explanation: Structures of Inquiry in Social Science,* New Haven, Conn., Yale University Press, 1981.

[R3.30] George, F. H., *Philosophical Foundations of Cybernetics*, Tunbridge Well, Great Britain, 1979.

[R3.31] Gillam, B., "Geometrical Illusions," *Scientific American*, January, 1980, pp.102-111.

[R3.32] Gödel , Kurt., "What is Cantor's Continuum Problem?" in Benecerraf, P. and H. Putnam (eds.), *Philosophy of Mathematics: Selected Readings*, Cambridge, Cambridge University Press, 1983, pp.470-486.

[R3.33] Gorsky, D.R., *Definition,* Moscow, Progress Publishers, 1974.

[R3.34] Gray, William and Nicholas D. Rizzo(eds.), *Unity Through Diversity*. New York, Gordon and Breach, 1973.

[R3.35] Hart, W. D. (ed.), *The Philosophy of Mathematics*, Oxford, Oxford University Press, 1996.

[R3.36] Hartkamper, a and H.- Schmidt, Structure and Approximation in Physical Theories, New York, Plenum Press, 1981.

[R3.37] Hausman, David, M., *The Exact and Separate Science of Economics,* Cambridge, Cambridge University Press, 1992.

[R3.38] Helmer, Olaf and Nicholar Rescher, *On the Epistemology of the Inexact Sciences*, P-1513, Santa Monica, CA, Rand Corporation, October 13, 1958.

[R3.39] Hempel, C. G., "Studies in the Logic of Confirmation," *Mind,* Vol.54, Part I , 1945, pp. 1-26.

[R3.40] Hempel, Carl G., "The Theoretician's Dilemma," in Herbert Feigl and M. Scriven (eds.) *Minnesota Studies in the Philosophy of Science*, Vol.II, 1958, pp. 37-98.

[R3.41] Hempel, C. G. and P. Oppenheim, "Studies in the Logic of Explanation," *Philosophy of Science*, Vol. 15, 1948, pp. 135-175. [also, in Brody, Baruch A. (ed.) *Reading in the Philosophy of Science*, Englewood Cliffs, NJ., Prentice-Hall,1970, pp. 8-27.

[R3. 42] Heyting, A., *Intuitionism: An Introduction,* Amsterdam: North-Holland, 1971.

[R3.43] Hintikka, Jackko (ed.), *The Philosophy of Mathematics*, London, Oxford University Press, 1969.

[R3.44] Hockney D. et al. (eds.), *Contemporary Research in Philosophical Logic and Linguistic Semantics*, Dordrecht-Holland, Reidel Pub., Co. 1975.

[R3.45] Hoyninggen-Huene, Paul and F. M. Wuketits, (eds.), *Reductionism and Systems Theory in the Life Science: Some Problems and Perspectives*, Dordrencht, Kluwer Academic Pub. 1989.

[R3.46] Ilyenkov, E.V., *Dialectical Logic: Essays on Its History and Theory*, Moscow, Progress Publishers, 1977.

[R3.47] Kedrov, B. M., "Toward the Methodological Analysis of Scientific Discovery," *Soviet Studies in Philosophy*, Vol. 11962, pp. 45 – 65.

[R3.48] Kemeny, John G, and P Oppenheim , "On Reduction," in Brody, Baruch A. (ed.) *Reading in the Philosophy of Science*, Englewood Cliffs, NJ., Prentice-Hall,1970, pp. 307-318.

[R3.49] Klappholz, K., "Value Judgments of Economics," *British Jour. of Philosophy*, Vol. 15, 1964, pp. 97-114.

[R 3.50] Kleene, S.C.,"On the Interpretation of Intuitionistic Number Theory," Journal of Symbolic Logic, Vol 10, 1945, pp. 109-124.

[R3.51] Kmita, Jerzy, "The Methodology of Science as a Theoretical Discipline," *Soviet Studies in Philosophy*, Spring, 1974, pp. 38 –49

[R3.52] Krupp, Sherman R.,(ed.), *The Structure of Economic Science*, Englewood Cliff, N. J., Prentice-Hall, 1966.

[R3.53] Kuhn, T., *The Structure of Scientific Revolution,* Chicago, University of Chicago Press, 1970.

[R3.54] Kuhn, Thomas, "The Function of Dogma in Scientific Research," in Brody, Baruch A. (ed.) *Reading in the Philosophy of Science*, Englewood Cliffs, NJ., Prentice-Hall,1970 pp. 356-374.

[R3.55] Kuhn, Thomas, *The Essential Tension: Selected Studies in Scientific Tradition and Change,* Chicago, University of Chicago Press, 1979.

[R3.56] Lakatos, I. (ed.), *The Problem of Inductive Logic,* Amsterdam, North Holland,1968.

[R3.57] Lakatos, I., *Proofs and Refutations: The Logic of Mathematical Discovery*, Cambridge, Cambridge University Press, 1976.

[R3.58] Lakatos, I., *Mathematics, Science and Epistemology: Philosophical Papers Vol. 2*, edited by J. Worrall and G. Currie, Cambridge, Cambridge Univ. Press, 1978.

[R3.59] Lakatos, I., *The Methodology of Scientific Research Programmes*, Vol 1, New York, Cambridge University Press, 1978.

[R3.60] Lakatos, Imre and A. Musgrave (eds.), *Criticism and the Growth of Knowledge*, New York, Cambridge University Press, 1979. Holland, 1979, pp.

153 – 164.

[R3.61] Lawson, Tony, *Economics and Reality*, New York, Routledge, 1977.

[R3.62] Lenzen, Victor, "Procedures of Empirical Science," in Neurath, Otto et al. (eds.), *International Encyclopedia of Unified Science, Vol. 1 – 10,* Chicago, University of Chicago Press, 1955, pp. 280-338.

[R3.63] Levi, Isaac, "Must the Scientist make Value Judgments?" in Brody, Baruch A. (Ed.) *Reading in the Philosophy of Science*, Englewood Cliffs, NJ., Prentice-Hall,1970, pp.559-570.

[R3.64] Tse-tung, *Mao, On Practice and Contradiction*, in *Selected works of Mao Tse-tung*, Piking, 1937. Also, London, Revolutions, 2008.

[R3.65] Lewis, David, *Convention: A Philosophical Study,* Cambridge, Mass., Harvard University Press, 1969.

[R3.66] Mayer, Thomas, *Truth versus Precision in Economics*, London, Edward Elgar 1993.

[R3.67] Menger, Carl, *Investigations into the Method of the Social Sciences with Special Reference to Economics*, New York, New York University Press, 1985.

[R3.68] Mirowski, Philip (ed.), *The Reconstruction of Economic Theory*, Boston, Mass. Kluwer Nijhoff, 1986.

[R3.69] Mueller, Ian, *Philosophy of Mathematics and Deductive Structure in Euclid's Elements*, Cambridge, Mass., MIT Press, 1981.

[R3.70] Nagel, Ernest, "Review: Karl Niebyl, Modern Mathematics and Some Problems of Quantity, Quality, and Motion in Economic Analysis," *The Journal of Symbolic Logic* , 1940, p.74.

[R3.71] Nagel, E. et al. (ed.), *Logic, Methodology, and the Philosophy of Science*, Stanford, Stanford University Press 1962.

[R3.72] Narens, Louis, "A Theory of Belief for Scientific Refutations," *Synthese,* Vol.145, 2005, pp. 397-423.

[R3.73] Narskii, I. S., "On the Problem of Contradiction in Dialectical Logic," *Soviet Studies in Philosophy,* Vol. vi, #4 pp.3-10, 1965

[R3.74] Neurath, Otto et al. (eds.), *International Encyclopedia of Unified Science, Vol. 1 – 10,* Chicago, University of Chicago Press, 1955.

[R3.75] Neurath Otto, "Unified Science as Encyclopedic," in Neurath, Otto et al. (eds.), *International Encyclopedia of Unified Science, Vol. 1 – 10,* Chicago, University of Chicago Press, 1955, pp.1-27.

[R3.76] Planck, Max, *Scientific Autobiography and Other Papers*, Westport, Conn. Greenwood 1971.

[R3.77] Planck, Max, "The Meaning and Limits of Exact Science," in Max Planck, *Scientific Autobiography and Other Papers,* Westport, Conn. Greenwood, 1971, pp. 80 -120.

[R3.78] Polanyi, Michael, "Genius in Science," in Robert S. Cohen, and Marx W. Wartofsky (eds.), *Methodological and Historical Essays in the Natural and Social Sciences*, Dordrecht, D. Reidel Publishing Co. 1974, pp.57-71.

[R3.79] Popper, Karl, *The Nature of Scientific Discovery*, New York, Harper and Row, 1968.

[R3.80] Putnam, Hilary., "Models and Reality," in Benecerraf, P. and H. Putnam (eds.), *Philosophy of Mathematics: Selected Readings* , Cambridge, Cambridge University Press, 1983. pp. 421- 444.

[R3.81] Reise, S., *The Universe of Meaning*, New York, The Philosophical Library, 1953.

[R3.82] Robinson, R., *Definition*, Oxford, clarendon Press, 1950.

[R3.83] Rudner, Richard, "The Scientist qua Scientist Makes Value Judgments," *Philosophy of Science*, Vol. 20, 1953, pp 1-6.

[R3.84] Russell, B., *Our Knowledge of the External World*, New York, Norton, 1929.

[R3.85] Russell, B., *Human Knowledge, Its Scope and Limits*, London, Allen and Unwin, 1948.

[R3.86] Russell, B., *Logic and Knowledge: Essays 1901-1950,* New York, Capricorn Books, 1971.

[R3.87] Russell, B., *An Inquiry into Meaning and Truth*, New York, Norton, 1940.

[R3.88] Russell, Bertrand, *Introduction to Mathematical Philosophy*, London, George Allen and Unwin, 1919.

[R3.89] Russell, Bertrand, *The Problems of Philosophy*, Oxford, Oxford University Press, 1978.

[R3.90] Rutkevih, M. N., "Evolution, Progress, and the Law of Dialectics," *Soviet Studies in Philosophy,* Vol. IV, #3, PP. 34-43, 1965.

[R3.91] Ruzavin, G. I., "On the Problem of the Interrelations of Modern Formal Logic and Mathematical Logic," *Soviet Studies in Philosophy*, Vol. 3, #1, 1964, pp.34- 44.

[R3.92] Scriven, Michael, "Explanations, Predictions, and Laws," in Brody, Baruch A. (ed.) *Reading in the Philosophy of Science, Englewood Cliffs*, NJ., Prentice- Hall, 1970, pp.88-104.

[R3.93] Sellars, Wilfrid, The Language of Theories," in Brody, Baruch A. (ed.) *Reading in the Philosophy of Science,* Englewood Cliffs, NJ., Prentice-Hall,1970 pp.343-353.

[R3.94a] Sterman, John, "The Growth of Knowledge: Testing a Theory of Scientific Revolutions with a Formal Model," *Technological Forecasting and Social Change*, Vol. 28, 1995, pp. 93-122.

[R3.94b] Tsereteli, S. B. "On the Concept of Dialectical Logic,", *Soviet Studies in Philosophy*, Vol. V, #2, 1966, pp. 15-21.

[R3.95] Tullock, Gordon, *The Organization of Inquiry,* Indianapolis, Indiana, Liberty Fund Inc. 1966.

[R3.96] Van Fraassen, B., *Introduction to Philosophy of Space and Time*, New York, Random House, 1970.

[R3.97] Veldman, W., "A Survey of Intuitionistic Descriptive Set Theory," in P.P. Petkov (ed.), Mathematical Logic: Proceedings of the Heyting Conference, New York, Plenum Press, 1990, pp. 155-174.

[R3.98] Vetrov, A. A., "Mathematical Logic and Modern Formal Logic," *Soviet Studies in Philosophy*, Vol. 3, #1, 1964 pp. 24 – 33.

[R3.99] von Mises, Ludwig, *Epistemological Problems in Economics*, New York, New York University Press, 1981.

[R3.100] Wang, Hao, *Reflections on Kurt* Gödel, Cambridge, Mass. MIT Press,1987

[R3.101] Watkins, J. W. N., "The Paradoxes of Confirmation"," in Brody, Baruch A. (ed.) *Reading in the Philosophy of Science*, Englewood Cliffs, NJ., Prentice-Hall, 1970, pp. 433-438.

[R3.102] Whitehead, Alfred North, *Process and Reality*, New York, The Free Press,

1978.
[R3.103] Wittgenstein, Ludwig, *Ttactatus Logico-philosophicus,* Atlantic Highlands, N.J., The Humanities Press Inc.1974.

[R3.104] Woodger, J. H., *The Axiomatic Method in Biology*, Cambridge, Cambridge University Press, 1937.
[R3.105] Zeman, Jiři', "Information, Knowledge and Time," in Libor Kubát, and J. Zeman (eds.), *Entropy and Information in the Physical Sciences*, Amsterdam, Elsevier, 1975, pp. 245-260.

R4. FUZZY LOGIC, INFORMATION AND KNOWLEDGE-PRODUCTION

[R4.1] Baldwin, J.F., "A New Approach to Approximate Reasoning Using a Fuzzy Logic," *Fuzzy Sets and Systems*, Vol. 2, #4, 1979, pp. 309-325.

[R4.2] Baldwin, J.F., "Fuzzy Logic and Fuzzy Reasoning," *Intern. J. Man-Machine Stud.*, Vol. 11, 1979, pp. 465-480.

[R4.3] Baldwin, J.F., "Fuzzy Logic and Its Application to Fuzzy Reasoning," in. M. M. Gupta et al. (eds.), *Advances in Fuzzy Set Theory and Applications*, New York, North-Holland, 1979, pp.96-115.

[R4.4] Baldwin, J.F. et al., "Fuzzy Relational Inference Language," *Fuzzy Sets and Systems*, Vol. 14, #2, 1984, pp. 155-174.

[R4.5] Baldsin, J. and B.W. Pilsworth, "Axiomatic Approach to Implication For Approximate Reasoning With Fuzzy Logic," *Fuzzy Sets and Systems*, Vol. 3, #2, 1980, pp. 193-219.

[R4.6] Baldwin, J.F. et. al., "The Resolution of Two Paradoxes by Approximate Reasoning Using A Fuzzy Logic," *Synthese*, Vol. 44, 1980, pp. 397-420.

[R4.7] Dompere, K. K., *Fuzzy Rationality: Methodological Critique and Unity of Classical, Bounded and Other Rationalities*, (Studies in Fuzziness and Soft Computing, vol.235) New York, Springer, 2009.

[R4.8] Dompere Kofi K., *Epistemic Foundations of Fuzziness*, (Studies in Fuzziness and Soft Computing, vol. 236), New York, Springer, 2009.

[R4.9] Dompere Kofi K., *Fuzziness and Approximate Reasoning: Epistemics on Uncertainty, Expectation and Risk in Rational Behavior*, (Studies in Fuzziness and Soft Computing, vol. 237), New York, Springer, 2009.

[R4.10] Dompere, Kofi K., *The Theory of the Knowledge Square: The Fuzzy Rational Foundations of Knowledge-Production Systems,* New York, Springer, 2013.

[R4.11] Dompere, Kofi K., "Cost-Benefit Analysis, Benefit Accounting and Fuzzy Decisions: Part I, Theory", *Fuzzy Sets and Systems*, Vol. 92, 1997, pp. 275-287.

[R4.12] Dompere, Kofi K., "The Theory of Social Cost and Costing for Cost-Benefit Analysis in a Fuzzy Decision Space", *Fuzzy Sets and Systems*. Vol. 76, 1995, pp. 1-24.

[R4.13] Dompere, Kofi K., *Fuzzy Rational Foundations of Exact and Inexact Sciences,* New York, Springer, 2013.

[R4.14] Gaines, B.R., "Foundations of Fuzzy Reasoning," *Inter. Jour. of Man-Machine Studies*, Vol. 8, 1976, pp. 623-668.

[R4.15] Gaines, B.R., "Foundations of Fuzzy Reasoning," in Gupta, M.M. et. al. (eds.), *Fuzzy Information and Decision Processes*, New York North-Holland, 1982, pp. 19-75.

[R4.16] Gaines, B. R., "Precise Past, Fuzzy Future," *International Journal. Of Man-Machine Studies.*, Vol. 19, #1,1983, pp.117-134.

[R4.17] Giles, R., "Lukasiewics Logic and Fuzzy Set Theory," *Intern. J. Man-Machine Stud.*, Vol. 8, 1976, pp. 313-327.

[R4.18] Giles, R., "Formal System for Fuzzy Reasoning," *Fuzzy Sets and Systems*, Vol. 2, #3, 1979, pp. 233-257.

[R4.19] Ginsberg, M. L.(ed.), *Readings in Non-monotonic Reason*, Los Altos, Ca., Morgan Kaufman, 1987.

[R4.20] Goguen, J.A., "The Logic of Inexact Concepts," *Synthese*, Vol. 19, 1969, pp. 325-373.

[R4.21] Gottinger, H.W., "Towards a Fuzzy Reasoning in the Behavioral Science," *Cybernetica*, Vol. 16, #2, 1973, pp. 113-135.

[R4.22] Gupta, M.M. et. al., (eds.), *Approximate Reasoning In Decision Analysis,* North Holland, New York, 1982.

[R4.23] Höhle Ulrich . and E.P. Klement , *Non-Clasical Logics and their Applications to Fuzzy Subsets: A Handbook of the Mathematical Foundations of Fuzzy Set Theory*, Boston, Mass. Kluwer, 1995.

[R4.24] Kaipov, V., Kh. et. al., "Classification in Fuzzy Environments," in M. M. Gupta et al. (eds.), *Advances in Fuzzy Set Theory and Applications,* New York, North-Holland, 1979, pp. 119-124,

[R4.25] Kaufman, A., "Progress in Modeling of Human Reasoning of Fuzzy Logic" in M. M. Gupta et al. (eds.), *Fuzzy Information and Decision Process*, New York, North-Holland, 1982, pp. 11-17.

[R4.26] Lakoff, G., "Hedges: A Study in Meaning Criteria and the Logic of Fuzzy Concepts," *Jour. Philos. Logic,* Vol. 2, 1973, pp. 458-508.

[R4.27] Lee, R.C.T., "Fuzzy Logic and the Resolution Principle," *Jour. of Assoc. Comput.* Mach., Vol. 19, 1972, pp. 109-119.

[R4.28] LeFaivre, R.A., "The Representation of Fuzzy Knowledge", *Jour. of Cybernetics*, Vol. 4, 1974, pp. 57-66.

[R4.29] Negoita, C.V., "Representation Theorems for Fuzzy Concepts,"

Kybernetes, Vol. 4, 1975, pp. 169-174.

[R4.30] Nowakowska, M., "Methodological Problems of Measurements of Fuzzy Concepts in Social Sciences", *Behavioral Sciences*, Vol. 22, #2, 1977, pp. 107-115.

[R4.31] Skala, H.J., "On Many-Valued Logics, Fuzzy Sets, Fuzzy Logics and Their Applications," *Fuzzy Sets and Systems*, Vol. 1, #2, 1978, pp. 129-149.

[R4.32] Van Fraassen, B.C., "Comments: Lakoff's Fuzzy Propositional Logic," in D. Hockney et. al., (Eds.), *Contemporary Research in Philosophical Logic and Linguistic Semantics* Holland, Reild, 1975, pp. 273-277.

[R4.33] Yager, R.R. et. al. (eds.), An Introduction to Fuzzy Logic Applications in

Intelligent Systems, Boston, Mass., Kluwer 1992.

[R4.34] Zadeh, L.A., "Quantitative Fuzzy Semantics," *Inform. Science*, Vol. 3, 1971, pp. 159-176.

[R4.35] Zadeh, L.A., "A Fuzzy Set Interpretation of Linguistic Hedges," *Jour. Cybernetics*, Vol. 2, 1972, pp. 4-34.

[R4.36] Zadeh, L.A, "The Concept of a Linguistic Variable and Its Application to Approximate Reasoning," in K.S. Fu et. al. (eds.), *Learning Systems and Intelligent Robots*, Plenum Press, New York, 1974, pp. 1-10.

[R4.37] Zadeh, L.A., et. al., (eds.), *Fuzzy Sets and Their Applications to Cognitive and Decision Processes*, New York, Academic Press, 1974.

[R4.38] Zadeh, L.A., "The Birth and Evolution of Fuzzy Logic," *Intern. Jour. of General Systems*, Vol. 17, #(2-3) 1990, pp. 95-105.

R5. FUZZY MATHEMATICS AND PARADIGM OF APPROXIMATE REASONING UNDER CONDITIONS OF INEXACTNESS AND VAGUENESS

[R5.1] Bellman, R.E., "Mathematics and Human Sciences," in J. Wilkinson et. al. (eds.), *The Dynamic Programming of Human Systems*, New York, MSS Information Corp., 1973, pp. 11-18.

[R5.2] Bellman, R.E and Glertz, M., "On the Analytic Formalism of the Theory of Fuzzy Sets," Information *Science*, Vol. 5, 1973, pp. 149-156.

[R5.3] Butnariu, D., Fixed Points for Fuzzy Mapping," *Fuzzy Sets and Systems*, Vol. 7, #2, pp. 191-207, 1982.

[R5.4] Butnariu, D., "Decompositions and Range for Additive Fuzzy Measures", *Fuzzy Sets and Systems*, Vol. 10, #2, pp. 135-155, 1983.

[R5.5] Chang, C.L., "Fuzzy Topological Spaces," *J. Math. Anal. and Applications*, Vol. 24, 1968, pp. 182-190.

[R5.6] Chang, S.S.L., "Fuzzy Mathematics, Man and His Environment", *IEEE Transactions on Systems, Man and Cybernetics*, SMC-2 1972, pp. 92-93,

[R5.7] Chang, S.S., "Fixed Point Theorems for Fuzzy Mappings," Fuzzy *Sets and Systems*, Vol. 17, 1985, pp. 181-187.

[R5.8] Chapin, E.W., "An Axiomatization of the Set Theory of Zadeh," *Notices, American Math. Society*, 687-02-4 754, 1971.

[R5.9] Chaudhury, A. K. and P. Das, "Some Results on Fuzzy Topology on Fuzzy Sets," *Fuzzy Sets and Systems,* Vol. 56, 1993, pp. 331-336.

[R5.10] Chitra, H., and P.V. Subrahmanyam, "Fuzzy Sets and Fixed Points," *Jour. of Mathematical Analysis and Application*, Vol. 124, 1987, pp. 584-590.

[R5.11] Czogala, J. et. al., Fuzzy Relation Equations on a Finite Set," *Fuzzy Sets and Systems,* Vol. 7, #1,1982. pp. 89-101.

[R5.12] DiNola, A. et. al., (eds.), *The Mathematics of Fuzzy Systems*, Koln, Verlag TUV Rheinland, 1986.

[R5.13] Dompere, Kofi K., *Cost-Benefit Analysis and the Theory of Fuzzy Decisions: Identification and Measurement Theory* (Series: Studies in Fuzziness and Soft Computing, Vol. 158), Berlin, Heidelberg, Springer, 2004.

[R5.14] Dompere, Kofi K., *Cost-Benefit Analysis and the Theory of Fuzzy Decisions: Fuzzy Value Theory* (Series: Studies in Fuzziness and Soft Computing, Vol.160), Berling, Heidelberg, Springer, 2004.

[R5.15] Dubois, D. and H. Prade, *Fuzzy Sets and Systems*, New York, Academic Press, 1980.

[R5.16] Dubois, "Fuzzy Real Algebra: Some Results," *Fuzzy Sets and Systems*, Vol. 2, #4, pp. 327-348, 1979.

[R5.17] Dubois, D. and H. Prade, Possibility Theory: An Approach to Computerized Processing of Uncertainty, New York, Plenum Press, 1988

[R5.18] Dubois, D. and H. Prade , "Gradual Inference Rules in Approximate Reasoning," *Information Sciences*, Vol. 61(1-2), 1992, pp. 103-122.

[R5.19] Dubois, D. and H. Prade, "On the Combination of Evidence in various Mathematical Frameworks," In: Flamm. J. and T. Luisi, (eds.), *Reliability Data Collection and Analysis*. Kluwer, Boston, 1992, pp. 213-241.

[R5.20] Dubois, D. and H. Prade, "Fuzzy Sets and Probability: Misunderstanding, Bridges and Gaps." *Proc. Second IEEE Intern. Conf. on Fuzzy Systems*, San Francisco, 1993, pp. 1059-1068.

[R5.21] Dubois, D. and H. Prade "A Survey of Belief Revision and Updating Rules in Various Uncertainty Models," *Intern. J. of Intelligent Systems*, Vol. 9, #1, 1994, pp. 61-100.

[R5.22] Filev, D.P. et. al., "A Generalized Defuzzification Method via Bag Distributions," *Intern. Jour. of Intelligent Systems*, Vol. 6, #7, 1991, pp. 687-697.

[R5.23] Goetschel, R. Jr., et. al., "Topological Properties of Fuzzy Number," *Fuzzy Sets and Systems*, Vol. 10, #1, pp. 87-99, 1983.

[R5.24] Goodman, I.R., "Fuzzy Sets as Equivalence Classes of Random Sets" inYager, R.R. (ed.), *Fuzzy Set and Possibility Theory: Recent Development*, New York, Pergamon Press 1992. pp. 327-343.

[R5.25] Gupta, M.M. et. al., (eds.), *Fuzzy Automata and Decision Processes*, New York, North-Holland, 1977.

[R5.26] Gupta, M.M. and E. Sanchez (eds.), *Fuzzy Information and Decision Processes*, New York, North-Holland, 1982.

[R5.27] Higashi, M. and G.J. Klir, "On measure of fuzziness and fuzzy complements," Intern. *J. of General Systems*, Vol. 8 #3, 1982, pp. 169-180.

[R5.28] Higashi, M. and G.J. Klir, "Measures of uncertainty and information based on possibility distributions," *International Journal of General Systems*, Vol. 9 #1, 1983, pp. 43-58.

[R5.29] Higashi, M. and G.J. Klir, "On the notion of distance representing information closeness: Possibility and probability distributions," *Intern. J. of General Systems,* Vol. 9 #2, 1983, pp. 103-115.

[R5.30] Higashi, M. and G.J. Klir , "Resolution of finite fuzzy relation equations," *Fuzzy Sets and Systems*, Vol. 13, #1,1984, pp. 65-82.

[R5.31] Higashi, M. and G.J. Klir , "Identification of fuzzy relation systems," *IEEE Trans. on Systems, Man, and Cybernetics*, Vol.14 #2, 1984, pp. 349-355.

[R5.32] Jin-wen, Z., "A Unified Treatment of Fuzzy Set Theory and Boolean Valued Set theory: Fuzzy Set Structures and Normal Fuzzy Set Structures," *Jour. Math. Anal. and Applications*, Vol. 76, #1, 1980, pp. 197-301.

[R5.33] Kandel, A. and W.J. Byatt, "Fuzzy Processes," *Fuzzy Sets and Systems*, Vol. 4, #2, 1980, pp. 117-152.

[R5.34] Kaufmann, A. and M.M. Gupta, *Introduction to fuzzy arithmetic: Theory and applications*, New York Van Nostrand Rheinhold,. 1991.

[R5.35] Kaufmann, A., *Introduction to the Theory of Fuzzy Subsets*, Vol. 1, New York, Academic Press, 1975.

[R5.36] Klement, E.P. and W. Schwyhla, "Correspondence Between Fuzzy Measures and Classical Measures," *Fuzzy Sets and Systems,* Vol. 7, #1, 1982.pp. 57-70.

[R5.37] Klir, George and Bo Yuan, *Fuzzy Sets and Fuzzy Logic*, Upper Saddle River, NJ Prentice Hall, 1995.

[R5.38] Kruse, R. et al., *Foundations of Fuzzy Systems*, New York, John Wiley and Sons, 1994.

[R5.39] Lasker, G.E. (ed.), *Applied Systems and Cybernetics, Vol. VI: Fuzzy Sets and Systems*, Pergamon Press, New York, 1981.

[R5.40] Lientz, B.P., "On Time Dependent Fuzzy Sets", *Inform, Science*, Vol. 4, 1972, pp. 367-376.

[R5.41] Lowen, R., "Fuzzy Uniform Spaces," *Jour. Math. Anal. Appl.*, Vol. 82, #21981, pp. 367-376.

[R5.42] Michalek, J., "Fuzzy Topologies," *Kybernetika*, Vol. 11, 1975, pp. 345-354.

[R5.43] Negoita, C.V. et. al., *Applications of Fuzzy Sets to Systems Analysis*, Wiley and Sons, New York, 1975.

[R5.44] Negoita, C.V., "Representation Theorems for Fuzzy Concepts," *Kybernetes*, Vol. 4, 1975, pp. 169-174.

[R5.45] Negoita, C.V. et. al., "On the State Equation of Fuzzy Systems," *Kybernetes*, Vol. 4, 1975, pp. 231-214.

[R5.46] Netto, A.B., "Fuzzy Classes," *Notices, American Mathematical Society,* Vol.68T-H28, 1968, pp.945.

[R5.47] Pedrycz, W., "Fuzzy Relational Equations with Generalized Connectives and Their Applications," *Fuzzy Sets and Systems*, Vol. 10, #2, 1983, pp. 185-201.

[R5.48] Raha, S. et. al., "Analogy Between Approximate Reasoning and the Method of Interpolation," *Fuzzy Sets and Systems*, Vol. 51, #3, 1992, pp. 259-266.

[R5.49] Ralescu, D., "Toward a General Theory of Fuzzy Variables," *Jour. of Math. Analysis and Applications*, Vol. 86, #1, 1982, pp. 176-193.

[R5.50] Rodabaugh, S.E., "Fuzzy Arithmetic and Fuzzy Topology," in G.E. Lasker, (ed.), *Applied Systems and Cybernetics, Vol. VI: Fuzzy Sets and Systems*, Pergamon Press, New York, 1981 pp. 2803-2807.

[R5.51] Rosenfeld, A., "Fuzzy Groups," *Jour. Math. Anal. Appln.*, Vol. 35, 1971, pp. 512-517.

[R5.52] Ruspini, E.H., "Recent Developments in Mathematical Classification Using Fuzzy Sets," in G.E. Lasker, (ed.), *Applied Systems and Cybernetics, Vol. VI: Fuzzy Sets and Systems*, Pergamon Press, New York, 1981. pp. 2785-2790.

[R5.53] Santos, E.S., "Fuzzy Algorithms," *Inform. and Control*, Vol. 17, 1970, pp. 326-339.

[R5.54] Stein, N.E. and K Talaki, "Convex Fuzzy Random Variables," *Fuzzy Sets and Systems*, Vol. 6, #3, 1981, pp. 271-284.

[R5.55] Triantaphyllon, E. et. al., "The Problem of Determining Membership Values in Fuzzy Sets in Real World Situations," in D.E. Brown et. al. (eds), *Operations Research and Artificial Intelligence: The Integration of Problem-solving Strategies*, Boston, Mass., Kluwer, 1990, pp. 197-214.

[R5.56] Tsichritzis, D., "Participation Measures," *Jour. Math. Anal. and Appln.*, Vol. 36, 1971, pp. 60-72.

[R5.57] Turksens, I.B., "Four Methods of Approximate Reasoning with Interval-Valued Fuzzy Sets," *Intern. Journ. of Approximate Reasoning*, Vol. 3, #2, 1989, pp. 121-142.

[R5.58] Turksen, I.B., "Measurement of Membership Functions and Their Acquisition," *Fuzzy Sets and Systems*, Vol. 40, #1, 1991, pp. 5-38.

[R5.59] Wang, P.P. (ed.), *Advances in Fuzzy Sets, Possibility Theory, and Applications*, New York, Plenum Press, 1983.

[R5.60] Wang, Zhenyuan, and George Klir, *Fuzzy Measure Theory,* New York, Plenum Press, 1992.

[R5.61] Wang, P.Z. et. al. (eds.), *Between Mind and Computer: Fuzzy Science and Engineering*, Singapore, World Scientific Press, 1993.

[R5.62] Wang, S., "Generating Fuzzy Membership Functions: A Monotonic Neural Network Model," *Fuzzy Sets and Systems*, Vol. 61, #1, 1994, pp. 71-82.

[R5.63] Wong, C.K., "Fuzzy Points and Local Properties of Fuzzy Topology," *Jour. Math. Anal. and Appln.*, Vol. 46, 19874, pp. 316-328.

[R5.64] Wong, C.K., "Categories of Fuzzy Sets and Fuzzy Topological Spaces," *Jour. Math. Anal. and Appln.*, Vol. 53, 1976, pp. 704-714.

[R5.65] Yager, R.R. et. al., (Eds.), *Fuzzy Sets, Neural Networks, and Soft Computing*, New York, Nostrand Reinhold, 1994.

[R5.66] Zadeh, L.A., "A Computational Theory of Decompositions," *Intern. Jour. of Intelligent Systems*, Vol. 2, #1, 1987, pp. 39-63.

[R5.67] Zimmerman, H.J., *Fuzzy Set Theory and Its Applications,* Boston, Mass, Kluwer, 1985.

R6. FUZZY OPTIMIZATION, INFORMATION, DECISION-CHOICE THEORY AND THE SCIENCE OF KNOWING

[R6.1] Bose, R.K. and Sahani D, "Fuzzy Mappings and Fixed Point Theorems," *Fuzzy Sets and Systems*, Vol. 21, 1987, pp. 53-58.

[R6.2] Butnariu D. "Fixed Points for Fuzzy Mappings," *Fuzzy Sets and Systems*, Vol. 7, 1982, pp.191-207.

[R6.3] Dompere, Kofi K., "Fuzziness, Rationality, Optimality and Equilibrium in Decision and Economic Theories" in Weldon A. Lodwick and Janusz Kacprzyk (Eds.), *Fuzzy Optimization: Recent Advances and Applications* (Series: Studies in Fuzziness and Soft Computing, Vol. 254), Berlin, Heidelberg, Springer, 2010.

[R6.4] Eaves, B.C., "Computing Kakutani Fixed Points," *Journal of Applied Mathematics*, Vol. 21, 1971,pp. 236-244.

[R6.5] Heilpern, S. "Fuzzy Mappings and Fixed Point Theorem," *Journal of Mathematical Analysis and Applications*, Vol. 83, 1981, pp.566-569.

[R6.6] Kacprzyk, J. et. al., (eds.), *Optimization Models Using Fuzzy Sets and Possibility Theory*, Boston, Mass., D. Reidel, 1987.

[R6.7] Kaleva, O. "A Note on Fixed Points for Fuzzy Mappings", *Fuzzy Sets and Systems,* Vol. 15, 1985, pp. 99-100.

[R6.8] Lodwick, Weldon A and Janusz Kacprzyk (eds.), *Fuzzy Optimization: Recent Advances and Applications,* (Studies in Fuzziness and Soft Computing, Vol. 254), Berlin Heidelberg, Springer, 2010,

[R6.9] Negoita, C.V., "The Current Interest in Fuzzy Optimization," *Fuzzy Sets and Systems*, Vol. 6, #3, 1981, pp. 261-270.

[R6.10] Negoita, C.V., et. al., "On Fuzzy Environment in Optimization Problems," in J. Rose et. al., (eds.), *Modern Trends in Cybernetics and Systems,* Springer, Berlin, 1977, pp.13-24.

[R6.11] Zimmerman, H.-J., "Description and Optimization of Fuzzy Systems," *Intern. Jour. Gen. Syst.* Vol. 2, #4, 1975, pp. 209-215.

R7. IDEOLOGY, DISINFORMATION, MISINFORMATION AND PROPAGANDA IN INTRA- AND INTER-EPISTEMOLOGICAL MASSEGING SYSTEMS

[R7.1] Abercrombie, Nicholas et al., *The Dominant Ideology Thesis*, London, Allen and Unwin, 1980.

[R7.2] Abercrombie, Nicholas, *Class, Structure, and Knowledge: Problems in the Sociology of Knowledge*, New York, New York University Press, 1980.

[R7.3] Aron, Raymond, *The Opium of the Intellectuals*, Lanham, MD, University Press of America, 1985.

[R7.4] Aronowitz, Stanley, *Science as Power: Discourse and Ideology in Modern Society,* Minneapolis, University of Minnesota Press, 1988.

[R7.5] Barinaga, M. and E. Marshall, *Confusion on the Cutting Edge*, Science, Vol. 257, July 1992, pp. 616-625.

[R7.6] Barnett, Ronald, *Beyond All Reason: Living with Ideology in the University,* Philadelphia, PA., Society for Research into Higher Education and Open University Press, 2003.

[R7.7] Barth, Hans, *Truth and Ideology*, Berkeley, University of California Press,1976.

[R7.8] Basin, Alberto, and Thierry Verdie, "The Economics of Cultural Transmission and the Dynamics of Preferences," *Journal of Economic Theory*, Vol. 97, 2001, pp. 298-319.

[R7.9] Bikhchandani, Sushil et al., "A Theory of Fads, Fashion, Custom, and Cultural Change," *Journal of political Economy*, Vol. 100 1992, pp 992-1026.

[R7.10] Boyd Robert and Peter J Richerson, *Culture and Evolutionary Process*, Chicago, University of Chicago Press, 1985.

[R7.11] Buczkowski, Piotr and Andrzej Klawiter, *Theories of Ideology and Ideology of Theories*, Amsterdam, Rodopi, 1986.

[R7.12] Chomsky, Norm, *Manufacturing Consent*, New York, Pantheo Pess, 1988.

[R7.13] Chomsky, N., *Problem of Knowledge and Freedom*, Glasgow, Collins, 1972.

[R7.14] Cole, Jonathan, R., "Patterns of Intellectual influence in Scientific Research," *Sociology of Education*, Vol. 43, 1968, pp.377-403.

[R7.15] Cole Jonathan, R. and Stephen Cole, *Social Stratification in Science*, Chicago, University of Chicago Press, 1973.

[R7.16] Debackere, Koenraad and Michael A. Rappa, "Institutioal Variations in Problem Choice and Persistence among Scientists in an Emerging Fields," *Research Policy,* Vol. 23, 1994, pp425- 441.

[R7.17] Dompere, K. K., "Ideology and Decision-Choice Rationalities" in Dompere, Kofi. Kissi., *Fuzzy Rationality: Methodological Critique and Unity of Classical, Bounded and Other Rationalities*, (Studies in Fuzziness and Soft Computing, vol.235) New York, Springer, 2009, pp.143-165.

[R7.18A] Dompere, K. K., "Culture, Ideology and Categorial Conversion of Social Polarities in Systems Dynamics," in Dompere, K. K., *The Theory of Philosophical Consciencism*; Practice Foundations of Nkrumaism in Social Systemicity, London, Adonis & Abbey Pub. 2017, pp.124 – 158.

[R7.18B] Dompere, K. K., "Abstract Ideas and Practice of Ideas in Social Settings Extensions and Reflections on Nkrumah and Africa under Systems Thinking," in Dompere, K. K., *The Theory of Categorial Conversion: Rational Foundations of Nkrumaism in Socio-natural Systemicity and Complexity*; London, Adonis & Abbey Pub. 2017, pp.61 85 .

[R7.19] Dompere, K. K., "Zones of Thought: Reflections on the theories of Thought," in Dompere, K. K, *Fuzziness and Foundations of Exact and Inexact Sciences*, (Studies in Fuzziness and Soft Computing, vol.290) New York, Springer, 2013, pp.103-116.

[R7.20] Fraser, Colin and George Gaskell (eds.), *The Social Psychological Study of Widespread Beliefs*, Oxford, Clarendon Press, 1990.

[R7.21] Gieryn, Thomas, F. "Problem Retention and Problem Change in Science," *Sociological Inquiry*, Vol. 48, 1978, pp96-115.

[R7.22] Harrington, Joseph E. Jr, "The Rigidity of social Systems," *Journal of Political Economy*, Vol. 107, pp. 40-64.

[R7.23] Hinich, Melvin and Michael Munger, *Ideology and the Theory of Political Choice*, Ann Arbor University of Michigan Press, 1994.

[R7.24] Hull, D. L., *Science as a Process: An Evolutionary Account of the Social and Conceptual Development of Science,* Chicago, University of Chicago Press, 1988.

[R7.25] Marx, Karl and Friedrich Engels, *The German Ideology*, New York, International Pub, 1970

[R7.26] Mészáros, István , *Philosophy, Ideology and Social Science*: Essay in Negation and Affirmation, Brighton, Sussex, Wheatsheaf, 1986.

[R7.27] Mészáros, István *The Power of Ideology*, New York, New York University Press, 1989.

[R7.28] Newcomb, Theodore M. et. al., *Persistence and Change*, New York, John Wiley, 1967

[R7.29] Pickering, Andrew, *Science as Practice and Culture,* Chicago, University of Chicago Press, 1992.

[R7.30] Therborn, Göran , *The Ideology of Power and the Power of Ideology*, London, NLB Publications, 1980.

[R7.31] Thompson, Kenneth, *Beliefs and Ideology*, New York, Tavistock Publication,1986.

[R7.32] Trine, Ralph W., *In Tune with the Infinite*: A Modern Classic of Spiritual Inspiration, New York, The Bobbs Merrill Company. 1970.

[R7.33] Ziman, John, "The Problem of 'Problem Choice'," *Minerva*, Vol. 25, 1987, pp. 92-105.

[R7.34] Ziman, John, *Public Knowledge: An Essay Concerning the Social Dimension of Science*, Cambridge, Cambridge University Press, 1968.

[R7.35] Zuckerman, Harriet, "Theory Choice and Problem Choice in Science," *Sociological Inquiry*, Vol.48, 1978, pp. 65-95.

R 8. INFORMATION, THOUGHT, KNOWING AND KNOWLEDGE

[R8.1] Aczel, J. and Z. Daroczy, *On Measures of Information and their Characterizations*, New York, Academic Press, 1975

[R8.2] Afanasyev, Social Information and Regulation of Social Development, Moscow, Progress,

1978

[R8.3] Anderson, J. R., *The Architecture of Cognition*, Cambridge, Mass., Harvard University Press, 1983.

[R8.4] Angelov, Stefan and Dimitr Georgiev, "The Problem of Human Being in Contemporary Scientic Knowledge," *Soviet Studies in Philosophy*, Summer, 1974, pp. 49-66.

[R8.5] Ash, Robert, *Information Theory*, New York, John Wiley and Sons, 1965.

[R8.6] Bergin, J., "Common Knowledge with Monotone Statistics," *Econometrica*, Vol. 69, 2001, pp. 1315-1332.

[R8.7] Bestougeff, Hélène and Gerard Ligozat, *Logical Tools for Temporal Knowledge Representation*, New York, Ellis Horwood, 1992.

[R8.8] Brillouin, L., *Science and information Theory,* New York, Academic Press, 1962.

[R8.9] Bruner, J. S., et. al., *A Study of Thinking*, New York, Wiley, 1956.

[R8.10] Brunner, K. and A. H. Meltzer (eds.), *Three Aspects of Policy and Policy Making: Knowledge, Data and Institutions*, Carnegie-Rochester Conference Series, Vol. 10, Amsterdam, North-Holland, 1979.

[R8.11] Burks, A. W., *Chance, Cause, Reason: An Inquiry into the Nature of Scientific Evidence*, Chicago, University of Chicago Press, 1977.

[R8.12] Calvert, Randall, *Models of Imperfect Information in Politics*, New York, Hardwood Academic Publishers, 1986.

[R8.13] Cornforth, Maurice, *The Theory of Knowledge*, New York, International Pub. 1972

[R8.14] Cornforth, Maurice, *The Open Philosophy and the Open Society,* New York, International Pub. 1970

[R8.15] Coombs, C. H., *A Theory of Data*, New York, Wiley, 1964.

[R8.16a] Dompere, Kofi K., *The Theory of Problem-Solution Dualities and Polarities: Information-Decision-Choice Foundations of Unity of Knowing and Unity of Science*, (Studies in Systems, Decision and Control, Vol. 405) Springer, New York, 2022.

[R8.16b] Dompere, Kofi K., *The Theory of Epistemic fields in the Problem-Solution Dualities and Polarities: Fuzzy-Semantic Foundations of Interdisciplinary Categories for the Diversity and the Unity of Knowing and the Unity of Science*, An Unpublished Monograph, (Research Institute of Information, Decision and Economic Sciences (RIIDES), Washington, DC., 2022.

[R8.17] Dretske, Fred. I., *Knowledge and the Flow of Information*, Cambridge, Mass., MIT Press 1981.

[R8.18] Dreyfus, Hubert L., "A Framework for Misrepresenting Knowledge," in Martin Ringle (ed.) *Philosophical Perspectives in Artificial Intelligence,* Atlantic Highlands, N.J., Humanities press, 1979.

[R8.19] Fagin R. et al., *Reasoning About Knowledge,* Cambridge, Mass, MIT Press, 1995.

[R8.20] George, F. H., *Models of Thinking*, London, Allen and Unwin, 1970.

[R8.21] George, F. H., "Epistemology and the problem of perception," *Mind*, Vol.66, 1957, pp.491-506.

[R8.22] Harwood, E. C., *Reconstruction of Economics,* Great Barrington, Mass, American Institute for Economic Research, 1955

.

[R8.23] Hintikka, J., *Knowledge and Belief,* Ithaca, N. Y., Cornell University Press, 1962.

[R8.24] Hirshleifer, Jack., "The Private and Social Value of Information and Reward to inventive activity," *American Economic Review,* Vol. 61, 1971, pp.561-574.

[R8.25] Kapitsa, P. L., "The Influence of Scientific Ideas on Society," *Soviet Studies in Philosophy*, Fall, 1979, pp.52-71.

[R8.26] Kedrov, B. M., "The Road to Truth," *Soviet Studies in Philosophy*, Vol. 4, 1965, pp 3 – 53.

[R8.27a] Klatzky, R. L., *Human Memory: Structure and Processes*, San Francisco, Ca., W. H. Freeman Pub., 1975.

[R8.27b] Klix, Friedhart,(ed.) Human and Artfificial Intelligence, (Fundamental Studies in Computer Science, Vol. 8),New York, North-Holland, 1979

[R8.28] Kreps, David and Robert Wilson, "Reputation and Imperfect Information," *Journal of Economic Theory,* Vol. 27. 1982, pp253-279.

[R8.29] Kubát, Libor and J. Zeman (eds.), *Entropy and Information*, Amsterdam, Elsevier, 1975.

[R8.30] Kurcz, G. and W. Shugar et al (eds.), *Knowledge and Language*, Amsterdom, North-

Holland, 1986.

[R8.31] Lakemeyer, Gerhard,and Bernhard Nobel (eds.), *Foundations of Knowledge Representation and Reasoning*, Berlin, Springer-Verlag, 1994.

[R8.32] Lektorskii, V. A., "Principles involved in the Reproduction of Objective in Knowledge,", *Soviet Studies in Philosophy*, Vol. 4, #4, 1967, pp. 11-21.

[R8.33] Levi, I., *The Enterprise of Knowledge*, Cambridge, Mass. MIT Press1980.

[R8.34] Levi, Isaac, "Ignorance, Probability and Rational Choice", *Synthese* Vol. 53, 1982, pp. 387-417.

[R8.35] Levi, Isaac, "Four Types of Ignorance," *Social Science*, Vol. 44, pp745-756.

[R8.36] Marschak, Jacob, *Economic Information, Decision and Prediction: Selected Essays*, Vol. II, Part II, Boston, Mass. Dordrecnt-Holland, 1974.

[R8.37] Menges, G. (ed.), *Information, Inference and Decision*, Dordrecht, Holland, D. Reidel Pub.,1974.

[R8.38] Michael Masuch and Lászlò Pólos (eds.), *Knowledge Representation and Reasoning Under Uncertainty*, New York, Springer-Verlag, 1994.

[R8.39] Moses, Y. (ed.), *Proceedings of the Fourth Conference of Theoretical Aspects of Reasoning about Knowledge*, San Mateo, Morgan Kaufmann, 1992.

[R8.40] Nielsen, L.T. et al., "Common Knowledge of Aggregation Expectations," *Econometrica*, Vol. 58, 1990, pp. 1235-1239.

[R8.41] Newell, A., *Unified Theories of Cognition*, Cambridge, Mass. Harvard University Press, 1990.

[R8.42] Newell, A and Herbert A. Simon., *Human Problem Solving*, Englewood Cliff, N.J. Prentice- Hall, 1972.

[R8.43] Ogden, G. K. and I. A., *The Meaning of Meaning*, New York, Harcourt-Brace Jovanovich, 1923.

[R8.44] Planck, Max, Scientific Autobiography and Other Papers, Westport, Conn., Greenwood, 1968.

[R8.45] Pollock, J., *Knowledge and Justification*, Princeton, Princeton University Press, 1974.

[R8.46] Polanyi, M., *Personal Knowledge*, London, Routledge and Kegan Paul, 1958.

[R8.47] Popper, K. R., *Objective Knowledge*, London, Macmillan, 1949.

[R8.48] Popper, K. R., *Open Society and it Enemies, Vols. 1 and 2* Princeton, Princeton Univ. Press, 2013.

[R8.49] Popper, K. R., *The Poverty of Historicism* New York, Taylor and Francis, 2002.

[R8.50] Price, H. H., *Thinking and Experience*, London, Hutchinson, 1953.

[R8.51] Putman, H., *Reason, Truth and History*, Cambridge, Cambridge University Press, 1981.

[R8.52] Putman. H., *Realism and Reason*, Cambridge, Cambridge University Press, 1983.

[R8.53] Putman, H., *The Many Faces of Realism*, La Salle, Open Court Publishing Co., 1987.

[R8.54] Russell, B., *Human Knowledge, its Scope and Limits*, London, Allen and Unwin, 1948.

[R8.55] Russell, B., *Our Knowledge of the External World,* New York, Norton, 1929.

[R8.56] Samet, D., "Ignoring Ignorance and Agreeing to Disagree," *Journal of Economic Theory*, Vol. 52, 1990, pp. 190-207.

[R8.57] Schroder, Harold, M. and Peter Suedfeld (eds.), *Personality Theory and Information Processing*, New York, Ronald Pub. 1971.

[R8.58] Searle J., *Minds, Brains and Science*, Cambridge, Mass., Harvard University Press, 1985.

[R8.59] Shin, H., "Logical Structure of Common Knowledge," *Journal of Economic Theory*, Vol. 60, 1993, pp. 1-13.

[R8.60] Simon, H. A., *Models of Thought*, New Haven, Conn., Yale University Press,1979.

[R8.61]
Smithson, M., *Ignorance and Uncertainty, Emerging Paradigms*, New York, Springer-Verlag, 1989.

[R8.62] Sowa, John F., *Knowledge Representation: Logical, Philosophical, and Computational Foundations*, Pacific Grove, Brooks Pub., 2000.

[R8.63] Stigler, G. J., The Economics of Information," *Journal of Political Economy*, Vol.69, 1961, pp.213-225.

[R8.64] Tiukhtin, V. S., "How Reality Can be Reflected in Cognition: Reflection as a Property of All Matter," *Soviet Studies in Philosophy*, Vol.3 #1, 1964, pp.3-12.

[R8.65] Tsypkin, Ya Z., *Foundations of the Theory of Learning Systems*, New York, Academic Press, 1973.

[R8.66] Ursul, A. D., "The Problem of the Objectivity of Information," in Libor Kubát , and J.

Zeman (eds.), *Entropy and Information*, Amsterdam, Elsevier, 1975, pp. 187 – 230.

[R8.67] Vardi, M. (ed.), *Proceedings of Second Conference on Theoretical Aspects of Reasoning about Knowledge*, Asiloman, Ca., Los Altos, Ca, Morgan Kaufman1988.

[R8.68] Vazquez, Mararita, et al., "Knowledge and Reality: Some Conceptual Issues in System Dynamics Modeling," *Systems Dynamics Review,* Vol. 12, 1996, pp. 21-37.

[R8.69] Zadeh, L. A., "A Theory of Commonsense Knowledge," in Skala, Heinz J. et al., (eds.), *Aspects of Vagueness*, Dordrecht, D. Reidel Co. 1984. pp 257 – 295.

[R8.70] Zadeh, L. A., "The Concept of Linguistic Variable and its Application to Approximate reasoning," *Information Science*, Vol. 8, 1975, pp. 199 – 249(Also in Vol. 9, pp. 40 – 80).

R9. LANGUAGES AND INFORMATION

[R9.1] Agha, Agha, *Language and Social Relations,* Cambridge, Cambridge University Press 2006.

[R9.2]Aitchison, Jean (ed.), *Language Change: Progress or Decay?* Cambridge, New York, Melbourne: Cambridge University Press, 2001.

[R9.3] Anderson, Stephen, *Languages: A Very Short Introduction.* Oxford: Oxford University Press (2012).

[R9.4] Aronoff, Mark and Fudeman, Kirsten, *What is Morphology. New York, John Wiley & Sons,* 2011.

[R9.5] Bauer, Laurie (ed.), *Introducing linguistic morphology* Washington, D.C.: Georgetown University Press, 2003.

[R9.6] Barber Alex and Robert J Stainton *(eds.). Concise Encyclopedia of Philosophy of Language and Linguistics.* New York, Elsevier, 2010.

[R9.7] Brown, Keith; Ogilvie, Sarah, (eds.), *Concise Encyclopedia of Languages of the World,* New York Elsevier Science 2008.

[R9.8] Campbell, Lyle_ (ed.), *Historical Linguistics: an Introduction Cambridge,* MASS, MIT Press, 2004.

[R9.9] Chomsky, Noam, *Syntactic Structures.* The Hague: Mouton,1957.

[R9.10] Chomsky, Noam, *The Architecture of Language*. Oxford, Oxford University Press, 2000.

[R9.11] Clarke, David S. *Sources of semiotic: readings with commentary from antiquity to the present* Carbondale: Southern Illinois University Press, 1990.

[R9.12] Collinge, N.E. (ed.), *An Encyclopedia of Language.* London: New York: Routledge (1989).

[R9.13] Comrie, Bernard (ed.), *Language universals and linguistic typology: Syntax and morphology,* Oxford, Blackwell,1989.

[R9.14] Comrie, Bernard (ed.), *The World's Major Languages.* New York: Routledge, 2009.

[R9.15] Coulmas, Florian, *Writing Systems: An Introduction to Their Linguistic Analysis.* Cambridge, Cambridge University Press, 2002.

[R9.16] Croft, William, Cruse, D. Alan, *Cognitive Linguistics.* Cambridge, Cambridge University Press, 2004.

[R9.17] Croft, William, "Typology". In Mark Aronoff; Janie Rees-Miller(ed.), *The Handbook of Linguistics,* Oxford, Blackwell. pp. 81–105, 2001.

[R9.18] Crystal, David (ed.) *The Cambridge Encyclopedia of Language. Cambridge: Cambridge University Press,* 1997.

[R9.19] Deacon, Terrence , *The Symbolic Species: The Co-evolution of Language and the Brain,* New York: W.W. Norton & Company, 1997.

[R9.20] Deming W. Edwards, “On Errors in Surveys,” *American Sociological Rev.*, Vol. IX, 1944, pp. 359-369.

[R9.21] Devitt, Michael and Sterelny, Kim, *Language and Reality: An Introduction to the Philosophy of Language. Boston:* MIT Press 1999.

[R9.22] Duranti, Alessandro "Language as Culture in U.S. Anthropology: Three Paradigms" *Current Anthropology* Vol. 44 (3), pp. 323–348, 2003.

[R9.23] Evans, Nicholas and Levinson, Stephen C, "The myth of language universals: Language diversity and its importance for cognitive science," Vol. 32 *(5). Behavioral and Brain Sciences.* pp. 429–492, 2009.

[R9.24] Fitch, W. Tecumseh, *The Evolution of Language,* Cambridge, Cambridge University Press, 2010.

[R9.25] Foley, William A., *Anthropological Linguistics: An Introduction,* Oxford, Blackwell, 1997.

[R9.26] Ginsburg, Seymour, *Algebraic and Automata-Theoretic Properties of Formal Languages*, New York, North-Holland, 1973.

[R9.27] Goldsmith, John A, *The Handbook of Phonological Theory: Blackwell Handbooks in Linguistics.* Oxford, Blackwell Publishers, 1995.

[R9.28] Greenberg, Joseph, *Language Universals: With Special Reference to Feature Hierarchies.* The Hague, Mouton & Co., 1966.

[R9.29] Hauser, Marc D.; Chomsky, Noam; Fitch, W. Tecumseh, "The Faculty of Language: What Is It, Who Has It, and How Did It Evolve?". *Science,* 22 **298** (5598), 2002, pp. 1569–1579,.

[R9.30] Hörz, Herbert, "Information, Sign, Image," in Libor Kubát and Jiří Zeman (eds.*) Entropy and Information in Science and Philosophy*, New York, Elsevier, 1975.

[R9.31] International Phonetic Association, *Handbook of the International Phonetic Association: A guide to the use of the International Phonetic Alphabet. Cambridge,* Cambridge University Press (1999).

[R9.32] Katzner, Kenneth, *The Languages of the World,* New York: Routledge, 1999.

[R9.33] Labov, William, *Principles of Linguistic Change vol.I Internal Factors,* Oxford, Blackwell, 1994.

[R9.34] Labov, William, *Principles of Linguistic Change vol.II Social Factors,* Oxford, Blackwell, 2001.

[R9.35] Levinson, Stephen C. *Pragmatics.* Cambridge: Cambridge University Press, 1983.

[R9.36] Lewis, M. Paul (ed.), *Ethnologue: Languages of the World*, Dallas, Tex.: SIL International 2009.

[R9.37] Lyons, John, *Language and Linguistics,* Cambridge, Cambridge University Press, 1981.

[R9.38] MacMahon, *April M.S., Understanding Language Change,* Cambridge, Cambridge University Press, 1994.

[R9.39] Matras, Yaron; Bakker, Peter, (eds.). *The Mixed Language Debate: Theoretical and Empirical Advances.* Berlin: Walter de Gruyter, 2003.

[R9.40] Moseley, Christopher (ed.), *Atlas of the World's Languages in Danger, Paris: UNESCO Publishing* (2010).

[R9.41] Nerlich, B. "History of pragmatics". In L. Cummings (ed.), *The Pragmatics Encyclopedia.* New York: Routledge. pp. 192–93, 2010.

[R9.42] Newmeyer, Frederick J., *The History of Linguistics*. Linguistic Society of America 2005.

[R9.43] Newmeyer, Frederick J., *Language Form and Language Function* (PDF). Cambridge, MA: MIT Press 1998.

[R9.44] Nichols, Johanna, *Linguistic diversity in space and time.* Chicago, University of Chicago Press,1992.

[R9.45] *Nichols, Johanna. "Functional Theories of Grammar". Annual Review of Anthropology* Vol. **13**: 1949 pp. 7–117.

[R9.46] *Sandler, Wendy; Lillo-Martin, Diane,* "Natural Sign Languages", *In Mark Aronoff; Janie Rees-Miller (eds.). The Handbook of Linguistics.* Oxford, Blackwell. pp. 533–563, 2001.

[R9.47] Swadesh, Morris, *"The phonemic principle", Language,* Vol. **10** (2): 1934, pp.117–129.

[R9.48] Tomasello, Michael, "The Cultural Roots of Language". In B. Velichkovsky and D. Rumbaugh 9EDS.), *Communicating Meaning: The Evolution and Development of Language.* New York, Psychology Press. 1996, pp. 275–308.

[R9.49] Tomasello, Michael, *Origin of Human Communication.* Cambridge Mass., MIT Press 2008.

[R9.50] Thomason, Sarah G, *Language Contact – An Introduction,* Edinburgh, Edinburgh University Press,2001.

[R9.51] Ulbaek, Ib, "The Origin of Language and Cognition", In J. R. Hurford and C. Knight (eds.). *Approaches to the evolution of language.* Cambridge, Cambridge University Press. *1998,* pp. 30–43.

[R9.52] Van Valin, jr. Robert D., "Functional Linguistics," In Mark Aronoff; Janie Rees-Miller (eds.), *The Handbook of Linguistics.* Oxford, Blackwell. (2001). pp. 319–337.

R10. LANGUAGE, KNOWLEDGE-PRODUCTION PROCESS AND EPISTEMICS

[R10.1] Aho, A. V. "Indexed Grammar - An Extension of Context-Free Grammars" *Journal of the Association for Computing Machinery*, Vol. 15, 1968, pp. 647-671.

[R10.2] Black, Max (ed.), *The Importance of Language*, Englewood Cliffs, N.J, Prentice-Hall, 1962.

[R10.3] Carnap, Rudolff, Meaning and Necessity: A Study in Semantics and Modal Logic, Chicago, University of Chicago Press, 1956.

[R10.4] Chomsky, Norm, "Linguistics and Philosophy" in S. Hook (ed.) *Language and Philosophy*, New York, New York University Press, 1968, pp. 51-94.

[R10.5] Chomsky, Norm, *Language and Mind*, New York, Harcourt Brace Jovanovich, 1972.

[R10.6] Cooper, William S., *Foundations of Logico-Linguistics: A Unified Theory of Information, Language and Logic*, Dordrecht, D. Reidel, 1978.

[R10.7] Cresswell, M.J.., *Logics and Languages*, London, Methuen Pub. 1973.

[R10.8] Dilman, Ilham, *Studies in Language and Reason*, Totowa, N.J., Barnes and Nobles, Books, 1981.

[R10.9] Fodor, Jerry A., *The Language and Thought*, New York, Thom as Y. Crowell Co, 1975.

[R10.10] Givon, Talmy, *On Understanding Grammar*, New York, Academic Press, 1979.

[R10.11] Gorsky, D.R., *Definition,* Moscow, Progress Publishers, 1974.

[R10.12] Hintikka, Jaakko, The Game of Language, Dordrecht, D. Reidel Pub. 1983.

[R10.13] Johnson-Lair, Philip N. *Mental Models: Toward Cognitive Science of Language, Inference and Consciousness*, Cambridge, Mass, Harvard University Pres, 1983.

[R10.14] Kandel, A., "Codes Over Languages," *IEEE Transactions on Systems Man and Cybernetics*, Vol. 4, 1975, pp. 135-138.

[R10.15] Keenan, Edward L. and Leonard M. Faltz, *Boolean Semantics for Natural Languages*, Dordrecht, D. Reidel Pub.,1985.

[R10.16] Lakoff, G. Linguistics and Natural Logic, *Synthese*, Vol. 22, 1970, pp. 151-271.

[R10.17] Lee, E.T., et. al., "Notes On Fuzzy Languages," *Information Science*, Vol. 1, 1969, pp. 421-434.

[R10.18] Mackey, A. and D. Merrill (eds.) *Issues in the Philosophy of Language,* New Haven, CT. Yale University Press, 1976.

[R10.19] Nagel, T., "Linguistics and Epistemology" in S. Hook(ed.) *Language and Philosophy*, New York, New York University Press, 1969, pp. 180-184.

[R10.20] Pike, Kenneth, *Language in Relation to a Unified Theory of Structure of Human Behavior*, The Hague, Mouton Pub., 1969.

[R10.21] Quine, W.V. O. *Word and object,* Cambridge, Mass, MIT Press, 1960.

[R10.22] Russell, Bernard, *An Inquiry into Meaning and Truth,* Penguin Books, 1970.

[R10.23] Tarski, Alfred, *Logic, Semantics and Metamathematics,* Oxford, Clarendon Press, 1956.

[R10.24] Whorf, B.L. (ed.), *Language, Thought and Reality*, New York, Humanities Press, 1956.

R11 PHILOSOPHY OF ECONOMICS AND PRODUCTION-CONSUMPTION DUALITIES

[R11.1] Alchian, A., "Uncertainty, Evolution and Economic Theory", *Journal of Political Economy*, Vol. 57,1950: pp.211–21.

R11.2] Angyrous, G., "Refutation or Comparison?" *British Journal for the Philosophy of Science*, Vol. 17, 1967, pp.279–96.

[R11.3] Arrow, K., "Economic Equilibrium", pp. 376–89, *International Encyclopedia of the Social Sciences*, New York: Macmillan 1968.

[R11.3] Arrow, K, *The Limits of Organization*, New York: Norton, 1974.

[R11.4] Aruka, Y., *Evolutionary Foundations of Economic Science: How Can Scientists Study Evolving Economic Doctrines from the Last Centuries*, New York: Springer, 2015.

[R11.5] Ayer, A., 1936. *Language, Truth and Logic*, 2nd edition, reprinted New York: Dover, 1946.

[R11.6] Backhouse, R., *Truth and Progress in Economic Knowledge*, Cheltenham: Edward Elgar 1997.

[R11.7] Bavetta, Sebastiano, Pietro Navarra, and Dario Maimone, , *Freedom and the Pursuit of Happiness: An Economic and Political Perspective*, Cambridge: Cambridge University Press 2014.

[R11.8] Becker, G., *The Economic Approach to Human Behavior*, Chicago: University of

Chicago Press, 1976.

[R11.9] Berger, L., "Economics and Hermeneutics", *Economics and Philosophy*, Vol. 5: 1989, pp. 209–34.

[R11.10] Boehm, S. *et al* (eds.), *Is There Progress in Economics? Knowledge, Truth and the History of Economic Thought*, Cheltenham: Edward Elgar, 2002.

[R11.11] Boylan, T. and P. O'Gorman, *Beyond Rhetoric and Realism in Economics: Towards a Reformulation of Economic Methodology*, London: Routledge,1995.

[R11.12] Brousseau, E. and J. Glachant (eds.), *New Institutional Economics: A Guidebook*, Cambridge: Cambridge University Press, 2008.

[R11.13] Brunner, K., "'Assumptions' and the Cognitive Quality of Theories", *Synthese*, Vol. 20, 1969. pp.501–25.

[R11.14] Brzezinski, J., F. Coniglione, R. Kuipers, and L. Nowak (eds.), *Idealization I: General Problems* (Poznan Studies in the Philosophy of the Sciences and Humanities, Vol, 16), Amsterdam: Rodopi, 1990.

[R11.15] Camerer, C., *Behavioral Game Theory: Experiments in Strategic Interaction*, Princeton: Princeton University Press, 2003.

[R11.16] Camerer, C, "Neuroeconomics: Using Neuroscience to Make Economic Predictions", *Economic Journal*, Vol. 117, 2007. pp. C26–C42.

[R11.17] Camerer, C., G. Loewenstein, and M. Rabin (eds.), *Advances in Behavioral Economics*, Princeton: Princeton University Press, 2003.

[R11.18] Camerer, C., G. Loewenstein and D. Prelec, "Neuroeconomics: How Neuroscience Can Inform Economics", *Journal of Economic Literature*, Vol.43, 2005.: pp.9–64.

[R11.19] Clark, Andrew and Sarah Flèche, *The Origins of Happiness: The Science of Well-Being over the Life Course*, Princeton: Princeton University Press. 2018.

[R11.20] Cohen, I., N. Daniels, and N. Eyal (eds.), ‚*Identified Versus Statistical Lives: An Interdisciplinary Perspective*, New York: Oxford University Press, 2015.

[R11.21] Coleman, J.,. "Economics and the Law: A Critical Review of the Foundations of the Economic Approach to Law", *Ethics*, Vol. 94, 1984, pp. 649–79.

[R11.22] Davis, J. (ed.),. *Recent Developments in Economic Methodology*, Cheltenham: Edward Edgar 2006.

[R11.23] De Alessi, L., "Reversals of Assumptions and Implications", *Journal of Political Economy*, Vol. 79 1971, pp.867–77.

[R11.24] Deaton, A "Instruments, Randomization, and Learning about Development", *Journal of Economic Literature*, Vol. 48, 2010, pp. 424 - 55.

[R11.25] De Marchi, D. and M. Blaug (eds.),. *Appraising Modern Economics: Studies in the Methodology of Scientific Research Programs*, Cheltenham: Edward Elgar,1991.

[R11.26] Diesing, P., *Science and Ideology in the Policy Sciences*, New York: Aldine, 1982.

[R11.27] Dolan, Paul and Daniel Kahneman , *Happiness by Design: Change What You Do, Not How You Think* New York: Avery Press, 2014.

[R11.28] Engelskirchen, H., *Capital as a Social Kind: Definitions and Transformations in the Critique of Political Economy*, New York: Routledge, 2011.

[R11.29] Epstein, Brian, *The Ant Trap: Rebuilding the Foundations of the Social Sciences*, New York: Oxford University Press, 2015.

[R11.30] Fox, G., *Reason and Reality in the Methodologies of Economics*, Cheltenham: Edward

Elgar 1997.

[R11.31] Frey, B., *Economics as a Science of Human Behaviour: Towards a New Social Science Paradigm*, 2nd edition, Dordrecht: Kluwer, 1999.

[R11.32] Gaus, G., *On Philosophy, Politics, and Economics*, Belmont, CA: Wadsworth, 2008.

[R11.33] George, D. (ed.), *Issues in Heterodox Economics*, *Journal of Economic Surveys*, Volume 21, No. 3, Malden, MA and Oxford: Blackwell, 2007.

[R11.34 Glimcher, P., *Foundations of Neuroeconomic Analysis.* New York: Oxford University Press, 2010.

[R11.35] Glimcher, P., C. Camerer, R. Poldrack, and E. Fehr (eds.), *Neuroeconomics: Decision Making and the Brain*, Amsterdam: Elsevier, 2009.

[R11.36] Gustafsson, B., C. Knudsen, and U. Mäki (eds.), *Rationality, Institutions and Economic Methodology*, London: Routledge, 1993.

[R11.37] Henderson, W., T. Dudley-Evans, and R. Backhouse (eds.), *Economics and Language*, London: Routledge, 1993.

[R11.38] Hodgson, G., "What Is the Essence of Institutional Economics?" *Journal of Economic Issues*, Vol. 34, 2000, pp.317–29.

[R11.39] Hutchison, T., *Knowledge and Ignorance in Economics*, Chicago: University of Chicago Press1977.

[R11.40] Jalladeau, J., "Research Program versus Paradigm in the Development of Economics",*Journal of Economic Issues*, Vol. 12, 1978, pp. 583–608.

[R11.41] Katouzian, H., *Ideology and Method in Economics*, New York: New York University Press,1980.

[R11.42] Kincaid, H., *Philosophical Foundations of the Social Sciences: Analyzing Controversies in Social Research*, Cambridge: Cambridge University Press,1996.

[R11.43] Knight, F., 1935. "Economics and Human Action", from Knight 1935b; reprinted in Hausman, ed. 2008b, pp. 111–18.

[R11.44] Koopmans, T., *Three Essays on the State of Economic Science*, New York: McGraw-Hill,1957.

[R11.45] Krupp, S. (ed.), *The Structure of Economic Science*, Englewood Cliffs: Prentice-Hall 1966.

[R11.46] Kuipers, T. (ed.), *What is Closer-to-the Truth? A Parade of Approaches to Truthlikeness*(Poznan Studies in the Philosophy of the Sciences and Humanities, Volume 10), Amsterdam: Rodopi, 1987.

[R11.47] Lavoie, D. (ed.), *Economics and Hermeneutics*, London: Routledge, 1990.

[R11.48] Lawson, T., *Economics and Reality*, London: Routledge, 1997.

[R11.49] Lee, F. and B. Cronin (eds.), *Handbook of Research Methods and Applications in Heterodox Economics*, Cheltenham, UK: Elgar 2016.

[R11.50] Leontief, W., "Theoretical Assumptions and Non-observed Facts", *American Economic Review*, Vol. 61,1971. Pp. 1–7.

[R11.51] Loasby, B., *Choice, Complexity and Ignorance*, Cambridge: Cambridge University Press, 1976.

[R11.52] Loewenstein, G., S. Rick, and J.D. Cohen, "Neuroeconomics", *Annual Review of Psychology*, Vol. 59, 2008, pp. 647–672.

[R11.53] McCloskey, D., *The Rhetoric of Economics*, Madison: University of Wisconsin Press,

1985.

[R11.54] Marchionni, C. and J. Vromen (eds.), *Neuroeconomics: Hype or Hope?*, London, Routledge, 2014.

[R11.55] Mayer, T., *Truth Versus Precision in Economics*, Cheltenham: Edward Elgar, 1993.

[R11.56] Mises, L. von, *The Ultimate Foundation of Economic Science: An Essay on Method*, 2nd edition, Kansas City: Sheed Andrews, 1978.

[R11.57] Mishan, E., *Cost Benefit Analysis: An Introduction*, New York: Praeger, 1971.

[R11.58] Mishan, E, "Imagination and Imaging in Model Building", *Philosophy of Science*, Vol. 71, 2004, pp. 753–66.

[R11.59] Nagel, E., "Assumptions in Economic Theory", *American Economic Review Papers and Proceedings*, Vol. 53,1963, pp. 211–19.

[R11.60] Neuberg, L., *Conceptual Anomalies in Economics*, Cambridge: Cambridge University Press, 1988.

[R11.61] Oakley, A., *Reconstructing Economic Theory: The Problem of Human Agency*, Cheltenham, Edward Elgar, 2002.

[R11.62] Ochangco, A.,. *Rationality in Economic Thought.: Methodological Ideas on the History of Political Economy*, Cheltenham: Edward Elgar, 1999.

[R11.63] O'Sullivan, P., *Economic Methodology and Freedom to Choose*, London: Allen & Unwin, 1987.

[R11.64] Pratten, S. (ed.), *Social Ontology and Modern Economics*, London: Routledge, 2015.

[R11.65] Rabin, M., "Psychology and Economics", *Journal of Economic Literature*, Vol. 36,1998, pp. 11–46.

[R11.66] Radcliff, Benjamin, *The Political Economy of Human Happiness*, Cambridge: Cambridge University Press 2013.

[R11.67] Redman, D., *Economic Methodology: A Bibliography with References to Works in the Philosophy of Science, 1860–1988*, New York: Greenwood Press, 1989.

[R11.68] Reuter, M., and C. Montag (eds.), *Studies in Neuroscience, Psychology and Behavioral Economics*, New York: Springer, 2016.

[R11.69] Richter, R. (ed.), *Essays on New Institutional Economics.* New York: Springer, 2015.

[R11.70] Ross, D., *Economic Theory and Cognitive Science: Microexplanation*, Cambridge, MA: MIT Press, 2005.

[R11.71] Schlefer, J., *The Assumptions Economists Make*, Cambridge, MA: Harvard University Press, 2012.

[R11.72] Sen, A. and B. Williams (eds.), *Utilitarianism and Beyond*, Cambridge: Cambridge University Press1982.

[R11.73] Shrader-Frechette, K., *Science Policy, Ethics, and Economic Methodology: Some Problems of Technology Assessment and Environmental-Impact Analysis*, Dordrect: D. Reidel,1984.

[R11.74] Simon, H., "Theories of Decision-Making in Economics and Behavioral Science", *American Economic Review*, 49: 1959. 253–83.

[R11.75] Sowell, T., *Knowledge and Decisions*, New York: Basic Books,1980.

[R11.76] Vromen, J., *Economic Evolution: An Inquiry into the Foundations of Institutional Economics*, London: Routledge.1995.

[R11.77] Vromen, J. and C. Marchionni (eds.), *Neuroeconomics*, London: Routledge, 2018.

[R11.78] Weimann, Joachim and Andreas Knabe, *Measuring Happiness: The Economics of*

Well-Being, Cambridge, MA: MIT Press,2015.

[R11.79] Wilson, D. and A. Kirman (eds.), *Complexity and Evolution: Toward a New Synthesis for Economics*, Cambridge, MA: MIT Press,2016.

R12. PHILOSOPHY OF INFORMATION AND SEMANTIC INFORMATION

[R12.1] Aisbett, J., Gibbon, G.: "A practical measure of the information in a logical theory" *Journal of Experimental and Theoretical Artificial Intelligence* Vol. 11(2), 1999, pp. 201–218.

[R12.2] Badino, M.: "An Application of Information Theory to the Problem of the Scientific Experiment" *Synthese* Vol. 140, 2004, pp.355–389.

[R12.3] Bar-Hillel, Y. (ed.), *Language and Information: Selected Essays on Their Theory and Application,* Reading, Addison-Wesley, (1964)

[R12.4] Bar-Hillel, Y., Carnap, R. "An Outline of a Theory of Semantic Information," (1953); in Bar-Hillel, Y. (ed.), *Language and Information: Selected Essays on Their Theory and Application,* Reading, Addison-Wesley, (1964), pp. 221–274.

[R12.5] Barwise, J., Seligman, J.: *Information Flow: The Logic of Distributed Systems*, Cambridge, University Press, Cambridge (1997)

[R12.6] Braman, S.: "Defining Information," *Telecommunications Policy*, Vol. 13, pp. 233–242 (1989)

[R12.7] Bremer, M.E.: "Do Logical Truths Carry Information?" *Minds and Machines* Vol. 13(4), 2003, pp. 567–575.

[R12.8] Bremer, M. and Cohnitz, D., *Information and Information Flow: an Introduction*, Ontos Verlag, Frankfurt, Lancaster, 2004.

[R12.9] Chaitin, G.J., *Algorithmic Information Theory*, Cambridge, Cambridge University

Press, 1987.

[R12.10] Chalmers, D.J.: *The Conscious Mind: In Search of a Fundamental Theory*, New York, Oxford Univ. Press, (1996).

[R12.11] Cherry, C.: *On Human Communication: A Review, a Survey, and a Criticism*, Cambridge, MIT Press, 1978.

[R12.12] Colburn, T.R., *Philosophy and Computer Science*, Armonk, M.E. Sharpe, 2000.

[R12.13] Cover, T.M., Thomas, J.A.: *Elements of Information Theory*, New York, Wiley, 1991.

[R12.14] Dennett, D.C.: "Intentional Systems," *The Journal of Philosophy,* Vol. 68, 1971, pp. 87–106.

[R12.15] Deutsch, D., *The Fabric of Reality*, London Penguin, 1997.

[R12.16] Devlin, K.J., *Logic and Information.* Cambridge, Cambridge University Press, 1991.

[R12.17] Fetzer, J.H., "Information, Misinformation, and Disinformation" *Minds and Machines*, Vol. 14(2), 2004, pp. 223–229.

[R12.18] Floridi, L., *Philosophy and Computing: An Introduction*, London, Routledge, (1999)

[R12.19] Floridi, L., "What Is the Philosophy of Information?" *Metaphilosophy*, Vol. 33(1-2) (2002), pp. 123–145,

[R12.20] Floridi, L., "Two Approaches to the Philosophy of Information," *Minds and Machines,* Vol. 13(4), (2003), pp. 459–469.

[R12.21] Floridi, L.: "Open Problems in the Philosophy of Information," *Metaphilosophy* Vol. 35(4), 2004, pp. 554–582.

[R12.22] Floridi, L., "Outline of a Theory of Strongly Semantic Information," *Minds and Machines,* Vol. 14(2), 2004, pp. 197–222.

[R12.23] Floridi, L., "Is Information Meaningful Data?," *Philosophy and Phenomenological*

Research, Vol. 70(2), 2005, pp. 351–370.

[R12.24] Fox, C.J.: *Information and Misinformation: An Investigation of the Notions of Information, Misinformation, Informing, and Misinforming,* Westport Greenwood Press, 1983.

[R12.25] Frieden, B.R., *Science from Fisher Information: A Unification*, Cambridge University Press, Cambridge, 2004.

[R12.26] Golan, A., "Information and Entropy Econometrics - Editor's View". Journal of Econometrics, Vol. 107(1-2), 2002, pp.1–15

[R12.27] Graham, G., *The Internet: A Philosophical Inquiry.* , London, Routledge, 1999.

[R12.28] Grice, H.P. *Studies in the Way of Words,* Cambridge, Harvard University Press, 1989.

[R12.29] Hanson, P.P. (ed.), *Information, language, and cognition*, Vancouver, University of British Columbia Press, 1990.

[R12.30] Harms, W.F., "The Use of Information Theory in Epistemology," *Philosophy of Science*, Vol. 65(3), 472–501 (1998).

[R12.31] Heil, J., "Levels of Reality," *Ratio* Vol. 16(3), 2003, pp. 205–221

[R12.32] Hintikka, J., Suppes, P. (eds.), *Information and Inference*, Reidel, Dordrecht, 1970.

[R12.33] Kemeny, J., "A Logical Measure Function," *Journal of Symbolic Logic* Vol. 18, 1953, pp. 289–308.

[R12.34] Kolin K.K. "The Nature of Information and Philosophical Foundations of Informatics". *Open Education,* Vol.*2*, (2005) pp. 43-51.

[R12.35] Kolin K.K.). "The Evolution of Informatics," *Information Technologies,* Vol.1, 2005, pp. 2-16.

[R12.36] Kolin K.K., "The Formation of Informatics as Basic Science and Complex

Scientific Problems," In K.Kolin (Ed.), *Systems and Means of Informatics. Special Issue. Scientific and Methodological Problems of Informatics.* Moscow: IPI RAS, 2006, Pp. 7-57.

[R12.37] Kolin K.K., Fundamental Studies in Informatics: A General Analysis, Trends and Prospects. *Scientific and Technical Information,* Vol.1, (7), 2007, pp. 5-11.

[R12.38] Kolin K.K. "Structure of Reality and the Phenomenon of Information," *Open Education,* Vol. 5, 2008, pp. 56-61.

[R12.39] Losee, R.M., "A Discipline Independent Definition of Information," *Journal of the American Society for Information Science* Vol. 48(3), 1997, pp. 254–269.

[R12.40] Lozinskii, E. "Information and evidence in logic systems," *Journal of Experimental and Theoretical Artificial Intelligence*, Vol. 6, 1994, pp. 163–193.

[R12.41] Machlup, F., Mansfield, U. (eds.), *The Study of Information: Interdisciplinary Messages*, New York, Wiley, 1983.

[R12.42] MacKay, D.M., *Information, Mechanism and Meaning*, Cambridge MIT Press, 1969.

[R12.43] Marr, D., Vision: *A Computational Investigation into the Human Representation and Processing of Visual Information*. San Francisco, W.H. Freeman, 1982.

[R12.44] Mingers, J., "The Nature of Information and Its Relationship to Meaning," In: Winder, R.L., et al. (eds.) *Philosophical Aspects of Information Systems*, London, Taylor and Francis, 1997, pp. 73–84.

[R12.45] Nauta, D., *The Meaning of Information*, The Hague, Mouton, 1972.

[R12.46] Newell, A. "The Knowledge Level". *Artificial Intelligence* Vol. 18, 1982, pp. 87–127.

[R12.47] Newell, A., Simon, H.A., "Computer Science as Empirical Inquiry: Symbols and Search," *Communications of the ACM*, Vol. 19, 1976 pp.113–126.

[R12.48] Pierce, J.R., *An Introduction to Information Theory: Symbols, Signals and Noise*, New York, Dover Publications, 1980.

[R12.49] Poli, R., "The Basic Problem of the Theory of Levels of Reality" *Axiomathes*, Vol. 12, 2001, pp. 261–283.

[R12.50] Sayre, K.M.: *Cybernetics and the Philosophy of Mind*. London, Routledge and Kegan Paul, (1976).

[R12.51] Simon, H.A., *The Sciences of the Artificial*, Cambridge, MIT Press, 1996.

[R12.52] Smokler, H., "Informational Content: A Problem of Definition," *The Journal of Philosophy,* Vol. 63(8), 1966, pp. 201–211.

[R12.53] Ursul A.D., *The Nature of the Information. Philosophical Essay*. Moscow: Politizdat, (1968).

[R12.54] Ursul A.D., *Information. Methodological Aspects*. Moscow: Nauka, 1971.

[R12.55] Ursul A.D., *Reflection and Information.* Moscow: Nauka (1973).

[R12.56] Ursul A.D. *The Problem of Information in Modern Science: Philosophical Essays.* Moscow: Nauka. (1975).

[R12.57] Ursul, A. D. "The Problem of the Objectivity of Information" in Libor Kubát ř, and Jiří Zeman (eds.), *Entropy and Information in Science and Philosophy,* New York, Elsevier, 1975.

[R12.58] Weaver, W., "The Mathematics of Communication," *Scientific American* Vol. 181(1), 1949, pp. 11–15.

[R12.59] Winder, R.L., Probert, S.K., Beeson, I.A.: *Philosophical Aspects of Information Systems*, London, Taylor & Francis 1997.

R13 PLANNING, PRESCRIPTIVE SCIENCE AND INFORMATION IN COST-BENEFIT ANALYTICS

[R13.1] Alexander, Ernest R., *Approaches to Planning*, Philadelphia, Pa. Gordon and Breach, 1992.

[R13.2] Bailey, J., *Social Theory for Planning*, London, Routledge and Kegan Paul, 1975.

[R13.3] Burchell, R.W. and G. Sternlieb (eds.), *Planning Theory in the 1980's: A Search for Future Directions*, New Brunswick, N. J., Rutgers University Center for Urban and Policy Research, 1978.

[R13.4] Cambis, Marios, *Planning Theory and Philosophy*, London, Tavistock Publicationa, 1979.

[R13.5] Chadwick, G., *A Systems View of Planning,* Oxford, Pergamon, 1971.

[R13.6] Cooke, P., *Theories of Planning and Special Development*, London, Hutchinson,1983

[R13.7] Dompere, Kofi K., and Taresa Lawrence, "Planning," in Syed B Hussain, *Encyclopedia of Capitalism,* Vol. II, New York, Facts On File, Inc., 2004, pp. 649-653.

[R13.8] Dompere, Kofi K., *Social Goal-Objective Formation, Democracy and National Interest: A Theory of Political Economy under Fuzzy Rationality*, (Studies in Systems, Decision and Control, Vol. 4), New York, Springer, 2014

[R13.9] Dompere, Kofi K., *Fuzziness, Democracy, Control and Collective Decision-Choice System: A Theory on Political Economy of Rent-Seeking and Profit-Harvesting,* (Studies In Systems, Decision and Control, Vol. 5), New York, Springer, 2014.

[R13.10] Dompere, Kofi K., *The Theory of Aggregate Investment in Closed Economic Systems*, Westport, CT, Greenwood Press, 1999.

[R13.11] Dompere, Kofi K., *The Theory of Aggregate Investment and Output Dynamics in Open Economic Systems*, Westport, CT, Greenwood Press, 1999.

[R13.12] Faludi, A., *Planning Theory*, Oxford, Pergamon, 1973.

[R13.13] Faludi, A.(ed.), *A Reader in Planning Theory,* Oxford, Pergamon, 1973.

[R13.14] Harwood, E.C. (ed.), Reconstruction of Economics, American Institute For Economic Research, Great Barrington, Mass, 1955., Also in John Dewey and Arthur Bently, 'Knowing and the known', Boston, Beacon Press, 1949, p. 269.

[R13.15] Kickert, W.J.M., *Organization of Decision-Making A Systems-Theoretic Approach*, New York, North-Holland, 1980.

[R13.16] Knight, Frank H. *Risk, Uncertainty and Profit,* Chicago, University of

Chicago Press, 1971.

[R13.17] Knight, Frank H. *On History and Method of Economics,* Chicago, University of Chicago Press, 1971.

R.14. POSSIBLE-ACTUAL WORLDS AND INFORMATION ANALYTICS

[R14.1] Adams, Robert M., "Theories of Actuality," *Noûs*, Vol. 8, 1974, pp211-231.

[R14.2] Allen, Sture (ed.) *Possible Worlds in Humanities, Arts and Sciences*, Proceedings of Nobel Symposium, Vol. 65, New York, Walter de Gruyter Pub., 1989.

[R14.3] Armstrong, D. M., *A Combinatorial Theory of Possibility.* Cambridge University Press, 1989.

[R14.4] Armstrong, D.M *A World of States of Affairs*, Cambridge, Cambridge University Press 1997.

[R14.5] Bell, J.S., "Six Possible Worlds of Quantum Mechanics" in Allen, Sture (Ed.) *Possible Worlds in Humanities, Arts and Sciences*, Proceedings of Nobel Symposium, Vol. 65, New York, Walter de Gruyter Pub., 1989.pp. 359-373.

[R14.6] Bigelow, John. "Possible Worlds Foundations for Probability", *Journal of Philosophical Logic,* 5 (1976), pp. 299-320.

[R14.7] Bradley, Reymond and Norman Swartz, *Possible World: An Introduction to Logic and its Philosophy,* Oxford, Bail Blackwell, 1997.

[R14.8] Castañeda, H.-N. "Thinking and the Structure of the World", *Philosophia*, 4 (1974), pp. 3-40.

[R14.9] Chihara, Charles S. *The Worlds of Possibility: Modal Realism and the Semantics of Modal Logic*, Clarendon, 1998.

[R14.10] Chisholm, Roderick. "Identity through Possible Worlds: Some Questions", *Noûs,* 1 1967, pp. 1-8; reprinted in Loux, *The Possible and the Actual*[R14.21].

[R14.11] Divers, John, *Possible Worlds*, London: Routledge, 2002.

[R14.12] Forrest, Peter. "Occam's Razor and Possible Worlds", *Monist,* 65 (1982), pp. 456-64.

[R14.13] Forrest, Peter, and Armstrong, D. M. "An Argument Against David Lewis' Theory of Possible Worlds", *Australasian Journal of Philosophy*, 62 (1984), pp. 164-168.

[R14.14] Grim, Patrick, "There is No Set of All Truths", *Analysis*, Vol. 46, 1986, pp. 186-191.

[R14.15] Heller, Mark. "Five Layers of Interpretation for Possible Worlds", *Philosophical Studies*, 90 (1998), pp. 205-214.

[R14.16] Herrick, Paul, *The Many Worlds of Logic,* Oxford: Oxford University Press1999.

[R14.17] Krips, H. "Irreducible Probabilities and Indeterminism", *Journal of Philosophical Logic*, Vol. 18, 1989, pp. 155-172.

[R14.18] Kuhn, Thomas S., "Possible Worlds in History of Science" in Allen, Sture (ed.) *Possible Worlds in Humanities, Arts and Sciences,* Proceedings of Nobel Symposium, Vol. 65, New York, Walter de Gruyter Pub. , 1989 pp. 9-41.

[R14.19] Kuratowski, K. and Mostowski, A. *Set Theory: With an Introduction to Descriptive Set Theory*, New York: North-Holland, 1976.

[R14.20] Lewis, David, *On the Plurality of Worlds*, Oxford , Basil Blackwell, 1986.

[R14.21] Loux, Michael J. (ed.) *The Possible and the Actual*: *Readings in the Metaphysics of Modality,* Ithaca & London: Cornell University Press,1979.

[R14.22] Parsons, Terence, *Nonexistent Objects,* New Haven, Yale University Press, 1980.

[R14.23] Perry, John, "From Worlds to Situations", *Journal of Philosophical Logic*, Vol. 15, 1986, pp. 83-107.

[R14.24] Rescher, Nicholas and Brandom, Robert. *The Logic of Inconsistency: A Study in Non-Standard Possible-World Semantics and Ontology*, Rowman and Littlefield, 1979.

[R14.25] Skyrms, Brian. "Possible Worlds, Physics and Metaphysics", *Philosophical Studies,* Vol. 30, 1976, pp. 323-32.

[R14.26] Stalmaker, Robert C. "Possible World", *Noûs* , Vol. 10, 1976, pp. 65-75

[R14.27] Quine, W.V.O. *Word and Object*, M.I.T. Press, 1960.

[R14.28] Quine, W.V.O "Ontological Relativity", *Journal of Philosophy*, 65 (1968), pp. 185-212.

R15. RATIONALITY, INFORMATION, GAMES, CONFLICTS AND EXACT REASONING

[R15.1] Border, Kim, *Fixed Point Theorems with Applications to Economics and Game Theory,* Cambridge, Cambridge University Press 1985.

[R15.2] Brandenburger, Adam, "Knowledge and Equilibrium Games," *Journal of Economic Perspectives*, Vol.6, 1992, pp. 83-102.

[R15.3] Campbell, Richmond and Lanning Sowden, *Paradoxes of Rationality and Cooperation: Prisoner's Dilemma and Newcomb's Problem*, Vancouver, University of British Columbia Press, 1985.

[R15.4] Gates Scott and Brian Humes, *Games, Information, and Politics: Applying Game Theoretic Models to Political Science*, Ann Arbor, University of Michigan Press, 1996.

[R15.5] Gjesdal, Froystein, "Information and Incentives: The Agency Information Problem," *Review of Economic Studies*, Vol.49, 1982, pp373-390.

[R15.6] Harsanyi, John, "Games with Incomplete Information Played by 'Bayesian' Players I: The Basic Model," *Management Science,* Vol.14, 1967, pp.159-182.

[R15.7] Harsanyi, John, "Games with Incomplete Information Played by 'Bayesian' Players II: Bayesian Equilibrium Points," *Management Science*, Vol.14, 1968, pp.320-334.

[R15.8] Harsanyi, John, "Games with Incomplete Information Played by 'Bayesian' Players III: The Basic Probability Distribution of the Game," *Management Science*, Vol.14, 1968, pp.486-502.

[R15.9] Harsanyi, John, *Rational Behavior and Bargaining Equilibrium in Games and Social Situations*, New York Cambridge University Press, 1977.

[R15.10] Krasovskii, N.N. and A.I. Subbotin, *Game-theoretical Control Problems*, New York, Springer-Verlag, 1988.

[R15.11] Lagunov, V. N., *Introduction to Differential Games and Control Theory*, Berlin, Heldermann Verlag, 1985.

[R15.12] Maynard Smith, John, *Evolution and the Theory of Games,* Cambridge, Cambridge University Press, 1982.

[R15.13] Myerson, Roger, *Game Theory: Analysis of Conflict*, Cambridge, Mass. Harvard University Press, 1991.

[R15.14] Rapoport, Anatol and Albert Chammah, *Prisoner's Dilemma: A Study in Conflict and Cooperation*, Ann Arbor, University of Michigan Press, 1965.

[R15.15] Roth, Alvin E., "The Economist as Engineer: Game Theory, Experimentation, and Computation as Tools for Design Economics," *Econometrica,* Vol.70, 2002, pp1341-1378.

[R15.16] Shubik, Martin, *Game Theory in the Social Sciences: Concepts and Solutions*, Cambridge, Mass., MIT Press, 1982.

R16. SOCIAL SCIENCES, MATHEMATICS AND THE PROBLEMS OF EXACT AND INEXACT INFORMATION

[R16.1] Ackoff, R.L., *Scientific Methods: Optimizing Applied Research Decisions*, New York, John Wiley, 1962.

[R16.2] Angyal, A. "The Structure of Wholes," *Philosophy of Sciences*, Vol.6, #1, 1939, pp 23-37.

[R16.3] Bahm, A.J., "Organicism: The Philosophy of Interdependence" *International Philosophical Quarterly*, Vol. VII # 2, 1967.

[R16.4] Bealer, George, *Quality and Concept*, Oxford, Clarendon Press, 1982.

[R16.5] Black, Max, *Critical Thinking*, Englewood Cliffs, N.J., Prentice-Hall, 1952.

[R16.6] Brewer, Marilynn B., and Barry E Collins (eds.) *Scientific Inquiry and Social Sciences,* San Francisco, Ca, Jossey-Bass Pub., 1981.

[R16.7] Campbell, D.T., "On the Conflicts Between Biological and Social Evolution and Between Psychology and Moral Tradition", *American Psychologist*, Vol. 30, 1975, pp1103-1126.

[R16.8] Churchman, C. W. and P. Ratoosh (eds.) *Measurement: Definitions and Theories*, New York, John Wiley, 1959.

[R16.9] Foley, Duncan, "Problems versus Conflicts Economic Theory and Ideology" American Economic Association Papers and Proceedings, Vol. 65, May 1975, pp. 231-237.

[R16.10] Garfinkel, Alan, *Forms of Explanation: Structures of Inquiry in Social Science,* New Haven, Conn., Yale University Press, 1981.

[R16.11] Georgescu-Roegen, Nicholas, *Analytical Economics*, Cambridge, Harvard University Press, 1967.

[R16.12] Gillespie, C., *The Edge of Objectivity*, Princeton, Princeton University press, 1960.

[R16.13] Hayek, F.A., *The Counter-Revolution of Science*, New York, Free Press of Glencoe Inc, 1952.

[R16.14] Laudan, L., *Progress and Its Problems: Towards a Theory of Scientific Growth,* Berkeley, CA, University of California Press, 1961.

[R16.15] Marx, Karl, *The Poverty of Philosophy,* New York, International Pub. 1971.

[R16.16] Miller, John W., *The Philosophy of History with Reflection and Aphorisms,* Norton, New York, 1951.

[R16.17] Phillips, Denis C., *Holistic Thought in Social Sciences*, Stanford, CA, Stanford University Press, 1976.

[R16.18] Popper, K., *Objective Knowledge*, Oxford, Oxford University Press, 1972.

[R16.19] Rashevsky, N. "Organismic Sets: Outline of a General Theory of Biological and Social Organism," *General Systems*, Vol XII, 1967, pp. 21-28.

[R16.20] Roberts, Blaine, and Bob Holdren, *Theory of Social Process*, Ames, Iowa University Press, 1972.

[R16.21] Rudner, Richard S., *Philosophy of Social Sciences*, Englewood Cliff, N.J., Prentice Hall, 1966.

[R16.22] Simon, H. A., "The Structure of Ill-Structured Problems," *Artificial Intelligence,* Vol. 4, 1973, pp. 181-201 .

[R16.23] Toulmin, S., *Foresight and understanding: An Enquiry into the Aims of Science,* New York, Harper and Row, 1961.

[R16.24] Winch, Peter, *The Idea of a Social Science*, New York, Humanities Press, 1958.

17 TRANFORMATIONS, DECISIONS, POLARITY, DUALITY AND CONFLICTS

[R17.1a] Abraham, W.E., *The Mind of Africa,*Chicago, The University of Chicago Press,1962.

[R17.1b] Amen, Ra Un Nefer, *Metu Neter Vol.1*, Khamit Corp., Bronx, NY, 1990.

[R17.2a] Anovsky, 0mely M.E., *Linin and Modern Natural Science, Moscow*, Progress Pub. 1978

[R17.2b] Arrow, Kenneth J., "Limited Knowledge and Economic Analysis", American Economic Review, Vol. 64, 1974, pp. 1-10.

[R17.3a] Asante, Molefi K, *Revolutionary Pedagogy,* Chicago African American Image Pub., 2003

[R17.3b] Asante, Molefi K, *Afrocentricity:The Theory of Social Change*, African American Image. Pub. Chicago, 2003

[R17.4a] Ben-Jochannan, Yosef A, A., *Cultural Genocide in the Black and African Studies Curriculum,* Baltimore, U.S.A..,Black Classic Press,2004 .

[R17.4b] Ben-Jochannan, Yosef A, A., *Africa: Mother of Western Civilization,* Baltimore, MD, U.S.A., Black Classic Press, 1997.

[R17.5a] Berkeley, George, *Treatise Concerning the Principles of Human Knowledge, Works*, Vol. I (edited by A. Fraser), Oxford, Oxford University Press, 1871-1814.

[R17.5b] Berkeley, George, "Material Things are Experiences of Men or God" in [R1.5], 1967, pp. 658-668.

[R17.6a] Boulding, Kenneth E., *A New Theory of Societal Evolution,* Sage Pub. Beverly Hills Ca, 1978.

[R17.6b] Brody, Baruch A. (ed.), *Readings in the Philosophy of Science*, Englewood Cliffs, NJ., Prentice-Hall Inc., 1970.

[R17.7a] Brouwer, L.E.J., "Consciousness, Philosophy, and Mathematics," in Benecerraf, P. and H. Putnam (eds.), *Philosophy of Mathematics: Selected Readings* , Cambridge, Cambridge University Press, 1983. pp90-96.

[R17.7b] Brown, B. and J Woods (eds.), *Logical Consequence; Rival Approaches and New Studies in exact Philosophy: Logic, Mathematics and Science*, Vol. II Oxford, Hermes, 2000.

[R17.8] Cornforth, Maurice, *Dialectical Materialism and Science*, New York, International Pub.1960.

[R17.9] Cornforth, Maurice, *Materialism and Dialectical Method*, New York, International Pub.1953.

[R17.10] Cornforth, Maurice, *Science and Idealism: An Examination of "Pure Empiricism"*, New York International Pub.1947.

[R17.11a] Cornforth, Maurice, *The Open Philosophy and the Open Society: A Reply to Dr. Karl Popper's Refutations of Marxism* New York, International Pub.1968.

[R17.11b] Cornforth, Maurice, *The Theory of Knowledge*, New York, International Pub.1960.

[R17.12a] Diop, Cheikh A., *The African Origins of Civilization: Myth or Reality*, Brooklyn, New York, Lawrence Hill, 1974.

[R17.12b] Diop, Cheikh A., *Civilization or Barbarism*, Brooklyn, New York, Lawrence Hill, 1991.

[R17.13] Dompere, Kofi K., "On Epistemology and Decision-Choice Rationality" in R. Trapple (ed.), *Cybernetics and System Research*, New York, North Holland, 1982, pp. 219-228.

[R17.14] Dompere, Kofi K. and M. Ejaz, *Epistemics of Development Economics: Toward a Methodological Critique and Unity*, Westport, CT, Greenwood Press, 1995.

[R17.15] Dompere, Kofi K., *The Theory of Categorial Conversion: Rational Foundations of Nkrumaism in socio-natural Systemicity and Complexity,* London, Adonis-Abbey Pubs. 2016-2017.

[R17.16] Dompere, Kofi K., *The Theory of Philosophical Consciencism: Practice Foundations of Nkrumaism in Social Systemicity,,* London, Adonis-Abbey Pubs. 2016-2017.

[R17.17] Dompere, Kofi K., *The Theory of Info-Statics: Conceptual Foundations of Information and Knowledge*, New York, Springer (Series: Studies in Systems, Decision and Control, Vol. 112), 2017.

[R17.18] Dompere, Kofi K., *The Theory of Info-Dynamics: Rational Foundations of Information-Knowledge Dynamics,* New York, Springer (Series: Studies in Systems, Decision and Control, Vol. 114), 2017.

[R17.19a] Dompere, Kofi K., *Polyrhythmicity: Foundations of African Philosophy*, London, Adonis, and Abbey Pub, 2006.

[R17.19b] Dubois, W. E. B., *Dubois Speaks: Speeches and Addresses 1890-1919*, New York, Pathfinder,1970.

[R17.20] Engels, Frederick, *Dialectics of Nature*, New York, International Pub.,1971

[R17.21] Engels, Frederick, *Origin of the Family, Private Property and State*, New York, International Pub.,1971.

[R17.22] Ewing, A.C., "A Reaffirmation of Dualism" in [R1.5], pp. 454-461.

[R17.23] Fedoseyer, P.N. *et al., Philosophy in USSR: Problems of Dialectical Materialism*, Moscow, Progress Pub., 1977.

[R17.24] Kedrov, B. M., "On the Dialectics of Scientific Discovery," *Soviet Studies in Philosophy* , Vol. 6 1967, pp16 – 27.

[R17.25] Lenin, V. I. *Materialism and Empirio-Criticism: Critical Comments on Reactionary Philosophy*, New York, International Pub., 1970.

[R17.26] Lenin, V. I. *Collected Works Vol. 38: Philosophical Notebooks*, New York, International Pub., 1978.

[R17.27] Lenin, V. I., *On the National Liberation Movement*, Peking, Foreign Language Press, 1960.

[R17.28] Hegel, George, *Collected Works*, Berlin, Duncher und Humblot, 1832 – 1845 [also

Science of Logic, translated by W. H. Johnston and L. G. Struther , London, 1951].

[R17.29] Hempel, Carl G. and P. Oppenheim, "Studies in the Logic of Explanation," in [R15.5], pp. 8 – 27.

[R17.30] Ilyenkov, E.V., *Dialectical Logic: Essays on its History and Theory*, Moscow, Progress Pub. 1977.

[R17.31] Keirstead, B.S., "The Conditions of Survival," American Economic Review, Vol. 40, #2, pp.435- 445.

[R17.32] Kühne, Karl, *Economics and Marxism, Vol.I: The Renaissance of the Marxian System*, New York, St Martin's Press, 1979.

[R17.33] Kühne, Karl, *Economics and Marxism, Vol.II: The Dynamics of the Marxian System*, New York, St Martin's Press, 1979.

[R17.34] March, J. C., "Bounded Rationality, Ambiguity and Engineering of Choice," *The Bell Journal of Economics*, Vol. 9 (2), 1978.

[R17.35] Marx, Karl, *Contribution to the Critique of Political Economy*, Chicago, Charles H. Kerr and Co. 1904.

[R17.36] Marx, Karl, *Economic and Philosophic Manuscripts of 1884,* Moscow, Progress Pub., 1967.

[R17.37] Marx, Karl, *The Poverty of Philosophy*, New York, International Publishers, 1963.

[R17.38] Marx, Karl, Economic and Philosophic Manuscripts of 1844, Moscow, Progress Pub, 196

[R17.39] Massey, Gerald, *A Book of the Beginnings*, Vols. 1-2, London, William and Norgate, 1881.

[R17.40] Massey, Gerald, *Pyramid Text*, Vols. 1-4, New York, Longmans Green, 1952.

[R17.41] Massey, Gerald, *The Natural Genesis*, Vols. 1 and 2, Baltimore, MD, Black Classic Press, 1998 (First published 1883).

[R17.42] Massey, Gerald, *Ancient Egypt: The Light of the World*, Baltimore, MD, Black Classic Press, 1992.

[R17.43] Niebyl, Karl, H., "Modern Mathematics and Some Problems of Quantity, Quality and Motion in Economic Analysis," *Philosophy of Science*, Vol 7, # 1, January,1940, pp. 103 – 120.

[R17.44] Nkrumah, Kwame, *Consciencism*, London, Heinemann, 1964.

[R17.45] Obenga, Theophile T, *African Philosophy During The Period Of The Pharaohs, 2800-300BC*, Popenguine, Senegal, W.A., Per Ankh Pub., 2006.

[R17.46] Obenga, Theophile T, *African Philosophy in World History,* Popenguine, Senegal, W.A., Per Ankh Pub., 1998.

[R17.47] Price, H. H., *Thinking and Experience*, London, Hutchinson, 1953.

[R17.48] Putman, H., *Reason, Truth and History*, Cambridge, Cambridge University Press, 1981.

[R17.49] Putman. H., *Realism and Reason*, Cambridge, Cambridge University Press, 1983.

[R17.50] Quiggin, John, *Zombie Economics: How Dead Ideas Still Walk Among Us, Princeton* University Press, Prince ton, NJ. 2010.

[R17.51] Robinson, Joan, *Economic Philosophy.* New York, Anchor Books, 1962.

[R17.52] Robinson, Joan, *Freedom and Necessity: An Introduction to the Study of Society,* New York, Vintage Books, 1971.

[R17.53] Robinson, Joan, *Economic Heresies: Some Old-Fashioned Questions in Economic Theory,* New York, Basic Books, 1973.

[R17.54] Schumpeter, Joseph A., *The Theory of Economic Development*, Cambridge, Mass. Harvard University Press, 1934.

[R17.55] Schumpeter, Joseph A., *Capitalism, Socialism and Democracy*, New York, Harper & Row, 1950.

[R17.56] Schumpeter, Joseph A., "March to Socialism," *American Economic Review*, Vol. 40 May 1950, pp. 446 456.

[R17.57] Schumpeter, Joseph A., "Theoretical Problems of Economic Growth" *Journal of Economic History* Vol. 8, Supplement 1947, pp. 1-9.

[R17.58] Schumpeter, Joseph A., "The Analysis of Economic Change," *Review of Economic Statistics*, Vol. 17, 1935, pp 2-10.

[R17.59] Schwaller de Lubicz, R.A., *The Egyptian Miracle: An Introduction to the Wisdom of the Temple*, Rochester (Vermont) Inner Traditions International,1985.

[R17.60] Schwaller de Lubicz, R.A., *A Study of Numbers: A Guide to The constant creation of The Universe*, Rochester, (Vermont) Inner Traditions International, 1986.

[R17.61] Schwallier de Lubicz, R.A., *The Temple in Man: The Secrets of Ancient Egypt*, Brookline,

(Massachusetts), Autumn Press, 1977.

[R17.62] Schaller de Lubicz, R.A., *The Temple In Man: Sacred Architecture and The Perfect Man.*, Rochester, (Vermont), Inner Traditions Published, 1981.

[R17.63] Schwaller de Lubicz, R.A., *Symbol and the Symbolic: Egypt, Science, and The Evolution of Consciousness*, Brookline, (Massachusetts) Autumn Press, 1978.

[R17.64] Schwaller de Lubicz, R.A., *The Temple of Man: Apet of The South at Luxor*, Rochester, (Vermont) Inner Traditions, 1998.

[R17.65] Wright, Richard A. (ed.), *African Philosophy: An Introduction*, New York, University Press of America, 1984.

[R17.66] Weber, Luc E., and James J. Duderswtadt (Eds.), Reinventing the Research University, Economica, London, 2004.

[R17.67] Welsing, Frances, C., *The Isis Papers The Key to the Colors*, Chicago, Third World Pres, 1991

[R17.68] Woodson, Carter G, *Miseducation of the Negro,* Washington, D.C, The Associated Pub.1969.

[R17.69] Yesufu, T. M. (ed.) *Creating the African University: Emerging Issues of the 1970s*, Oxford, Oxford University Press, 1973.

R18. VAGUENESS, THINKING, APPROXIMATION AND REASONING IN THE INFORMATION-KNOWLEDGE PROCESS

[R18.1] Adams, E. w., and H. F. Levine, "On the Uncertainties Transmitted from Premises to Conclusions in deductive Inferences," *Synthese* Vol. 30, 1975, pp. 429 – 460.

[R18.2] Arbib, M. A., *The Metaphorical Brain*, New York, McGraw-Hill, 1971.

[R18.3] Bečvář, Jiří , "Notes on Vagueness and Mathematics," in Skala, Heinz J. et al., (eds.), *Aspects of Vagueness*, Dordrecht, D. Reidel Co. 1984, pp.1-11.

[R18.4] Black, M, "Vagueness: An Exercise in Logical Analysis," *Philosophy of Science*, Vol. 17, 1970, pp141-164.

[R18.5] Black, M. "Reasoning with Loose Concepts," *Dialogue*, Vol. 2, 1973, pp. 1-12.

[R18.6] Black, Max, *Language and Philosophy*, Ithaca, N.Y.: Cornell University

Press. 1949.

[R18.7] Black, Max , *The Analysis of Rules*, in Black, Max [R18.8] *Models and Metaphors: Studies in Language and Philosophy*, Ithaca, NewYork: Cornell University Press 1962 pp. 95-139.

[R18.8] Black, Max, *Models and Metaphors: Studies in Language and Philosophy*, Ithaca, NewYork: Cornell University Press. 1962.

[R18.9] Black, Max, *Margins of Precision*, Ithaca: Cornell University Press,1970.

[R18.10] Boolos, G. S. and R. C. Jeffrey, *Computability and Logic,* New York, Combridge University Press, 1989.

[R18.11] Cohen, P. R., *Heuristic Reasoning about uncertainty: An Artificial Intelligent Approach*, Boston, Pitman, 1985.

[R18.12] Darmstadter, H., "Better Theories," *Philosophy of Science*, Vol. 42, 1972, pp. 20 – 27.

[R18.13] Davis, M., *Computability and Unsolvability*, New York, McGraw-Hill, 1958.

[R18.14] Dummett, M., "Wang's Paradox," *Synthese,* Vol. 30, 1975, pp301 – 324.

[R18.15] Dummett, M., *Truth and Other Enigmas*, Cambridge, Mass. Harvard University Press, 1978.

[R18.16] Endicott, Timothy, *Vagueness in the Law*, Oxford, Oxford University Press, 2000.

[R18.17] Evans, Gareth, "Can there be Vague Objects?" *Analysis,* Vol. 38, 1978, p. 208.

[R18.18] Fine, Kit, "Vagueness, Truth and Logic," *Synthese,* Vol.54, 1975, pp. 235-259.

[R18.19] Gale, S., "Inexactness, Fuzzy Sets and the Foundation of Behavioral Geography," *Geographical Analysis*, Vol.4, #4, 1972, pp.337-349.

[R18.20] Ginsberg, M. L. (ed.), *Readings in Non-monotonic Reason*, Los Altos, Ca., Morgan Kaufman, 1987.

[R18.21] Goguen, J. A., "The Logic of Inexact Concepts," *Synthese*, Vol. 19, 1968/69, pp. 325 – 373.

[R18.22] Grafe, W., "Differences in Individuation and Vagueness," in A. Hartkamper and H. –J. Schmidt, *Structure and Approximation in Physical Theories*, New York, Plenum Press, 1981. pp.113–122.

[R18.23] Goguen, J. A, "The Logic of Inexact Concepts" *Synthese*, Vol.19, 1968-1969.

[R18.24] Graff, Delia and Timothy (eds.), *Vagueness*, Aldershot, Ashgate Publishing, 2002.

[R18.25] Hartkä Amper and H.J. Schmidt (eds.), *Structure and Approximation in Physical Theories*, New York, Plenum Press, 1981.

[R18.26] Hersh, H.M. et. al., "A Fuzzy Set Approach to Modifiers and Vagueness in Natural Language," *J. Experimental*, Vol. 105, 1976, pp. 254-276.

[R18.27] Hilpinen, R., "Approximate Truth and Truthlikeness," in M. Pprelecki et al. (eds.) *Formal Methods in the Methodology of Empirical Sciences*, Wroclaw, Reidel, Dordrecht and Ossolineum, 1976 pp. 19 – 42.

[R18.28] Hockney D. et al. (eds.), *Contemporary Research in Philosophical Logic and Linguistic Semantics*, Dordrecht-Holland, Reidel Pub. Co. 1975.

[R18.29] Höhle Ulrich et al (eds.), *Non-Clasical Logics and their Applications to Fuzzy Subsets: A Handbook of the Mathematical Foundations of Fuzzy Set Theory*, Boston, Mass. Kluwer, 1995.

[R18.30] Katz, M., "Inexact Geometry," *Notre-Dame Journal of Formal Logic*, Vol.21, 1980, pp. 521-535.

[R18.31] Katz, M., "Measures of Proximity and Dominance," *Proceedings of the Second World Conference on Mathematics at the Service of Man*, Universidad Politecnica de Las Palmas, 1982, pp. 370 – 377.

[R18.32] Katz, M., "The Logic of Approximation in Quantum Theory," *Journal of Philosophical Logic*, Vol. 11, 1982, pp. 215 – 228.

[R18.33] Keefe, Rosanna, *Theories of Vagueness*, Cambridge, Cambridge University Press, 2000.

[R18.34] Keefe, Rosanna and Peter Smith (eds.) *Vagueness: A Reader*, Cambridge, MIT Press, 1996.

[R18.35] Kling, R., "Fuzzy Planner: Reasoning with Inexact Concepts in a Procedural Problem-solving Language," *Jour. Cybernetics*, Vol. 3, 1973, pp. 1-16.

[R18.36] Kruse, R.E. et. al., *Uncertainty and Vagueness in Knowledge Based Systems*: *Numerical Methods*, New York, Springer-Verlag 1991.

[R18.37] Ludwig, G., "Imprecision in Physics," in A. Hartkämper and H.J. Schmidt (eds.), *Structure and Approximation in Physical Theories*, New York, Plenum Press, 1981, pp. 7 – 19.

[R18.38] Kullback, S. and R. A. Leibler, "Information and Sufficiency," *Annals of Math. Statistics*, Vol. 22, 1951, pp. 79 – 86.

[R18.39] Lakoff, George, "Hedges: A Study in Meaning Criteria and Logic of Fuzzy Concepts," in, Hockney D. et al. (eds.), *Contemporary Research in Philosophical Logic and Linguistic Semantics*, Dordrecht-Holland, Reidel Pub. Co. 1975, pp.221-271.

[R18.40] Lakoff, G., "Hedges: A Study in Meaning Criteria and the Logic of Fuzzy Concepts," *Jour. Philos. Logic,* Vol. 2, 1973, pp. 458-508.

[R18.41] Levi, I., *The Enterprise of Knowledge*, Cambridge, Mass. MIT Press 1980.

[R18.42] Łucasiewicz, J., *Selected Works: Studies in the Logical Foundations of Mathematics,* Amsterdam, North-Holland, 1970.

[R18.43] Machina, K.F., "Truth, Belief and Vagueness," *Jour. Philos. Logic*, Vol. 5, 1976, pp. 47-77.

[R18.44] Menges, G., et. al., "On the Problem of Vagueness in the Social Sciences," in Menges, G. (ed.), *Information, Inference and Decision*, D. Reidel Pub., Dordrecht, Holland, 1974, pp. 51-61.

[R18.45] Merricks, Trenton, "Varieties of Vagueness," *Philosophy and Phenomenological Research*, Vol.53, 2001, pp. 145-157.

[R18.46] Mycielski, J., "On the Axiom of Determinateness," *Fund. Mathematics*, Vol. 53, 1964, pp. 205 – 224.

[R18.47] Mycielski, J., "On the Axiom of Determinateness II," *Fund. Mathematics,* Vol. 59, 1966, pp. 203 – 212.

[R18.48] Naess, A., "Towards a Theory of Interpretation and Preciseness," in L. Linsky (ed.) *Semantics and the Philosophy of Language*, Urbana, Ill. Univ. of Illinois Press, 1951.

[R18.49] Narens, Louis, "The Theory of Belief," *Journal of Mathematical Psychology*, Vol. 49, 2003, pp1-31.

[R18.50] Narens, Louis, "A Theory of Belief for Scientific Refutations," *Synthese,* Vol.145, 2005, pp397-423.

[R18.51] Netto, A. B., "Fuzzy Classes," *Notices, Amar, Math. Society*, Vol. 68T- H28,

1968, pp.945.

[R18.52] Neurath, Otto et al. (eds.), *International Encyclopedia of Unified Science, Vol. 1 – 10,* Chicago, University of Chicago Press, 1955.

[R18.53] Niebyl, Karl, H., "Modern Mathematics and Some Problems of Quantity, Quality and Motion in Economic Analysis," *Science*, Vol 7, # 1, January,1940, pp. 103 – 120.

[R18.54] Orlowska, E., "Representation of Vague Information," *Information Systems*, Vol. 13, #2, 1988, pp. 167-174.

[R18.55] Parrat, L. G., *Probability and Experimental Errors in Science*, New York, John Wiley and Sons, 1961.

[R18.56] Raffman. D., "Vagueness and Context-sensitivity," *Philosophical Studies*, Vol. 81, 1996, pp.175-192.

[R18.57] Reiss, S., *The Universe of Meaning*, New York, The Philosophical Library, 1953.

[R18.58] Russell, B., "Vagueness," *Australian Journal of Philosophy*, Vol.1, 1923 pp. 84-92.

[R18.59] Russell, B., *An Inquiry into Meaning and Truth*, New York, Norton, 1940.

[R18.60] Shapiro, Stewart, *Vagueness in Context*, Oxford, Oxford University Press, 2006.

[R18.61] Skala, H. J. "Modelling Vagueness," in M. M. Gupta and E. Sanchez, *Fuzzy Information and Decision Processes*, Amsterdam North-Holland, 1982, pp 101 – 109.

[R18.62] Skala, Heinz J. et al., (eds.), *Aspects of Vagueness*, Dordrecht, D. Reidel Co. 1984.

[R18.63] Sorensen, Roy, *Vagueness and Contradiction*, Oxford, Oxford University Press, 2001.

[R18.64] Tamburrini, G. and S. Termini, "Some Foundational Problems in Formalization of Vagueness," in M. M. Gupta et al (eds.), *Fuzzy Information and Decision Processes*, Amsterdam, North Holland, 1982, pp. 161-166.

[R18.65] Termini, S. "Aspects of Vagueness and Some Epistemological Problems Related to their Formalization," in Skala, Heinz J. et al., (eds.), *Aspects of Vagueness,* Dordrecht, D. Reidel Co. 1984, pp.205 – 230.

[R18.66] Tikhonov, Andrey N. and Vasily Y. Arsenin., *Solutions of Ill-Posed Problems*, New York, John Wiley and Sons, 1977.

[R18.67] Tversky, A. and D. Kahneman, "Judgments under Uncertainty: Heuristics and

Biases," *Science,* Vil 185 September 1974, pp. 1124-1131.

[R18.68] Ursul, A. D., "The Problem of the Objectivity of Information," in Kuba't, Libor and J. Zeman (eds.), *Entropy and Information*, Amsterdam, Elsevier, 1975, pp. 187-230.

[R18.69] Vardi, M. (ed.), *Proceedings of Second Conference on Theoretical Aspects of Reasoning about Knowledge*, Asiloman, Ca, Los Altos, Ca, Morgan Kaufman, 1988.

[R18.70] Verma, R.R., "Vagueness and the Principle of the Excluded Middle," *Mind*, Vol. 79, 1970, pp. 66-77.

[R18.71] Vetrov, A. A., "Mathematical Logic and Modern Formal Logic," *Soviet Studies in Philosophy*, Vol. 3, #1, 1964 pp. 24 – 33.

[R18.72] von Mises, Richard, *Probability, Statistics and Truth,* New York, Dover Pub. 1981.

[R18.73] Williamson, Timothy, *Vagueness,* London, Routledge, 1994.

[R18.74] Wiredu, J.E., "Truth as a Logical Constant with an Application to the Principle of the Excluded Middle," *Philos. Quart.*, Vol. 25, 1975, pp. 305-317.

[R18.75] Wright, C., "On Coherence of Vague Predicates," *Synthese*, Vol. 30, 1975. pp. 325 – 365.

[R18.76] Wright, Crispin, "The Epistemic Conception of Vagueness," *Southern Journal of Philosophy,* Vol. 33, Supplement, 1995, pp. 133-159.

[R18.77] Zadeh, L. A., A Theory of Commonsense Knowledge," in Skala, Heinz J. et al., (eds.), *Aspects of Vagueness*, Dordrecht, D. Reidel Co. 1984., pp 257 – 295.

[R18.78] Zadeh, L. A., "The Concept of Linguistic Variable and its Application to Approximate reasoning," *Information Science*, Vol. 8, 1975, pp. 199 – 249 (Also, in Vol. 9, pp. 40 – 80).

R19 VAGUENESS, DISINFORMATION, MISINFORMATIOMN AND FUZZY-GAME THEORY IN SOCIO-NATURAL TRANSFORMATIONS

[R19.1] Aubin, J.P. "Cooperative Fuzzy Games", Mathematics of Operations Research, Vol.6, 1981, ppl – 13.

[R19.2] Aubin, J.P. Mathematical Methods of Game and Economics Theory, New York, North Holland. 1979.

[R19.3] Barone, J.M. "Morphisms in Set H and Fuzzy Move Games," *Annual meeting of the North American Fuzzy Information Processing Society, NAFIPS* , 2006, pp.290-295.

[R19.4] Butnaria, D., "Fuzzy Games: A description pf the concepts," *Fuzzy sets and systems* Vol.1, 1978, pp. 181 – 192.

[R19.5] Butnaria, D., "Stability and shapely value for a n – persons Fuzzy Games," *Fuzzy sets and systems,* Vol. 4, #1, 1980, pp 63 – 72.

[R19.6] Chakeri, A., Dariani, A.N., Lucas, C. "How can fuzzy logic determine game equilibriums better?" *Intelligent Systems (IS'08), 4th International IEEE Conference,* Volume 1, 2008, pp.2-56.

[R19.7] Ereshko F.I. "Reflexive games with uncertain and fuzzy interests," *Proceedings of the IEEE International Conference on Fuzzy Systems, FUZZIEEE'02,* Volume 2, 2002, pp.1608-1612.

[R19.8] Li, K.W., Karray, F., Hipel, K.W. and Kilgour, D.M.). "Fuzzy approaches to the game of Chicken," IEEE Transactions on Fuzzy Systems, Volume 9, Issue 4, 2001, pp.608-623

[R19.9] Michael Aristidou, Sudipta Sarangi. Games in fuzzy environments. Southern Economic Journal, Vol. 72, #3, 2006, pp.645-659.

[R19.10] P.K.Nayak, M. Pal. Bi-matrix games with intuitionistic fuzzy goals. Iranian Journal of Fuzzy Systems. Vol. 7, No. 1, 2010. pp. 65-79.

[R19.11] Nurmi, H.., "A Fuzzy Solution to a Majority Voting Game, "Fuzzy sets and systems, Vol.5, 1981 pp187-198.

[R19.12] Al-Mutairi, M.S. "Two-decision-maker cooperative games with fuzzy preferences," *IEEE International Conference on Industrial Engineering and Engineering Management (IEEM)*, (2010), pp.6-12.

[R19.13] Qian Song, and Kandel, A. A fuzzy approach to strategic games. *IEEE Transactions on Fuzzy Systems*, vol.7, no.6, (1999), pp.634-642.

[R19.14] Regade., R. K., "Fuzzy Games in the Analysis of Options," jour. Of Cybernetics, Vol. 6, 1976, pp 213 – 221.

[R19.15] Rodica Branzei, Dinko Dimitrov, Stef Tijs. Models in Cooperative Game Theory: Crisp, Fuzzy and Multichoice Games. Springer, 2005.

[R19.16] Russell, S.,Lodwick, W.A. Fuzzy game theory and Internet commerce: e-strategy and metarationality. *Proceedings of the Annual Meeting of the North American Fuzzy Information Processing Society*, 2002 NAFIPS, 2002, pp. 93- 98.

[R19.17] Sharifian, S., Chakeri, A., Sheikholeslam, F. "Linguistic representation of Nash equilibriums in fuzzy games." *Annual Meeting of the North American Fuzzy Information Processing Society (NAFIPS)*, (2010). pp.1-6.

[R19.18] Sharma, R., Gopal, M. "Hybrid Game Strategy in Fuzzy Markov-Game Based Control." *IEEE Transactions on Fuzzy Systems*, Volume 16, Issue.5, (2008), pp.1315- 1327.

[R19.19] Shih-Hung Wu, Von Wun-Soo. A Fuzzy Game Theoretic Approach to Multi-Agent Coordination. Multiagent Platforms. Lecture Notes in Computer Science, Volume 1599, 1998. pp.76-87.

[R19.20] Spillman, B. et al., "Coalition Analysis with Fuzzy Sets," *Kybernetes*, Vol. 8, 1979, pp. 203-211.

[R19.21] Wernerfelt, B., "Semifuzzy Games" *Fuzzy sets and systems*, Vol. 19, 1986, pp. 21 – 28.

[R19.22] Xiaohui Yu, Qiang Zhang. (2010). Fuzzy Nash equilibrium of fuzzy n-person non-cooperative game. Journal of Systems Engineering and Electronics, Vol. 21, No. 1, February 2010, pp.47–56.

[R19.23] Yan-Qing Zhang, Kandel, A., Henne, M. "On normal fuzzy reasoning and fuzzy moves" *Proceedings of the Sixth IEEE International Conference on Fuzzy Systems*, Volume 1, (1997). pp.549-553.

[R19.24] Zhaoming Wang, Yonghui Ling.. "A Static Negotiation Model of Ekectronic Commerce," *International Journal of Nonlinear Science*,Volume 5, #1, 2008, pp.43-50.

R20 WEAPON FOUNDATIONS FOR INFORMATION SYSTEM

[R20.1] Forte, B., "On a System of Functional Equation in Information Theory," *Aequationes*

Math. Vol5, 1970, pp. 202-211.

[R20.2] Gallick, James, *The Information: A History, a Theory, a Flood*. Pantheon, New York, NY, 2011.

[R20.3] Hopcroft, John, E. Rajeev Motwani and D. Ljeffrey, Ullman, *Introduction to automata Theory, Languages, and Computation*, Pearson Education, 2000.

[R20.4] Howard, N., *Paradoxes of Rationality*, Cambridge, Mass., MIT Press, 1972.

[R20.5] Ingarden, R.S, "A Simplified Axiomatic Definition of Information," *Bull. Acad. Polo Sci. Ser, Sci. Math Astronomy Phys.*, Vol.11, 1963, pp. 209-212.

[R20.6] Lee, P. M., "On the Axioms of Information Theory," *Annals of Math. Statistics*, vol.35, 1964, pp 415-418.

[R20.7] Luce, R. D. (ed.) , *Development in Mathematical Psychology*, Westport, Greenwood Press, 1960.

[R20.8] Luciano, Floridi, "Is Information Meaningful Data?," *Philosophy and Phenomenological Research*, 70 (2), 2005 pp. 351 – 370.

[R20.9] Meyer, L, "Meaning in Music and Information Theory," *Journal of Aesthetics and Art Criticism,* Vol. 15, 1957, pp412 – 424.

[R20.10] Rich, Elaine, *Automata, Computability, and Complexity: Theory and Applications*, Pearson, 2008.

[R20.11] Shannon, Claude E., "The Mathematical Theory of Communication," *Bell System Technical Journal,* Vol. 27,#3, 1945, pp.379-423 and Vol. 27, #4, 1948, pp623-666.

[R20.12] *Shannon, Claude E.* and Warren Weaver, *The Mathematical Theory of Communication.* University of Illinois Press,1949.

[R20.13] Theil, Henri, *Statistical Decomposition Analysis*, Amsterdam, North-Holland, 1974.

[R20.14] Vigo, R. "Representational information: a new general notion and measure of Information". *Information Sciences.* Vol.18, *(*2011*)*, pp. *4847–4859.*

[R20.15] Vigo, R "Complexity over Uncertainty in Generalized Representational Information Theory (GRIT): A Structure-Sensitive General Theory of Information". *Information.* **4** (1), 2013, pp. 1–30.

[R20.16] Vigo, R. *Mathematical Principles of Human Conceptual Behavior: The Structural Nature of Conceptual Representation and Processing*, Routledge, New York and London, 2014.

[R20.17] von Mises, R., *Probability, Statistics and Truth*, New York, Dover, 1981.

[R20.18] Wicker Stephen B. and Saejoon Kim*, Fundamentals of Codes, Graphs, and Iterative Decoding. Springer*, New York, 2003.

[R20.19] Wiener, N., Cybernetics, New York, John Wiley and Sons, 1948.

[R20.20] Wiener, N., The Human use of Human Beings, Boston, Mass, Houghton, 1950.

[R20.21] Young, Paul. *The Nature of Information,* Greenwood Publishing Group, Westport, Ct, 1987.

21. WEAPON FOUNDATIONS FOR FUZZY INFORMATION AND ENTROPY

[R21.1] Belis, M. and S. Guiasu, "A quantitative–qualitative measure of information in cybernetic systems," *IEEE Trans. Inform. Theory* Vol.14, 1968, pp. 593–594.

[R21.2] Burillo, P. and H. Bustince, "Entropy on intuitionistic fuzzy sets and on interval-valued fuzzy sets," *Fuzzy Sets and Systems*, vol. 78, no. 3, pp. 305–316, 1996.

[R21.3] Ceng, H.D., Chen, Y.H. & Sun,Y. "A novel fuzzy entropy approach to image enhancement and thresholding," *Signal Processing, 75,* 1999, pp.277-301.

[R21.4] De Luca, A. S. Termini, "A definition of non-probabilistic entropy in setting of fuzzy set theory", *Inform. Control.,* Vol. 20, 1972, pp. 301–312.

[R21.5] Dompere, Kofi K., *A General Theory of Entropy: Rational Foundations of Information-Knowledge Dynamics,* New York, Springer (Series: Studies in Systems, Decision and Control, Vol. 114), 2019.

[R21.6] Dumitrescu, D., "Fuzzy measures and the entropy of fuzzy partitions," *J. Math. Anal. Appl.*, Vol.176, 1993, pp. 359–373.

[R21.7] Dumitrescu, D., "Entropy of fuzzy process," *Fuzzy Sets Syst.*, Vol.55, 1993, pp. 169–177.

[R21.8] Dumitrescu, D., "Entropy of a fuzzy dynamical system," *Fuzzy Sets Syst.*, Vol. 70, 1995, pp.45–57.

[R21.9] Garbaczewski, P., "Differential entropy and dynamics of uncertainty," *J. Stat. Phys.* Vol.123, 2006, pp. 315–355.

[R21.10] Hu, Q. and Yu, D., "Entropies of fuzzy indiscernibility relation and its operations," *International Journal of Uncertainty, Fuzziness and Knowledge-Based Systems* Vol.12, 2004, pp.575-589.

[R21.11] Hu, Q., Yu, D., Xie, Z. and Liu, J. "Fuzzy probabilistic approximation spaces and their information measures," *IEEE Trans. Fuzzy Syst.*, Vol. 14, 2006, pp.191–201.

[R21.12] Hudetz, T. "Space-time dynamical entropy for quantum systems," *Lett. Math. Phys.*, Vol.16, 1988, pp.151–161.

[R21.14] Hung, W. L. and M. S. Yang, "Fuzzy entropy on intuitionistic fuzzy sets," *International Journal of Intelligent Systems,* vol. 21, no. 4, 2006, pp. 443–451.

[R21.15] Kapur, J. N., "Four Families of Measures of Entropy," *Indian J. Pure Appl. Math.* Vol.17, 1986, pp. 429-449.

[R21.16] Kapur, J.N. *Measures of Fuzzy Information*. Mathematical Sciences Trust Society, New Delhi 1997.

[R21.17] Kasko, B. "Fuzzy Entropy and Conditioning," *Information Sciences,* Vol.40, 1986, pp. 165-174.

[R.21.18] Klir, G.J. "Generalized information theory : aims, results and open problems," *Reliab. Eng. Syst. Safety*, Vol. 85 (1–3), 2004, pp. 21–38.

[R21.19] Kolmogorov, A.N. *Foundations of the Theory of Probability*; Chelsea Publishing Company: New York, NY, 1950.

[R21.20] Liu, B. and Y.K. Liu, "Expected value of fuzzy variable and fuzzy expected value models," *IEEE Transactions on Fuzzy Systems,* Vol.10, #4, 2002, pp. 445–450.

[R21.21] S.G. Loo, "Measures of fuzziness," *Cybernetica,* Vol. 20, 1977, pp. 201–210.

[R21.22] Markechová, D. "The entropy of fuzzy dynamical systems and generators," *Fuzzy Sets Syst.* 1992, Vol. 48, pp. 351–363.

[R21.23] Mesiar, R. and Rybárik, J. "Entropy of Fuzzy Partitions: A General Model," *Fuzzy Sets Syst.*, 99, 1998, pp. 73–79.

[R21.24] Mesiar, R. "The Bayes principle and the entropy on fuzzy probability spaces," Int. J. Gen. Syst., Vol. 20, 1991, pp. 67–72.

[R21.25] Parkash, O. "A new parametric measure of fuzzy entropy," *Inform. Process. Manage. Uncertain.* Vol.2, 1998, pp. 1732–1737.

[R21.26] Parkash, O. and Sharma, P.K. "Measures of fuzzy entropy and their relations," *International Journal of Management & Systems,* Vol. 20, 2004, pp. 65-72.

[R21.27] Rahimi, M.and Riazi, A., "On local entropy of fuzzy partitions," *Fuzzy Sets Syst.*, Vol. 234, 2014, pp. 97–108.

[R21.28] Rieˇcan, B. "An entropy construction inspired by fuzzy sets," Soft Comput., Vol.7, 2003, pp. 486–488.

[R21.29] Szmidt, E. and J. Kacprzyk, "Entropy for intuitionistic fuzzy sets," *Fuzzy Sets and Systems,* vol. 118, no. 3, 2001, pp. 467–477.

[R21.30] Verma R. and B. D. Sharma, "On generalized exponential fuzzy entropy," *Engineering and Technology,* vol. 5, 2011, pp. 956–959.

[R21.31] Zeng W. and H. Li, "Relationship between similarity measure and entropy of interval valued fuzzy sets," *Fuzzy Sets and Systems*, vol. 157, #11, 2006 pp. 1477–1484.

R22 WEAPON FOUNDATIONS FOR KNOWLEDGE BY ACQUAINTANCE AND KNOWLEDGE BY DESCRIPTION

[R22.1] Balog, K. "Acquaintance and the Mind-Body Problem." In Simone Gozzano and Christopher Hill (eds.), *New Perspectives on Type Identity: The Mental and the Physical* Cambridge: Cambridge University Press 2012 (pp. 16-43).

[R22.2] BonJour, L., "Toward a Defense of Empirical Foundationalism." In Michael Raymond DePaul (ed.), *Resurrecting Old-Fashioned Foundationalism,* Lanham: Rowman and Littlefield, 2001, pp. 21-38.

[R22.3] Brewer, B. *Perception and its Objects*. Oxford: Oxford University Press. 2011.

[R22.4] Chalmers, D. *The Conscious Mind.* Oxford: Oxford University Press,1996.

[R22.5] Chisolm, Roderick, *Acquaintance and the Mind-Body Problem*, Oxford, Oxford University Press, (2008).

[R22.6] Churchman, C. W., and Ratoosh, P., *Measurement, Definitions and Theories*, Wiley and Sons, 1959.

[R22.7] Fales, E. *A Defense of the Given.* Lanham: Rowman and Littlefield1996.

[R22.8] Gertler, B. "A Defense of the Knowledge Argument," *Philosophical Studies,* Vol. 93, 1999., pp. 317-336.

[R22.9] Gertler, B. "Introspecting Phenomenal States." *Philosophy and Phenomenological*

Research Vol. 63, 2001, (pp. 305–328).

[R22.10a] Gertler, B. "Renewed Acquaintance." In Declan Smithies and Daniel Stoljar (eds.), *Introspection and Consciousness*, Oxford, Oxford University Press, 2012, pp. 93-128.

[R22.10b] Harrison, Glenn W. and John A List, "*Field Experiments,*" *Journal of Economic Literature,* Vol. 42, Dec. 2004.

[R22.11] Hasan, Ali and Fumerton, Richard, "Knowledge by Acquaintance vs. Description," *The Stanford Encyclopedia of Philosophy*, Edward N. Zalta (ed.), (Spring 2014 Edition).

[R22.12] Lazerowitz, M., "Knowledge by Description", *The Philosophical Review*, Vol.46, No.4, July 1937, pp. 402–415.

[R22.13] Morgenstern, O., *On the Accuracy of Economic Observation*, Princeton, Princeton University Press, 1973.

[R22.14] Parker, D.H., "Knowledge by Acquaintance", *The Philosophical Review*, Vol.54, No.1, January 1945A, (pp. 1–18).

[R22.15] Parker, D.H., "Knowledge by Description", *The Philosophical Review*, Vol.54, No.5, September1945B, pp. 458–488.

[R22.16] Poston, T., "Similarity and Acquaintance." *Philosophical Studies* Vol. 147, 2010, pp. 369-378.

[R22.17] Russell, B. "On Denoting." *Mind,* Vol.14, 479-493, 1905.

[R22.18] Russell, B. "Knowledge by Acquaintance and Knowledge by Description." *Proceedings of the Aristotelian Society* Vol.11, 1910-1911, pp.108-128.

[R22.19] Russell, B., "On the Nature of Acquaintance," in *Bertrand Russell: Logic and Knowledge: Essays 1901-1950*, R. C. Marsh (ed.) New York, Capricorn Books,1971, pp.,127-174.

[R22.20] Russell, B., *Logic and Knowledge: Essays 1901-1950*, R. C. Marsh (ed.) New York, Capricorn Books,1971.

R23 WRITTEN AND AUDIO LANGUAGES AND INFORMATION

[R23.1] Agha, Agha, *Language and Social Relations,* Cambridge, Cambridge University Press, 2006.

[R23.2] Aitchison, Jean (ed.), *Language Change: Progress or Decay?* Cambridge, New York, Melbourne: Cambridge University Press 2001.

[R23.3] *Allerton, D. J.* "Language as Form and Pattern: Grammar and its Categories". In Collinge, N.E. (ed.) *An Encyclopedia of Language.* London and New York: Routledge, 1989.

[R23.4] Anderson, Stephen, *Languages: A Very Short Introduction.* Oxford: Oxford University Press, 2012.

[R23.5] Aronoff, Mark; Fudeman, Kirsten, *What is Morphology.* New York, John Wiley & Sons, 2011.

[R23.6] Barber Alex & Robert J Stainton (eds.), *Concise Encyclopedia of Philosophy of Language and Linguistics.* New York, Elsevier, 2010.

[R23.7] Bauer, Laurie (ed.), *Introducing linguistic morphology* Washington, D.C., Georgetown University Press, 2003.

[R23.8] Brown, Keith; Ogilvie, Sarah, (eds.), *Concise Encyclopedia of Languages of the World,* New York Elsevier Science, 2008.

[R23.9] Campbell, Lyle (ed.), *Historical Linguistics: an Introduction,* Cambridge, MA: MIT Press, 2004.

[R23.10] Chandler, W.L. "Abstract Families of Deterministic Languages," *Proceedings of the First ACM Symposium on Theory of Computing,* 1969, pp21-30.

[R23.11] Chao, Yuen Ren, *Language and Symbolic Systems,* Cambridge, Cambridge University Press. 1968.

[R23.12] *Chomsky, Noam, Syntactic Structures. The Hague: Mouton,* 1957.

[R23.13] Chomsky, Noam, *The Architecture of Language*. Oxford: Oxford University Press 2000.

[R23.14] Clarke, David S., *Sources of semiotic: readings with commentary from antiquity to the present.* Carbondale: Southern Illinois University Press, 1990.

[R23.15] Comrie, Bernard (ed.), *Language universals and linguistic typology: Syntax and morphology*. Oxford: Blackwell, 1989.

[R23.16] Comrie, Bernard, (ed.), *The World's Major Languages.* New York: Routledge, 2009.

[R23.17] Coulmas, Florian, *Writing Systems: An Introduction to Their Linguistic Analysis.* Cambridge University Press, 2002.

[R23.18] Croft, William, Cruse, D. Alan, *Cognitive Linguistics. Cambridge, Cambridge University Press,* 2004.

[R23.19] Crystal, David, *The Cambridge Encyclopedia of Language.* Cambridge: Cambridge University Press, 1997.

[R23.19] Culik, K. H., "Abstract Families of Relations," *Journal. of Computer and System Science,* Vol.5, 1971, PP596-606.

[R23.20] Deacon, Terrence , *The Symbolic Species: The Co-evolution of Language and the Brain,* New York: W.W. Norton & Company, 1997.

[R23.21] Devitt, Michael and Sterelny, Kim *Language and Reality: An Introduction to the Philosophy of Language. Boston:* MIT Press, 1999.

[R23.22] Evans, Nicholas; Levinson, Stephen C.. "The myth of language universals: Language diversity and its importance for cognitive science". *Behavioral and Brain Sciences***32** (5), 2009, pp. 429–492.

[R23.23] Fitch, W. Tecumseh, *The Evolution of Language. Cambridge: Cambridge University Press,* 2010.

[R23.24] Foley, William A., *Anthropological Linguistics: An Introduction,* Oxford, Blackwell, 1997.

[R23.25] Ginsburg, S, *Algebraic and Automata-Theoretic Properties of Formal Languages,* (Fundamental Studies in Computer Science, Vol. 2), New Amsterdam, North- Holland, 1975.

[R23.26] Greenberg, Joseph, *Language Universals: With Special Reference to Feature Hierarchies. The Hague: Mouton & Co.* 1966.

[R23.27] Hauser, Marc D.; Chomsky, Noam; Fitch, W. Tecumseh, "The Faculty of Language: What Is It, Who Has It, and How Did It Evolve," *Science 22,* ***298,*** (5598) , 2002, pp. 1569–1579

[R23.28] Hopcroft, J, E. and J. D. Ullman, *Formal Languages and their relation to Automata*, Reading, Mass, Addison-Wesley, 1959

[R23.29] International Phonetic Association, *Handbook of the International Phonetic Association: A guide to the use of the International Phonetic Alphabet.* Cambridge: Cambridge University Press, 1999.

[R23.30] Katzner, Kenneth, *The Languages of the World,* New York: Routledge, 1999.

[R23.31] Labov, William, *Principles of Linguistic Change vol. I: Internal Factors,* Oxford, Blackwell, 1994.

[R23.32] Labov, William, *Principles of Linguistic Change vol. II: Social Factors,* Oxford, Blackwell 2001.

[R23.33] Ladefoged, Peter, Maddieson, Ian, *The sounds of the world's languages*, Oxford: Blackwell., 1996, pp. 329–330.

[R23.34] Levinson, Stephen C. *Pragmatics.* Cambridge: Cambridge University Press 1983.

[R23.35] Lewis, M. Paul (ed.), *Ethnologue: Languages of the World*, Dallas, Tex.: SIL International 2009.

[R23.36] Lyons, John, *Language and Linguistics,* Cambridge, Cambridge University Press 1981.

[R23.37] MacMahon, *April M.S., Understanding Language Change,* Cambridge, Cambridge University Press, 1994.

[R23.38] Matras, Yaron; Bakker, Peter, (eds). *The Mixed Language Debate: Theoretical and Empirical Advances.* Berlin: Walter de Gruyter 2003.

[R23.39] Moseley, Christopher (ed.), *Atlas of the World's Languages in Danger,* Paris: UNESCO Publishing, 2010.

[R23.40] Nerlich, B. "History of pragmatics". In L. Cummings (ed.), *The Pragmatics Encyclopedia.* New York: Routledge. 2010, pp. 192–93.

[R23.41] Newmeyer, Frederick J., *The History of Linguistics*., Linguistic Society of America, 2005.

[R23.42] Newmeyer, Frederick J., *Language Form and Language Function,* Cambridge, MA: MIT Press, 1998.

[R23.43] Nichols, Johanna *Linguistic diversity in space and time.* Chicago: University of Chicago Press,1992.

[R23.44] Nichols, Johanna "Functional Theories of Grammar". *Annual Review of Anthropology,* Vol. **13,** 1984, pp.97–117.

[R23.45] Saloman, A., Formal Languages, New York, Academic Press, 1973

[R23.46] Senft, Gunter, *Systems of Nominal Classification.* Cambridge University Press. (ed.), 2008.

[R23.47] Swadesh, Morris, "The phonemic principle", *Language,* Vol. **10** (2): (1934), 117–129.

[R23.48] Tomasello, Michael "The Cultural Roots of Language". In B. Velichkovsky and D. Rumbaugh (Eds.), *Communicating Meaning: The Evolution and Development of Language.* Psychology Press, 1996, pp. 275–308.

[R23.49] Tomasello, Michael, *Origin of Human Communication.* MIT Press, 2008.

[R23.50] Ulbaek, Ib, "The Origin of Language and Cognition", In J. R. Hurford & C. Knight (eds.). *Approaches to the evolution of language.* Cambridge University Press. 1998, pp. 30–43.

[R 24] ZONES OF EPISTEMIC CONFLICTS ON POSSIBILITY AND PROBABILITY

.

[R24.1] Alberoni, F. Contribution to the study of subjective probability. Part I. *Journal of General Psychology,* 1962, 66, 241-264.

[R24.2] Berger, James O., *Statistical Decision Theory and Bayesian Analysis* (Springer Series in Statistics). New York, Springer-Verlag 1985.

[R24.3] Benacerraf, P.: "Mathematical Truth," *Journal of Philosophy* Vol. 70, 1973, pp.661-79.

[R24.4] de Finetti, Bruno, *Theory of Probability. A Critical Introductory Treatment,* (translation by A.Machi and AFM Smith of 1970 book) 2 volumes. New York, Wiley, 1974 – 75.

[R.24.5] Dompere, Kofi K., "Savage Axioms, Ellsberg's Paradox and Fuzzy Optimal Decision-Choice Rationality, *Fuzzy Economic Review,* Vol 17, #1, 2012, pp. 15 – 51

[R24.6] DeGroot, Morris, *Optimal Statistical Decisions.* Wiley Classics Library, New York, Wiley, 2004.

[R24.7] Edwards, W. "Conservatism in human information processing," In B. Kleinmuntz (ed.),

Formal representation of human judgment. New York: Wiley, 1968. Pp. 17-52.

[R24.8] Eells, E. and E. Sober, "Probabilistic Causality and the Question of Transitivity," *Philosophy of Science* Vol.50, 1983, 35-57.

[R24.9] Feller, W., *An introduction to probability theory and its applications*. Vol.1 New York: Wiley, 1968.

[R24.10] Fetzer, J. H., "Dispositional Probabilities," in R. C. Buck and R. S. Cohen (eds.), *Boston Studies in the Philosophy of Science VIII* 1971, Reidel, Dordrecht, pp. 473-82.

[R24.11] Fetzer, J. H., "Statistical Probabilities: Single Case Propensities vs. Long-Run Frequencies," in W. Leinfellner and E. Kohler (eds.), *Developments in the Methodology of Social Science,* Reidel, Dordrecht,1974, pp. 387-97.

[R24.12] Fetzer, J. H. and D. E. Nute: 1979, "Syntax, Semantics, and Ontology: A Probabilistic Causal Calculus," *Synthese* Vol. 40, 1979, 453-95.

[R24.13a] Gärdenfors, P. and Sahlin, Nils-Eric. *Decision, Probability, and Utility: Selected Readings*. Cambridge, Cambridge University Press1988.

[R24.13b] Good, Irving, J. Good, *Thinking: Foundations of Probability and its Applications*, New York, Dover, 1983.

[R24.14] Hacking, Ian "Slightly More Realistic Personal Probability". *Philosophy of Science.* Vol. **34** (4): (December 1967). 311–325.

[R24.15] Hazewinkel, M, (ed.), "Bayesian approach to statistical problems, " in *Encyclopedia of Mathematics,* New York, Springer Science, 2001.

[R24.16a] James H. Fetzer (ed.), *Probability and Causality,* Reidel, Dordrecht, 1988.

[R24.16b] Jazwinski, A. H., *Stochastic Processes and Filtering Theory,*New York, Dover, 1998

[R24.17] Jeffrey, R. C., *The Logic* of *Decision,* University of Chicago Press, Chicago 1990.

[R24.18] Kolmogorov, A.N. *Foundations of the Theory of Probability*; Chelsea Publishing Company: New York, NY, 1950.

[R24.19] Kyburg, H. E., Jr. "Propensities and Probabilities," *British Journal for the Philosophy of Science,* Vol. 25, 1974 pp359-75.

[R24.20a] Lindley, Dennis, V. *Introduction to Probability and Statistics: From a Bayesian Viewpoint Part I, Probability*, Cambridge, Cambridge Univ. Press, 1965.

[R24.20b] Lindley, Dennis, V. *Introduction to Probability and Statistics: From a Bayesian Viewpoint Part 2, Inference,* Cambridge, Cambridge Univ. Press, 1965.

[R24.21] Morgenstern, Oskar *(1978). "Utility". In Andrew Schotter. Selected Economic Writings of Oskar Morgenstern.* New York, New York University Press.1978, pp. 65–70.

[R24.22] Ozga, Andrew S. *Expectations in Economic Theory,* Chicago, Aldine Pub., 1965.

[R24.23] Pfanzagl, J., "Subjective Probability Derived from the Morgenstern-von Neumann Utility Theory". In Martin Shubik.(ed.) *Essays in Mathematical Economics In Honor of Oskar Morgenstern.* Princeton, N.J., Princeton University Press. 1967, pp. 237–251.

[R24.24] Popper, K. R., "The Propensity Interpretation of the Calculus of Probability, and the Quantum Theory", in S. Korner (ed.), *Observation and Interpretation in the Philosophy* of *Physics,* New York, Dover Publications, 1957, pp. 65-70.

[R24.25] Popper, K. R., 'The Propensity Interpretation of Probability', *British Journal for the Philosophy of Science* Vol. 10, pp. 25- 42, 1959.

[R24.26] Salmon, W. C., *The Foundations of Scientific Inference,* Pittsburg, University of Pittsburgh Press1967.

[R24.27] Savage, L. J., *The Foundations of Statistics*, New York, Dover Pub, 1972.

[R24.28] Savage, L. J., "Subjective Probability and Statistical Practice," in Robert E Machol and Paul Gray (eds.), *Recent Developments in Decision and Information Processes*, New York, Macmillan, 1962.

[R24.29] Savage, L. J., "Difficulties in the theory of Personal Probability," *Philosophy of Science*, Vol.34, 1967, pp. 305 – 310.

[R24.30] Sklar, L.: "Is Probability a Dispositional Property?," *Journal of Philosophy* Vol.67, 1970, pp.355-66.

[R24.31] Skyrms, B.: *Causal Necessity,* New Haven, Yale University Press, 1980.

[R24.32] Stigler, SM. *The History of Statistics: The Measurement of Uncertainty before 1900.*

Cambridge, Mass., Belknap Press/Harvard University *Press* 1990.

[R24.33] Suppes, P., "New Foundations of Objective Probability: Axioms for Propensities," in P. Suppe, L. Henkin, A. Joja, and G. C. Moisil (eds.), *Logic, Methodology and Philosophy of Science IV,* North-Holland, Amsterdam, London, 1973, pp. 515-29.

[R24.34] Tversky, A., & Kahneman, D. The belief in the law of small numbers. Psychological Bulletin, 1971, 76, 105-110.

[R24.35] Von Mises, R., *Probability, Statistics and Truth,* Macmillan, New York 1957.

[R24.36] von Neumann, John and Oskar Morgenstern, *Theory of Games and Economic Behavior,* Princeton University Press, 1943, 1947.

[R24.37] Wagenaar, W. A. "Subjective randomness and the capacity to generate information," In A. F. Sanders (ed.) *Attention and performance III*, Acta Psychologica, Vol. 33,1970, pp. 233-242.

R25. ZONES OF EPISTEMIC CONFLICTS IN UNITY OF SCIENCE AND KNOWING.

[R25.1] Andler, Daniel, "Unity Without Myths" In John Symons, Juan Manuel Torres & Olga Plomb (eds.), *New approaches to the Unity of Science, vol. 1: Otto Neurath and the Unity of Science.* New York, Springer. 201.

[R25.2] Anton, J. P., "The Unity of Scientific Inquiry and Categorial Theory in Aristotle," *Boston Studies in the Philosophy of Science* 1990, pp. 121:29-43.

[R25.3] Barrow, John D. *New Theories of Everything: The Quest for Ultimate Explanation,* Oxford, Oxford University Press, 2007.

[R25.4] Bechtel, William P. & Andrew Hamilton "Reduction, Integration, and the Unity of Science: Natural, Behavioral, and Social Sciences and the Humanities," In T. Kuipers (ed.), *Philosophy of Science: Focal Issues (Volume 1 of the Handbook of the Philosophy of Science).* New York, Elsevier, 2007.

[R25.5] Boersema, David , "Metaphysics, Mind, and the Unity of Science," *Behavioral and Brain Sciences* 27 (5), 2004 pp. 627-628.

[R25.6] Briskman, Larry, "Three Views Concerning the Unity of Science," In Gerard Radnitzky (ed.), *Centripetal Forces in the Sciences*, New York, Paragon House Publishers, 1987. pp. 1--105.

[R25.7] Bunge Mario Augusto (ed.) *The Methodological Unity of Science.*, Boston, Reidel1973.

[R25.8] Burian, Richard M. "Conceptual Change, Cross-Theoretical Explanation, and the Unity of Science" *Synthese* 32 (#1-2), 1975, pp. 1 - 28.

[R25.9]Byrne, Lee "An Educational Application of Resources of the Unity of Science Movement," - *Philosophy of Science,* Vol. 7 (#2): 1940, pp. 241-262.

[R25.10] Carnap, Rudolf, *The Unity of Science*, London: K. Paul, Trench, Trubner & Co.. 1934.

[R25.11] Carrier, Martin, "The Unity of Science," *International Studies in the Philosophy of Science* Vol. 4 (1): 1990, pp. 17-31.

[R25.12] Causey, Robert L., "Unity of Science," *Philosophical Review* Vol. 90 (1) 1977, pp. 150-153.

[R25.13] Crave, Carl F. r, "Beyond Reduction: Mechanisms, Multifield Integration and the Unity of Neuroscience," *Studies in History and Philosophy of Science Part C* ,Vol. 36 (#2), 2005 . pp. 373-395.

[R25.14] Damasio, Antonio R. (ed.) *Unity of Knowledge: The Convergence of Natural and Human Science.* New York, New York Academy of Sciences, 2001.

[R25.15] Davies, David "Explanatory Disunities and the Unity of Science," *International Studies in the Philosophy of Science* Vol. 10 (1): 1996, pp.5 – 21.

[R25.16] de Santillana, George & Edgar Zilsel "Foundations of the Unity of Science. II 8: The Development of Rationalism and Empiricism". *Philosophical Review*, 1943 Vol. 52 (1):87-87.

[R25.17] Dewey, John "Foundations of the Unity of Science," Vol. II, No. 4: Theory of Valuation.

Philosophical Review Vol. 50 (4): 1941, pp. 443-446.

[R25.18] Donovan, Rickard, "Science Without Unity," *International Philosophical Quarterly* Vol. 30 (1): 1990, pp. 122-125.

[R25.19] Dupre J. "Against Scientific Imperialism," - 1994 - *PSA: Proceedings of the Biennial Meeting of the Philosophy of Science Association* 1994, pp. 374 - 381.

[R25.20] Dupré, John "The Disunity of Science," *Mind* Vol. 92 (367): 1983, pp. 321-346.

[R25.21] Finch III, Charles, S, *The Star of Deep Beginnings : The Genesis of African sciences and Technology,Decatur,* GA, USA, Khenti INC, 2001.

[R25.22] Galison, Peter & David J. Stump (eds.), *The Disunity of Science: Boundaries, Contexts, and Power*. Stanford, Ca. Stanford University Press. 1996.

[R25.23] Neurath, Otto et al. (eds.), *International Encyclopedia of Unified Science, Vol.*

1 – 10, Chicago, University of Chicago Press, 1955.

[R25.24] Neurath, Otto, "Unified Science as Encyclopedic," in Neurath, Otto et al. (eds.),

International Encyclopedia of Unified Science, Vol. 1 – 10, Chicago, University of Chicago Press, 1955, pp.1-27.

[R25.25] Sertima, Ivan Van (ed.) *Blacks in Science: ancient and modern,* New Brunswick, NJ. USA, Transaction Pub., 1998

[R26] ZONES OF EPISTEMIC CONFLICTS ON DECIDABILITY, COMPUTABILITY AND PREDICTABILITY IN THE SPACE OF THE PROBLEM-SOLUTION DUALITIES AND POLARITIES

[R26.1] Davis, Martin, *The Undecidable: Basic Papers on Undecidable Propositions, Unsolvable Problems and Computable Functions*, Dover, Dover Books, 2004.

[R26.2] Gödel, Kurt, *On formally undecidable Propositions of Principia Mathematica and Related Systems*, Dover, Dover Pub.,1992.

[R26.3] Gödel, K. Kurt, *Gödel, Collected Works, Vols.1- 3.* Oxford University Press, Oxford, 1995

[R26.4] Hartshorne, C. *The Logic of Perfection and Other Essays in Neoclassical Metaphysics*. Open Court Press, Lasalle, Illinois,1962.

[R26.5] Tarski, Alfred, A Mostowski et al., *Undecidable Theories: Studies in Logic and Foundation of Mathematics*, Dover, Dover Books, 2010.

[R26.6] Turing, A. M., "Solvable and Unsolvable Problems"., *UroBooks,* pp 576-595, 1954.

[R26.7] Turing, A. M., "Intelligent Machinery, A Heretical Theory", in *The Essential Turing*, 2004, pp465-475.

[R26.6] Wang, Hao, *Reflections on Kurt* Gödel, Cambridge, Mass. MIT Press,1987.

[R27] ZONE OF SYSMMERY, ADINKRA, ADINKRAMATICS, ADINKRA GRAPHS AND POLYRHYTHMS

[R27.1] Douglas B. L.,[1]S. James Gates,[2]B. L. Segler,[1]and J. B. Wang *Automorphism Properties and Classification of Adinkras,* Vol. 2015| Article IID584542 htt://doi.org/10.1155/2016/584542 ,

[R27.2] Doran C. F., M. G. Faux, S. J. Gates, Jr., T. Hübsch, K. M. Iga and G. D. Landweber,"Relating doubly-even error-correcting codes, graphs, and irreducible representations of *N*-supersymmetry," in: *Discrete and Computational Mathematics,* pp. 53–71 (Nova Sci. Publ., 2008

[R27.3] Doran C. F., M. G. Faux, S. J. Gates, Jr., T. Hübsch, K. M. Iga and G. D. Landweber, "On Graph-Theoretic Identifications of Adinkras, Supersymmetry Representations and Superfields," *Int. J. Mod. Phys.* A **22**, pp.869–930, 2012

[R27.4] Faux, M. and S. J. Gates, Jr., "Adinkras: a graphical technology for supersymmetric representation theory," *Phys. Rev.* D **71**(3), 065002 (2005).

[R27.5] . Gates, S, K. Iga,, K. Stiffler *Generating All 36, 864 Four-Color Adinkras via Signed Permutations and Organizing into ℓ- and ℓ ˜ -Equivalence Classes,* Published 21 December 2017.

[R27.6] Hrnciar P. and A. Haviar, "All trees of diameter five are graceful," ˇ Discrete Mathematics, vol. 233, no. 1–3, 2001, pp. 133–150.

[R27.7] Huang, C., A. Kotzig, and A. Rosa, "Further results on tree Labellings," Utilitas Mathematica, vol. 21, pp. 31–48, 1982.

[R27.8] Iga, Kevin and Yan Zhang, "Structural Theory and Classification of 2D Adinkras," *High Energy Physics,* Vol, 2016, 2016 12pages, (SCOAP3)

[R27.9] Korankye, Charles M., *Adinkra Alphabets: The Adinkra Symbols as Alphabets and Meanings*, Washington, USA, Korankye 2017.

[R27.10] Marcolli, M., Zolman, N. Adinkras, Dessins, Origami, and Supersymmetry Spectral Triples. *P-Adic Num Ultrametr Anal Appl* **11**, 223–247 (2019).

[R27.11] Ofori-Ansah, Kwaku, P, *Symbols of Adinkra Cloth,* Hyattsville, MD, USA Sankofa Pub. 1995.

[R27.12] Ofori-Ansah, Kwaku, P, *Weights of Wisdom: History,Uses and Sybolism of Akan Goldweights*, Hyattsville, MD, USA, Sankofa Pub. 1997.

[R27.13] Owusu-Ansah, J.V., Nana. *New Versions of the Traditional Motifs*, Kumasi, Ghana, Degraft Graphics ang Pub. 1992.

[R27.14] A. Panpa and T. Poomsaard "On Graceful Spider Graphs with at Most Four Legs of Lengths Greater than One" , *Hindawi Journal of Applied Mathematics Volume 2016*, Article ID 5327026, 5 pages.

[R27.15] Rosa, A. "On certain valuations of the vertices of a graph," in *Theory of Graphs,*

International Symposium, Rome, July 1966, New York, NY, Gordon and Breach 1967, pp. 349–355.

[R27.16] Sarpong, P., *The Sacred Stools of the Akan*, Tema, Ghana, Ghana Pub. Co, 1971.

[R27.17] Willis, W. Bruce, *The Adinkra dictionary: A visual primer on the language of Adinkra,* Washington, D. C., Pyramid Complex, 1998.

[R27.18] Zhang Yan X *Adinkras for Mathematicians,* Massachusetts Institute of Technology November28, 201.

[R28] ZONE OF SYSMMERY, FRATAL GRAPH, MATHEMATICS, ART, RHYTHMS AND GEOMETRY

[R28.1] Bangura, Abdul K., *African Mathematics: From Bones to Computers*, Lanham, MD, University Press of America, 2012.

[R28.2] Burkle-Elizondo, Gerardo; Valdéz-Cepeda, Ricardo David "Fractal analysis of Mesoamerican pyramids". *Nonlinear Dynamics, Psychology, and Life Sciences.* Vol. 10 (1): (2006). pp,105–122.

[R28.3] Cannon, W. J. Floyd, W. R. Parry. "*Finite subdivision rules*", *Conformal Geometry and Dynamics*, vol. 5, 2001, pp. 153–196.

[R28.4] Cheng, Qiuming *"Multifractal Modeling and Lacunarity Analysis". Mathematical Geology.* Vol. 29 (7): (1997), pp. 919–932.

[R28.5] Cross, Simon S. "Fractals in Pathology". *The Journal of Pathology. 182 (1):* 1997, pp.1–8.

[R28.6] Edgar, Gerald, *Measure, Topology, and Fractal Geometry,* New York, *Springer Science & Business Media., 2007.*

[R28.7] Eglash, Ron, *African Fractals: Modern Computing and Indigenous Design.,* New Brunswick: Rutgers University Press. (Archived from the original on January 3, 2018),

[R28.8] Edgar, Gerald *Measure, topology, and fractal geometry. New York: Springer-Verlag.* 2008.

[R28.9] Enright, Matthew B.; Leitner, David M. *"Mass fractal dimension and the compactness of proteins". Physical Review E. 71 (1)* , January 27, 2005

[R28.10] Falconer, Kenneth *Fractals, A Very Short Introduction,* New York, Oxford University Press2013.

[R28.11] Falconer, Kenneth *Fractal Geometry: Mathematical Foundations and Applications.* New York, John Wiley & Sons. (2003).

[R28.12] Falconer, Kenneth *Fractals, A Very Short Introduction,* New York,. Oxford University Press2013.

[R28.13] Gouyet, Jean-François *physics and fractal structures. new* york: masson springer, 1996.

[R28.14] Hahn, Horst K.; Georg, Manfred; Peitgen, Heinz-Otto "Fractal aspects of three-dimensional vascular constructive optimization". In Losa, Gabriele A. and Nonnenmacher, Theo F. (eds.). *Fractals in biology and medicine,* New York, Springer. (2005), pp. 55–66.

[R28.15] Hu, Shougeng; Cheng, Qiuming; Wang, Le; Xie, Shuyun "Multifractal characterization of urban residential land price in space and time". *Applied Geography.* Vol34: (2012), pp. 161–170.

[R28.16] Jelinek, Herbert F; Fernandez, Eduardo "Neurons and fractals: how reliable and useful are calculations of fractal dimensions?". *Journal of Neuroscience Methods. Vol 81 (1–2):* (June 1998). pp. 9–18.

[R28.17] Krapivsky, P. L.; Ben-Naim, E., "Multiscaling in Stochastic Fractals". *Physics Letters A. 196 ,1994 ,* pp*3–4: 168.*

[R28.18] Karperien, Audrey L.; Jelinek, Herbert F.; Buchan, Alastair M. "Box-Counting Analysis of Microglia Form in Schizophrenia, Alzheimer's Disease and Affective Disorder". *Fractals.* Vol.16 (2): 2008. p.103.

[R28.19] Landini, Gabriel *"Fractals in microscopy". Journal of Microscopy.* Vol.241 (1*):* 2011 pp. 1–8.

[R28.20] Lee, S.; Olsen, S.; Gooch, B. "Simulating and Analyzing Jackson Pollock's Paintings". *Journal of Mathematics and the Arts*. Vol.1 (2): (2007). Pp.73–83.

[R28.21] Liu, Jing Z.; Zhang, Lu D.; Yue, Guang H. "Fractal Dimension in Human Cerebellum Measured by Magnetic Resonance Imaging". *Biophysical Journal. 85 (6):* (2003). Pp. 4041–4046.

[R28.22] Losa, Gabriele A., Nonnenmacher, Theo F. *Fractals in biology and medicine*. New York, *Springer*. 2005.

[R28.23] Mandelbrot, Benoît B. *Fractals and Chaos. Berlin: Springer*. 2004.

[R28.24] Mandelbrot, B., and Michael Frame, *Fractals, Graphics. And Mathematics Education,* Mathematical Association of America, Notes Series Number 58, *(Translation, Mathematical Monograph)* 2002

[R28.25] Mandelbrot, Benoît B., *The fractal geometry of nature*. New York, *Macmillan. 1983.*

[R28.26] Mandelbrot, B. B.: *The Fractal Geometry of Nature*. W. H. Freeman and Company, New York 1982.

[R28.27] Nelson, Bryn; "Sophisticated Mathematics Behind African Village Designs Fractal patterns use repetition on large, small scale," *San Francisco Chronicle*, February 23, 2009.

[R28.28] Ofori-Ansah, Kwaku, P, *Kente is More than A Cloth: History and Significance of Ghana's Kente Cloth,* Hyattsville, MD, USA Sankofa Pub. 1995.

[R28.29] Ostwald, Michael J., and Vaughan, Josephine, *The Fractal Dimension of Architecture*. Birhauser, Basel. 2016.

[R28.30] Ozhovan M. I., Dmitriev I. E., Batyukhnova O. G. "Fractal structure of pores of clay soil". *Atomic Energy*, Vol. 74, 1993, pp.241–243.

[R28.31] Peters, Edgar, *Chaos and order in the capital markets: a new view of cycles, prices, and market volatility*. New York, Wiley. *(1996).*

[R28.32] Tan, Can Ozan; Cohen, Michael A.; Eckberg, Dwain L.; Taylor, J. Andrew "Fractal properties of human heart period variability: Physiological and methodological implications". *The Journal of Physiology. 587* (15), 2009, pp, 3929–41.

[R28.33] Russ, John C. *Fractal surfaces. Vol. 1.* New York, *Springer*. (1994).

[R28.34] D. Seekell, M. L. Pace, L. J. Tranvik, C. Verpoorter, "A fractal-based approach to lake size-distributions" *Geophysical Research Letters,*. Vol 40 (3), 2013 pp. 517–521.

- **[R28.35]** Shamar, L "What Makes a Pollock Pollock: A Machine Vision Approach," *International Journal of Arts and Technology*. Vol 8: . (2015). 1–10

[R28.36] Sornette, Didier, *Critical phenomena in natural sciences: chaos, fractals, selforganization, and disorder: concepts and tools*. New York, *Springer*. 2004. *pp. 128–140.*

[R28.37] Spencer, John; Thomas, Michael S. C.; McClelland, James L. *Toward a unified theory of development: connectionism and dynamic systems theory re-considered.* New York, Oxford University Press, 2009.

[R28.38] Sreenivasan, K. R.; Meneveau, C. "The Fractal Facets of Turbulence". *Journal of Fluid Mechanics. 173:* 1986. *357–386.*

[R28.39] Stumpff, Andrew "The Law is a Fractal: The Attempt to Anticipate Everything". *Loyola University Chicago Law Journal Vol. 44.*, 2013, p. 649.

[R28.40] Takayasu, H. *Fractals in the physical sciences,* Manchester, Manchester University Press. 1990.

[R28.41] Tan, Can Ozan; Cohen, Michael A.; Eckberg, Dwain L.; Taylor, J. Andrew "Fractal properties of human heart period variability: Physiological and methodological implications". *The Journal of Physiology. 587 (15),* 2009, pp,3929–41.

[R28.42] Taylor, Richard P. "Fractal Fluency: An Intimate Relationship Between the Brain and Processing of Fractal Stimuli". In Di Ieva, Antonio (ed.). *The Fractal Geometry of the Brain,* Springer Series in Computational Neuroscience. Springer. *(2016).* pp. 485–496.

[R28.43] Taylor, R. P.; Spehar, B.; Van Donkelaar, P.; Hagerhall, C. M. "Perceptual and Physiological Responses to Jackson Pollock's Fractals". *Frontiers in Human Neuroscience.*Vol. *5:* (2011). 1–13.

[R28.44] Zhiwei Huang; Yunho Hwang; Vikrant Aute; Reinhard Radermacher "Review of Fractal Heat Exchangers" *International Refrigeration and Air Conditioning Conference. Paper 1725,* 2016.

[R29] ALGEBRAIC-GEEOMETRIC THEORY OF SPIDER GRAPH, SPIDERMATICS AND ANANSIMATICS

[R29.1] Barwise, J., Etchemendy, J.: “Visual information and valid reasoning.” In: Allwein, G., Barwise, J. (eds.) *Logical Reasoning with Diagrams*, pp. 3–25. Oxford University Press, New York (1996)

[R29.2] Bottoni, P., Fish, A.: "Coloured Euler diagrams: a tool for visualizing dynamic systems and structured information." In: *Proceeding Diagrams 2010, LNCS*, (Low Noise Cable Sensors) Vol. 6170, pp. 39–53 (2010)

[R29.3] Bottoni, P., Andrew Fish & Francesco Parisi Presicce Spider Graphs: a graph transformation system for spider diagrams *Software & Systems Modeling* **volume 14**, pages1421–1453 (2015)

[R29.4] Bottoni, P., Koch, M., Parisi-Presicce, F., Taentzer, G.: "Consistency checking and visualization of OCL constraints." In: *Proceedings of the UML 2000, LNCS,* (Low Noise Cable Sensors), vol. 1939, pp. 294–308 (2000)

[R29.5] Bottoni, P., Koch, M., Parisi-Presicce, F., Taentzer, G.: "A visualization of OCL using collaborations." In: Gogolla, M., Kobryn, C. (eds.) *Proceedings of the UML (unified Model Language) 2001, LNCS,* Vol. 2185 (2001) , pp. 257–271

[R29.6] Bottoni, P., Schürr, A., Taentzer, G.: "Efficient parsing of visual languages based on critical pair analysis (and contextual layered graph transformation." In: *Proceedings of the IEEE-VL'00,* IEEE CS Press (2000), pp. 59-61.

[R29.7] Cabot, J., Clariso, R., Riera, D.: "Verification of UML/OCL class diagrams using constraint programming." *In: Proceedings of the ICSTW '08,* IEEE CS Press (2008) pp. 73–80.

[R29.8] Cordasco, G., De Chiara, R., Fish, A.: "Interactive visual classification with Euler diagrams." In*: Proceedings of the VL/HCC 2009*, IEEE CS Press (2009) pp. 185–192.

[R29.9] Ehrig, H., Ehrig, K., Habel, A., Pennemann, K.H.: "Theory of constraints and application conditions: from graphs to high-level structures." *Fundam. Inform.* **74**(1), (2006) pp.135–166

[R29.10] Ehrig, H., Ehrig, K., Prange, U., Taentzer, G.: "Fundamental theory for typed attributed graphs and graph transformation based on adhesive HLR categories." *Fundam. Inform.* **74**(1), 31–61 (2006)

[R29.11] Ehrig, H., Ehrig, K., Prange, U., Taentzer, G.: *Fundamentals of Algebraic Graph Transformation*. Berlin, Springer, (2006)

[R29.12] Ehrig, H., Heckel, R., Korff, M., Löwe, M., Ribeiro, L., Wagner, A., Corradini, A.: "Algebraic approaches to graph transformation—part II: single pushout approach and comparison with double pushout approach." In: Rozenberg, G. (ed.) *Handbook of Graph Grammars and Computing by Graph Transformations, Volume 1: Foundations*, World Scientific, Singapore (1997) pp. 247–312.

[R29.13] Fish, A., Flower, J., Howse, J.: The semantics of augmented constraint diagrams. (JVLC) *Journal of Visual Language and Computing* Vol.**16** (2005) , pp.541–573

[R29.14] Fish, A., John, C., Taylor, J.: "A normal form for Euler diagrams with shading." *In: Proceedings of the diagrams 2008, LNCS, vol. 5223,*. Berlin, Springer, (2008) pp. 206–221

[R29.15] Flower, J., Masthoff, J., Stapleton, G.: "Generating proofs with spider diagrams using heuristics." *In: Proceedings of DMS-VLC,* Knowledge Systems Institute (2004), pp. 279–285.

[R29.16] Flower, J., Masthoff, J., Stapleton, G.: "Generating readable proofs: a heuristic approach to theorem proving with spider diagrams." In*: Proceedings of Diagrams 2004, LNAI, vol. 2980*, Berlin, Springer, (2004) pp. 166–181.

[R29.17] Flower, J., Stapleton, G.: "Automated theorem proving with spider diagrams." In: *Proceedings of Computing: The Australasian Theory Symposium, ENTCS,* vol. 91, Elsevier (2004) pp. 116–132

[R29.18] Goedicke, M., Meyer, T., Taentzer, G.: "Viewpoint-oriented software development by distributed graph transformation: towards a basis for living with inconsistencies." In*: Proceedings of IEEE RE'99*, (1999), pp. 92–99

[R29.18] Habel, A., Heckel, R., Taentzer, G.: "Graph grammars with negative application conditions." *Fundam. Inform.* **26**(3,4), (1996), pp. 287–313

[R29.19] Habel, A., Pennemann, K.H.: "Correctness of high-level transformation systems relative to nested conditions." *Math. Structructures in. Computer. Sci.* **19**(2), (2009), pp. 245–296

[R29.20] Hammer, E., Shin, S.J.: "Euler's visual logic." *History and Philosophy of Logic,* (1998) pp. 1–29

[29.21] Howse, J., Molina, F., Taylor, J., Kent, S., Gil, J.: Spider diagrams: a diagrammatic reasoning system. JVLC (Journal of Visual Language and Computing)**, Vol. 12**(3), 299–324 (2001)

[R29.22] Howse, J., Stapleton, G., Taylor, J.: "Spider diagrams." *LMS J. Comput. Math.* Vol. **8**, (2005), 145–194

[R29.23] Kreowski, H.J., Kuske, S., Schürr, A.: "Nested graph transformation units." (Int. J. SEKE) *International Journal of Software Engineering and Knowledge Engineering*, Vol. 7(4), (1997) pp.479–502

[R29.24] Shin, S.J., *The Logical Status of Diagrams*., Cambridge, Cambridge University Press (1994)

[R29.25] Sowa, J.: *Conceptual Structures: Information Processing in Mind and Machine.* London, Addison-Wesley, (1984)

[R29.26] Stapleton, G., Howse, J., Taylor, J., "A decidable constraint diagram reasoning system." *J. Logic and Computaion.* Vol. **15**(6), (2005), pp. 975–1008.

[R29.27] Stapleton, G., Masthoff, J., Flower, J., Fish, A., Southern, J.: "Automated Theorem Proving in Euler Diagram systems." *J. Automated. Reason.* Vol. **39**(4), (2007), 431–470.

[R29.28] Stapleton, G., Thompson, S., Howse, J., Taylor, J.: "The expressiveness of spider diagrams." *J. Logic and. Computation*. Vol. **14**(6), (2004), pp. 857–880.

[R29.29] Urbas, M., Jamnik, M., Stapleton, G., Flower, J., "Speedith: a diagrammatic reasoner for spider diagrams." In: Proceedings of Diagrams 2012, LNCS, (Low Noise Cable Sensors). Vol. 7352, pp. 163–177. Springer, Berlin (2012)

[R29.30] Warmer, J., Kleppe, A.: *The Object Constraint Language: Precise Modeling with UML*. London Addison-Wesley, (1999)

[R29.31] Zykov, A.: Hypergraphs. Russian. Math. Survs. Vol. **29**(6), (1974) pp. 89–156

Index

A
Abraham, W.E., 310
Absoluteness-relativity duality, 220, 260, 419, 420
Abstract language, 101, 319, 331
Acceptance-rejection dualities, 121, 128, 203, 244, 299, 318, 327, 328, 330, 356, 366
Accident-risk dualities, 276
Acquaintance-description dualities, 46, 90, 120, 124, 161, 162, 172, 235, 427
Acquaintance space, 39, 98, 153, 191, 278, 304, 347, 350, 355, 404, 419
Actual-potential dualities, 6, 11, 17, 20, 34, 37–40, 43, 82, 85, 87, 100, 111, 116–118, 120–122, 126, 131, 132, 135, 138–140, 150, 153–155, 167, 169, 171–173, 177, 178, 181, 185, 187, 188, 205, 206, 213, 216, 222, 226, 230, 233, 234, 247, 248, 250, 252, 255, 262, 267–269, 271, 273, 276, 279–281, 285, 286, 304, 322, 328, 330, 351, 352, 358, 360, 377, 381, 384, 387, 390, 393, 402–405, 412
Adinkra, 211, 218, 235, 295, 370, 388, 398, 421–424
Adinkralogy, 235, 236, 238, 388, 398, 421, 424
Adinkramatics, 235, 236, 238, 388, 398, 421, 423
Adinkra tradition, 65
African system of thought, 65
Algebraic analytics, 112, 247
Algebraic-geometric systems, 235, 236
Algebraic systems, 227
Alternative fact, 113, 153, 247
Alternative reality, 153, 182, 247
Anansimatics, 235, 236, 238, 398, 421, 424
Anansi stories, 398, 421, 424
Anasiology
Anoma-kokone-kone, 47, 65, 87, 424, 425
Anomalies, 111–113, 116, 117, 129, 153, 154, 182, 247, 248, 283, 284, 291, 292, 295, 301, 302, 431
Anticipation, 145, 152, 266
Anticipation-expectation dualities, 115, 124, 132, 135, 141, 148, 167, 169, 219, 258, 265, 280, 361, 362, 366
Anticipation operator, 130, 147, 193
Anticipations, 6, 34, 95, 116–118, 130, 135, 141, 142, 148, 149, 153, 160, 166, 167, 169, 193, 199, 204–207, 209, 214, 259, 260–262, 265, 266, 269, 270, 280, 284, 305, 310, 323, 335, 351, 362, 366, 406, 422, 431
Antonyms, 407
Applied problem-solution dualities, 18, 21, 22, 30, 31, 33, 67, 68
Applied problem-solution polarity, 67
Apprenticeship, 94, 95
A-prior knowledge, 41, 42
Arrogance of knowledge, 225
Asante, M., 309
Asantrofi-anoma, 47, 99, 141, 214, 297, 316, 358–361, 425, 426
Asantrofi-anoma problem, 316, 359–361, 370, 425
Asantrofi-anoma rationality, 211, 358–361, 364, 425
Attractive-repulsive dualities, 361, 363
Axiomatic information, 39, 42

K. K. Dompere, *Theories of Knowing and Knowledge Production*, Studies in Systems, Decision and Control 501, https://doi.org/10.1007/978-3-031-42863-0

Axiomatic-information-support, 42, 43

B

BA, 221
Basic institutions primary-derived dualities, 22, 25
Basic organization, 22, 27, 29, 32, 55, 56, 59
Basic structure, 22, 25, 27, 28, 35, 219, 374, 391
Beginning disposition, 48, 132, 134, 221, 223, 224, 229–231, 271, 272, 389, 401, 403, 409
Beginning-end duality, 132, 133, 165, 177, 218, 223–226, 229–231, 240, 242, 268, 272, 292, 303, 370, 388, 392, 401, 402, 409, 426, 427
Beginning-end problem, 350
Belief disposition, 234
Belief functions, 260
Benefit-risk duality, 133
Ben-Jocannan, Y., 309
Bennu-anoma, 235, 389, 426
Bennu-anoma connectivity structure, 230
Bennu-anoma-gyenyame, 177, 426
Bennu anoma principle, 230, 231, 234
Body-mind duality, 227, 254
Body sovereignty, 61, 62, 273
Bounded rationality, 359
Brouwer, 365

C

Capital stock-flows, 333, 363, 417
Categorial analytics, 227, 229, 235, 385
Categorial converters, 2, 62, 136, 137, 138, 161, 174, 192, 193, 195, 228, 249, 250, 255, 256, 271, 274, 281, 302, 368, 379, 394, 411, 413–415, 417, 418, 420, 431
Categorial diversity, 7, 8, 66, 97, 105
Category of non-science, 71
Category of sciences, 71
Central analytical concept, 371, 379
Certainty, 35, 45, 46, 49, 50–53, 55, 173, 214, 261, 296, 305, 325, 327, 328, 405, 406
Certainty dominance
Certainty-uncertainty dualities, 2, 11, 35, 38, 43, 46, 49, 50, 52, 53, 56, 71, 76, 80, 98, 113, 120, 121, 124, 130, 133, 135, 139, 153, 172, 174, 179, 193, 201, 203, 208, 214, 216, 220, 247, 249, 259–263, 265, 272, 284, 294, 296, 301, 303, 320, 325, 326, 345, 347, 348, 351, 357, 358, 366, 369, 406, 419
Characteristic disposition, 11, 23, 39, 41, 42, 44, 50, 51, 60, 71, 83, 88, 96, 98, 100–102, 116, 118, 119, 120–127, 129, 133, 136, 137, 141–143, 147, 153, 156, 159, 160, 163, 167–169, 170–174, 178, 179, 183, 184, 186–190, 192, 194, 200, 201, 204, 212, 214, 238, 243, 251–253, 262, 277, 278, 280, 284–290, 292, 293, 298, 301, 303, 304, 317–320, 322, 323, 325, 326, 328, 329, 331, 334, 336, 339, 341, 343, 344, 345, 347–350, 359, 360, 365, 368, 369, 371, 379, 380, 381, 387, 396, 405, 412, 414, 417, 418
Characteristic-signal decomposition
Characteristic-signal disposition, 39, 100, 142, 143, 168, 170, 181, 273, 275, 278, 283, 292, 304, 319, 323, 325, 329, 348, 368
Chemical information, 321
Chicken-egg problem, 350
Classical indicator function
Classical information, 245, 355, 367
Classical mathematics, 101, 102
Classical paradigm of thought, 49, 102, 114, 240, 242, 244, 245, 337, 354, 355, 361, 362, 367
Classical true-false analytics, 354
Climatological changes, 387
Closed set, 127, 371, 373, 383, 385
Cognitive agents, 2, 16, 17, 19, 26, 28, 31, 40, 48, 51, 80, 82, 84, 88, 89, 91–93, 96, 115, 120, 124, 125, 132, 134, 139, 172, 182, 237, 268, 271, 278, 280, 281, 297, 298, 304, 306, 308, 329, 330, 349, 355, 356, 365, 373–375, 378–380, 385, 388–390, 393, 394, 396–399, 402, 403, 407
Cognitive capacity limitations, 8–11, 23, 26, 30, 38, 46, 48, 49, 51, 52, 54, 55, 62, 66, 68, 72–77, 79, 87, 95, 98, 107, 109, 113, 115, 121, 123, 127, 130, 133, 139, 150, 170, 175, 179, 193, 220, 222, 234, 235, 249, 259, 262, 263, 265, 276, 279, 280, 281, 283, 305, 307, 311, 325, 326, 338, 348, 355–357, 361, 368, 370, 377,

388, 394, 397, 399, 403, 405, 406, 408, 414, 417, 418, 419, 431
Cognitive geometry fuzzy paradigm
Cognitive imbecility, 4, 219
Cognitive preferences, 70, 82
Cognitive principle epistemic creational structure, 233
Cognitive processers
Cognitive state of being, 128
Cognitive transversality, 3, 295, 329, 331
Collaboration, 74, 106, 108, 109
Collective preference, 4, 13, 23, 82, 83, 203, 363, 377, 425
Collective skill, 95
Collective vision, 94, 95, 127, 174
Comparative analytics, 1, 58, 85, 147, 151, 177, 189, 247, 272, 339, 340, 375, 377, 378, 402, 407
Completeness-incompleteness dualities, 262
Computable-incomputable dualities, 49
Concept of policy, 57
Concept of uncertainty, 260
Conceptual-philosophical systems, 238
Conceptual zombies, 163
Confidence interval, 209
Conflict-cooperation dualities, 127, 174
Conflict-resolution dualities, 69, 88, 129
Conformal mapping, 113, 125, 138, 173, 182, 189, 203, 218, 220, 224, 249, 260, 267, 272, 329
Constructionism-reductionism dualities, 328
Control variable, 317, 318
Conversion moments, 256, 257
Converter-enhancement-operator function, 184
Correct-wrong dualities, 249
Cost-benefit distribution, 26, 36, 61, 73, 74, 108, 164, 273, 291, 294
Cost-benefit dualities, 1, 2, 4, 6–10, 19, 20, 23, 29, 30, 34–36, 42, 46, 59, 62, 63, 64, 66, 68, 69, 76, 77, 79, 80, 82–86, 88, 91–94, 100, 105, 107, 113, 126, 127, 139, 141, 144, 149, 151, 156, 158, 160, 162, 164, 166, 173, 178, 179, 182, 186, 194, 199, 203, 207, 212, 215, 218, 239, 248, 249, 255–258, 279, 282, 283, 290, 291, 297, 298, 301, 311, 329, 330, 335–338, 344, 346, 355, 358, 360–363, 373, 374, 390, 393, 394, 396, 403, 407, 409, 426, 431
Courage, 111, 165, 174, 183, 186, 189, 198–204, 212, 213, 215, 217, 221, 222, 247, 248, 250, 258, 260, 267, 282, 301, 305, 359, 408
Courage operator function, 198, 202
Creative function, 180, 187, 188
Creative innovation, 340, 342
Creativity, 2, 7, 87, 111, 116, 125–128, 153, 165, 174, 175, 177, 179, 180, 182–187, 189–204, 212, 213, 215–218, 221, 222, 247–250, 256–258, 260, 267, 276, 280, 282, 284, 301, 305, 308, 310, 339–342, 358, 366, 405, 408, 411, 417, 420, 431
Creativity discovery, 339, 340, 342, 405
Creativity disposition, 190, 202
Creativity index, 180
Creativity operator function, 191, 192, 193
Curiosity, 2, 7, 111, 125, 127–129, 148, 149, 165, 174, 175, 177, 179, 180, 183–192, 198–204, 213, 215–217, 221, 247, 248, 250, 258, 260, 282, 301, 305, 308, 358, 366, 408, 411, 417, 420, 431
Curiosity disposition, 190, 201, 202
Curiosity function, 180, 187, 188, 190, 213
Curiosity index, 180
Curriculum development, 93

D

Data, 49, 52, 54, 98, 261, 264, 272, 273, 292, 298–301, 319, 320, 329, 351, 367
Decidability-undecidability dualities, 45, 56, 142, 272, 290, 328, 330, 335, 345, 349, 355, 357, 359, 366, 423
Decision-choice action, 1, 3–10, 12, 14–18, 21–23, 26, 28–30, 32, 35, 57–60, 62, 76–80, 82, 83, 91, 95, 100, 130, 172, 187, 201, 233, 266, 292, 293, 295, 297, 329, 331, 345, 349, 350, 372, 392, 426, 431
Decoding function, 121, 292, 293
Decomposition-composition dualities, 21
Defuzzification, 356, 357
Degrees of doubt, 296
Degrees of surety, 296
Delusions, 156, 182
Demand-supply chain, 47
Departments, 89, 90, 92, 319, 418

Derived category, 22, 27, 28, 67, 82, 83, 85, 90, 190, 252, 259, 263, 278, 303, 304, 351, 365, 378, 389, 400, 404
Derived category superstructure, 22
Design quantitative measures, 329
Destruction-consumption dynamics
Destruction-consumption process, 306
Destruction-creation dualities
Diop, C. A., 309
Discovery creativity, 191–194
Disequilibrium dynamics, 1, 3, 17, 19, 28, 88, 89, 97, 103, 122, 155, 176, 218, 250, 301, 304, 323, 333, 368, 369, 375, 384, 394, 432
Disequilibrium path, 34
Disequilibrium process, 63, 92, 96, 121, 124, 172
Disequilibrium system, 63, 64
Disinformation, 4, 62, 91, 112, 113, 125, 160, 161, 164, 273, 290, 293–295, 305, 306, 323, 331, 344, 417
Distributive justice, 77
Dividedness, 3, 7–9, 84, 85, 88, 105, 106, 266, 276, 278, 279, 404, 423, 428
Dompere, K.K., 310
Doubt-surety, 30, 38, 45, 46, 49, 56, 120, 124, 142, 172, 258, 272, 294, 330, 345, 347, 361, 366, 406, 407
Dualistic dominance, 147, 387, 396
Dualistic efficiency
Dualistic-polar categories, 138
Dualistic-polar conditions, 57, 58, 64, 65, 84, 86, 89, 118, 128, 131, 136, 148, 154, 164, 168, 181, 185, 188, 213, 217, 222–224, 226, 229, 230, 232, 233, 256, 258, 263, 265, 276, 279, 286, 291, 335, 337, 349, 350, 353, 356, 360–362, 375, 392, 421, 423, 424
Dualistic-polar continuum, 6, 136, 171
Dualistic-polar decision, 340, 342
Dualistic-polar force, 324
Dualistic-polar games, 68, 107, 236, 237, 257, 294, 316, 317, 327, 332, 335, 339, 348, 358, 387, 397, 411, 413, 414
Dualistic-polar indispensability condition, 358
Dualistic-polar structure, 5, 74, 85, 86, 94, 101, 105, 112, 116, 125, 132–134, 140, 150, 153, 155, 163, 171, 190, 218, 225, 272, 278, 295, 297, 303, 313, 316, 327, 338, 343, 344, 346, 349, 350, 353, 365, 371, 374, 377, 378, 381, 383, 384, 387, 396, 404, 407, 414
Dualistic-polar zones, 140, 336, 342
Duals, 3, 10, 17, 69, 84, 121–124, 129, 161, 171, 172, 182, 188, 194, 197, 210, 214, 217, 223, 224, 229, 236, 263, 274, 294, 295, 297, 315, 316, 322, 324, 327, 334, 335, 343–346, 348, 356–358, 360, 362, 364, 367, 368, 380, 382, 387, 391, 396, 412, 414, 420–422, 425–429
Du Bois, W.E.B., 309
Dynamic problem-solution, 325
Dynamic stability, 126

E

Each-for-all-and-all-for each, 19, 72, 77, 85, 424, 429
Each-for-himself, 72
Earthquakes, 307, 387–389
Economic institutions, 27, 32
Economics of specialization, 92
Economic sovereignty, 56, 61
Economic structure, 27, 32, 56, 61, 104, 164, 219
Education, 21, 90, 92–97, 99, 108, 295, 310, 345, 431
Effort-reward dualities, 113
Emptiness disposition, 227, 228, 231–233
Emptiness-fullness dualities, 218, 229, 231–234, 241, 242
End disposition, 48, 134, 223, 224, 229, 230, 272, 389, 401, 403, 409
Endo-categorial convertor, 137, 138
Endo-functor, 137
Engineering, 43, 64, 127, 217, 235, 332, 369, 404
Engineering-economic system, 23
Entropic conditionality, 170
Epigenetics, 23
Epistemic anomalies, 111–113, 129, 247, 431
Epistemic characteristic disposition, 41, 44, 120, 122, 134, 135, 173, 220, 301, 305, 326, 403, 405
Epistemic conditionality, 121, 296, 327, 355
Epistemic creativity, 124, 173
Epistemic decomposition, 269
Epistemic distance, 41, 51, 52, 71, 98, 130, 133, 134, 139, 193, 216, 220, 226,

234, 262, 265, 326, 349, 369, 405, 406
Epistemic emptiness, 233
Epistemic geometry, 24, 43, 196, 374, 386, 391, 395, 416
Epistemic identification transversality, 328, 333, 344, 346
Epistemicity, 177, 431
Epistemic journey, 134, 226, 315–317, 336
Epistemic Ponzi schemes, 344
Epistemic space, 98, 173, 179, 220–222, 234, 269, 284, 369, 371, 380, 381, 392, 405, 424
Epistemic toolboxes, 86
Epistemic transformation transversality, 328, 333, 346
Epistemic transversality, 294, 295, 341, 342, 356, 369
Epistemic transversality conditions, 3, 106, 156, 233, 259, 295, 312, 315, 316, 321, 322, 324–328, 332, 333, 335–337, 343–347, 349, 353, 362, 366, 431
Epistemic varieties, 50, 51, 127, 140, 172, 220, 222, 262, 283, 285, 301, 302, 326, 349, 355, 395, 405
Epistemological information, 39, 221, 364, 368, 402
Epistemological space, 19, 20, 23, 36, 39, 41, 42, 44, 46–48, 50–152, 80, 82, 100, 103, 125, 126, 128, 132, 134, 135, 151, 175, 179, 215, 219–225, 227, 229, 231, 233–235, 237, 238, 241–243, 247, 249, 256, 262, 263, 267, 268, 271, 277, 279, 280, 282–286, 289, 290, 298, 300, 303–305, 307, 311, 322, 325, 338, 345, 347, 350–352, 356, 362, 365, 371, 373–375, 377, 378, 380, 381, 385, 390, 391, 393–395, 397–399, 402–406, 412, 413, 418
Epistemological time, 8, 225, 227, 228, 312, 339, 351, 391, 401, 409
Epistemological universe, 125
Epistemological variety, 134, 135, 140, 141, 181, 226, 239, 262, 285, 286, 374, 390, 396–400, 403
Equilibrium path, 34
Espionage, 70
Essence of life, 5, 7, 8, 94, 282
Evidence, 49, 52, 261, 272, 273, 292, 298–301, 319, 320, 329, 367
Exact information, 102, 355, 356
Exactness-inexactness, 30, 64
Excluded middle, 52, 102, 114, 196, 240, 242, 244, 245, 344, 354, 356, 357, 359, 362, 367, 368, 419, 422
Exoplanetary, 223, 307, 389, 403, 404
Exoplanetary system, 223, 307, 389, 403, 404
Expansion of universe, 125, 308
Expectation operator, 130, 141–143, 146–149, 156, 169, 193
Expectations mega-phenomenon, 260
Experiential information, 38, 39, 41, 76, 101, 107, 191, 255, 264, 365, 367, 368, 419
Explanatory-information-support, 42
Explanatory conditions, 35, 39, 40, 44
Explanatory science, 40, 270, 275, 332
Explanatory system
Explanatory theory, 36–42, 44, 270, 275
External dualistic-polar condition, 84
External duality, 421, 426

F

Fact, 7, 103, 113, 271, 273, 295, 298, 299, 300, 319, 320, 329
Fake news, 113, 182, 247, 289, 290, 295, 306, 344
Fantasies, 182, 282
Finite-infinite dualities, 19, 20, 64, 427
First principle, 229
Fix-level cuts, 222
Food chain, 367
Frequency function, 260
Full ignorance, 354
Fullness disposition, 227, 228, 231, 233
Functor, 2, 6, 37, 42, 43, 71, 86, 121, 130, 136–138, 141, 142, 153, 156, 159–161, 167, 173, 174, 180, 183, 184, 190–195, 197, 199, 228, 249, 255, 256, 264, 266, 271, 274, 276, 281, 285, 336, 379, 394, 408, 411, 413–415, 417, 418, 420, 431
Functor moment, 137, 138
Functor space, 36, 62, 117, 130, 136–141, 143, 146–148, 169, 179, 192–196, 255, 431
Fundamental problem-solution dualities, 18, 21, 24, 31, 67, 68, 90
Fundamental problem-solution polarities, 67
Funtum-mereku-denkyemmereku, 47
Fuzzification, 356, 357

Fuzzy analytics, 352, 362
Fuzzy conditionality, 115, 119, 123, 142, 147, 150, 168, 185, 188, 208, 213, 229, 230, 231, 233, 251, 253, 259, 356, 358
Fuzzy constrained set, 363
Fuzzy decision, 194, 195, 207, 254, 317, 334, 339
Fuzzy decomposition, 84, 88, 162, 200, 222, 254
Fuzzy goals, 297
Fuzzy indicator function, 148, 168, 222, 254, 265
Fuzzy information, 5, 245, 349, 353–357, 361, 368, 419, 420
Fuzzy logic, 419
Fuzzy logical support principle, 355
Fuzzy mathematical problem, 419
Fuzzy mathematical programming, 195, 208, 334
Fuzzy mathematics, 419
Fuzzy optimal control, 317
Fuzzy optimality, 158, 318, 358
Fuzzy optimal problem, 196, 208
Fuzzy paradigm of thought, 2, 5, 112, 244, 245, 316, 334, 337, 342, 349, 352, 354–356, 358, 359, 361, 411, 412, 419, 420, 432
Fuzzy parameters, 318
Fuzzy partitions, 92, 105, 106
Fuzzy process, 209, 317
Fuzzy set, 191, 205, 206, 207, 317, 318, 334, 339, 344, 353, 354, 356
Fuzzy set of despair, 206
Fuzzy set of hope, 205
Fuzzy set of ignorance, 339, 353
Fuzzy set of knowledge, 339, 353
Fuzzy space, 212, 244, 349
Fuzzy-stochastic, 212, 305, 327, 328, 357, 362, 366, 367
Fuzzy-stochastic conditionality, 212, 276, 318, 327, 329, 338, 357, 362, 366
Fuzzy-stochastic information, 347, 348, 365
Fuzzy true-false analytics, 355
Fuzzy variable, 205, 352

G

Galaxies, 126, 307, 389
Game space, 16, 113, 237, 327, 345, 387, 397, 414, 415, 416, 420
Game zone, 295, 296
General principle of opposites, 4, 47, 126, 196, 208, 255, 259, 295, 344, 358, 359, 369, 414
Generational cognitive capacity limitations, 72, 74
Generational lifetime, 72
Geomorphological changes, 240, 387
Gigaverse, 408
Give-and-take sharing, 7–9, 12–14, 16, 19, 28, 74, 76, 77, 90, 92, 95, 96, 107, 120, 122, 123, 126, 131, 154, 171, 189, 211, 248, 251, 253, 302, 383, 393, 412–414, 421, 422, 426
Give-and-take-sharing principle, 357
Global collaborations, 36, 106
Global system, 70, 108, 223
Goal-facilitator, 281
Goal-objective actions
Goal-objective element, 6, 91, 106, 127, 200, 337, 344
Gyenyame creational structure, 230
Gyenyame principle, 224, 230, 231

H

Happiness-sadness dualities, 47, 173, 291, 310
Hate, 105, 234, 300
Hesitation interval, 45, 209
Hope-despair duality, 2, 56, 177, 193, 201, 204, 207, 208, 215, 216, 218, 219, 248, 258
Hope indicator, 205, 206
Human atrocities, 108
Human right, 61, 62, 164
Hybrid index, 180

I

Identifiability, 106
Identification game, 343
Identification problem-solution, 55, 323, 325
Identification-transformation dualities, 1, 6–8, 37, 38, 41, 47, 50, 51, 92, 97, 112, 116, 120, 126, 128, 153, 156, 157, 175, 178, 179, 186–188, 190, 191, 196, 198, 200, 213, 215, 217, 238, 239, 248, 254, 255, 262–264, 272, 274, 280, 284, 298, 300, 301, 308, 309, 312, 315, 319, 321, 328, 336, 338, 339, 341, 342, 347, 364, 374, 377, 384, 387, 390, 392–394,

396, 398, 403–405, 407, 408, 411, 413–415, 418–420, 424
Identification-transformation process, 17, 283, 348, 380, 384, 389, 391, 399, 403, 413
Identification transversality, 315, 326
Ideological belief, 62
Ideological structures, 59
Ignorance-benefit dualities, 266
Ignorance-knowledge dualities, 3, 4, 19, 21, 37, 38, 40, 41, 43, 45, 47, 48, 63, 66, 71, 82, 103, 107, 112, 115, 127, 128, 133, 134, 156, 177, 180, 182, 184, 188, 192, 195, 196, 199, 204–206, 214–218, 226, 232, 233, 235, 252, 255, 259, 267, 275, 283, 284, 291, 295, 297, 301, 307, 309, 312, 313, 318, 321–323, 364–366, 378, 394, 409, 412–414, 418–420, 429
Ill-posed problem, 92, 112, 113, 225, 247, 287, 288, 408
Imagination dual, 123, 129, 171, 185, 213
Imagination-reality dualities, 11, 17, 20, 26, 34, 35, 37–39, 42, 43, 74, 83, 85, 87, 92, 99, 100, 111, 115–124, 126, 129, 132–135, 139–142, 148, 150, 151, 153–155, 157, 168, 170–174, 177, 179, 181–185, 187–190, 192, 199, 200, 203–206, 213, 215, 218, 222, 223, 225, 226, 231–233, 247, 248, 250–253, 255, 260, 262, 266–268, 279–281, 298, 304, 311, 320, 322, 323, 325, 328, 330, 332, 337, 338, 346, 358–361, 364, 369, 370, 377, 393, 394, 396, 402, 404, 405, 412, 417, 418, 421
Imbalance-balance dualities, 223
Impossible world, 111, 112, 116, 119, 128, 129, 131, 140–143, 146–150, 153, 156, 166, 168, 181, 272, 274, 278, 279, 285, 287, 288, 406
Impossible world intentionality
Improbability space, 143, 147, 167, 277
Improbable world, 119, 128, 150, 166, 168, 406
Incentive-disincentive dualities, 46, 158, 162, 164, 179, 193, 203, 239, 255, 274, 301, 311, 329, 330, 363, 374, 393, 399, 403, 407
Incubator of ideas, 274
Indicator conflict function
Individual-collective, 2, 8, 75, 77, 310
Individual-collective dualities, 8, 9, 21, 29, 30, 33, 34, 45, 48, 62, 63, 68, 73, 78, 79, 86, 87, 90, 91, 93, 97, 105, 117, 125, 127, 128, 129, 161–163, 165, 174, 180, 191, 203, 275, 291, 332, 393, 417, 420, 429
Individual-collective successes, 78
Individual community, 26, 33, 35
Individual-community dualities, 4, 13, 55–57, 61, 65, 66, 69, 75–77, 82, 85, 87, 88, 94, 96, 112, 118, 156, 158, 161, 164, 175, 199, 240, 282, 291, 303, 424, 425
Individual goals, 18, 20, 27, 94, 127
Individual preference, 13, 69, 425
Individual vision, 127, 174
Inductive-deductive duality, 65
Inexact information, 102, 367
Infinite ontological time, 384
Infinite stock-flow, 17, 322
Infinite universe, 150, 226
Infitiase-she-nawieyemu-nawieye-she-nefitiasemu, 370
Information field, 1, 13, 17, 63, 77–81, 84, 88, 89, 90, 92, 93, 96, 101, 104, 111, 131, 137, 154, 187, 199, 216, 300, 302, 303, 417, 418, 420
Information-knowledge disequilibrium, 28, 62, 69, 70, 78, 121, 124, 140, 172, 301, 374, 391, 394, 408, 428
Information-knowledge equality, 50, 51, 132, 173, 178, 220, 222, 263, 312, 325, 352, 371, 374, 381, 383, 397, 405, 414
Information-knowledge inequality, 51, 173, 179, 220, 222, 325, 351, 352, 405, 414
Information-knowledge production, 1, 2, 11, 19–22, 28, 29, 32, 33, 36, 44, 49, 54, 66, 69, 71, 80, 83, 91, 99, 103, 111, 117, 124, 126, 128, 139, 154, 161, 172, 174, 175, 186, 187, 188, 190, 191, 203, 213, 216, 222, 233, 237, 245, 247, 248, 255, 258, 261, 263, 266, 267, 272, 278, 296, 299, 332, 338, 367, 369, 379, 382, 390, 394, 397, 405, 415, 418, 420, 431
Information-knowledge structure, 28, 71, 79, 163, 181, 283, 293, 298, 300, 325, 327, 335, 365, 424
Information-knowledge system, 1, 3, 6, 7, 10, 14, 18, 21–23, 25, 29, 30–33, 35,

54, 55, 60–63, 65–69, 70, 74, 75, 77, 78–81, 86, 88–90, 94, 96, 97, 99, 102–107, 109, 125, 176, 191, 247, 260, 298, 300, 302, 328, 330, 378, 399, 403, 424
Information-time principle, 229, 231, 232
Information-time trinity, 175, 224, 234
Initial condition, 38, 40, 48, 117, 130, 133, 135, 140–142, 149, 152, 160, 161, 166, 167, 193, 199, 204–206, 217, 228, 241–243, 261, 262, 266, 268–271, 278, 279, 284, 290, 310, 323, 335, 338, 341, 350, 351, 366, 380, 388, 422
Innovation creativity, 191–194
Input dual, 382, 391
Input-output dualities, 3, 6, 9, 10, 18, 19, 23, 26, 32, 36, 47, 52, 62, 68, 74, 80, 82, 83, 86, 97, 138, 149, 154, 156–160, 165, 175, 176, 181, 184, 186, 189, 191, 201, 218, 222, 238, 239, 255, 257–259, 273, 278, 291–294, 300, 310, 311, 321, 323, 330, 346, 368, 373, 374, 377, 379, 381, 383, 384, 387, 390, 393, 397, 398, 402, 404, 412, 414, 418
Input-output process, 2, 29, 31, 32, 47, 50, 52, 64, 70, 80, 83, 92, 96, 109, 128, 149, 155, 240, 248, 279, 291, 298, 301, 306, 308, 309, 311, 317, 323, 330, 398, 399, 417
Input-output structure, 47, 60, 69, 155, 330, 363
Input supply, 311
Institutional channel, 9
Institutional diversity
Institution of research, 54, 81, 105
Insufficient information, 351
Intellectual capital stocks, 17, 20, 28, 37, 41, 52, 73, 74, 112, 117, 128, 129, 131, 134, 154, 175, 177, 180, 183, 186, 199, 207, 211, 212, 216, 218, 225, 227, 234, 254, 258, 259, 264, 266, 267, 269, 274, 278, 283, 291, 295, 296, 300, 302, 315, 318, 324, 327, 345, 365, 378, 392, 406, 411, 417, 432
Intellectual factories, 82
Intellectual heritage of humanity, 73, 95, 102
Intellectual investment flows, 8, 17, 28, 37, 41, 52, 74, 107, 112, 117, 122, 128, 129, 131, 134, 150, 154, 175, 177, 180, 183, 186, 190, 199, 207, 211, 212, 216, 218, 223, 225, 226, 227, 234, 254, 258, 259, 264, 267, 274, 278, 279, 283, 291, 296, 300, 302, 315, 317, 322, 324, 327, 345, 365, 378, 392, 406, 411, 431
Intellectual-investment process, 124, 172
Intelligence-unintelligence dualities, 293
Interactive relation, 342
Interdisciplinary, 99, 100
Intergenerational dynamics, 34, 160
Intergenerational-knowledge transfer
Intergenerational philosophical consciencism, 141
Intergenerational time point, 202
Intergeneration preferences, 34, 36
Inter-individual structure, 30
Internal dualistic-polar condition, 11, 84
Internal duality, 369, 380, 421
Internal-external dualities, 12, 84, 96, 199, 301, 362, 374, 412
Internality-externality dualities, 239
Internal relational unity, 10
International collaboration, 107, 108
International cooperation, 77, 108
Inter-planetary, 223
Interpretational problems, 292
Inter-supportiveness, 357
Inter-zone dynamics, 327
Intrageneration preferences exceptionalism
Intra-individual structure, 30
Intra-planetary, 223
Intra-zone dynamics, 327
Investment-capital process, 278, 417

J

Joy-sorrow dualities, 173, 291, 310
Judicial decision, 367
Justice-injustice dualities, 19, 21, 77, 86, 129, 158, 291, 407
Justification-trust conditions, 328
Justified true belief, 299
Justified true despair, 209
Justified true of hope, 209

K

KA, 221
Kente fractals, 236, 427
Kentelogy, 235, 236, 238, 427
Kentematics, 235, 236, 238, 427
Key opposites, 377
Kindness-wickedness dualities, 249

Knowability, 125, 224, 232, 234, 235, 397, 402
Knowability-unknowability dualities, 307, 309
Knowable universe, 125
Knowable-unknowable dualities telescopic structure, 224
Knowing-knowledge dualities, 40
Knowledge, 1–9, 11, 13–21, 26–28, 31–33, 35–38, 40, 41–43, 45, 46, 48, 50–52, 54–60, 62, 63, 64, 68, 70, 71, 73–75, 79, 82–86, 88, 91, 93–97, 103–109, 111–113, 116, 117, 120–122, 124, 125, 128, 129, 131, 133, 134, 149, 150, 153, 155, 159–162, 164–166, 169, 172, 175–177, 179, 180, 182, 183, 186, 187, 189–191, 197, 199–201, 204, 207, 209, 212, 213, 216–225, 227, 228, 234, 236, 238, 250, 254, 255, 257–259, 261, 263, 264, 267–269, 272–274, 278, 279, 283, 291–307, 309, 313, 315–320, 322–330, 332, 333, 335–359, 363–370, 375, 378–380, 392, 394, 395, 397, 400, 402–409, 411, 412, 417, 418, 419, 421, 422, 427–429, 431, 432
Knowledge by acquaintance, 30, 38, 39, 56, 76, 101, 111, 264, 300, 364, 365, 368, 369, 406
Knowledge by description, 30, 40, 41, 76, 111, 255, 264, 300, 364, 365, 369, 406
Kuhn's paradigm shifting, 114

L
Laws of dualities, 380
Laws of motion, 381
Laws of opposites, 34, 380, 422
Laws of organization, 10, 11
Laws of polarity, 380
Learning, 3, 21, 35, 36, 62, 64, 65, 74, 88, 89, 90, 92–95, 99, 101–103, 105, 108, 218, 293, 295, 309, 330, 331, 345, 380, 424, 429, 430
Learning-by-doing, 103, 125, 367
Learning without supervision, 306
Learning with supervision, 74
Legal institutions, 3, 27, 33
Legal rules misinformation, 62
Legal sovereignty, 57
Legal structure, 1, 27, 33, 57, 61, 78, 164, 219, 273
Legal structure economic structure, 78, 219
Limitationality, 79
Limitations, 72, 74, 75, 79, 101, 107, 109, 126, 139, 226, 338, 355, 366, 385, 419
Limitativeness, 79
Limits and non-limits, 10, 11
Logico-mathematical reasoning, 363
Love, 105, 234, 300

M
Macro units
Management system, 384
Mathematical system, 227, 388, 398
Matter-energy category, 17
Matter-energy disposition, 89
Matter-energy duality, 140, 157, 158, 227, 241, 250, 267, 350, 393
Matter-energy equilibrium, 24, 70, 131, 140, 155, 159, 301, 384, 387, 396
Matter-energy principle, 229, 232
Maximization, 211, 214, 318, 362
Meaning-essence dualities, 282
Meaning of life, 5, 282, 301
Medical information, 321
Mega-algorithms, xxxiii
Mega paradigm, 341
Megaverse, 408
Membership characteristic function, 194, 195, 204, 205, 206, 208, 327, 334, 341, 344, 348, 353, 354
Meta-algorithms, 114, 115
Meta-paradigm, 113–115
Methods, 59, 60, 70, 74, 83, 85, 90, 95, 101–103, 114, 125, 159, 160, 163, 191, 192, 330, 345, 349, 351–353, 366, 367, 375
Methods of research, 99
Micro-macro dualities, 5, 30, 408
Mini-max principle, 59
Minimization, 211, 214, 362
Misinformation, 4, 62, 91, 112, 113, 125, 160, 161, 164, 273, 290, 293–295, 305, 306, 323, 331, 344, 417
Mistake-correction dualities, 283
Multi-relationality, 85

N
National cognitive limitations, 109
National exceptionalism, 162, 164, 209, 212
National personality, 202, 266, 417

Natural categorial converter, 228, 276
Natural dualistic-polar structure, 85
Natural functors, 131, 138, 140, 222, 302, 368, 414
Natural laws of identity, 222
Natural organization, 17, 86, 238
Natural science, 60, 63, 101, 332
Natural technological process, 11, 71, 276
Necessity-freedom dualities, 21, 56, 87, 115, 116, 120, 124, 128, 132, 135, 141, 148, 153, 169, 172, 175, 181, 206, 209, 219, 258, 260, 265, 272, 277, 280, 310, 361
Negation function
Negation-negation, 137, 138, 154, 155, 196, 228, 236, 237, 260, 324
Negative-positive dualities, 7, 26, 30, 58, 86, 186, 189, 346, 360, 373, 383, 386, 387, 390, 394–396, 412
Negative-positive duals, 364
Negative-positive polarity, 413
Negative-sum game, 165
Negative system, 391
Negative variety, 30, 371, 374, 381, 387
Neocolonialism, 108, 164, 310, 415, 416
Neurobiology, 23
Neurochemistry, 23
Neuro-decision-choice, 1, 2, 4, 6–13, 15–18, 26–28, 30–34, 37–42, 45, 46–49, 52, 55–63, 65, 66–80, 82–91, 93–101, 104–106, 108, 112–115, 118, 119, 124, 126, 131, 132, 135, 139–141, 148, 151, 155, 158, 161, 162, 164, 166, 177, 180, 181, 182, 186–189, 191, 193–196, 198, 200, 201, 203, 204, 208, 209, 211–219, 235, 237, 239, 242, 244, 245, 247–249, 252, 254, 255, 257, 259–266, 270, 272, 275, 278, 280–282, 290, 291, 293, 294, 296–302, 304, 309–312, 318–320, 322–324, 326, 328–330, 332, 333, 335, 338, 344–350, 355–366, 368, 369, 372–374, 382, 390, 392–394, 396, 397, 403, 404, 407, 408, 412–415, 417, 421, 423, 425, 426, 429, 430
Neuro-decision-choice agents, 8, 17, 46–49, 52, 101, 114, 183, 207, 214, 250, 272, 293, 311, 348, 355, 402, 406
Neuro-decision-choice process, 26, 37, 55, 86, 92, 104, 165, 186, 211, 239, 259, 305, 309, 316, 318, 361, 362, 377, 379, 403
Neuro-decision-choice science, 23
Neuroeconomics, 23
Neuroscience, 23
Nkrumah, K., 309
Non-human resource, 18, 66
Non-ubuntu
Nonviolence-violence dualities, 165
Nothingness disposition, 221, 224, 227, 229–231, 270, 272
Nothingness-somethingness duality, 5, 204, 218, 224–230, 232, 234, 240, 242, 272, 283, 303, 389, 427, 428
Nyansapo, 218, 370, 427, 428

O

Objective probability, 98, 126
Object space, 119, 121, 168, 170, 185, 251, 253
Old-new dualities, 37, 412
Oneness, 3, 7–9, 276, 278, 279, 388, 423
Oniam-sua-ohu
Onim-sua-ohu, 3
Ontic characteristic disposition, 41, 44, 50, 51, 135, 172, 173, 220, 301, 303, 305, 403, 405
Ontic-epistemic dualities, 2, 38, 40, 41, 43, 46, 52, 71, 83, 87, 98, 101, 115, 126, 130, 139, 153, 173, 177, 181, 193, 203, 209, 216, 220, 222, 226, 234, 260, 262, 263, 265, 268, 296, 301, 303, 320, 325, 326, 338, 347, 348, 350–352, 355, 358, 405–407, 431
Ontic-epistemic variety, 51
Onticity, 177, 431
Ontic space, 173, 178, 220–222, 262, 349, 369, 371, 380, 381, 405
Ontic varieties, 50, 127, 222, 262, 283, 285, 302, 326, 355, 405
Ontological-epistemological dualities, 20, 39, 43, 46, 82, 84, 88, 91, 97, 122, 139, 173, 177, 183, 229, 237, 260, 262, 263, 273, 280, 302, 304, 305, 310, 312, 319, 320, 322, 324, 338, 344, 347, 350, 371, 372, 375, 378, 380, 397, 402, 405, 406, 419
Ontological how, 388, 398
Ontological information, 39, 125, 384
Ontological potential, 123
Ontological space, 19, 39, 46, 50–52, 100, 120, 131–135, 140, 169, 178, 215,

216, 219, 220–223, 225, 226, 230, 231, 233–235, 237, 238, 240, 242, 243, 247, 262, 263, 268, 269, 276, 278, 279, 282, 283, 285, 286, 289, 300, 303–305, 311, 312, 325, 350, 352, 358, 371, 373–375, 377, 378, 380–386, 388, 390, 392–394, 396, 397, 399, 401–406, 408
Ontological system, 388
Ontological time, 223, 224, 227, 228, 312, 350, 351, 390, 392, 394, 401, 409, 412
Ontological universe, 125, 151
Ontological variety, 121, 131, 136, 140, 169, 181, 183, 234, 238, 245, 262, 285, 303, 304, 312, 382–384, 387–390, 398, 402, 412
Ontological why, 388
Open set, 127, 373
Opportunity benefit, 370
Opportunity cost, 294, 311, 369, 370
Ordinary language, 186, 319, 321, 331
Organicity, 26, 35, 64, 118
Organization, 1–5, 8–16, 18, 20, 21–24, 26, 28, 29, 31–35, 40, 48, 57–59, 62, 65–69, 71, 73–80, 83–89, 91, 92, 94–97, 106–108, 111, 112, 115, 117, 137, 160, 177, 219, 227, 229, 238, 303, 316, 319, 379, 385, 411, 418, 420, 431
Organizational analytics, 1, 3, 35, 45, 53, 55, 59, 63, 65, 66, 68, 83, 108, 112, 116, 117, 162, 177, 219, 234, 240, 242, 344, 378, 382, 391, 404, 405
Organizational complexity, 91
Organizational cost-shifting
Organizational embedding, 91
Organizational objectives, 77
Organizational paremiology, 82, 221
Organizational rationality, 61
Organization-disorganization dualities, 3, 15
Organized algorithms, 73
Organized-disorganized dualities, 9
Output dual, 382, 391
Output supply, 1

P

Paradigm of thought, 30, 56, 111, 114, 115, 156, 276, 283, 353, 365, 366, 368, 406
Paradigm shifting, 114
Paradox, 92, 101, 112, 113, 115, 225, 226, 358, 408
Parametric shifter, 342
Paremiological geometry, 55, 67, 265, 266, 408, 422
Paremiological relation, 238
Parent-offspring dynamics, 94, 115
Parent-offspring process, 115, 148, 218, 228, 308, 370, 390, 422
Past-future dualities, 134, 157, 228, 229, 231, 233, 234, 235, 241, 242, 304, 338, 370, 385, 392, 393, 401, 402, 429
Past-present dualities, 150, 151, 157, 177, 218, 224, 225, 228, 231, 234, 235, 270, 275, 304, 307, 335, 338, 378, 426, 429
Past-present-future, 47, 165, 175, 178, 179, 224, 268, 283, 284, 298, 389, 390
Past-present telescopic structure, 40
Past-present varieties, 270
Peace-war dualities, 21, 218, 291, 330
Penumbral space, 361
Perception-imperfection dualities
Perfect recycling, 398
Pesemekomiya
Petaverse, 408
Phantom problem, 113, 225, 226, 247, 284–287, 290, 388
Philosophy of language, 319
Planet, 126, 307, 308, 328
Planning, 43, 127, 271, 332
Point of essentiality, 346
Polar dominance, 223, 244, 317, 346, 387, 396, 411
Polarities, 5, 7, 10–12, 14, 16, 17, 23, 30, 48, 55, 64, 67, 72, 75, 77, 84, 85, 94, 106, 112, 128, 155, 163, 175, 177, 182, 195, 200, 207, 213, 217, 221, 223, 235, 236, 244, 247, 257, 258, 276, 300, 315, 316, 322, 324, 327, 328, 330, 332, 334, 343, 344, 347, 356–358, 359, 360, 362, 364, 367, 369, 370, 373, 374, 377, 378, 380, 381, 383, 387, 390, 392–394, 396, 411, 413, 414, 420–422, 424, 426–429
Political institutions, 27, 33
Political-legal constraints, 62
Political sovereignty, 57
Political structure, 27, 32, 56, 61, 164, 219
Politico-legal restrictions, 87
Polyrhythmic integrity, 223

Positive-sum game, 108, 165, 414
Positive system, 94
Positive variety, 30, 374
Possibilistic information, 304
Possibilistic reasoning, 130
Possibility, 29, 34, 35, 38, 43, 46, 52, 71, 76, 104, 111, 115–118, 126, 128, 129, 133, 134–136, 139, 140, 143, 147, 148, 150, 151, 153, 155, 156, 159, 160, 167, 181, 184, 189, 192, 193, 204, 209, 221, 225, 234, 259, 260–264, 266, 268, 270–274, 278, 279, 284, 329, 331, 335, 351, 355, 362, 367, 406, 422, 431
Possibility-impossibility, 38, 41, 87, 115, 117, 118, 123, 130, 132, 135, 139, 141, 144, 146, 167, 171, 174, 179, 199, 248, 258–261, 267, 275, 279, 280, 320
Possibility space, 87, 98, 116, 117, 130, 136, 139, 142, 143, 145, 147–150, 153, 155, 156, 166, 175, 192, 193, 199, 204–207, 247, 263–265, 273, 274, 276, 279, 304, 305, 331, 350, 352
Possible universe, 133, 268, 271
Possible world, 54, 111, 112, 116, 118, 128, 129, 131, 132, 134, 135, 140–143, 146–153, 155, 156, 166, 168, 181, 247, 259, 262, 267–275, 278–280, 287, 288, 406
Possible-world semantics, 276
Possible-world transformation, 274, 280
Pragmatics, 132, 135
Predictability, 56, 106
Predictability-unpredictability dualities, 106, 358
Preference scheme, 45, 61, 63, 68, 86, 100, 310, 345, 362, 390
Preference space, 29, 424
Prescriptive science, 45, 270, 275, 332
Prescriptive system
Prescriptive theory, 36, 37, 39, 41–43, 270, 274, 275
Prescriptivity, 56
Present-future dualities, 19, 151, 157, 224, 228, 232, 234, 235, 271, 275, 304, 307, 338, 378, 424, 426, 429
Present-future varieties, 284
Primary category, 21, 22, 27, 29, 56, 59, 82, 83, 85, 88, 90, 190, 252, 259, 263, 273, 274, 275, 299, 301, 303, 350, 351, 365, 389, 390, 400, 404
Primary-derived, 190, 255, 302
Primary-derived organizations, 2, 21
Primary identity, 50, 233, 234, 240, 268, 279, 405
Primitive axiom, 303
Principle of diversity, 60, 65, 331
Principle of doubt, 94
Principle of indifference, 208, 356, 358
Principle of insufficient reasoning, 358
Principle of opposites, 2, 4, 5, 10, 11, 19, 23, 85, 99, 106, 118–120, 122, 131, 135, 138, 142, 150, 152, 154, 155, 168, 177, 178, 190, 194, 199, 204, 206, 207, 211, 220, 222, 223, 226, 235–238, 240–244, 251, 253, 255–257, 263, 264, 278, 279, 286, 297, 303, 308, 316, 317, 324, 337, 342, 347, 351, 353–355, 358, 362, 367, 370–375, 377, 378, 380–384, 386–389, 391–393, 395–398, 400, 404, 407–409, 411–413, 419–421, 424, 426, 428, 431
Principle of supersymmetry, 223
Principle of unity, 60, 65
Probabilistic information, 305, 367
Probabilistic reasoning, 366
Probability-improbability, 38, 41, 87, 115, 117, 118, 123, 130, 132, 135, 141, 144, 146, 167, 171, 174, 179, 248, 258–260, 320
Probability space, 35, 117, 129, 130, 136, 139, 142, 143, 148–150, 166, 175, 193, 199, 204–207, 247, 264, 273, 274, 276, 304, 305, 350
Probable world, 46, 119, 128, 148, 150, 166, 262, 406
Problem-solution dualities, 1–20, 22, 23, 26–33, 35, 40, 42, 45–50, 55–63, 65–73, 75–80, 82–101, 103–108, 112, 114–116, 125, 126, 128, 139, 143, 150–153, 155–157, 164, 166, 172, 174, 175, 177, 180, 182–184, 186, 188, 189, 192, 198–205, 207, 211, 212, 214–218, 222, 232, 235, 239, 247, 248, 250, 252, 254, 255, 257–259, 264, 266, 269, 274–277, 280–282, 284, 290, 291, 294–297, 300, 302, 304, 305, 307, 309, 313, 322–326, 328–332, 336, 337, 339–343, 345, 346, 348, 349, 353, 355, 358, 360, 361, 363, 364–366, 373, 374, 390, 393, 394, 397, 399, 412, 417, 431

Processes, 1–3, 5, 6, 8, 9, 14, 15, 17–20, 26, 28, 32, 33, 35–37, 39–41, 43, 46, 48, 55, 58, 62–65, 68, 71, 73, 76, 77, 79, 80, 82, 83, 85, 88–90, 92–96, 99, 101–107, 113–120, 122, 124, 125, 128, 131–133, 148, 152, 154–160, 166, 169, 172–175, 177–179, 187, 190, 196, 211, 215–217, 220, 227–229, 234, 236–243, 245, 248, 249, 252, 254, 256–258, 263, 264, 266, 268, 269, 274, 276, 277, 279, 281–283, 292–296, 298, 301–306, 308, 312, 315, 317, 318, 322, 323, 325–327, 330, 331, 333, 335–343, 345, 346, 348, 350–353, 356–359, 362–364, 367, 369, 378–381, 384, 387–389, 394–399, 403, 405, 408, 409, 412, 414, 415, 417, 420, 421, 423, 426, 427
Process-technology duality, 159, 160
Production-consumption dualities, 2, 6, 9, 10, 18, 19, 23, 26, 27, 31, 32, 34, 36, 50, 52, 62, 64, 67, 68, 70, 75, 80, 86, 91, 92, 97, 107–109, 117, 124, 125, 149, 154, 160, 172, 173, 175, 181, 184, 186, 190, 201, 204, 216–218, 222, 255, 256, 258, 259, 264, 267, 273, 278, 294, 298, 301, 310, 311, 312, 320, 323, 324, 330, 352, 368, 374, 397–399, 402, 407, 418, 420
Profit-loss duality, 73
Progress-retrogress dualities, 1, 46, 59, 173, 201
Propaganda, 4, 63, 91, 112, 113, 125, 160, 161, 163, 164, 166, 173, 273, 290, 293–295, 305, 306, 323, 331, 344, 417
Property-relation dualities, 21, 39, 40, 157, 158

Q

Qualitative motion, 349
Quantitative motion, 16, 128, 140, 158, 160, 381
Quantitative-qualitative dualities, 79
Question-answer dualities, 47, 50, 78, 80, 81, 83, 91, 112, 115, 126, 151, 172, 174, 182, 252, 307, 313, 397, 399

R

RA, 221
Racism, 108, 163, 164
Reality, 4, 30, 36, 37, 43, 48, 51, 54, 82, 104, 113, 116, 118–124, 126–130, 135, 139, 141–143, 145, 148, 153, 154, 156, 157, 167–174, 178, 180, 182, 185, 188, 193, 197, 204, 210, 215, 216, 221, 233, 234, 242, 249, 251–253, 256, 268–271, 273, 274, 281, 284, 287, 290, 298, 305, 323, 336, 360, 366, 391, 392, 394, 395
Reality dual, 129, 171, 185, 188, 213
Real-nominal dualities, 36, 116, 178, 328, 345
Reasoning, 21, 36, 60, 73, 92–95, 102, 111, 112, 114, 115, 162, 163, 176, 182, 193, 197, 210, 217, 247–249, 252–256, 267, 274, 276, 282, 283, 291–293, 295, 304, 317, 331, 332, 349, 352, 358, 362, 363, 389, 399, 421, 422, 428
Reasoning process, 217, 254, 295, 316, 330, 332
Relational continuum, 5, 7, 16, 28, 31, 33, 34, 47, 52, 64–67, 70, 76–78, 84, 85, 87–89, 99, 101, 102, 114, 118, 119, 126, 131, 142, 147, 154, 171, 174, 186, 204, 220–226, 228, 229, 233, 235, 236, 243, 244, 251, 253, 265, 272, 278, 279, 286, 295, 296, 306, 316, 317, 339, 342, 344, 347–351, 354, 355, 357, 359, 364, 368, 373, 383, 389, 392, 400, 412, 414, 418, 419, 421, 424, 426–428
Relational field, 13, 14, 26, 31, 78, 79, 81, 92, 93, 128, 137, 199, 284
Relationality, 11–14, 18, 26, 31, 35, 64, 66, 67, 72, 156, 171, 189, 256, 266, 281, 301, 302, 328, 362, 382, 387, 396, 411, 423, 429
Relational unity, 8, 11, 29, 79, 85, 89, 120, 121, 171, 232, 251, 253, 348, 354, 423
Relative frequency, 98, 126
Religious belief, 62, 161, 163
Resource state of being, 128
Restorative justice, 77
Risk-benefit condition, 266
Risk-reward duality, 116, 147, 148, 153, 203
Russell, 365

S

Sankofa-anoma, 47, 52, 133, 157, 175, 211, 220, 224, 234–236, 268, 270, 292, 304, 338, 399, 400, 429
Science, Technology, Engineering and Mathematics (STEM), 64
Scientific discovery, 114
Scientific efficiency, 59
Scientific practices, 59
Scientific sovereignty
Scientific theories, 59
Second principle, 229
Self-contained modules, 363
Self-correction system, 19, 63, 330
Self-destructive, 358
Self-learning, 306, 330
Semantic analytics, 153
Semantic reflection, 50
Semantics, 50, 132, 135, 143, 148, 151, 152, 156, 255, 267–271, 273–276, 287
Sending-interpretation process, 292
Sending problems, 292
Sending-receiving dualities, 113
Set of hope, 205
Sharing relation, 12, 14, 65, 66, 92, 107
Sickness-wellness dualities, 291
Signal decipher function, 121, 170
Signal disposition, 39, 50, 51, 81, 118, 120–122, 141, 167–170, 262, 278, 285, 289, 290, 292, 293, 301, 303, 304, 319, 323, 326, 331, 348, 349, 365, 369
Similarity principle, 86
Slavery, 61, 163, 415
Smart material, 20, 293
Social action, 26, 162, 219, 223, 311, 379
Social cognitive capacity limitations, 55, 72
Social decision-choice, 16
Social education, 95
Social functors, 134, 135, 140, 141, 183, 271, 368, 414
Social history, 158
Social organization, 1, 2, 9, 16, 18, 22, 27, 28, 30, 34, 56, 60, 61, 63, 65, 68, 69, 71, 75, 86, 238, 273, 425
Social philosophical consciencism, 2, 3, 6, 7, 15, 26, 27, 29–31, 34, 36, 40, 60, 62, 63, 66, 90, 104, 106–108, 129, 135, 150, 151, 160–166, 174, 180, 184, 189, 190, 200, 202, 203, 207, 209, 212–215, 217, 219, 221, 248–250, 255, 258, 284, 301, 323, 332, 376, 377, 379, 415, 417
Social policy, 59–62, 294, 344
Social progress, 4, 9, 46, 59, 68, 94, 129, 150, 151, 175, 189, 279, 330, 397, 399
Social rationality, 61, 156
Social science, 33, 60, 63, 64, 71, 101, 102, 332, 365
Social transformation, 32, 33, 130, 179, 193, 276, 424
Socioeconomic productions, 159
Socioeconomic transformation, 46, 408
Socio-material relations, 57
Socio-natural information, 60, 75, 104
Socio-psychological principle, 75, 186, 291
Somethingness disposition, 221, 224, 227, 229, 230, 270, 272
Source-destination dualities, 11, 39, 113, 116, 118, 153, 156, 167, 276, 280, 291–293, 298, 302, 303, 306, 320, 321, 331, 390
Source problems, 293
Sources of uncertainty, 35, 50, 260, 261, 263, 431
Sovereign power, 61
Space of absoluteness, 45, 52, 117, 140, 154, 243, 244, 337, 353–355, 357, 371, 379, 404, 419, 422
Space of acquaintance, 5, 38, 39, 42, 44, 51, 83, 87, 98, 118, 120, 122, 167, 169, 170, 188, 220, 226, 262, 265, 276, 308, 323, 351, 390, 405, 406
Space of actual, 11, 17, 20, 34, 37–39, 42, 43, 118–121, 124, 131, 133, 135, 136, 138, 140, 148, 167, 170, 172, 178, 185, 188, 213, 250–252, 262, 273, 274, 281, 283, 284, 360, 366, 385, 394, 403
Space of categorial convertors, 117, 130, 136–138, 146, 169, 179, 192
Space of creativities, 184, 187, 193, 281, 340
Space of curiosities, 187, 206
Space of epistemic characteristic disposition, 51, 54, 138, 220, 223, 395, 404
Space of identifications, 1, 7, 8, 38, 55, 71, 120, 183, 275, 386
Space of imaginations, 6, 11, 17, 20, 23, 26, 30, 34–39, 41, 43, 46, 48, 52, 71, 98, 103, 111, 115, 118–122, 125–127, 129, 133–136, 138–142, 144–146,

148–151, 155, 160, 166–170, 173–175, 178, 180–183, 185, 186, 192, 193, 197, 200, 203–206, 210, 213, 215, 216, 219, 232, 247, 256, 262, 264, 266–269, 271, 273–276, 278–281, 284–287, 289, 290, 305, 306, 308, 310, 323, 325, 331, 338, 351, 352, 355, 358, 360–362, 364, 366, 367, 369, 371, 379, 388, 391, 406, 414, 431
Space of knowability, 6, 125, 126, 224, 225, 228, 231, 233, 270, 271, 307, 397, 398, 400, 401, 409
Space of ontic characteristic disposition, 51, 54, 138, 220, 222, 286, 386, 404
Space of paradigms, 53, 80
Space of potentials, 6, 37, 38, 41, 43, 46, 51, 54, 88, 98, 103, 119, 121, 123, 124, 131, 132, 136, 138, 140, 148, 149, 151, 160, 166, 168, 170, 171, 174, 175, 182, 183, 185, 188, 213, 251, 252, 269, 273, 276, 278, 279, 281, 287, 325, 351, 352, 360, 385, 388, 394, 403
Space of realities, 37, 38, 40–43, 111, 115, 120, 121, 125, 126, 129, 134, 135, 138–142, 148–152, 167, 169, 172–175, 178, 180, 181, 183–185, 200, 203, 205, 206, 213, 215, 216, 219, 247, 250, 262, 268, 269, 271–276, 280, 281, 283–285, 287, 288, 305, 306, 325, 338, 352, 360–362, 364, 366, 369, 371, 379
Space of relativity, 5, 17, 45, 52, 117, 140, 154, 243, 244, 254, 258, 282, 306, 312, 337, 344, 353, 355, 357, 358, 361, 370, 379, 419, 422, 425
Space of transformations, 71, 120, 217, 275, 352
Space of varieties, 3, 4, 6, 8, 15, 23, 27–29, 31, 33, 37, 39, 40, 43, 47, 50–52, 55, 56, 58, 60, 62, 63, 68, 71, 72, 74, 76, 84, 88, 90, 92, 95, 102, 104, 105, 121, 123, 127, 139, 140, 157–159, 160, 161, 167, 173, 181, 184, 202, 220, 237, 256, 260, 263, 269, 278, 281, 298, 301, 305, 318, 319, 322, 326, 328, 332, 338, 343, 347, 363, 371, 373, 378, 381, 383, 385, 386, 392–394, 398, 402, 403, 407, 408, 413, 418, 423–425
Stability-instability dualities, 223, 308, 312, 388, 398
Star rebirth, 388
State variable, 317, 318
Static-dynamic dualities, 1, 7, 15, 28, 31, 32, 59, 70, 93, 122, 137, 150, 155–157, 199, 224, 225, 238, 239, 263, 279, 300, 308, 311, 325, 335, 361, 378, 392, 423, 427
Static equilibrium, 34
Static problem-solution, 325
Stock-flow, 1, 3, 8, 19, 20, 28, 64, 77, 79, 86, 89, 96, 97, 103, 107, 176, 185, 304, 368–371, 375, 383, 387, 391, 396, 408, 417, 428
Stock-flow systems, 6, 176
Stopping role fuzzy paradigm
Subjective matters
Subjective-objective dualities, 272, 300
Subjective probability, 98, 126
Subjective reality, 4
Success-failure dualities, 1, 2, 10, 11, 16, 38, 42, 43, 45, 46, 50, 71, 80, 103, 108, 141, 177, 182, 193, 201, 203, 207, 248, 258, 263, 266, 326, 345, 361, 362, 366, 408
Success-failure process, 4, 34, 48
Success-happiness dualities
Successor-predecessor, 184, 218, 350
Successor-predecessor process, 190
Sufficiency, 48, 117, 148, 160, 193, 199, 271, 305, 310, 406
Sufficient conditions, 149, 152, 161, 166, 167, 188, 199, 204–206, 261, 262, 266, 280, 284, 323, 351, 366
Superstructure institutions, 22
Superstructure organization, 22, 37, 45, 65, 71–73, 75, 77
Supersymmetry, 178, 211, 235, 236, 388, 398, 408, 422, 423, 427
Supply-demand chain, 109, 311, 370, 383, 418
Supporting opposites, 377
Sustainability, 223
Sustainability-unsustainability dualities, 223
Symmetric difference, 231, 385, 400
Symmetry, 178, 235, 236, 295, 398, 408, 422, 423, 427
Symmetry-asymmetry dualities, 223, 227, 236, 278
Synonym-antonym dualities, 226, 297, 303, 319
Synonyms, 407

T

Teaching, 17, 21, 35, 62, 65, 74, 88–90, 92, 93, 95, 99, 101, 105, 309, 345, 429
Technique, 59, 60, 70, 74, 83, 85, 95, 101–103, 114, 160, 163, 191, 192, 212, 330, 351–353, 366, 375
Technological process, 11, 23, 180, 245, 369
Technology, 5, 20, 63, 64, 92, 131, 152, 156, 157, 159, 160, 183, 292, 293, 298, 308, 351, 420
Telescopic disposition, 231, 232
Telescopic future, 19, 30, 51, 72, 73, 96, 139, 155, 228, 269–271, 307, 337
Telescopic past, 28, 30, 131, 133, 134, 139, 140, 151, 155, 228, 269, 270, 284, 307, 316, 337, 338
Telescopic present, 30, 139, 155, 269–271, 307, 337
Temporally equilibrium
Tension-change dualities, 88
Theory-applied dualities, 183
Theory of entropy, 300, 319
Theory of info-dynamics, 160, 298, 300
Theory of information, 27, 97, 319, 407
Theory of info-statics, 298, 300
Theory of knowing, 1, 36, 117, 124, 154, 161, 172, 267, 278, 283, 291, 322, 324, 325, 353, 361, 379, 402, 406, 417, 420, 421, 431
Theory of mastering, 283
Theory of possibility, 260, 264, 352
Theory of probability, 260, 264, 351
Theory-practice dualities, 48, 109, 328, 370
Thinking, 7, 21, 23, 36, 60, 73, 92–95, 102, 111–115, 118, 162, 163, 167, 176, 182, 187, 199, 217, 247–256, 260, 267, 274, 276, 282–284, 291–293, 295, 304, 317, 330–332, 353, 358, 363, 366, 403, 421, 422, 428
Thinking-algorithmic power, 187
Thinking process, 249
Thinking-reasoning dualities, 119, 124, 128, 183, 222, 249, 253–256, 261, 268, 271, 352, 359, 392, 419
Thinking-reasoning-learning storage, 331
Tipping points, 158, 327
Transformability, 106
Transformational dynamics, 131, 157
Transformation game, 5, 12, 305, 343, 345
Transformation problem-solution, 18, 23, 55, 69, 71, 96, 106, 323, 325, 353
Transformation transversality, 315, 326, 333
Transversality conditions, 5, 38, 315, 318, 327, 331, 337–339, 346, 347, 353, 369, 411, 420, 429
Transversality line, 161, 296, 334, 339, 342
Trilogy, 332
Truth-falsity dualities, 102
Try-and-error, 48, 367
Try-and-error process, 103, 125

U

Ubuntu, 19, 72, 424
Uncertainty, 41, 45, 46, 49–51, 53, 55, 101, 115, 126, 145, 173, 175, 201, 214, 220, 259, 260, 261, 263–265, 281, 325, 326, 349, 354, 355, 362, 366, 367, 405, 406, 414
Uncertainty dominance
Unintentionality, 113
Unity-disunity dualities, 8, 28, 83
Unity-disunity dynamics, 21
Universal oneness, 16
Universal principle of opposite
Universal system, 9, 322, 373, 383, 393
Universal variety, 69
Unknowability, 125, 224, 225, 228, 307, 388, 400–402, 409
Unknowable universe, 125, 126
Unsolvable problem, 113, 225, 226, 247, 289, 290, 295, 388

V

Vagueness, 5, 55, 56, 98, 101, 126, 209, 245, 263, 327, 347, 352, 353, 355, 361, 365–368, 419
Vagueness-precision
Variety-commonness dualities, 261
Variety curiosity, 189, 201, 211
Variety hope, 199, 211, 212, 215, 216
Variety imaginations, 30, 37, 43, 51, 88, 104, 118, 119, 122, 123, 124, 127, 129, 151, 159, 160, 167, 168, 171–174, 180, 184, 188, 204, 215, 216, 242, 249, 251, 253, 256, 281, 298, 366, 394
Variety potential, 43, 50, 118, 138, 160, 188, 249, 274, 276, 381
Violence-nonviolence dualities, 36, 163, 166
Volume incompleteness, 56, 98, 126, 263, 327, 347, 353, 365, 419

W

Well-posed problem, 92
Welsing, C., 309
What-it-is, 29, 69
What-it-is-not, 29, 69
What-there-is, 6, 29, 30, 40, 41, 50, 51, 55, 133, 134, 139, 151, 155, 220, 231, 275, 309, 325, 333, 335–337, 390, 397, 409, 429
What-there-was, 6, 30, 40, 41, 50, 51, 55, 133, 134, 139, 150, 151, 155, 220, 275, 325, 333, 335–337, 390, 397, 402, 409, 429
What-there-would-be, 30, 41, 50, 51, 133, 134, 139, 151, 155, 220, 275, 325, 390, 397, 429
What-was, 29, 335
What-was-not, 29
What-would-be, 29, 55, 333, 335–337, 403, 409
What-would-be-not, 29, 409
Wickedness-kindness dualities, 291
Wisdom knot, 218, 370, 427
Witness evidence, 367
Woodson, Cater, 309

Z

Zero-sum game, 108, 165, 336, 349, 415
Zone of ignorance, 134, 294, 296, 341–343
Zone of knowledge, 294, 296, 333, 341–343
Zone of positive deficiency, 343
Zone of variety ignorance, 294, 333, 335
Zone of variety knowledge, 294, 333, 335